STUDENT'S
SOLUTIONS MANUAL
JUDITH A. PENNA
Indiana University Purdue University Indianapolis

INTERMEDIATE ALGEBRA
GRAPHS & MODELS
SECOND EDITION

Marvin L. Bittinger
Indiana University Purdue University Indianapolis

David J. Ellenbogen
Community College of Vermont

Barbara L. Johnson
Indiana University Purdue University Indianapolis

PEARSON

Addison
Wesley

Boston   San Francisco   New York
London   Toronto   Sydney   Tokyo   Singapore   Madrid
Mexico City   Munich   Paris   Cape Town   Hong Kong   Montreal

Reproduced by Pearson Addison-Wesley from electronic files supplied by the author.

Copyright © 2004 Pearson Education, Inc.
Publishing as Pearson Addison-Wesley, 75 Arlington Street, Boston MA 02116

ISBN   0-321-16870-4

2 3 4 5 6 CRS 06 05 04

PEARSON
Addison
Wesley

Contents

Chapter 1

Basics of Algebra and Graphing

1. Substitute and carry out the operations indicated.
$$7x + y = 7 \cdot 3 + 4$$
$$= 21 + 4$$
$$= 25$$

2. 27

3. Substitute and carry out the operations indicated.
$$2c \div 3b = 2 \cdot 6 \div 3 \cdot 4$$
$$= 12 \div 3 \cdot 4$$
$$= 4 \cdot 4$$
$$= 16$$

4. 9

5. Substitute and carry out the operations indicated.
$$25 + r^2 - s = 25 + 3^2 - 7$$
$$= 25 + 9 - 7$$
$$= 34 - 7$$
$$= 27$$

6. 5

7. $3n^2p - 3pn^2 = 3 \cdot 5^2 \cdot 9 - 3 \cdot 9 \cdot 5^2$

Observe that $3 \cdot 5^2 \cdot 9$ and $3 \cdot 9 \cdot 5^2$ represent the same number, so their difference is 0.

8. 280

9. Substitute and carry out the operations indicated.
$$5x \div (2 + x - y) = 5 \cdot 6 \div (2 + 6 - 2)$$
$$= 5 \cdot 6 \div (8 - 2)$$
$$= 5 \cdot 6 \div 6$$
$$= 30 \div 6$$
$$= 5$$

10. 3

11. Substitute and carry out the operations indicated.
$$[10 - (a - b)]^2 = [10 - (7 - 2)]^2$$
$$= [10 - 5]^2$$
$$= 5^2$$
$$= 25$$

12. 144

13. Substitute and carry out the operations indicated.
$$m + [n(3 + n)]^2 = 9 + [2(3 + 2)]^2$$
$$= 9 + [2(5)]^2$$
$$= 9 + 10^2$$
$$= 9 + 100$$
$$= 109$$

14. 13

15. We substitute 5 for b and 7 for h and multiply:
$$A = \frac{1}{2} \cdot b \cdot h = \frac{1}{2} \cdot 5 \cdot 7 = 17.5 \text{ sq ft}$$

16. 3.045 sq m

17. We substitute 7 for b and 3.2 for h and multiply:
$$A = \frac{1}{2} \cdot b \cdot h = \frac{1}{2}(7)(3.2) = 11.2 \text{ sq m}$$

18. 7.2 sq ft

19. a) We replace y with 7.
$$\frac{12 - y = 5}{12 - 7 \ ? \ 5}$$
$$5 \ | \ 5 \quad \text{TRUE}$$
Since $5 = 5$ is true, 7 is a solution of the equation.

   b) We replace y with 5.
$$\frac{12 - y = 5}{12 - 5 \ ? \ 5}$$
$$7 \ | \ 5 \quad \text{FALSE}$$
Since $7 = 5$ is false, 5 is not a solution of the equation.

   c) We replace y with 12.
$$\frac{12 - y = 5}{12 - 12 \ ? \ 5}$$
$$0 \ | \ 5 \quad \text{FALSE}$$
Since $0 = 5$ is false, 12 is not a solution of the equation.

20. a) No
    b) no
    c) yes

. a) We replace x with 0.

$$5 - x \leq 2$$
$$5 - 0 \ ? \ 2$$
$$5 \ | \ 2 \quad \text{FALSE}$$

Since $5 \leq 2$ is false, 5 is not a solution of the inequality.

b) We replace x with 4.

$$5 - x \leq 2$$
$$5 - 4 \ ? \ 2$$
$$1 \ | \ 2 \quad \text{TRUE}$$

Since $1 \leq 2$ is true, 4 is a solution of the inequality.

c) We replace x with 3.

$$5 - x \leq 2$$
$$5 - 3 \ ? \ 2$$
$$2 \ | \ 2 \quad \text{TRUE}$$

Since $2 \leq 2$ is true, 3 is a solution of the inequality.

. a) No

b) no

c) yes

. a) We replace m with 6.

$$3m - 8 < 13$$
$$3 \cdot 6 - 8 \ ? \ 13$$
$$18 - 8$$
$$10 \ | \ 13 \quad \text{TRUE}$$

Since $10 < 13$ is true, 6 is a solution of the inequality.

b) We replace m with 7.

$$3m - 8 < 13$$
$$3 \cdot 7 - 8 \ ? \ 13$$
$$21 - 8$$
$$13 \ | \ 13 \quad \text{FALSE}$$

Since $13 < 13$ is false, 7 is not a solution of the inequality.

c) We replace m with 9.

$$3m - 8 < 13$$
$$3 \cdot 9 - 8 \ ? \ 13$$
$$27 - 8$$
$$19 \ | \ 13 \quad \text{FALSE}$$

Since $19 < 13$ is false, 9 is not a solution of the inequality.

24. a) Yes

b) no

c) no

25. List the letters in the set: {a,e,i,o,u}, or {a,e,i,o,u,y}

26. {Sunday, Monday, Tuesday, Wednesday, Thursday, Friday, Saturday}

27. List the numbers in the set: $\{1, 3, 5, 7, \ldots\}$

28. $\{2, 4, 6, 8, \ldots\}$

29. List the numbers in the set: $\{5, 10, 15, 20, \ldots\}$

30. $\{10, 20, 30, 40, \ldots\}$

31. Specify the conditions under which a number is in the set: $\{x | x$ is an odd number between 10 and 20$\}$

32. $\{x | x$ is a multiple of 4 between 22 and 35$\}$

33. Specify the conditions under which a number is in the set: $\{x | x$ is a whole number less than 5$\}$

34. $\{x | x$ is an integer greater than -4 and less than 3$\}$

35. Specify the conditions under which a number is in the set: $\{n | n$ is a multiple of 5 between 7 and 79$\}$

36. $\{x | x$ is an even number between 9 and 99$\}$

37. Since 9 is a natural number, the statement is true.

38. False

39. Since every member of the set of natural numbers is also a member of the set of whole numbers, the statement is true.

40. True

41. Since $\sqrt{8}$ is not a rational number, the statement is false.

42. False

43. Since every member of the set of irrational numbers is also a member of the set of real numbers, the statement is true.

44. True

45. Since 4.3 is not an integer, the statement is true.

46. True

47. Since every member of the set of rational numbers is also a member of the set of real numbers, the statement is true.

48. False

49.

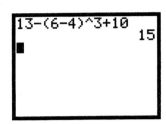

50. 143

51.

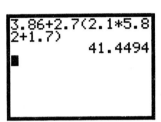

52. 7.978

53.

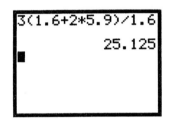

54. 24.54

55.

56. 97.25

57. *Writing Exercise*

58. *Writing Exercise*

59. *Writing Exercise*

60. *Writing Exercise*

61. The only whole number that is not also a natural number is 0. Using roster notation to name the set, we have {0}.

62. $\{-1, -2, -3, \ldots\}$

63. List the numbers in the set:

$\{5, 10, 15, 20, \ldots\}$

64. $\{3, 6, 9, 12, \ldots\}$

65. List the numbers in the set:

$\{1, 3, 5, 7, \ldots\}$

66. $\{\ldots, -4, -2, 0, 2, 4, \ldots\}$

67. Recall from geometry that when a right triangle has legs of length 2 and 3, the length of the hypotenuse is $\sqrt{2^2 + 3^2} = \sqrt{4 + 9} = \sqrt{13}$. We draw such a triangle:

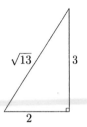

Exercise Set 1.2

1. $|-9| = 9$ -9 is 9 units from 0.

2. 7

3. $|6| = 6$ 6 is 6 units from 0.

4. 47

5. $|-6.2| = 6.2$ -6.2 is 6.2 units from 0.

6. 7.9

7. $|0| = 0$ 0 is 0 units from itself.

8. $3\frac{3}{4}$

9. $\left|1\frac{7}{8}\right| = 1\frac{7}{8}$ $1\frac{7}{8}$ is $1\frac{7}{8}$ units from 0.

10. 7.24

11. $|-4.21| = 4.21$ -4.21 is 4.21 units from 0.

12. 5.309

13. $-6 \leq -2$ is a true statement since -6 is to the left of -2

14. False

15. $-9 > 1$ is a false statement since -9 is to the left of 1.

16. True

7. $0 \geq -5$ is a true statement since -5 is to the left of 0.

. True

9. $-8 < -3$ is a true statement since -8 is to the left of -3.

0. True

1. $-4 \geq -4$ is a true statement since $-4 = -4$ is true.

2. False

3. $-5 > -5$ is a false statement since -5 does not lie to the left of itself.

4. True

5. $4 + 7$

Two positive numbers: Add the numbers, getting 11. The answer is positive, 11.

6. 11

7. $-4 + (-7)$

Two negative numbers: Add the absolute values, getting 11. The answer is negative, -11.

8. -11

9. $-3.9 + 2.7$

A negative and a positive number: The absolute values are 3.9 and 2.7. Subtract 2.7 from 3.9 to get 1.2. The negative number is farther from 0, so the answer is negative, -1.2.

0. 5.4

1. $\dfrac{2}{7} + \left(-\dfrac{3}{5}\right) = \dfrac{10}{35} + \left(-\dfrac{21}{35}\right)$

A positive and a negative number. The absolute values are $\dfrac{10}{35}$ and $\dfrac{21}{35}$. Subtract $\dfrac{10}{35}$ from $\dfrac{21}{35}$ to get $\dfrac{11}{35}$. The negative number is farther from 0, so the answer is negative, $-\dfrac{11}{35}$.

2. $-\dfrac{1}{40}$

3. $-4.9 + (-3.6)$

Two negative numbers: Add the absolute values, getting 8.5. The answer is negative, -8.5.

4. -9.6

5. $-\dfrac{1}{9} + \dfrac{2}{3} = -\dfrac{1}{9} + \dfrac{6}{9}$

A negative and a positive number. The absolute values are $\dfrac{1}{9}$ and $\dfrac{6}{9}$. Subtract $\dfrac{1}{9}$ from $\dfrac{6}{9}$ to get $\dfrac{5}{9}$. The positive number is farther from 0, so the answer is positive, $\dfrac{5}{9}$.

36. $\dfrac{3}{10}$

37. $0 + (-4.5)$

One number is zero: The sum is the other number, -4.5.

38. -3.19

39. $-7.24 + 7.24$

A negative and a positive number: The numbers have the same absolute value, 7.24, so the answer is 0.

40. 0

41. $15.9 + (-22.3)$

A positive and a negative number: The absolute values are 15.9 and 22.3. Subtract 15.9 from 22.3 to get 6.4. The negative number is farther from 0, so the answer is negative, -6.4.

42. -6.6

43. The opposite of 3.14 is -3.14, because $3.14 + (-3.14) = 0$.

44. -5.43

45. The opposite of $-4\dfrac{1}{3}$ is $4\dfrac{1}{3}$, because $-4\dfrac{1}{3} + 4\dfrac{1}{3} = 0$.

46. $-2\dfrac{3}{5}$

47. The opposite of 0 is 0, because $0 + 0 = 0$.

48. $2\dfrac{3}{4}$

49. If $x = 7$, then $-x = -7$. (The opposite of 7 is -7.)

50. -3

51. If $x = -2.7$, then $-x = -(-2.7) = 2.7$.
(The opposite of -2.7 is 2.7.)

52. 1.9

53. If $x = 1.79$, then $-x = -1.79$. (The opposite of 1.79 is -1.79.)

54. -3.14

55. If $x = 0$, then $-x = 0$. (The opposite of 0 is 0.)

56. 1

57. $9 - 2 = 9 + (-2)$ Change the sign and add.
  $= 7$

58. 7

59. $2 - 9 = 2 + (-9)$ Change the sign and add.

$\qquad = -7$

60. -7

61. $-6 - (-10) = -6 + 10$ Change the sign and add.

$\qquad\qquad = 4$

62. 6

63. $-5 - 14 = -5 + (-14) = -19$

64. -15

65. $2.7 - 5.8 = 2.7 + (-5.8) = -3.1$

66. 0

67. $-\dfrac{3}{5} - \dfrac{1}{2} = -\dfrac{3}{5} + \left(-\dfrac{1}{2}\right)$

$\qquad = -\dfrac{6}{10} + \left(-\dfrac{5}{10}\right)$ Finding a common denominator

$\qquad = -\dfrac{11}{10}$

68. $-\dfrac{13}{15}$

69. $-3.9 - (-3.9)$

Observe that we are subtracting -3.9 from itself. Thus, the difference is 0.

We could also do this exercise as follows:

$-3.9 - (-3.9) = -3.9 + 3.9 = 0$

70. -1.1

71. $0 - (-7.9) = 0 + 7.9 = 7.9$

72. -5.3

73. $(-5)6$

Two numbers with unlike signs: Multiply their absolute values, getting 30. The answer is negative, -30.

74. -28

75. $(-3)(-8)$

Two numbers with the same sign: Multiply their absolute values, getting 24. The answer is positive, 24.

76. 56

77. $(4.2)(-5)$

Two numbers with unlike signs: Multiply their absolute values, getting 21. The answer is negative, -21.

78. -28

79. $\dfrac{3}{7}(-1)$

Two numbers with unlike signs: Multiply their absolute values, getting $\dfrac{3}{7}$. The answer is negative, $-\dfrac{3}{7}$.

80. $-\dfrac{2}{5}$

81. $15.2 \times 0 = 0$

82. 0

83. $(-3.2) \times (-1.7)$

Two numbers with the same sign: Multiply their absolute values, getting 5.44. The answer is positive, 5.44.

84. 8.17

85. $\dfrac{-10}{-2}$

Two numbers with the same sign: Divide their absolute values, getting 5. The answer is positive, 5.

86. 5

87. $\dfrac{-100}{20}$

Two numbers with unlike signs: Divide their absolute values, getting 5. The answer is negative, -5.

88. -10

89. $\dfrac{73}{-1}$

Two numbers with unlike signs: Divide their absolute values, getting 73. The answer is negative, -73.

90. -62

91. $\dfrac{0}{-7} = 0$

92. 0

93. $\dfrac{-6}{0}$

Since division by 0 is undefined, this quotient is undefined.

94. Undefined

95. The reciprocal of 4 is $\dfrac{1}{4}$, because $4 \cdot \dfrac{1}{4} = 1$.

96. $\dfrac{1}{3}$

97. The reciprocal of -9 is $\dfrac{1}{-9}$, or $-\dfrac{1}{9}$, because

$-9\left(-\dfrac{1}{9}\right) = 1.$

8. $-\dfrac{1}{5}$

9. The reciprocal of $\dfrac{2}{3}$ is $\dfrac{3}{2}$, because $\dfrac{2}{3} \cdot \dfrac{3}{2} = 1$.

0. $\dfrac{7}{4}$

1. The reciprocal of $-\dfrac{3}{11}$ is $-\dfrac{11}{3}$, because
$-\dfrac{3}{11}\left(-\dfrac{11}{3}\right) = 1$.

2. $-\dfrac{3}{7}$

3. $\dfrac{2}{3} \div \dfrac{4}{5}$

$= \dfrac{2}{3} \cdot \dfrac{5}{4}$ Multiplying by the reciprocal of 4/5

$= \dfrac{10}{12}$, or $\dfrac{5}{6}$

4. $\dfrac{5}{21}$

5. $\left(-\dfrac{3}{5}\right) \div \dfrac{1}{2}$

$= -\dfrac{3}{5} \cdot \dfrac{2}{1}$ Multiplying by the reciprocal of 1/2

$= -\dfrac{6}{5}$

6. $-\dfrac{12}{7}$

7. $\left(-\dfrac{2}{9}\right) \div (-8)$

$= -\dfrac{2}{9} \cdot \left(-\dfrac{1}{8}\right)$ Multiplying by the reciprocal of -8

$= \dfrac{2}{72}$, or $\dfrac{1}{36}$

8. $\dfrac{1}{33}$

9. $\dfrac{12}{7} \div \left(-\dfrac{12}{7}\right) = \dfrac{12}{7} \cdot \left(-\dfrac{7}{12}\right)$ Multiplying by the

reciprocal of $-\dfrac{12}{7}$

$= -1$

0. $\dfrac{2}{7}$

111. $7-(8-3\cdot2^3) = 7 - (8 - 3 \cdot 8)$ Working within

$= 7 - (8 - 24)$ the parentheses

$= 7 - (-16)$ first

$= 7 + 16$

$= 23$

112. -3

113. $\dfrac{5 \cdot 2 - 4^2}{27 - 2^4} = \dfrac{5 \cdot 2 - 16}{27 - 16} = \dfrac{10 - 16}{11} = \dfrac{-6}{11}$, or $-\dfrac{6}{11}$

114. $-\dfrac{4}{17}$

115. $\dfrac{3^4 - (5-3)^4}{8 - 2^3} = \dfrac{3^4 - 2^4}{8 - 8} = \dfrac{81 - 16}{0}$

Since division by 0 is undefined, this expression is undefined.

116. $\dfrac{55}{2}$

117. $\dfrac{(2-3)^3 - 5|2-4|}{7 - 2 \cdot 5^2} = \dfrac{(-1)^3 - 5|-2|}{7 - 2 \cdot 25} = \dfrac{-1 - 5(2)}{7 - 50} =$

$\dfrac{-1 - 10}{-43} = \dfrac{-11}{-43} = \dfrac{11}{43}$

118. Undefined

119. $|2^2 - 7|^3 + 1 = |4 - 7|^3 + 1 = |-3|^3 + 1 =$
$3^3 + 1 = 27 + 1 = 28$

120. 79

121. $28 - (-5)^2 + 15 \div (-3) \cdot 2$

$= 28 - 25 + 15 \div (-3) \cdot 2$ Evaluating the exponential expression

$= 28 - 25 - 5 \cdot 2$ Dividing

$= 28 - 25 - 10$ Multiplying

$= -7$ Subtracting

122. -12

123. $12 - \sqrt{11 - (3+4)} \div [-5 - (-6)]^2$

$= 12 - \sqrt{11 - 7} \div [-5 + 6]^2$

$= 12 - \sqrt{4} \div [1]^2$

$= 12 - 2 \div 1$

$= 10 \div 1$

$= 10$

124. 11

125. Observe that $5.2(-1.7 - 3.8)^2 \cdot 0 = 0$ since this expression has a factor of 0. Thus, $-12.86 - 5.2(-1.7 - 3.8)^2 \cdot 0 = -12.86 - 0 = -12.86$.

126. 0

127. We would use the keystrokes in (a) to enter the given expression on a calculator.

128. (c)

129. We would use the keystrokes in (d) to enter the given expression on a calculator.

130. (b)

131. *Writing Exercise*

132. *Writing Exercise*

133. Substitute and carry out the indicated operations.
$$2(x + 5) = 2(3 + 5) = 2 \cdot 8 = 16$$
$$2x + 10 = 2 \cdot 3 + 10 = 6 + 10 = 16$$

134. 11; 11

135. *Writing Exercise*

136. *Writing Exercise*

137. $(8 - 5)^3 + 9 = 36$

138. $2 \cdot (7 + 3^2 \cdot 5) = 104$

139. $5 \cdot 2^3 \div (3 - 4)^4 = 40$

140. $(2 - 7) \cdot 2^2 + 9 = -11$

141. Any value of a such that $a \leq -6.2$ satisfies the given conditions. The largest of these values is -6.2.

Exercise Set 1.3

1. Using the commutative law of addition, we have
$$4a + 7b = 7b + 4a.$$
Using the commutative law of multiplication, we have
$$4a + 7b = a4 + 7b$$
$$\text{or} \quad 4a + 7b = 4a + b7$$
$$\text{or} \quad 4a + 7b = a4 + b7.$$

2. $xy + 6; 6 + yx$

3. Using the commutative law of multiplication, we have
$$(7x)y = y(7x)$$
$$\text{or} \quad (7x)y = (x7)y.$$

4. $(ab)(-9); -9(ba)$

5. $\quad (3x)y$
$$= 3(xy) \quad \text{Associative law of multiplication}$$

6. $(-7a)b$

7. $\quad x + (2y + 5)$
$$= (x + 2y) + 5 \quad \text{Associative law of addition}$$

8. $3y + (4 + 10)$

9. $\quad 3(a + 7)$
$$= 3 \cdot a + 3 \cdot 7 \quad \text{Using the distributive law}$$
$$= 3a + 21$$

10. $8x + 8$

11. $\quad 4(x - y)$
$$= 4 \cdot x - 4 \cdot y \quad \text{Using the distributive law}$$
$$= 4x - 4y$$

12. $9a - 9b$

13. $\quad -5(2a + 3b)$
$$= -5 \cdot 2a + (-5) \cdot 3b$$
$$= -10a - 15b$$

14. $-6c - 10d$

15. $\quad 9a(b - c + d)$
$$= 9a \cdot b - 9a \cdot c + 9a \cdot d$$
$$= 9ab - 9ac + 9ad$$

16. $5xy - 5xz + 5xw$

17. $5x + 25 = 5 \cdot x + 5 \cdot 5 = 5(x + 5)$

18. $7(a + b)$

19. $3p - 9 = 3 \cdot p - 3 \cdot 3 = 3(p - 3)$

20. $3(5x - 1)$

21. $7x - 21y + 14z = 7 \cdot x - 7 \cdot 3y + 7 \cdot 2z = 7(x - 3y + 2z)$

22. $3(2y - 3z - w)$

23. $255 - 34b = 17 \cdot 15 - 17 \cdot 2b = 17(15 - 2b)$

24. $33(4a + 1)$

25. $xy + x = x \cdot y + x \cdot 1 = x(y + 1)$

26. $b(a + 1)$

27. $4x - 5y + 3 = 4x + (-5y) + 3$
The terms are $4x$, $-5y$, and 3.

28. $2a, 7b, -13$

9. $x^2 - 6x - 7 = x^2 + (-6x) + (-7)$
The terms are x^2, $-6x$, and -7.

10. $-3y^2$, $6y$, 10

11. $3x + 7x = (3 + 7)x = 10x$

12. $12x$

13. $7rt - 9rt = (7 - 9)rt = -2rt$

14. $10ab$

15. $9t^2 + t^2 = (9 + 1)t^2 = 10t^2$

16. $8a^2$

17. $12a - a = (12 - 1)a = 11a$

18. $14x$

19. $n - 8n = (1 - 8)n = -7n$

20. $-5x$

21. $5x - 3x + 8x = (5 - 3 + 8)x = 10x$

22. $-6x$

23.
$4x - 2x^2 + 3x$
$= 4x + 3x - 2x^2$ Commutative law of
                    addition
$= (4 + 3)x - 2x^2$
$= 7x - 2x^2$

24. $13a - 5a^2$

25.
$6a + 7a^2 - a + 4a^2$
$= 6a - a + 7a^2 + 4a^2$ Commutative law of
                         addition
$= (6 - 1)a + (7 + 4)a^2$
$= 5a + 11a^2$

26. $14x + 2x^3 - 6x^2$

27.
$4x - 7 + 18x + 25$
$= 4x + 18x - 7 + 25$
$= (4 + 18)x + (-7 + 25)$
$= 22x + 18$

28. $9p + 12$

29.
$-7t^2 + 3t + 5t^3 - t^3 + 2t^2 - t$
$= (-7 + 2)t^2 + (3 - 1)t + (5 - 1)t^3$
$= -5t^2 + 2t + 4t^3$

30. $-12n + 6n^2 + 5n^3$

51.
$7a - (2a + 5)$
$= 7a - 2a - 5$
$= 5a - 5$

52. $-4x - 9$

53.
$m - (3m - 1)$
$= m - 3m + 1$
$= -2m + 1$

54. $a + 3$

55.
$3d - 7 - (5 - 2d)$
$= 3d - 7 - 5 + 2d$
$= 5d - 12$

56. $13x - 16$

57.
$-2(x + 3) - 5(x - 4)$
$= -2x - 6 - 5x + 20$
$= -7x + 14$

58. $-15y - 45$

59.
$4x - 7(2x - 3)$
$= 4x - 14x + 21$
$= -10x + 21$

60. $-11y + 24$

61.
$9a - [7 - 5(7a - 3)]$
$= 9a - [7 - 35a + 15]$
$= 9a - [22 - 35a]$
$= 9a - 22 + 35a$
$= 44a - 22$

62. $47b - 51$

63.
$5\{-2a + 3[4 - 2(3a + 5)]\}$
$= 5\{-2a + 3[4 - 6a - 10]\}$
$= 5\{-2a + 3[-6 - 6a]\}$
$= 5\{-2a - 18 - 18a\}$
$= 5\{-20a - 18\}$
$= -100a - 90$

64. $-721x - 728$

65.
$2y + \{7[3(2y - 5) - (8y + 7)] + 9\}$
$= 2y + \{7[6y - 15 - 8y - 7] + 9\}$
$= 2y + \{7[-2y - 22] + 9\}$
$= 2y + \{-14y - 154 + 9\}$
$= 2y + \{-14y - 145\}$
$= 2y - 14y - 145$
$= -12y - 145$

66. $-11b + 217$

67. Substitute each value in the first expression.

$$2x - 3(x + 5) = 2(1.3) - 3(1.3 + 5)$$
$$= 2(1.3) - 3(6.3)$$
$$= 2.6 - 18.9$$
$$= -16.3$$
$$2x - 3(x + 5) = 2(-3) - 3(-3 + 5)$$
$$= 2(-3) - 3(2)$$
$$= -6 - 6$$
$$= -12$$
$$2x - 3(x + 5) = 2 \cdot 0 - 3(0 + 5)$$
$$= 2 \cdot 0 - 3(5)$$
$$= 0 - 15$$
$$= -15$$

Now substitute each value in the second expression.

$$-x + 15 = -1.3 + 15 = 13.7$$
$$-x + 15 = -(-3) + 15 = 3 + 15 = 18$$
$$-x + 15 = -0 + 15 = 15$$

Since the values of the expression differ when one or more of the replacements are the same, the expressions are not equivalent.

68. $-3.9; 9; 0$

$9.1; -21; 0;$

Not equivalent

69. Substitute each value in the first expression.

$$4(x + 3) = 4(1.3 + 3) = 4(4.3) = 17.2$$
$$4(x + 3) = 4(-3 + 3) = 4 \cdot 0 = 0$$
$$4(x + 3) = 4(0 + 3) = 4 \cdot 3 = 12$$

Now substitute each value in the second expression.

$$4x + 12 = 4(1.3) + 12 = 5.2 + 12 = 17.2$$
$$4x + 12 = 4(-3) + 12 = -12 + 12 = 0$$
$$4x + 12 = 4 \cdot 0 + 12 = 0 + 12 = 12$$

Since the values of the expressions are the same for three replacements, we can be reasonably certain that the expressions are equivalent.

70. $10.1; -20; 1$

$16.1; -14; 7$

Not equivalent

71. *Writing Exercise*

72. *Writing Exercise*

73. $|-3| = 3$ -3 is 3 units from 0.

74. 3.59

75. The opposite of -35 is 35, because $-35 + 35 = 0$.

76. $-\dfrac{1}{35}$

77. *Writing Exercise*

78. *Writing Exercise*

79.
$$11(a-3)+12a-\{6[4(3b-7)-(9b+10)]+11\}$$
$$= 11(a - 3) + 12a - \{6[12b - 28 - 9b - 10] + 11\}$$
$$= 11(a - 3) + 12a - \{6[3b - 38] + 11\}$$
$$= 11(a - 3) + 12a - \{18b - 228 + 11\}$$
$$= 11(a - 3) + 12a - \{18b - 217\}$$
$$= 11a - 33 + 12a - 18b + 217$$
$$= 23a - 18b + 184$$

80. $-42x - 360y - 276$

81.
$$z - \{2z + [3z - (4z + 5z) - 6z] + 7z\} - 8z$$
$$= z - \{2z + [3z - 9z - 6z] + 7z\} - 8z$$
$$= z - \{2z - 12z + 7z\} - 8z$$
$$= z - \{-3z\} - 8z$$
$$= z + 3z - 8z$$
$$= -4z$$

82. $4x - f$

83.
$$x - \{x + 1 - [x + 2 - (x - 3 - \{x + 4 -$$
$$[x - 5 + (x - 6)]\})]\}$$
$$= x - \{x + 1 - [x + 2 - (x - 3 - \{x + 4 -$$
$$[x - 5 + x - 6]\})]\}$$
$$= x - \{x + 1 - [x + 2 - (x - 3 - \{x + 4 -$$
$$[2x - 11]\})]\}$$
$$= x - \{x + 1 - [x + 2 - (x - 3 - \{x + 4 - 2x + 11\})]\}$$
$$= x - \{x + 1 - [x + 2 - (x - 3 - \{-x + 15\})]\}$$
$$= x - \{x + 1 - [x + 2 - (x - 3 + x - 15)]\}$$
$$= x - \{x + 1 - [x + 2 - (2x - 18)]\}$$
$$= x - \{x + 1 - [x + 2 - 2x + 18]\}$$
$$= x - \{x + 1 - [-x + 20]\}$$
$$= x - \{x + 1 + x - 20\}$$
$$= x - \{2x - 19\}$$
$$= x - 2x + 19$$
$$= -x + 19$$

4.
$$5(a + bc)$$
$= 5a + 5(bc)$ Distributive law
$= 5(bc) + 5a$ Commutative law of addition
$= (bc)5 + 5a$ Commutative law of multiplication
$= (cb)5 + 5a$ Commutative law of multiplication
$= c(b5) + 5a$ Associative law of multiplication
$= c(b5) + a5$ Commutative law of multiplication

5. *Writing Exercise*

6. *Writing Exercise*

---

xercise Set 1.4

1. $5^6 \cdot 5^3 = 5^{6+3} = 5^9$

2. 6^8

3. $t^0 \cdot t^8 = t^{0+8} = t^8$

4. x^5

5. $6x^5 \cdot 3x^2 = 6 \cdot 3 \cdot x^5 \cdot x^2 = 18x^{5+2} = 18x^7$

6. $8a^{10}$

7. $(-3m^4)(-7m^9) = (-3)(-7)m^4 \cdot m^9 = 21m^{4+9} = 21m^{13}$

8. $-14a^9$

9. $(x^3y^4)(x^7y^6z^0) = (x^3x^7)(y^4y^6)(z^0) = x^{3+7}y^{4+6} \cdot 1 = x^{10}y^{10}$

10. $m^{10}n^{12}$

11. $\dfrac{a^9}{a^3} = a^{9-3} = a^6$

12. x^9

13. $\dfrac{12t^7}{4t^2} = \dfrac{12}{4} \cdot t^{7-2} = 3t^5$

14. $4a^{16}$

15. $\dfrac{m^7n^9}{m^2n^5} = m^{7-2} \cdot n^{9-5} = m^5n^4$

16. m^8n^3

17. $\dfrac{28x^{10}y^9z^8}{-7x^2y^3z^2} = \dfrac{28}{-7} \cdot x^{10-2}y^{9-3}z^{8-2} = -4x^8y^6z^6$

18. $-6x^6y^3z^6$

19. $-x^0 = -(-2)^0 = -(1) = -1$

20. 1

21. $(4x)^0 = (4(-2))^0 = (-8)^0 = 1$

22. 4

23. $(-3)^4 = -3(-3)(-3)(-3) = 81$

24. 64

25. $-3^4 = -3 \cdot 3 \cdot 3 \cdot 3 = -81$

26. -64

27. $(-4)^{-2} = \dfrac{1}{(-4)^2} = \dfrac{1}{-4(-4)} = \dfrac{1}{16}$

28. $\dfrac{1}{25}$

29. $-4^{-2} = -\dfrac{1}{4^2} = -\dfrac{1}{16}$

30. $-\dfrac{1}{25}$

31. $-1^{-4} = -\dfrac{1}{1^4} = -\dfrac{1}{1} = -1$

32. $\dfrac{1}{a^3}$

33. $n^{-6} = \dfrac{1}{n^6}$

34. 5^3, or 125

35. $\dfrac{1}{2^{-6}} = 2^6 = 64$

36. $\dfrac{4}{x^7}$

37. $2a^2b^{-6} = 2a^2 \cdot \dfrac{1}{b^6} = \dfrac{2a^2}{b^6}$

38. $\dfrac{1}{x^2y^5}$

39. We can move x^{-2} and z^{-3} to the other side of the fraction bar if we change the sign of the exponent.
$$\dfrac{y^4z^{-3}}{x^{-2}} = \dfrac{x^2y^4}{z^3}$$

40. $\dfrac{y^7z^4}{x^2}$

41. $\dfrac{1}{3^4} = 3^{-4}$

42. $(-8)^{-6}$

43. $x^5 = \dfrac{1}{x^{-5}}$

44. $\dfrac{1}{n^{-3}}$

45. $6x^2 = 6 \cdot \dfrac{1}{x^{-2}} = \dfrac{6}{x^{-2}}$, or $\dfrac{1}{6^{-1}x^{-2}}$

46. $\dfrac{1}{(-4y)^{-5}}$

47. $\dfrac{1}{(5y)^3} = (5y)^{-3}$

48. $\dfrac{y^{-4}}{3}$, or $3^{-1}y^{-4}$

49. $7^{-1} \cdot 7^{-6} = 7^{-1+(-6)} = 7^{-7}$, or $\dfrac{1}{7^7}$

50. 3^{-12}, or $\dfrac{1}{3^{12}}$

51. $b^2 \cdot b^{-5} = b^{2+(-5)} = b^{-3}$, or $\dfrac{1}{b^3}$

52. a

53. $a^{-3} \cdot a^4 \cdot a^2 = a^{-3+4+2} = a^3$

54. 1

55. $(5a^{-2}b^{-3})(2a^{-4}b) = 5 \cdot 2 \cdot a^{-2} \cdot a^{-4} \cdot b^{-3} \cdot b$
$$= 10a^{-2+(-4)}b^{-3+1}$$
$$= 10a^{-6}b^{-2}, \text{ or } \dfrac{10}{a^6b^2}$$

56. $6a^{-4}b^{-9}$, or $\dfrac{6}{a^4b^9}$

57. $\dfrac{10^{-3}}{10^6} = 10^{-3-6} = 10^{-9}$, or $\dfrac{1}{10^9}$

58. 12^{-12}, or $\dfrac{1}{12^{12}}$

59. $\dfrac{2^{-7}}{2^{-5}} = 2^{-7-(-5)} = 2^{-7+5} = 2^{-2}$, or $\dfrac{1}{2^2}$, or $\dfrac{1}{4}$

60. 9^2, or 81

61. $\dfrac{8^{-7}}{8^{-2}} = 8^{-7-(-2)} = 8^{-7+2} = 8^{-5}$, or $\dfrac{1}{8^5}$

62. 10^2, or 100

63. $\dfrac{-5x^{-2}y^4z^7}{30x^{-5}y^6z^{-3}} = \dfrac{-5}{30}x^{-2-(-5)}y^{4-6}z^{7-(-3)} =$
$$-\dfrac{1}{6}x^3y^{-2}z^{10}, \text{ or } -\dfrac{x^3z^{10}}{6y^2}$$

64. $\dfrac{1}{3}a^{10}b^{-9}c^{-2}$, or $\dfrac{a^{10}}{3b^9c^2}$

65. $(x^4)^3 = x^{4 \cdot 3} = x^{12}$

66. a^6

67. $(9^3)^{-4} = 9^{3(-4)} = 9^{-12}$, or $\dfrac{1}{9^{12}}$

68. 8^{-12}, or $\dfrac{1}{8^{12}}$

69. $(t^{-8})^{-5} = t^{-8(-5)} = t^{40}$

70. 6^{12}

71. $(a^3b)^4 = a^{3 \cdot 4}b^4 = a^{12}b^4$

72. $x^{15}y^5$

73. $5(x^2y^2)^{-7} = 5(x^2)^{-7}(y^2)^{-7} = 5x^{2(-7)}y^{2(-7)} =$
$$5x^{-14}y^{-14}, \text{ or } \dfrac{5}{x^{14}y^{14}}$$

74. $7a^{-15}b^{-20}$, or $\dfrac{7}{a^{15}b^{20}}$

75. $(a^{-5}b^2)^3(a^4b^{-1})^2 = (a^{-5})^3(b^2)^3(a^4)^2(b^{-1})^2$
$$= a^{-5 \cdot 3}b^{2 \cdot 3}a^{4 \cdot 2}b^{-1 \cdot 2}$$
$$= a^{-15}b^6a^8b^{-2}$$
$$= a^{-15+8}b^{6+(-2)}$$
$$= a^{-7}b^4, \text{ or } \dfrac{b^4}{a^7}$$

76. $x^{-2}y^{-7}$, or $\dfrac{1}{x^2y^7}$

77. $(5x^{-3}y^2)^{-4}(5x^{-3}y^2)^4 = (5x^{-3}y^2)^{-4+4} = (5x^{-3}y^2)^0 = 1$

78. $2^{-4}a^4b^{-12}$, or $\dfrac{a^4}{16b^{12}}$

79. $\dfrac{(3x^3y^4)^3}{6xy^3} = \dfrac{3^3(x^3)^3(y^4)^3}{6xy^3} = \dfrac{27x^9y^{12}}{6xy^3} =$
$$\dfrac{27}{6}x^{9-1}y^{12-3} = \dfrac{9}{2}x^8y^9, \text{ or } \dfrac{9x^8y^9}{2}$$

80. $\dfrac{5}{2}a^4b$

81. $\left(\dfrac{-4x^4y^{-2}}{5x^{-1}y^4}\right)^{-4} = \left(\dfrac{-4}{5}x^{4-(-1)}y^{-2-4}\right)^{-4} =$
$$\left(\dfrac{-4}{5}x^5y^{-6}\right)^{-4} = \dfrac{(-4)^{-4}(x^5)^{-4}(y^{-6})^{-4}}{5^{-4}} =$$
$$\dfrac{5^4x^{-20}y^{24}}{(-4)^4} = \dfrac{625x^{-20}y^{24}}{256}, \text{ or } \dfrac{625y^{24}}{256x^{20}}$$

$\dfrac{8x^9 y^3}{27}$

$\left(\dfrac{4a^3 b^{-9}}{2a^{-2} b^5}\right)^0 = 1$

(Any nonzero real number raised to the zero power is 1.)

. 1

$\left(\dfrac{6a^3 b^{-4}}{5a^{-2} b^2}\right)^{-2} = \left(\dfrac{6a^5 b^{-6}}{5}\right)^{-2} = \left(\dfrac{5}{6a^5 b^{-6}}\right)^2 =$

$\dfrac{25}{36a^{10} b^{-12}} = \dfrac{25}{36} a^{-10} b^{12}$, or $\dfrac{25 b^{12}}{36 a^{10}}$

$\dfrac{625}{81} m^{16} n^{-28}$, or $\dfrac{625 m^{16}}{81 n^{28}}$

$\left(\dfrac{8x^3 y^{-10}}{6x^{-5} y^{-6}}\right)^3 = \left(\dfrac{4x^8 y^{-4}}{3}\right)^3 = \dfrac{64 x^{24} y^{-12}}{27}$, or

$\dfrac{64 x^{24}}{27 y^{12}}$

$\dfrac{4}{9} x^{-4} y^4$, or $\dfrac{4y^4}{9x^4}$

.

$47,000,000,000$

$= \dfrac{47,000,000,000}{10^{10}} \cdot 10^{10}$ Multiplying by 1:

$\dfrac{10^{10}}{10^{10}} = 1$

$= 4.7 \times 10^{10}$ This is scientific notation.

. 2.6×10^{12}

.

0.000000016

$= \dfrac{0.000000016}{10^8} \cdot 10^8$ Multiplying by 1:

$\dfrac{10^8}{10^8} = 1$

$= \dfrac{1.6}{10^8}$

$= 1.6 \times 10^{-8}$ Writing scientific notation

2. 2.63×10^{-7}

3.

$407,000,000,000$

$= \dfrac{407,000,000,000}{10^{11}} \cdot 10^{11}$

$= 4.07 \times 10^{11}$

4. 3.09×10^{12}

---

95. 0.000000603

$= \dfrac{0.000000603}{10^7} \cdot 10^7$

$= \dfrac{6.03}{10^7}$

$= 6.03 \times 10^{-7}$

96. 8.02×10^{-9}

97. $4 \times 10^{-4} = 0.0004$ Moving the decimal point 4 places to the left

98. 0.00005

99. $6.73 \times 10^8 = 673,000,000$ Moving the decimal point 8 places to the right

100. $92,400,000$

101. $8.923 \times 10^{-10} = 0.0000000008923$ Moving the decimal point 10 places to the left

102. 0.07034

103. $9.03 \times 10^{10} = 90,300,000,000$ Moving the decimal point 10 places to the right

104. $90,010,000,000$

105. $(2.3 \times 10^6)(4.2 \times 10^{-11})$

$= (2.3 \times 4.2)(10^6 \times 10^{-11})$

$= 9.66 \times 10^{-5}$

106. 3.38×10^{-4}

107. $(2.34 \times 10^{-8})(5.7 \times 10^{-4})$

$= (2.34 \times 5.7)(10^{-8} \times 10^{-4})$

$= 13.338 \times 10^{-12}$

$= (1.3338 \times 10^1) \times 10^{-12}$

$= 1.3338 \times (10^1 \times 10^{-12})$

$= 1.3338 \times 10^{-11}$

108. 2.6732×10^{-11}

109. $(2.506 \times 10^{-7})(1.408 \times 10^{10})$

$= (2.506 \times 1.408)(10^{-7} \times 10^{10})$

$= 3.528448 \times 10^3$

110. 1.4663385×10^{-1}

111. $\dfrac{5.1 \times 10^6}{3.4 \times 10^3} = \dfrac{5.1}{3.4} \times \dfrac{10^6}{10^3}$

$= 1.5 \times 10^3$

112. 2.5×10^3

113. $\dfrac{7.5 \times 10^{-9}}{2.5 \times 10^{-4}} = \dfrac{7.5}{2.5} \times \dfrac{10^{-9}}{10^{-4}}$

$\qquad = 3.0 \times 10^{-5}$ (2 significant digits)

114. 3.0×10^{11}

115. $\dfrac{1.23 \times 10^{8}}{6.87 \times 10^{-13}} = \dfrac{1.23}{6.87} \times \dfrac{10^{8}}{10^{-13}}$

$\qquad \approx 0.1790393 \times 10^{21}$

$\qquad \approx (1.790393 \times 10^{-1}) \times 10^{21}$

$\qquad \approx 1.790393 \times (10^{-1} \times 10^{21})$

$\qquad \approx 1.790393 \times 10^{20}$

116. $3.0182927 \times 10^{-13}$

117. $\dfrac{780,000,000 \times 0.00071}{0.000005}$

$= \dfrac{(7.8 \times 10^{8}) \times (7.1 \times 10^{-4})}{5 \times 10^{-6}}$

$= \dfrac{7.8 \times 7.1}{5} \times \dfrac{10^{8} \times 10^{-4}}{10^{-6}}$

$= 11.076 \times 10^{10}$

$= 1.1076 \times 10^{1} \times 10^{10}$

$= 1.1076 \times (10^{1} \times 10^{10})$

$= 1.1076 \times 10^{11}$

118. 3.213×10

119. $\quad 5.9 \times 10^{23} + 2.4 \times 10^{23}$

$= (5.9 + 2.4) \times 10^{23}$

$= 8.3 \times 10^{23}$

120. 7.2×10^{-34}

121. *Writing Exercise*

122. *Writing Exercise*

123. $-\dfrac{5}{6} - \left(-\dfrac{3}{4}\right) = -\dfrac{5}{6} + \dfrac{3}{4} = -\dfrac{10}{12} + \dfrac{9}{12} = -\dfrac{1}{12}$

124. 30.96

(The product of a positive number and a negative number is negative.)

125. $-2(4x - 6y) = -8x + 12y$

126. $2(4x - 5)$

127. *Writing Exercise*

128. *Writing Exercise*

129. *Writing Exercise*

130. $4a^{-x-4}$

131. $\dfrac{-12x^{a+1}}{4x^{2-a}} = -3x^{a+1-(2-a)} = -3x^{a+1-2+a} = -3x^{2a-1}$

132. 3^{a^2+2a}

133. $(12^{3-a})^{2b} = 12^{(3-a)2b} = 12^{6b-2ab}$

134. $2x^{a+2}y^{b-2}$

135. $\dfrac{25x^{a+b}y^{b-a}}{-5x^{a-b}y^{b+a}} = -5x^{a+b-(a-b)}y^{b-a-(b+a)} =$

$-5x^{a+b-a+b}y^{b-a-b-a} = -5x^{2b}y^{-2a}$

136. T8 $\cdot 10^{-90}$ is larger than $9 \cdot 10^{-91}$. Scientific notation for the difference is 7.1×10^{-90}.

137. $\dfrac{1}{8.00 \times 10^{-23}} = \dfrac{1}{8.00} \times \dfrac{1}{10^{-23}} = 0.125 \times 10^{23} =$

1.25×10^{22}

138. 8

139. The unit's digit is the digit 3 raised to some power. Look for a pattern:

Exponent	1	2	3	4	5	6	7	8
Unit's digit	3	9	7	1	3	9	7	1

$(513)^{128} = (513^{4})^{32}$, so the unit's digit is 1.

140. Approximately 8×10^{18} grains of sand

Exercise Set 1.5

1. A is 5 units right of the origin and 3 units up, so its coordinates are $(5, 3)$.

B is 4 units left of the origin and 3 units up, so its coordinates are $(-4, 3)$.

C is 0 units right or left of the origin and 2 units up, so its coordinates are $(0, 2)$.

D is 2 units left of the origin and 3 units down, so its coordinates are $(-2, -3)$.

E is 4 units right of the origin and 2 units down, so its coordinates are $(4, -2)$.

F is 5 units left of the origin and 0 units up or down, so its coordinates are $(-5, 0)$.

2. $(2, 4); (-3, 1); (0, -2); (2, -2); (-5, -4); (4, 0)$

5.

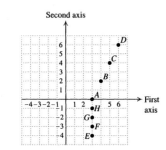

$A(3, 0)$ is 3 units right and 0 units up or down.

$B(4, 2)$ is 4 units right and 2 units up.

$C(5, 4)$ is 5 units right and 4 units up.

$D(6, 6)$ is 6 units right and 6 units up.

$E(3, -4)$ is 3 units right and 4 units down.

$F(3, -3)$ is 3 units right and 3 units down.

$G(3, -2)$ is 3 units right and 2 units down.

$H(3, -1)$ is 3 units right and 1 unit down.

6.

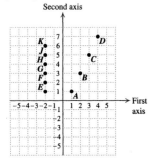

7.

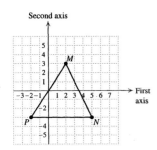

A triangle is formed. The area of a triangle is found by using the formula $A = \frac{1}{2}bh$. In this triangle the base and height are 7 units and 6 units, respectively.

$A = \frac{1}{2}bh = \frac{1}{2} \cdot 7 \cdot 6 = \frac{42}{2} = 21$ square units

8. A parallelogram is formed; 36 square units

7. Both coordinates are negative, so the point $(-4, -9)$ is in quadrant III.

8. I

9. The first coordinate is negative and the second positive, so the point $(-6, 1)$ is in quadrant II.

10. IV

11. Both coordinates are positive, so the point $\left(3, \frac{1}{2}\right)$ is in quadrant I.

12. III

13. The first coordinate is positive and the second negative, so the point $(6.9, -2)$ is in quadrant IV.

14. II

15.

$$y = 3x - 4$$

-1 ? $3 \cdot 1 - 4$ Substituting 1 for x and -1 for y

$3 - 4$ (alphabetical order of variables)

$-1 \mid -1$

Since $-1 = -1$ is true, $(1, -1)$ is a solution of $y = 3x - 4$.

16. Yes

17.

$$5s - t = 8$$

$5 \cdot 2 - 4$? 8 Substituting 2 for s and 4 for t

$10 - 4$ (alphabetical order of variables)

$6 \mid 8$

Since $6 = 8$ is false, $(2, 4)$ is not a solution of $5s - t = 8$.

18. No

19.

$$4x - y = 7$$

$4 \cdot 3 - 5$? 7 Substituting 3 for x and 5 for y

$12 - 5$ (alphabetical order of variables)

$7 \mid 7$

Since $7 = 7$ is true, $(3, 5)$ is a solution of $4x - y = 7$.

20. Yes

21.

$$6a + 5b = 3$$

$6 \cdot 0 + 5 \cdot \frac{3}{5}$? 3 Substituting 0 for a and $\frac{3}{5}$ for b

$0 + 3$ (alphabetical order of variables)

$3 \mid 3$

Since $3 = 3$ is true, $\left(0, \frac{3}{5}\right)$ is a solution of $6a + 5b = 3$.

22. Yes

23.
$$4r + 3s = 5$$
$$4\cdot2+3\cdot(-1) \ ? \ 5 \quad \text{Substituting 2 for } r \text{ and}$$
$$ -1 \text{ for } s$$
$$8 - 3 \quad\quad \text{(alphabetical order of}$$
$$ \text{variables)}$$
$$5 \ \big| \ 5$$

Since $5 = 5$ is true, $(2, -1)$ is a solution of $4r + 3s = 5$.

24. Yes

25. $x - 3y = -4$
$$5-3\cdot3 \ ? \ -4 \quad \text{Substituting 5 for } x \text{ and 3 for } y$$
$$5 - 9 \quad\quad \text{(alphabetical order of variables)}$$
$$-4 \ \big| \ -4$$

Since $-4 = -4$ is true, $(5, 3)$ is a solution of
$x - 3y = 4$.

26. No

27. $y = 3x^2$
$$3 \ ? \ 3(-1)^2 \quad \text{Substituting } -1 \text{ for } x \text{ and 3 for } y$$
$$3\cdot1 \quad\quad \text{(alphabetical order of variables)}$$
$$3 \ \big| \ 3$$

Since $3 = 3$ is true, $(-1, 3)$ is a solution of $y = 3x^2$.

28. No

29. $5s^2 - t = 7$
$$5(2)^2 - 3 \ ? \ 7 \quad \text{Substituting 2 for } s \text{ and 3 for } t$$
$$5\cdot4 - 3 \quad\quad \text{(alphabetical order of variables)}$$
$$20 - 3$$
$$17 \ \big| \ 7$$

Since $17 = 7$ is false, $(2, 3)$ is not a solution of
$5s^2 - t = 7$.

30. Yes

31. $y = x + 1$

To find an ordered pair, we choose any number for x and then determine y by substitution.

When $x = 0$, $y = 0 + 1 = 1$.

When $x = 3$, $y = 3 + 1 = 4$.

When $x = -2$, $y = -2 + 1 = -1$.

x	y	(x, y)
0	1	$(0, 1)$
3	4	$(3, 4)$
-2	-1	$(-2, -1)$

Plot these points, draw the line they determine, and label the graph $y = x + 1$.

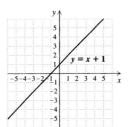

32.

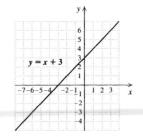

33. $y = -x$

To find an ordered pair, we choose any number for x and then determine y. For example, if we choose 1 for x, then $y = -1$. We find several ordered pairs, plot them, and draw the line.

x	y	(x, y)
1	-1	$(1, -1)$
2	-2	$(2, -2)$
-1	1	$(-1, 1)$
-3	3	$(-3, 3)$

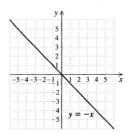

34.

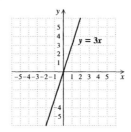

35. $y = -2x + 3$

To find an ordered pair, we choose any number for x and then determine y. For example, if $x = 1$, then $y = -2\cdot1 + 3 = -2 + 3 = 1$. We find several ordered pairs, plot them, and draw the line.

x	y
1	1
3	-3
-1	5
0	3

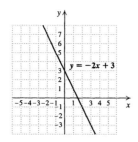

$y = -2x + 3$

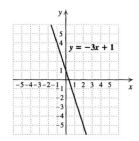

$y = -3x + 1$

7. $y + 2x = 3$

$y = -2x + 3$ Solving for y

Observe that this is the equation that was graphed in Exercise 35. The graph is shown above.

8. Since $y + 3x = 1$ is equivalent to $y = -3x + 1$, this is the equation that was graphed in Exercise 36. The graph is shown above.

9. $y = \dfrac{2}{3}x + 1$

To find an ordered pair, we choose any number for x and then determine y. For example, if $x = 3$, then $y = \dfrac{2}{3} \cdot 3 + 1 = 2 + 1 = 3$. We find several ordered pairs, plot them, and draw the line.

x	y
3	3
0	1
-3	-1

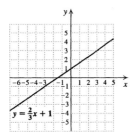

$y = \dfrac{2}{3}x + 1$

10.

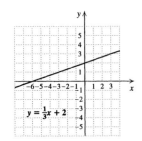

$y = \dfrac{1}{3}x + 2$

41.

X	Y₁
-3	5.5
-2	4
-1	2.5
0	1
1	-.5
2	-2
3	-3.5

X = -3

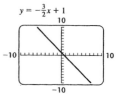

$y = -\dfrac{3}{2}x + 1$

42.

X	Y₁
-3	0
-2	-.6667
-1	-1.333
0	-2
1	-2.667
2	-3.333
3	-4

X = -3

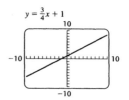

$y = -\dfrac{2}{3}x - 2$

43.

X	Y₁
-3	-1.25
-2	-.5
-1	.25
0	1
1	1.75
2	2.5
3	3.25

X = -3

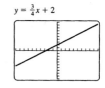

$y = \dfrac{3}{4}x + 1$

44.

X	Y₁
-3	-.25
-2	.5
-1	1.25
0	2
1	2.75
2	3.5
3	4.25

X = -3

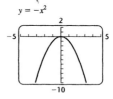

$y = \dfrac{3}{4}x + 2$

45.

X	Y₁
-3	-9
-2	-4
-1	-1
0	0
1	-1
2	-4
3	-9

X = -3

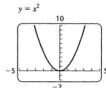

$y = -x^2$

46.

X	Y₁
-3	9
-2	4
-1	1
0	0
1	1
2	4
3	9

X = -3

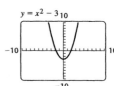

$y = x^2$

47.

X	Y₁
-3	6
-2	1
-1	-2
0	-3
1	-2
2	1
3	6

X = -3

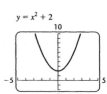

$y = x^2 - 3$

48.

X	Y₁
-3	11
-2	6
-1	3
0	2
1	3
2	6
3	11

X = -3

$y = x^2 + 2$

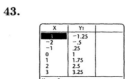

49.

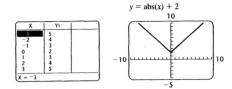

$y = \text{abs}(x) + 2$

50.

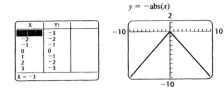

$y = -\text{abs}(x)$

51.

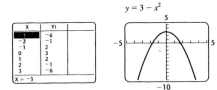

$y = 3 - x^2$

52.

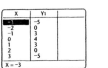

$y = 4 - x^2$

53.

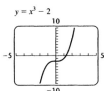

$y = x^3$

54.

$y = x^3 - 2$

55. Only window (b) shows where the graph crosses the x- and y-axes.

56. (b)

57. Only window (a) shows the shape of the graph and where it crosses the y-axis.

58. (a)

59. Only window (b) shows where the graph crosses the x- and y-axes.

60. (b)

61. The equations in the odd-numbered exercises 31-43 and in Exercise 55 have graphs that are straight lines, so they are linear equations.

62. 32-44 and 56

63. *Writing Exercise*

64. *Writing Exercise*

65. We replace x with 5.

$$3x - 5 = 10$$

$$3 \cdot 5 - 5 \ ? \ 10$$
$$15 - 5$$
$$10 \ | \ 10 \quad \text{TRUE}$$

Since $10 = 10$ is true, 5 is a solution of the equation.

66. Yes

67. We replace n with 4.

$$3n - 7 < 5$$

$$3 \cdot 4 - 7 \ ? \ 5$$
$$12 - 7$$
$$5 \ | \ 5 \quad \text{FALSE}$$

Since $5 < 5$ is false, 4 is not a solution of the inequality.

68. No

69. $5s - 3t = 5 \cdot 2 - 3 \cdot 4 = 10 - 12 = -2$

70. 49

71. $(5 - x)^4(x + 2)^3 = (5 - (-2))^4(-2 + 2)^3$

Observe that $-2 + 2 = 0$, so the product is 0.

72. -3

73. *Writing Exercise*

74. *Writing Exercise*

75. *Writing Exercise*

76. *Writing Exercise*

77. Substitute $-\dfrac{1}{3}$ for x and $\dfrac{1}{4}$ for y in each equation.

a)
$$-\frac{3}{2}x - 3y = -\frac{1}{4}$$

$$-\frac{3}{2}\left(-\frac{1}{3}\right) - 3\left(\frac{1}{4}\right) \ ? \ -\frac{1}{4}$$
$$\frac{1}{2} - \frac{3}{4}$$
$$-\frac{1}{4}$$

Since $-\dfrac{1}{4} = -\dfrac{1}{4}$ is true, $\left(-\dfrac{1}{3}, \dfrac{1}{4}\right)$ is a solution.

b)
$$8y - 15x = \frac{7}{2}$$

$$8\left(\frac{1}{4}\right) - 15\left(-\frac{1}{3}\right) \;?\; \frac{7}{2}$$

$$2 + 5$$

$$7$$

Since $7 = \frac{7}{2}$ is false, $\left(-\frac{1}{3}, \frac{1}{4}\right)$ is not a solution.

c)
$$0.16y = -0.09x + 0.1$$

$$0.16\left(\frac{1}{4}\right) \;?\; -0.09\left(-\frac{1}{3}\right) + 0.1$$

$$0.04 \;\Big|\; 0.03 + 0.1$$

$$0.13$$

Since $0.04 = 0.13$ is false, $\left(-\frac{1}{3}, \frac{1}{4}\right)$ is not a
solution.

d)
$$2(-y + 2) - \frac{1}{4}(3x - 1) = 4$$

$$2\left(-\frac{1}{4} + 2\right) - \frac{1}{4}\left[3\left(-\frac{1}{3}\right) - 1\right] \;?\; 4$$

$$2\left(\frac{7}{4}\right) - \frac{1}{4}(-2)$$

$$\frac{7}{2} + \frac{1}{2}$$

$$\frac{8}{2}$$

$$4$$

Since $4 = 4$ is true, $\left(-\frac{1}{3}, \frac{1}{4}\right)$ is a solution.

. $(2, 4)$; $(-5, -3)$; 49 square units

. Plot $(-10, -2)$, $(-3, 4)$, and $(6, 4)$, and sketch a parallelogram.

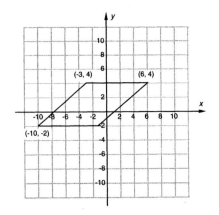

Since $(6,4)$ is 9 units directly to the right of $(-3,4)$, a
fourth vertex could lie 9 units directly to the right of
$(-10, -2)$. Then its coordinates are $(-10 + 9, -2)$, or
$(-1, -2)$.

If we connect the points in a different order, we get a second parallelogram.

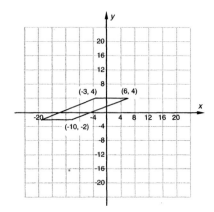

Since $(-3, 4)$ is 9 units directly to the left of $(6,4)$, a fourth
vertex could lie 9 units directly to the left of $(-10, -2)$.
Then its coordinates are $(-10 - 9, -2)$, or $(-19, -2)$.

If we connect the points in yet a different order, we get a
third parallelogram.

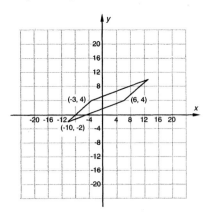

Since $(6, 4)$ lies 16 units directly to the right of and 6 units
above (-10,-2), a fourth vertex could lie 16 units to the
right of and 6 units above $(-3, 4)$. Its coordinates are
$(-3 + 16, 4 + 6)$, or $(13, 10)$.

80. -2

81. Equations (a), (c), and (d) appear to be linear.

82. $[0, 30, -50, 10]$, Xscl $= 10$, Yscl $= 10$, is a good choice.

Exercise Set 1.6

1. Three less than some number

Let n represent the number. Then we have

$n - 3$.

2. Let n represent the number; $n + 4$, or $4 + n$

3. Twelve times a number

Let t represent the number. Then we have

$12t$.

4. Let x represent the number; $2x$

5. Sixty-five percent of some number

Let x represent the number. Then we have

$0.65x$, or $\dfrac{65}{100}x$.

6. Let x represent the number; $0.39x$, or $\dfrac{39}{100}x$.

7. Ten more than twice a number

Let y represent the number. Then we have

$2y + 10$.

8. Let y represent the number; $\dfrac{1}{2}y - 6$

9. Eight more than ten percent of some number

Let s represent the number. Then we have

$0.1s + 8$, or $\dfrac{10}{100}s + 8$

10. Let s represent the number; $0.06s - 5$, or $\dfrac{6}{100}s - 5$

11. One less than the difference of two numbers

Let m and n represent the numbers. Then we have

$m - n - 1$.

12. Let m and n represent the numbers; $mn + 2$

13. Ninety miles per every four gallons of gas

We have

$90 \div 4$, or $\dfrac{90}{4}$.

14. $100 \div 60$, or $\dfrac{100}{60}$

15. **Familiarize**. There are two numbers involved, and we want to find both of them. We can let x represent the first number and note that the second number is 7 more than the first. Also, the sum of the numbers is 65.

Translate. The second number can be named $x + 7$. We translate to an equation:

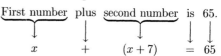

16. Let x and $x + 11$ represent the numbers;

$x + (x + 11) = 83$

17. **Familiarize**. Let $x =$ the number.

Translate.

128 is 0.4 of what number?
$128 = 0.4 \times x$

18. Let $n =$ the number; $456 = \dfrac{1}{3}n$

19. **Familiarize**. Let $n =$ the number. Then the quotient will be expressed as $\dfrac{n}{4}$.

Translate.

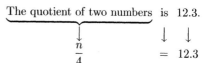

20. Let $x =$ the larger number; then $x - 65 =$ the smaller number; $x + (x - 65) = 92$

21. **Familiarize**. Recall that the perimeter P of a rectangle with length l and width w is given by the formula $P = 2l + 2w$. Let $x =$ the width of the rectangle. Then $2x =$ the length.

Translate. We substitute in the formula.

$$P = 2l + 2w$$
$$21 = 2 \cdot 2x + 2 \cdot x$$

22. Let $y =$ the length; then $\dfrac{1}{3}y =$ the width.

$2 \cdot y + 2 \cdot \dfrac{1}{3}y = 32$

23. **Familiarize**. Let $p =$ the original price. Then 35% of the original price is $35\%p$, or $0.35p$.

Translate.

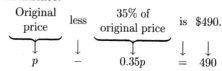

l. Let s = the number of students who would be expected to live in apartment communities; $0.3(1250) = s$

5. *Familiarize.* Let l = the length of an American bullfrog, in millimeters.

Translate.

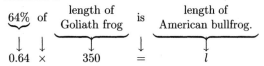

6. Let p = the price the store paid for the book.

$$p + 0.8p = 22$$

7. *Familiarize.* Recall that Distance = Rate × Time. Let t = the length of time Rhonda bicycled, in hours.

Translate.

$$\begin{array}{ccccc}
\text{Distance} & = & \text{Rate} & \times & \text{Time} \\
\downarrow & & \downarrow & \downarrow & \downarrow \\
25 & = & 15 & \times & t
\end{array}$$

8. Let t = the number of minutes Logan ran; $9 = \dfrac{6}{45}t$

9. *Familiarize.* There are three angle measures involved, and we want to find all three. We can let x represent the smallest angle measure and note that the second is one more than x and the third is one more than the second, or two more than x. We also note that the sum of the three angle measures must be $180°$.

Translate. The three angle measures are x, $x + 1$, and $x + 2$. We translate to an equation:

$$\begin{array}{ccccccc}
\text{First} & \text{plus} & \text{second} & \text{plus} & \text{third} & \text{is} & 180°. \\
\downarrow & \downarrow & \downarrow & \downarrow & \downarrow & \downarrow & \downarrow \\
x & + & (x+1) & + & (x+2) & = & 180
\end{array}$$

10. Let w = the wholesale price; $w + 0.5w + 0.25 = 1.99$, or $1.5w + 0.25 = 1.99$

11. *Familiarize.* Let c = the original cost of the order. Then the discounted cost is $c - 10\%c$, or $c - 0.1c$.

Translate.

$$\begin{array}{ccc}
\text{Discounted price} & \text{is} & \$279. \\
\downarrow & \downarrow & \downarrow \\
c - 0.1c & = & 279
\end{array}$$

12. Let c = the cost without the 5% discount, in dollars; $c - 0.05c = 142.50$

13. *Familiarize.* Let t represent the time required. Note that the plane must climb $29{,}000 - 8000$, or $21{,}000$ ft.

Translate.

$$\begin{array}{ccccc}
\text{Speed} & \times & \text{Time} & = & \text{Distance} \\
\downarrow & & \downarrow & & \downarrow \\
3500 & \times & t & = & 21{,}000
\end{array}$$

34. Let x represent the longer length; $x + \dfrac{2}{3}x = 10$

35. *Familiarize.* Let x represent the measure of the second angle. Then the first angle is three times x, and the third is $12°$ less than twice x. The sum of the three angle measures is $180°$.

Translate. The first angle is $3x$, the second is x, and the third is $2x - 12$. Translate to an equation:

$$\begin{array}{ccccccc}
\text{First} & \text{plus} & \text{second} & \text{plus} & \text{third} & \text{is} & 180°. \\
\downarrow & \downarrow & \downarrow & \downarrow & \downarrow & \downarrow & \downarrow \\
3x & + & x & + & (2x-12) & = & 180
\end{array}$$

36. Let x represent the measure of the second angle;

$$4x + x + (2x + 5) = 180$$

37. *Familiarize.* Note that each even integer is 2 more than the one preceding it. If we let n represent the first even integer, then $n + 2$ represents the next even integer.

Translate.

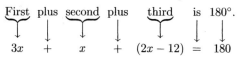

38. Let n represent the first odd integer;

$$n + 2(n + 2) + 3(n + 4) = 70$$

39. *Familiarize.* The perimeter of an equilateral triangle is 3 times the length of a side. Let s = the length of a side of the smaller triangle. Then $2s$ = the length of a side of the larger triangle. The sum of the two perimeters is 90 cm.

Translate.

$$\begin{array}{ccccc}
\text{Perimeter of} & \text{plus} & \text{perimeter of} & \text{is 90 cm.} \\
\text{smaller triangle} & & \text{larger triangle} & \\
\downarrow & \downarrow & \downarrow & \downarrow \\
3s & + & 3 \cdot 2s & = & 90
\end{array}$$

40. Let x represent the length of one piece;

$$\left(\frac{x}{4}\right)^2 = \left(\frac{100 - x}{4}\right)^2 + 144$$

41. *Familiarize.* After the next test there will be six test scores. The average of the six scores is their sum divided by 6. We let x represent the next test score.

Translate.

$$\begin{array}{cc}
\text{The average of the six scores} & \text{is } 88. \\
\downarrow & \downarrow \downarrow \\
\dfrac{93 + 89 + 72 + 80 + 96 + x}{6} & = 88
\end{array}$$

42. Let x represent the price of the least expensive set;

$$(x + 20) + (x + 6 \cdot 20) = x + 12 \cdot 20$$

43. Locate the point that is directly above 225. Then estimate its second coordinate by moving horizontally from the point to the vertical axis. The rate is about 75 heart attacks per 10,000 men.

44. 125 heart attacks per 10,000 men

45. Locate the point on the graph that is directly above '60. Then estimate its second coordinate by moving horizontally from the point to the vertical axis. In 1960, about 56% of Americans were willing to vote for a woman for president.

46. About 92%

47. Plot and connect the points, using body weight as the first coordinate and the corresponding number of drinks as the second coordinate.

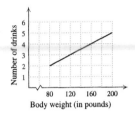

To estimate the number of drinks that a 140-lb person would have to drink to be considered intoxicated, first locate the point that is directly above 140. Then estimate its second coordinate by moving horizontally from the point to the vertical axis. Read the approximate function value there. The estimated number of drinks is 3.5.

48. 3 drinks

49. Enter the data and select the line graph option as described in Example 6. Then select a viewing window that will display all of the data points. The years range from 1980 through 2001 and the numbers of endangered species range from 32 through 64. One choice is [1979, 2010, 0, 70], Xscl = 10, Yscl = 10. Then press GRAPH to display the graph.

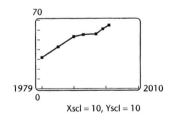

50.

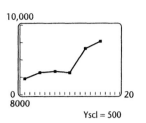

51. Replace f with y and t with x and graph $y = -\dfrac{1}{100}x + 2$ in an appropriate window. One good choice is $[0, 30, 0,$ with Xscl = 5. Note that $1985 - 1970 = 15$, so 1985 is years after 1970. Using the Value feature from the CAL menu we see that $y = 2.35$ when $x = 15$.

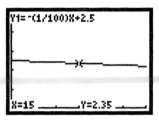

Thus, there were 2.35 million farms in the United State in 1985.

52. 2020

53. Replace R with y and t with x and graph $y = 0.3x^2$ $0.1x + 57.84$ in an appropriate window. One good choi is $[0, 15, 0, 100]$, Yscl = 10. Using the TRACE feature find that $x \approx 3$ when $y \approx 60$.

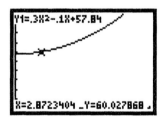

Thus, about 3 years after 1990, or in 1993, the averag room rate was $60.

54. $99.84

55. *Writing Exercise*

56. *Writing Exercise*

57. $(-2)^4 = (-2)(-2)(-2)(-2) = 16$

58. -16

59. $z^{-2} \cdot z^6 = z^{-2+6} = z^4$

60. z^{-8}, or $\dfrac{1}{z^8}$

1. $2(x^2y^3)^4 = 2(x^2)^4(y^3)^4 = 2x^{2\cdot4}y^{3\cdot4} = 2x^8y^{12}$

2. $16x^8y^{12}$

3. *Writing Exercise*

4. *Writing Exercise*

5. a) Graph III seems most appropriate for this situation. It reflects less than 40 hours of work per week until September, 40 hours per week from September until December, and more than 40 hours per week in December.

   b) Graph II seems most appropriate for this situation. It reflects 40 hours of work per week until September, 20 hours per week from September until December, and 40 hours per week again in December.

   c) Graph I seems most appropriate for this situation. It reflects more than 40 hours of work per week until September, 40 hours per week from September until December, and more than 40 hours per week again in December.

   d) Graph IV seems most appropriate for this situation. It reflects less than 40 hours of work per week until September, approximately 20 hours per week from September until December, and about 40 hours per week in December.

6. a) IV

   b) III

   c) I

   d) II

Chapter 2

Functions, Linear Equations, and Models

1. The correspondence is not a function, because a member of the domain (3) corresponds to more than one member of the range.

2. Yes

3. The correspondence is a function, because each member of the domain corresponds to just one member of the range.

4. Yes

5. The correspondence is a function, because each member of the domain corresponds to just one member of the range.

6. No

7. This correspondence is a function, because each Christmas tree has only one price.

8. Function

9. The correspondence is not a function, since it is reasonable to assume that at least one member of a rock band plays more than one instrument.

   The correspondence is a relation, since it is reasonable to assume that each member of a rock band plays at least one instrument.

10. Function

11. This correspondence is a function, because each number in the domain, when squared and then increased by 4, corresponds to only one number in the range.

12. Function

13. a) Locate 1 on the horizontal axis and then find the point on the graph for which 1 is the first coordinate. From that point, look to the vertical axis to find the corresponding y-coordinate, -2. Thus, $f(1) = -2$.

    b) The domain is the set of all x-values in the graph. It is $\{x | -2 \le x \le 5\}$.

    c) To determine which member(s) of the domain are paired with 2, locate 2 on the vertical axis. From there look left and right to the graph to find any points for which 2 is the second coordinate. One such point exists. Its first coordinate is 4. Thus, the x-value for which $f(x) = 2$ is 4.

    d) The range is the set of all y-values in the graph. It is $\{y | -3 \le y \le 4\}$.

14. a) -1

    b) $\{x | -4 \le x \le 3\}$

    c) -3

    d) $\{y | -2 \le y \le 5\}$

15. a) Locate 1 on the horizontal axis and then find the point on the graph for which 1 is the first coordinate. From that point, look to the vertical axis to find the corresponding y-coordinate, 3. Thus, $f(1) = 3$.

    b) The set of all x-values in the graph extends from -1 to 4, so the domain is $\{x | -1 \le x \le 4\}$.

    c) To determine which member(s) of the domain are paired with 2, locate 2 on the vertical axis. From there look left and right to the graph to find any points for which 2 is the second coordinate. One such point exists. Its first coordinate is 3. Thus, the x-value for which $f(x) = 2$ is 3.

    d) The set of all y-values in the graph extends from 1 to 4, so the range is $\{y | 1 \le y \le 4\}$.

16. a) 1

    b) $\{x | -3 \le x \le 5\}$

    c) -1

    d) $\{y | 0 \le y \le 4\}$

17. a) Locate 1 on the horizontal axis and the find the point on the graph for which 1 is the first coordinate. From that point, look to the vertical axis to find the corresponding y-coordinate. It appears to be -2. Thus, $f(1) = -2$.

    b) The set of all x-values in the graph extends from -4 to 2, so the domain is $\{x | -4 \le x \le 2\}$.

    c) To determine which member(s) of the domain are paired with 2, locate 2 on the vertical axis. From there look left and right to the graph to find any points for which 2 is the second coordinate. One such point exists. Its first coordinate is -2, so the x-value for which $f(x) = 2$ is -2.

    d) The set of all y-values in the graph extends from -3 to 3, so the range is $\{y | -3 \le y \le 3\}$.

a) 3

b) $\{x| -4 \leq x \leq 3\}$

c) 0

d) $\{y| -5 \leq y \leq 4\}$

a) Locate 1 on the horizontal axis and then find the point on the graph for which 1 is the first coordinate. From that point, look to the vertical axis to find the corresponding y-coordinate, 3. Thus, $f(1) = 3$.

b) The set of all x-values in the graph extends from -4 to 3, so the domain is $\{x| -4 \leq x \leq 3\}$.

c) To determine which member(s) of the domain are paired with 2, locate 2 on the vertical axis. From there look left and right to the graph to find any points for which 2 is the second coordinate. One such point exists. Its first coordinate is -3. Thus, the x-value for which $f(x) = 2$ is -3.

d) The set of all y-values in the graph extends from -2 to 5, so the range is $\{y| -2 \leq y \leq 5\}$.

a) 4

b) $\{x| -3 \leq x \leq 4\}$

c) -1

d) $\{y| 0 \leq y \leq 5\}$

a) Locate 1 on the horizontal axis and then find the point on the graph for which 1 is the first coordinate. From that point, look to the vertical axis to find the corresponding y-coordinate, 1. Thus, $f(1) = 1$.

b) The domain is the set of all x-values in the graph. It is $\{-3, -1, 1, 3, 5\}$.

c) To determine which member(s) of the domain are paired with 2, locate 2 on the vertical axis. From there look left and right to the graph to find any points for which 2 is the second coordinate. One such point exists. Its first coordinate is 3. Thus, the x-value for which $f(x) = 2$ is 3.

d) The range is the set of all y-values in the graph. It is $\{-1, 0, 1, 2, 3\}$.

a) 3

b) $\{-4, -3, -2, -1, 0, 1, 2\}$

c) $-2, 0$

d) $\{1, 2, 3, 4\}$

a) Locate 1 on the horizontal axis and then find the point on the graph for which 1 is the first coordinate. From that point, look to the vertical axis to find the corresponding y-coordinate, 4. Thus, $f(1) = 4$.

b) The set of all x-values in the graph extends from -3 to 4, so the domain is $\{x| -3 \leq x \leq 4\}$.

c) To determine which member(s) of the domain are paired with 2, locate 2 on the vertical axis. From there look left and right to the graph to find any points for which 2 is the second coordinate. There are two such points, $(-1, 2)$ and $(3, 2)$. Thus, the x-values for which $f(x) = 2$ are -1 and 3.

d) The set of all y-values in the graph extends from -4 to 5, so the range is $\{y| -4 \leq y \leq 5\}$.

24. a) 2

b) $\{x| -5 \leq x \leq 2\}$

c) $-5, 1$

d) $\{y| -3 \leq y \leq 5\}$

25. a) Locate 1 on the horizontal axis and then find the point on the graph for which 1 is the first coordinate. From that point, look to the vertical axis to find the corresponding y-coordinate, 1. Thus, $f(1) = 1$.

b) The set of all x-values in the graph extends from -4 to 5, so the domain is $\{x| -4 \leq x \leq 5\}$.

c) To determine which member(s) of the domain are paired with 2, locate 2 on the vertical axis. From there look left and right to the graph to find any points for which 2 is the second coordinate. All points in the set $\{x| 2 < x \leq 5\}$ satisfy this condition. These are the x-values for which $f(x) = 2$.

d) The domain is the set of all y-values in the graph. It is $\{-1, 1, 2\}$.

26. a) 2

b) $\{x| -4 \leq x \leq 4\}$

c) $\{x| 0 < x \leq 2\}$

d) $\{1, 2, 3, 4\}$

27. We can use the vertical line test:

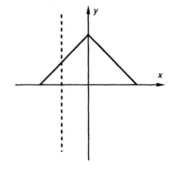

Visualize moving this vertical line across the graph. No vertical line will intersect the graph more than once. Thus, the graph is a graph of a function.

28. No

29. We can use the vertical line test:

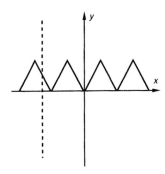

Visualize moving this vertical line across the graph. No vertical line will intersect the graph more than once. Thus, the graph is a graph of a function.

30. No

31. We can use the vertical line test.

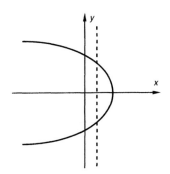

It is possible for a vertical line to intersect the graph more than once. Thus this is not the graph of a function.

32. Yes

33. We can use the vertical line test.

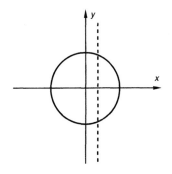

It is possible for a vertical line to intersect the graph more than once. Thus this is not a graph of a function.

34. Yes

35. $g(x) = 2x + 3$

    a) $g(0) = 2 \cdot 0 + 3 = 0 + 3 = 3$

    b) $g(-4) = 2(-4) + 3 = -8 + 3 = -5$

    c) $g(-7) = 2(-7) + 3 = -14 + 3 = -11$

    d) $g(8) = 2 \cdot 8 + 3 = 16 + 3 = 19$

    e) $g(a + 2) = 2(a + 2) + 3 = 2a + 4 + 3 = 2a + 7$

    f) $g(a) + 2 = (2a + 3) + 2 = 2a + 5$

36. a) 10

    b) 22

    c) -11

    d) -14

    e) $3a - 5$

    f) $3a - 3$

37. $f(n) = 5n^2 + 4n$

    a) $f(0) = 5 \cdot 0^2 + 4 \cdot 0 = 0 + 0 = 0$

    b) $f(-1) = 5(-1)^2 + 4(-1) = 5 - 4 = 1$

    c) $f(3) = 5 \cdot 3^2 + 4 \cdot 3 = 45 + 12 = 57$

    d) $f(t) = 5t^2 + 4t$

    e) $f(2a) = 5(2a)^2 + 4 \cdot 2a = 5 \cdot 4a^2 + 8a = 20a^2 + 8a$

    f) $2 \cdot f(a) = 2(5a^2 + 4a) = 10a^2 + 8a$

38. a) 0

    b) 5

    c) 21

    d) $3t^2 - 2t$

    e) $12a^2 - 4a$

    f) $6a^2 - 4a$

39. $f(x) = \dfrac{x - 3}{2x - 5}$

    a) $f(0) = \dfrac{0 - 3}{2 \cdot 0 - 5} = \dfrac{-3}{0 - 5} = \dfrac{-3}{-5} = \dfrac{3}{5}$

    b) $f(4) = \dfrac{4 - 3}{2 \cdot 4 - 5} = \dfrac{1}{8 - 5} = \dfrac{1}{3}$

    c) $f(-1) = \dfrac{-1 - 3}{2(-1) - 5} = \dfrac{-4}{-2 - 5} = \dfrac{-4}{-7} = \dfrac{4}{7}$

    d) $f(3) = \dfrac{3 - 3}{2 \cdot 3 - 5} = \dfrac{0}{6 - 5} = \dfrac{0}{1} = 0$

    e) $f(x + 2) = \dfrac{x + 2 - 3}{2(x + 2) - 5} = \dfrac{x - 1}{2x + 4 - 5} = \dfrac{x - 1}{2x - 1}$

. a) $\dfrac{26}{25}$

b) $\dfrac{2}{9}$

c) $-\dfrac{5}{12}$

d) $-\dfrac{7}{3}$

e) $\dfrac{3x+5}{2x+11}$

. $A(s) = s^2 \dfrac{\sqrt{3}}{4}$

$A(4) = 4^2 \dfrac{\sqrt{3}}{4} = 4\sqrt{3} \approx 6.93$

The area is $4\sqrt{3}$ cm$^2 \approx 6.93$ cm^2.

. $9\sqrt{3}$ in$^2 \approx 15.59$ in^2

. $V(r) = 4\pi r^2$

$V(3) = 4\pi(3)^2 = 36\pi$

The area is 36π in$^2 \approx 113.10$ in^2.

. 100π cm$^2 \approx 314.16$ cm^2

. $F(C) = \dfrac{9}{5}C + 32$

$F(-10) = \dfrac{9}{5}(-10) + 32 = -18 + 32 = 14$

The equivalent temperature is $14°$F.

. $41°$F

. $H(x) = 2.75x + 71.48$

$H(32) = 2.75(32) + 71.48 = 159.48$

The predicted height is 159.48 cm.

. 167.73 cm

. Plot and connect the points, using the year as the first coordinate and the corresponding number of reported cases of AIDS as the second coordinate.

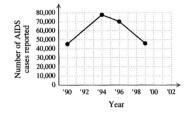

To estimate the number of cases of AIDS reported in 2002, extend the graph and extrapolate. It appears that about 24,000 cases of AIDS were reported in 2002.

. About 61,000 cases

51. Plot and connect the points, using the year as the first coordinate and the population as the second.

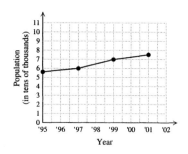

To estimate what the population was in 1998, first locate the point that is directly above 1998. Then estimate its second coordinate by moving horizontally from the point to the vertical axis. Read the approximate function value there. The population was about 65,000.

52. About 80,000

53. Plot and connect the points, using the year as the first coordinate and the sales total as the second coordinate.

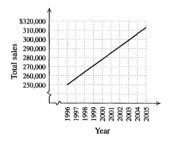

To predict the total sales for 2005, first locate the point directly above 2005. Then estimate its second coordinate by moving horizontally to the vertical axis. Read the approximate function value there. The predicted 2005 sales total is about $313,000.

54. About $271,000

55. *Writing Exercise*

56. *Writing Exercise*

57.
$3x - 5 - 9x + 15$
$= 3x - 9x - 5 + 15$
$= (3 - 9)x - 5 + 15$
$= -6x + 10$

58. $-8x - 3$

59. $2x + 4 - 6(5 - 7x) = 2x + 4 - 30 + 42x =$
$44x - 26$

60. $3y - 21$

61. $3 - 2[5(x - 7) + 1] = 3 - 2[5x - 35 + 1] =$
$3 - 2[5x - 34] = 3 - 10x + 68 = -10x + 71$

62. $-11x - 6$

63. $\dfrac{10 - 3^2}{9 - 2 \cdot 3} = \dfrac{10 - 9}{9 - 6} = \dfrac{1}{3}$

64. -1

65. *Writing Exercise*

66. *Writing Exercise*

67. To find $f(g(-4))$, we first find $g(-4)$:
$g(-4) = 2(-4) + 5 = -8 + 5 = -3$.
Then $f(g(-4)) = f(-3) = 3(-3)^2 - 1 = 3 \cdot 9 - 1 = 27 - 1 = 26$.
To find $g(f(-4))$, we first find $f(-4)$:
$f(-4) = 3(-4)^2 - 1 = 3 \cdot 16 - 1 = 48 - 1 = 47$.
Then $g(f(-4)) = g(47) = 2 \cdot 47 + 5 = 94 + 5 = 99$.

68. $26; 9$

69. Graph $y = (4/3)\pi x^3$ in an appropriate window such as $[0, 3, 0, 60]$, Yscl $= 10$. Then use the TRACE feature to find the value of x that corresponds to $y = 50$.

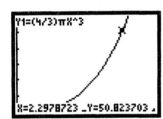

The radius is about 2.3 cm when the volume is 50 cm^3.

70. About 0.659 in.

71. Locate the highest point on the graph. Then move horizontally to the vertical axis and read the corresponding pressure. It is about 22 mm.

72. About 2 minutes, 50 seconds.

73. *Writing Exercise*

74. 1 every 3 minutes

75.

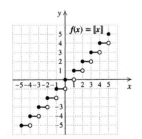

76. $g(x) = \dfrac{15}{4}x - \dfrac{13}{4}$

77. Graph the energy expenditures for walking and for bicycling on the same axes. Using the information given we plot and connect the points $\left(2\frac{1}{2}, 210\right)$ and $\left(3\frac{3}{4}, 300\right)$ for walking. We use the points $\left(5\frac{1}{2}, 210\right)$ and $(13, 660)$ for bicycling.

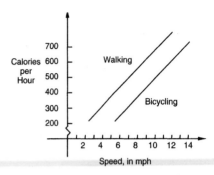

From the graph we see that walking $4\frac{1}{2}$ mph burns about 350 calories per hour and bicycling 14 mph burns about 725 calories per hour. Walking for two hours at $4\frac{1}{2}$ mph, then, would burn about $2 \cdot 350$, or 700 calories. Thus bicycling 14 mph for one hour burns more calories than walking $4\frac{1}{2}$ mph for two hours.

Exercise Set 2.2

1. The solution of the equation is the x-coordinate of the point of intersection of the graphs. This x-value appears to be -2.

2. 4

3. The solution of the equation is the x-coordinate of the point of intersection of the graphs. This x-value appears to be -4.

4. 1

5. The solution of the equation is the x-coordinate of the point of intersection of the graphs. This x-value appears to be 8.

6. 3

7. The solution of the equation is the x-coordinate of the point of intersection of the graphs. This x-value appears to be 0.

8. 2

9. The solution of the equation is the x-coordinate of the point of intersection of the graphs. This x-value appears to be 5.

10. 1

11. $x - 3 = 4$

Graph $f(x) = x - 3$ and $g(x) = 4$ on the same set of axes.

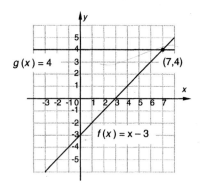

The lines appear to intersect at $(7, 4)$, so the solution is apparently 7.

Check: $x - 3 = 4$

$$7 - 3 \ ? \ 4$$
$$4 \ | \ 4 \quad \text{TRUE}$$

The solution is 7.

12. 2

13. $2x + 1 = 7$

Graph $f(x) = 2x + 1$ and $g(x) = 7$ on the same grid.

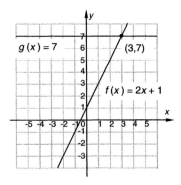

The lines appear to intersect at $(3, 7)$, so the solution is apparently 3.

Check: $2x + 1 = 7$

$$2 \cdot 3 + 1 \ ? \ 7$$
$$6 + 1 \ |$$
$$7 \ | \ 7 \quad \text{TRUE}$$

The solution is 3.

14. 2

15. $\frac{1}{3}x - 2 = 1$

Graph $f(x) = \frac{1}{3}x - 2$ and $g(x) = 1$ on the same grid.

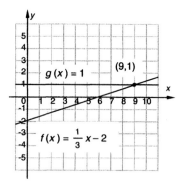

The lines appear to intersect at $(9, 1)$, so the solution is apparently 9.

Check: $\frac{1}{3}x - 2 = 1$

$$\frac{1}{3} \cdot 9 - 2 \ ? \ 1$$
$$3 - 2$$
$$1 \ | \ 1 \quad \text{TRUE}$$

The solution is 9.

16. -8

17. $x + 3 = 5 - x$

Graph $f(x) = x + 3$ and $g(x) = 5 - x$ on the same grid.

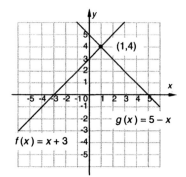

The lines appear to intersect at $(1, 4)$, so the solution is apparently 1.

Check: $x + 3 = 5 - x$

$$1 + 3 \ ? \ 5 - 1$$
$$4 \ | \ 4 \quad \text{TRUE}$$

The solution is 1.

18. -2

19. $5 - \dfrac{1}{2}x = x - 4$

Graph $f(x) = 5 - \dfrac{1}{2}x$ and $g(x) = x - 4$ on the same grid.

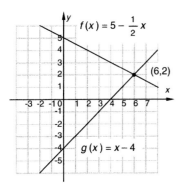

The lines appear to intersect at $(6, 2)$, so the solution is apparently 6.

Check: $5 - \dfrac{1}{2}x = x - 4$

$$5 - \dfrac{1}{2} \cdot 6 \ ?\ 6 - 4$$
$$5 - 3 \ \big|\ 2$$
$$2 \ \big|\ 2 \qquad \text{TRUE}$$

The solution is 6.

20. 4

21. $2x - 1 = -x + 3$

Graph $f(x) = 2x - 1$ and $g(x) = -x + 3$ on the same grid.

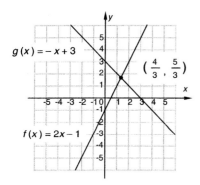

The lines appear to intersect at $\left(\dfrac{4}{3}, \dfrac{5}{3}\right)$, so the solution is apparently $\dfrac{4}{3}$.

Check: $2x - 1 = -x + 3$

$$2 \cdot \dfrac{4}{3} - 1 \ ?\ -\dfrac{4}{3} + 3$$
$$\dfrac{8}{3} - 1 \ \Big|\ \dfrac{5}{3}$$
$$\dfrac{5}{3} \ \Big|\ \dfrac{5}{3} \qquad \text{TRUE}$$

The solution is $\dfrac{4}{3}$. If we solve the equation using a graphing calculator, we get 1.33333.

22. $\dfrac{4}{3}$

23. Familiarize. A monthly fee is charged after the purchase of the phone. After one month of service, the total cost will be $50 + 25 = \$75$. After two months, the total cost will be $50 + 2 \cdot \$25 = \$100$. We can generalize this with a model, letting $C(t)$ represent the total cost, in dollars, for t months of service.

Translate. We reword the problem and translate.

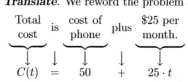

Carry out. First write the model in slope-intercept form: $C(t) = 25t + 50$. The vertical intercept is $(0, 50)$ and the slope, or rate, is $25 per month. Plot $(0, 50)$ and from there go *up* $25 and to the *right* 1 month. This takes us to $(1, 75)$. Draw a line passing through both points.

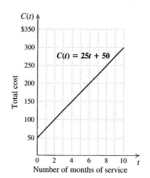

To estimate the time required for the total cost to reach $150, we are estimating the solution of $25t + 50 = 150$. We do this by graphing $y = 150$ and finding the point of intersection of the graphs.

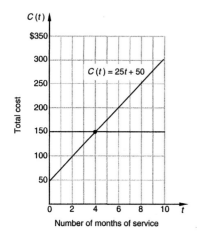

Number of months of service

This point appears to be $(4, 150)$. Thus, we estimate that it takes 4 months for the total cost to reach \$150.

Check. We evaluate.

$$C(4) = 25 \cdot 4 + 50$$
$$= 100 + 50$$
$$= 150$$

The estimate is precise.

State. It takes 4 months for the total cost to reach \$150.

, 5 months

. **Familiarize.** After an initial \$3.00 parking fee, an additional 50¢ fee is charged for each 15-min unit of time. After one 15-min unit of time the cost is \$3.00 + \$0.50, or \$3.50. After two 15-min units, or 30 min, the cost is \$3.00 + 2(\$0.50), or \$4.00. We can generalize this with a model if we let $C(t)$ represent the total cost, in dollars, for t 15-min units of time.

Translate. We reword the problem and translate.

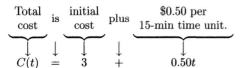

$$C(t) = 3 + 0.50t$$

Carry out. First write the model in slope-intercept form: $C(t) = 0.50t + 3$. The vertical intercept is $(0, 3)$ and the slope, or rate, is 0.50, or $\frac{1}{2}$. Plot $(0, 3)$ and from there go *up* \$1 and to the *right* 2 15-min units of time. This takes us to $(2, 4)$. Draw a line passing through both points.

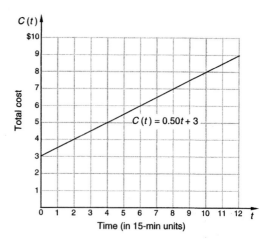

Time (in 15-min units)

To estimate how long someone can park for \$7.50, we are estimating the solution of $0.50t + 3 = 7.50$. We do this by graphing $y = 7.50$ and finding the point of intersection of the graphs.

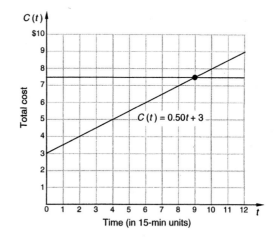

Time (in 15-min units)

The point appears to be $(9, 7.50)$. Thus, we estimate that someone can park for nine 15-min units of time, or 2 hr, 15 min, for \$7.50.

Check. We evaluate:

$$C(9) = 3 + 0.50(9)$$
$$= 3 + 4.50$$
$$= 7.50$$

The estimate is precise.

State. Someone can park for 2 hr, 15 min for \$7.50.

26. 1 hr

27. Familiarize. After a cost of \$18.75 for the first 6 lb, another \$0.75 is charged for each additional pound. Since 7 lb − 6 lb = 1 lb, the shipping cost for a 7-lb package is \$18.75 + (7 − 6)(\$0.75), or \$18.75 + \$0.75, or \$19.50. For an 8-lb package the cost is \$18.75 + (8 − 6)(\$0.75), or \$18.75 + 2(\$0.75), or \$20.25. We can generalize this with a model if we let $C(p)$ represent the total shipping cost, in dollars, for a package weighing p pounds.

Translate. We reword the problem and translate.

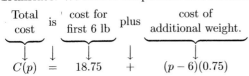

Total cost	is	cost for first 6 lb	plus	cost of additional weight.
$C(p)$	$=$	18.75	$+$	$(p-6)(0.75)$

Carry out. First write the model in slope-intercept form.

$$C(p) = 18.75 + (p-6)(0.75)$$
$$C(p) = 18.75 + 0.75p - 4.5$$
$$C(p) = 0.75p + 14.25$$

The vertical intercept is $(0, 14.25)$ and the slope, or rate, is 0.75, or $\dfrac{0.75}{1}$. Plot $(0, 14.25)$ and from there go *up* 0.75 unit and to the *right* 1 unit. This takes us to $(1, 15)$. Draw a line passing through both points.

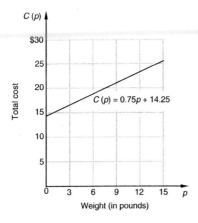

To estimate the weight of a package that costs \$24 to ship, we are estimating the solution of $0.75p + 14.25 = 24$. We do this by graphing $y = 24$ and finding the point of intersection of the graphs.

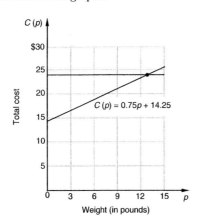

The point appears to be $(13, 24)$. Thus, we estimate that a package that weights 13 lb costs \$24 to ship.

Check. We evaluate:
$$C(13) = 18.75 + (13-6)(0.75)$$
$$= 18.75 + 7(0.75)$$
$$= 18.75 + 5.25$$
$$= 24$$

The estimate is precise.

State. A package that costs \$24 to ship weighs 13 lb.

28. 25 pages

29. Graph $y = 219 + 0.035x$. Use the Intersect feature find the x-coordinate that corresponds to the y-coordina 370.03. We find that the sales total was about \$4315.14

30. 249 shirts

31. $3x = 15$ and $5x = 25$

The equation $3x = 15$ is true only when $x = 5$. Similar $5x = 25$ is true only when $x = 5$. Since both equatio have the same solution, they are equivalent.

32. Yes

33. $t + 5 = 11$ and $3t = 18$

Each equation has only one solution, the number 6. Th the equations are equivalent.

34. No

35. $12 - x = 3$ and $2x = 20$

When x is replaced by 9, the first equation is true, b the second equation is false. Thus the equations are n equivalent.

36. Yes

37. $5x = 2x$ and $\dfrac{4}{x} = 3$

When x is replaced by 0, the first equation is true, but t second equation is not defined. Thus the equations are n equivalent.

38. No

39.
$$x - 2.9 = 13.4$$
$$x - 2.9 + 2.9 = 13.4 + 2.9 \quad \text{Addition principle;}$$
$$\text{adding } 2.9$$
$$x + 0 = 13.4 + 2.9 \quad \text{Law of opposites}$$
$$x = 16.3$$

Check:
$$x - 2.9 = 13.4$$
$$\overline{16.3 - 2.9 \ ? \ 13.4}$$
$$13.4 \mid 13.4 \qquad \text{TRUE}$$

The solution is 16.3.

40. 6.9

41.

$$8t = 72$$

$$\frac{1}{8} \cdot 8t = \frac{1}{8} \cdot 72 \quad \text{Multiplication principle;}$$
$$\text{multiplying by } \frac{1}{8}, \text{ the}$$
$$\text{reciprocal of 8}$$

$$1t = 9$$

$$t = 9$$

Check:

$$\frac{8t = 72}{8 \cdot 9 \,\, ? \,\, 72}$$
$$72 \mid 72 \quad \text{TRUE}$$

The solution is 9.

42. 7

43.

$$4x - 12 = 60$$

$$4x - 12 + 12 = 60 + 12$$

$$4x = 72$$

$$\frac{1}{4} \cdot 4x = \frac{1}{4} \cdot 72$$

$$1x = \frac{72}{4}$$

$$x = 18$$

Check:

$$\frac{4x - 12 = 60}{4 \cdot 18 - 12 \,\, ? \,\, 60}$$
$$72 - 12 \mid$$
$$60 \mid 60 \quad \text{TRUE}$$

The solution is 18.

44. 19

45.

$$3n + 5 = 29$$

$$3n + 5 + (-5) = 29 + (-5)$$

$$3n = 24$$

$$\frac{1}{3} \cdot 3n = \frac{1}{3} \cdot 24$$

$$1n = 8$$

$$n = 8$$

Check:

$$\frac{3n + 5 = 29}{3 \cdot 8 + 5 \,\, ? \,\, 29}$$
$$24 + 5 \mid$$
$$29 \mid 29 \quad \text{TRUE}$$

The solution is 8.

46. 9

47.

$$3x + 5x = 56$$

$$8x = 56$$

$$\frac{1}{8} \cdot 8x = \frac{1}{8} \cdot 56$$

$$x = 7$$

Check:

$$\frac{3x + 5x = 56}{3 \cdot 7 + 5 \cdot 7 \,\, ? \,\, 56}$$
$$21 + 35 \mid$$
$$56 \mid 56 \quad \text{TRUE}$$

The solution is 7.

48. 12

49.

$$9y - 7y = 42$$

$$2y = 42$$

$$\frac{1}{2} \cdot 2y = \frac{1}{2} \cdot 42$$

$$y = 21$$

Check:

$$\frac{9y - 7y = 42}{9 \cdot 21 - 7 \cdot 21 \,\, ? \,\, 42}$$
$$189 - 147 \mid$$
$$42 \mid 42 \quad \text{TRUE}$$

The solution is 21.

50. 13

51.

$$-6y - 10y = -32$$

$$-16y = -32$$

$$-\frac{1}{16} \cdot (-16y) = -\frac{1}{16} \cdot (-32)$$

$$y = 2$$

Check:

$$\frac{-6y - 10y = -32}{-6 \cdot 2 - 10 \cdot 2 \,\, ? \,\, -32}$$
$$-12 - 20 \mid$$
$$-32 \mid -32 \quad \text{TRUE}$$

The solution is 2.

52. −2

53.

$$2(x + 6) = 8x$$

$$2x + 12 = 8x$$

$$2x + 12 - 2x = 8x - 2x$$

$$12 = 6x$$

$$\frac{1}{6} \cdot 12 = \frac{1}{6} \cdot 6x$$

$$2 = x$$

Check:
$$
\begin{array}{c}
2(x+6) = 8x \\
\hline
\end{array}
$$

$$
\begin{array}{c|c}
2(2+6) \ ? \ 8 \cdot 2 & \\
2 \cdot 8 & 16 \\
16 & 16 \qquad \text{TRUE}
\end{array}
$$

The solution is 2.

54. 3

55.
$$70 = 10(3t - 2)$$
$$70 = 30t - 20$$
$$70 + 20 = 30t - 20 + 20$$
$$90 = 30t$$
$$\frac{1}{30} \cdot 90 = \frac{1}{30} \cdot 30t$$
$$3 = t$$

Check:
$$
\begin{array}{c}
70 = 10(3t - 2) \\
\hline
\end{array}
$$

$$
\begin{array}{c|c}
70 \ ? \ 10(3 \cdot 3 - 2) & \\
& 10(9 - 2) \\
& 10 \cdot 7 \\
70 & 70 \qquad \text{TRUE}
\end{array}
$$

The solution is 3.

56. 1

57.
$$180(n - 2) = 900$$
$$180n - 360 = 900$$
$$180n - 360 + 360 = 900 + 360$$
$$180n = 1260$$
$$\frac{1}{180} \cdot 180n = \frac{1}{180} \cdot 1260$$
$$n = 7$$

Check:
$$
\begin{array}{c}
180(n - 2) = 900 \\
\hline
\end{array}
$$

$$
\begin{array}{c|c}
180(7 - 2) \ ? \ 900 & \\
180 \cdot 5 & \\
900 & 900 \quad \text{TRUE}
\end{array}
$$

The solution is 7.

58. 7

59.
$$5y - (2y - 10) = 25$$
$$5y - 2y + 10 = 25$$
$$3y + 10 = 25$$
$$3y + 10 - 10 = 25 - 10$$
$$3y = 15$$
$$\frac{1}{3} \cdot 3y = \frac{1}{3} \cdot 15$$
$$y = 5$$

Check:
$$
\begin{array}{c}
5y - (2y - 10) = 25 \\
\hline
\end{array}
$$

$$
\begin{array}{c|c}
5 \cdot 5 - (2 \cdot 5 - 10) \ ? \ 25 & \\
25 - (10 - 10) & \\
25 - 0 & \\
25 & 25 \quad \text{TRUE}
\end{array}
$$

The solution is 5.

60. 7

61.
$$7y - 1 = 23 - 5y$$
$$7y - 1 + 5y = 23 - 5y + 5y$$
$$12y - 1 = 23$$
$$12y - 1 + 1 = 23 + 1$$
$$12y = 24$$
$$\frac{1}{12} \cdot 12y = \frac{1}{12} \cdot 24$$
$$y = 2$$

Check:
$$
\begin{array}{c}
7y - 1 = 23 - 5y \\
\hline
\end{array}
$$

$$
\begin{array}{c|c}
7 \cdot 2 - 1 \ ? \ 23 - 5 \cdot 2 & \\
14 - 1 & 23 - 10 \\
13 & 13 \qquad \text{TRUE}
\end{array}
$$

The solution is 2.

62. -7

63.
$$\frac{1}{5} + \frac{3}{10}x = \frac{4}{5}$$
$$\frac{1}{5} + \frac{3}{10}x - \frac{1}{5} = \frac{4}{5} - \frac{1}{5}$$
$$\frac{3}{10}x = \frac{3}{5}$$
$$\frac{10}{3} \cdot \frac{3}{10}x = \frac{10}{3} \cdot \frac{3}{5}$$
$$x = 2$$

Check:
$$
\begin{array}{c}
\frac{1}{5} + \frac{3}{10}x = \frac{4}{5} \\
\hline
\end{array}
$$

$$
\begin{array}{c|c}
\frac{1}{5} + \frac{3}{10} \cdot 2 \ ? \ \frac{4}{5} & \\
\frac{1}{5} + \frac{3}{5} & \\
\frac{4}{5} & \frac{4}{5} \quad \text{TRUE}
\end{array}
$$

The solution is 2.

64. $\dfrac{37}{5}$

5.
$$0.9y - 0.7 = 4.2$$
$$0.9y - 0.7 + 0.7 = 4.2 + 0.7$$
$$0.9y = 4.9$$
$$\frac{1}{0.9}(0.9y) = \frac{1}{0.9}(4.9)$$
$$y = \frac{4.9}{0.9}$$
$$y = \frac{49}{9}$$

Check:

$$\frac{0.9y - 0.7 = 4.2}{}$$
$$0.9\left(\frac{49}{9}\right) - 0.7 \ ? \ 4.2$$
$$4.9 - 0.7$$
$$4.2 \ \bigg| \ 4.2 \quad \text{TRUE}$$

The solution is $\dfrac{49}{9}$.

6. 13

7.
$$4.23x - 17.898 = -1.65x - 42.454$$
$$5.88x - 17.898 = -42.454$$
$$5.88x = -24.556$$
$$x = -\frac{24.556}{5.88}$$
$$x \approx -4.176190476$$

The check is left to the student. The solution is approximately -4.176190476.

8. 0.21402

9.
$$7r - 2 + 5r = 6r + 6 - 4r$$
$$12r - 2 = 2r + 6$$
$$12r - 2 - 2r = 2r + 6 - 2r$$
$$10r - 2 = 6$$
$$10r - 2 + 2 = 6 + 2$$
$$10r = 8$$
$$\frac{1}{10} \cdot 10r = \frac{1}{10} \cdot 8$$
$$r = \frac{8}{10}$$
$$r = \frac{4}{5}$$

Check:

$$\frac{7r - 2 + 5r = 6r + 6 - 4r}{}$$
$$7\left(\frac{4}{5}\right) - 2 + 5\left(\frac{4}{5}\right) \ ? \ 6\left(\frac{4}{5}\right) - 6 - 4\left(\frac{4}{5}\right)$$
$$\frac{28}{5} - \frac{10}{5} + \frac{20}{5} \ \bigg| \ \frac{24}{5} + \frac{30}{5} - \frac{16}{5}$$
$$\frac{38}{5} \ \bigg| \ \frac{38}{5} \qquad \text{TRUE}$$

The solution is $\dfrac{4}{5}$.

70. 7

71.
$$\frac{1}{4}(16y + 8) - 17 = -\frac{1}{2}(8y - 16)$$
$$4y + 2 - 17 = -4y + 8$$
$$4y - 15 = -4y + 8$$
$$4y - 15 + 4y = -4y + 8 + 4y$$
$$8y - 15 = 8$$
$$8y - 15 + 15 = 8 + 15$$
$$8y = 23$$
$$\frac{1}{8} \cdot 8y = \frac{1}{8} \cdot 23$$
$$y = \frac{23}{8}$$

Check:

$$\frac{\frac{1}{4}(16y + 8) - 17 = -\frac{1}{2}(8y - 16)}{}$$
$$\frac{1}{4}\left(16 \cdot \frac{23}{8} + 8\right) - 17 \ ? \ -\frac{1}{2}\left(8 \cdot \frac{23}{8} - 16\right)$$
$$\frac{1}{4}(46 + 8) - 17 \ \bigg| \ -\frac{1}{2}(23 - 16)$$
$$\frac{1}{4}(54) - 17 \ \bigg| \ -\frac{1}{2}(7)$$
$$\frac{27}{2} - \frac{34}{2} \ \bigg| \ -\frac{7}{2}$$
$$-\frac{7}{2} \ \bigg| \ -\frac{7}{2} \qquad \text{TRUE}$$

The solution is $\dfrac{23}{8}$.

72. $\dfrac{22}{5}$

73.
$$5 + 2(x - 3) = 2[5 - 4(x + 2)]$$
$$5 + 2x - 6 = 2[5 - 4x - 8]$$
$$2x - 1 = 2[-4x - 3]$$
$$2x - 1 = -8x - 6$$
$$2x - 1 + 1 = -8x - 6 + 1$$
$$2x = -8x - 5$$
$$2x + 8x = -8x - 5 + 8x$$
$$10x = -5$$
$$\frac{1}{10} \cdot 10x = \frac{1}{10}(-5)$$
$$x = -\frac{1}{2}$$

Check:

$$\frac{5 + 2(x - 3) = 2[5 - 4(x + 2)]}{}$$

$$5 + 2\left(-\frac{1}{2} - 3\right) \ ? \ 2\left[5 - 4\left(-\frac{1}{2} + 2\right)\right]$$

$$5 + 2\left(-\frac{7}{2}\right) \ \bigg| \ 2\left[5 - 4\left(\frac{3}{2}\right)\right]$$

$$\begin{array}{c|c} 5 - 7 & 2[5 - 6] \\ -2 & 2[-1] \\ -2 & -2 \end{array} \qquad \text{TRUE}$$

The solution is $-\frac{1}{2}$.

74. $\frac{23}{8}$

75. $5x + 7 - 3x = 2x$

$$2x + 7 = 2x$$

$$2x + 7 - 2x = 2x - 2x$$

$$7 = 0$$

Since the original equation is equivalent to the false equation $7 = 0$, there is no solution. The solution set is $\emptyset$. The equation is a contradiction.

76. All real numbers; identity

77. $1 + 9x = 3(4x + 1) - 2$

$$1 + 9x = 12x + 3 - 2$$

$$1 + 9x = 12x + 1$$

$$1 + 9x - 1 = 12x + 1 - 1$$

$$9x = 12x$$

$$9x - 9x = 12x - 9x$$

$$0 = 3x$$

$$\frac{1}{3} \cdot 0 = \frac{1}{3} \cdot 3x$$

$$0 = x$$

The solution set is $\{0\}$. The equation is a conditional equation.

78. $\emptyset$; contradiction

79. $-9t + 2 = -9t - 7(6 \div 2(49) + 8)$

Observe that $-7(6 \div 2(49) + 8)$ is a negative number. Then on the left side we have $-9t$ plus a positive number and on the right side we have $-9t$ plus a negative number. This is a contradiction, so the solution set is $\emptyset$.

80. $\emptyset$; contradiction

81. $2\{9 - 3[-2x - 4]\} = 12x + 42$

$$2\{9 + 6x + 12\} = 12x + 42$$

$$2\{21 + 6x\} = 12x + 42$$

$$42 + 12x = 12x + 42$$

$$42 + 12x - 12x = 12x + 42 - 12x$$

$$42 = 42$$

The original equation is equivalent to the equation $42 = 42$, which is true for all real numbers. Thus the solution set is the set of all real numbers. The equation is an identity.

82. $\{0\}$; conditional

83. *Writing Exercise*

84. *Writing Exercise*

85. Roster notation: List the numbers in the set.

$$\{1, 2, 3, 4, 5, 6, 7, 8, 9\}$$

Set-builder notation: Specify the conditions under which a number is in the set.

$$\{x | x \text{ is a positive integer less than } 10\}$$

86. $\{-8, -7, -6, -5, -4, -3, -2, -1\}$;

$\{x | x \text{ is a negative integer greater than } -9\}$

87. Let x and y represent the numbers; $xy - 3$

88. Let m and n represent the numbers; $m - n + 10$

89. Let n represent the number. Then we have

$$2n + 9, \quad \text{or} \quad 9 + 2n.$$

90. Let n represent the number; $0.42\left(\frac{n}{2}\right)$

91. *Writing Exercise*

92. *Writing Exercise*

93. $4x - \{3x - [2x - (5x - (7x - 1))]\} = 4x + 7$

$$4x - \{3x - [2x - (5x - 7x + 1)]\} = 4x + 7$$

$$4x - \{3x - [2x - (-2x + 1)]\} = 4x + 7$$

$$4x - \{3x - [2x + 2x - 1]\} = 4x + 7$$

$$4x - \{3x - [4x - 1]\} = 4x + 7$$

$$4x - \{3x - 4x + 1\} = 4x + 7$$

$$4x - \{-x + 1\} = 4x + 7$$

$$4x + x - 1 = 4x + 7$$

$$5x - 1 = 4x + 7$$

$$x - 1 = 7$$

$$x = 8$$

The check is left to the student. The solution is 8.

4. 4

5. $17 - 3\{5 + 2[x - 2]\} + 4\{x - 3(x + 7)\} =$
$$9\{x + 3[2 + 3(4 - x)]\}$$
$$17 - 3\{5 + 2x - 4\} + 4\{x - 3x - 21\} =$$
$$9\{x + 3[2 + 12 - 3x]\}$$
$$17 - 3\{1 + 2x\} + 4\{-2x - 21\} = 9\{x + 3[14 - 3x]\}$$
$$17 - 3 - 6x - 8x - 84 = 9\{x + 42 - 9x\}$$
$$-14x - 70 = 9\{-8 + 42\}$$
$$-14x - 70 = -72x + 378$$
$$58x - 70 = 378$$
$$58x = 448$$
$$x = \frac{448}{58}, \text{ or } \frac{224}{29}$$

The check is left to the student. The solution is $\frac{224}{29}$.

6. $\frac{19}{46}$

7. Graph $y_1 = 2x$ and $y_2 = \text{abs}(x + 1)$ and find the first coordinate of the point of intersection of the graphs.

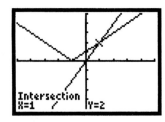

The solution is 1.

8. 1.6666667, 3; or $\frac{5}{3}$, 3

9. Graph $y_1 = (1/2)x$ and $y_2 = 3 - \text{abs}(x)$ and find the first coordinates of the points of intersection of the graphs.

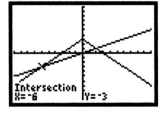

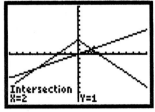

The solutions are −6 and 2.

100. −0.25, or $-\frac{1}{4}$

101. Graph $y_1 = x^2$ and $y_2 = x + 2$ and find the first coordinates of the points of intersection of the graphs.

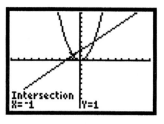

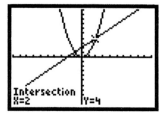

The solutions are −1 and 2.

102. 0, 1

103. We graph $C(t) = 0.50t + 3$, where t represents the number of 15-min units of time, as a series of steps. The cost is constant within each 15-min unit of time. Thus,

for $0 < t \le 1$, $C(t) = 0.5(1) + 3 = \$3.50$;

for $1 < t \le 2$, $C(t) = 0.5(2) + 3 = \$4.00$;

for $2 < t \le 3$, $C(t) = 0.5(3) + 3 = \$4.50$;

and so on. We draw the graph. An open circle at a point indicates that the point is not on the graph.

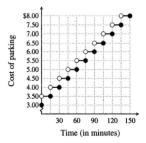

104.

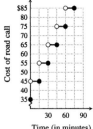

105. *Writing Exercise*

Exercise Set 2.3

1. **Familiarize.** Let x = the smaller number. Then $x + 7$ = the larger number, and $x + (x + 7)$ = the sum of the numbers.

Translate.

$$\underbrace{\text{The sum of two numbers}}_{\downarrow} \; \text{is} \; 65.$$
$$x + (x + 7) \quad\quad = \quad 65$$

Carry out. We solve the equation.

$$x + (x + 7) = 65$$
$$2x + 7 = 65 \quad \text{Combining like terms}$$
$$2x = 58 \quad \text{Subtracting 7 from both sides}$$
$$\frac{2x}{2} = \frac{58}{2} \quad \text{Dividing both sides by 2}$$
$$x = 29 \quad \text{Simplifying}$$

If $x = 29$, then $x + 7 = 29 + 7 = 36$.

Check. The number 36 is 7 more than 29, and $29 + 36 = 65$. The answer checks.

State. The numbers are 29 and 36.

2. 36, 47

3. **Familiarize.** Let n = the unknown number. The product of the numbers is $48 \cdot n$.

Translate.

$$\underbrace{\text{The product of two numbers}}_{\downarrow} \; \text{is} \; 72.$$
$$48 \cdot n \quad\quad = \quad 72$$

Carry out. We solve the equation.

$$48 \cdot n = 72$$
$$\frac{48 \cdot n}{48} = \frac{72}{48} \quad \text{Dividing both sides by 48}$$
$$n = \frac{3}{2} \quad \text{Simplifying}$$

Check. Since $48 \cdot \dfrac{3}{2} = \dfrac{48 \cdot 3}{2} = \dfrac{2 \cdot 24 \cdot 3}{2} = \dfrac{2}{2} \cdot \dfrac{24 \cdot 3}{1} = 72$, the answer checks.

State. The other number is $\dfrac{3}{2}$.

4. 894

5. **Familiarize.** Let x = the width of the rectangle. Then $2x$ = the length. Recall that the perimeter P of a rectangle with length l and width w is $P = 2l + 2w$.

Translate.

$$\underbrace{\text{The perimeter}}_{\downarrow} \; \text{is} \; \underline{15 \text{ ft.}}$$
$$2(2x) + 2x \quad = \quad 15$$

Carry out. We solve the equation.

$$2(2x) + 2x = 15$$
$$4x + 2x = 15$$
$$6x = 15 \quad \text{Combining like terms}$$
$$\frac{6x}{6} = \frac{15}{6} \quad \text{Dividing both sides by 6}$$
$$x = \frac{5}{2} \quad \text{Simplifying}$$

If $x = \dfrac{5}{2}$, then $2x = 2 \cdot \dfrac{5}{2} = 5$.

Check. The length, 5 ft, is twice the width, $\dfrac{5}{2}$ ft. The perimeter is $2 \cdot 5 + 2 \cdot \dfrac{5}{2}$, or $10 + 5$, or 15 ft. The answer checks.

State. The length of the rectangle is 5 ft, and the width is $\dfrac{5}{2}$ ft.

6. Length: 20 m; width: 5 m

7. **Familiarize.** We will use the formula Distance = Speed · Time. Let t = the time, in hours, it takes the Delta Queen to cruise 2 mi upstream. The speed of the paddle boat going upstream is $7 - 3$, or 4 mph.

Translate. We substitute in the formula.

$$2 = 4t$$

Carry out. We solve the equation.

$$2 = 4t$$
$$\frac{2}{4} = \frac{4t}{4} \quad \text{Dividing both sides by 4}$$
$$\frac{1}{2} = t$$

Check. At the rate of 4 mph, in $\dfrac{1}{2}$ hr the Delta Queen will travel $4 \cdot \dfrac{1}{2}$, or 2 mi. The answer checks.

State. It will take the boat $\dfrac{1}{2}$ hr to cruise 2 mi upstream.

8. $t = \dfrac{2}{3}$ hr

9. **Familiarize.** Since the sidewalk's speed is 5 ft/sec and Alida's walking speed is 4 ft/sec, Alida will move at a speed of $5 + 4$, or 9 ft/sec on the sidewalk. Let t = the time, in seconds, it takes her to walk the length of the moving sidewalk, 300 ft.

Translate. We will use the formula Distance = Speed × Time.

$$\underbrace{\text{Distance}}_{300} = \underbrace{\text{Speed}}_{9} \times \underbrace{\text{Time}}_{t}$$

Carry out. We solve the equation.

$$300 = 9t$$

$$\frac{300}{9} = \frac{9t}{9} \quad \text{Dividing both sides by 9}$$

$$\frac{100}{3} = t \quad \text{Simplifying}$$

Check. At a rate of 9 ft/sec, in $\frac{100}{3}$ sec Alida walks $9 \cdot \frac{100}{3}$, or $\frac{900}{3}$, or 300 ft. The answer checks.

State. It will take Alida $\frac{100}{3}$ sec to walk the length of the moving sidewalk.

. $\frac{29}{13}$ hr

. **Familiarize**. Using the labels on the drawing in the text, we let x, x, and $2x$ represent the measures of the angles. Recall that the sum of the measures of the angles in a triangle is 180°.

Translate.

$$\underbrace{\text{The sum of the}}_{x + x + 2x} \underbrace{\text{angle measures}}_{} \text{ is } \underbrace{180°.}_{= \quad 180}$$

Carry out. We solve the equation.

$$x + x + 2x = 180$$

$$4x = 180$$

$$\frac{4x}{4} = \frac{180}{4}$$

$$x = 45$$

If $x = 45$, then $2x = 2 \cdot 45 = 90$.

Check. The measure of the third angle, 90°, is twice the measure of the other angles. Also $45° + 45° + 90° = 180°$. The answer checks.

State. The measures of the angles are 45°, 45°, and 90°.

. 25°, 50°, 105°

. **Familiarize**. Let $p =$ the previous power usage record.

Translate.

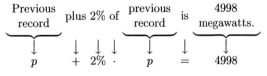

$$\underbrace{\text{Previous}}_{p} \underbrace{\text{plus 2\% of}}_{+ \quad 2\% \cdot} \underbrace{\text{previous}}_{p} \underbrace{\text{is}}_{=} \underbrace{\begin{array}{c}4998\\ \text{megawatts.}\end{array}}_{4998}$$

Carry out. We solve the equation, expressing 2% as 0.02.

$$p + 0.02p = 4998$$

$$1.02p = 4998 \quad \text{Adding}$$

$$\frac{1.02p}{1.02} = \frac{4998}{1.02}$$

$$p = 4900$$

Check. 2% of 4900 is 0.02(4900), or 98, and $4900 + 98 = 4998$. The answer checks.

State. The previous power usage record was 4900 megawatts.

14. 8.6 in.

15. **Familiarize**. Let $f =$ the amount of fudge Kelly got, in pounds. Then $\frac{2}{3}f =$ the amount Kim got and the total amount of fudge is $f + \frac{2}{3}f$.

Translate.

$$\underbrace{\begin{array}{c}\text{The total amount}\\ \text{of fudge}\end{array}}_{f + \frac{2}{3}f} \underbrace{\text{is}}_{=} \underbrace{\frac{1}{2} \text{ lb.}}_{\frac{1}{2}}$$

Carry out. We solve the equation.

$$f + \frac{2}{3}f = \frac{1}{2}$$

$$\frac{5}{3}f = \frac{1}{2} \quad \text{Adding}$$

$$\frac{3}{5} \cdot \frac{5}{3}f = \frac{3}{5} \cdot \frac{1}{2}$$

$$f = \frac{3}{10}$$

If $f = \frac{3}{10}$, then $\frac{2}{3}f = \frac{2}{3} \cdot \frac{3}{10} = \frac{1}{5}$.

Check. $\frac{1}{5}$ lb is $\frac{2}{3}$ of $\frac{3}{10}$ lb, and $\frac{1}{5} + \frac{3}{10} = \frac{2}{10} + \frac{3}{10} = \frac{5}{10} = \frac{1}{2}$ lb. The answer checks.

State. Kim got $\frac{1}{5}$ lb of fudge.

16. 15 gal

17. **Familiarize**. Let $s =$ the amount of herbal supplement sales in 1997, in millions of dollars. A 110% increase in these sales is 110%s, or 1.1s.

Translate.

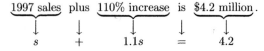

$$\underbrace{\text{1997 sales}}_{s} \underbrace{\text{plus}}_{+} \underbrace{\text{110\% increase}}_{1.1s} \underbrace{\text{is}}_{=} \underbrace{\$4.2 \text{ million}.}_{4.2}$$

Carry out. We solve the equation.

$$s + 1.1s = 4.2$$
$$2.1s = 4.2$$
$$\frac{2.1s}{2.1} = \frac{4.2}{2.1}$$
$$s = 2$$

Check. If sales in 1997 were $2 million, then a 110% increase in sales is 1.1(2), or $2.2 million. Since
$2 million + $2.2 million = $4.2 million, the sales in 2002, the answer checks.

State. Herbal supplement sales in 2002 were $2 million.

18. 54,000 employees

19. Familiarize. Let $p =$ the number of prints. Then the total charge is $6.30 + $0.22p$.

Translate.

$$\underbrace{\text{The total charge}}_{6.30 + 0.22p} \text{ is } \underbrace{\$14.66.}_{= \quad 14.66}$$

Carry out. We solve the equation.

$$6.30 + 0.22p = 14.66$$
$$0.22p = 8.36 \qquad \text{Subtracting 6.30 on both}$$
$$\qquad\qquad\qquad \text{sides}$$
$$\frac{0.22p}{0.22} = \frac{8.36}{0.22}$$
$$p = 38$$

Check. If there are 38 prints, the total charge will be $6.30 + $0.22(38), or $14.66. The answer checks.

State. LaKenya received 38 prints.

20. 750 cubic feet

21. Familiarize. Let $p =$ the price of the compact disk player. The amount owed after a year is the price plus 12% of the price or $p + 12\% p$, or $p + 0.12p$.

Translate.

$$\underbrace{\text{The amount owed}}_{p + 0.12p} \text{ is } \underbrace{\$504.}_{= \quad 504}$$

Carry out. We solve the equation.

$$p + 0.12p = 504$$
$$1.12p = 504 \qquad \text{Adding}$$
$$\frac{1.12p}{1.12} = \frac{504}{1.12}$$
$$p = 450$$

Check. 12% of $450, or 0.12($450), is $54, and $450 + $54 = $504. The answer checks.

State. The price of the compact disc player was $450.

22. $1095

23. Familiarize. The distance Venus travels in one orbit can be approximated by the circumference of a circle whose radius is the average distance from the sun to Venus, 1.08×10^8 km. Recall that the formula for the circumference of a circle is $C = 2\pi r$, where r is the radius.

Translate. Substitute 1.08×10^8 for r in the formula

$$C = 2\pi r$$
$$C = 2\pi \times 1.08 \times 10^8$$

Carry out. Do the calculation. Use a calculator with π key.

$$C = 2\pi \times 1.08 \times 10^8$$
$$\approx 6.79 \times 10^8$$

Check. Repeat the calculation.

State. Venus travels about 6.79×10^8 km in one orbit.

24. 4.5 g

25. Familiarize. First we will find the number n of $5 bills in $4,540,000 worth of $5 bills. Then we will find the weight w of a $5 bill. Recall that 1 ton = 2000 lb.

Translate. To find the number of $5 bills in $4,540,000 worth of $5 bills we divide:

$$n = \frac{4,540,000}{5}.$$

Then we divide again to find the weight w of a $5 bill:

$$w = \frac{2000}{n}.$$

Carry out. We begin by finding n.

$$n = \frac{4,540,000}{5} = \frac{4.54 \times 10^6}{5} = 0.908 \times 10^6 =$$
$$(9.08 \times 10^{-1}) \times 10^6 = 9.08 \times 10^5$$

Now we find w.

$$w = \frac{2000}{n} = \frac{2000}{9.08 \times 10^5} = \frac{2 \times 10^3}{9.08 \times 10^5} \approx$$
$$0.22 \times 10^{-2} = (2.2 \times 10^{-1}) \times 10^{-2} =$$
$$2.2 \times 10^{-3}$$

Check. We recheck the translation and calculations.

State. A $5 bill weighs about 2.2×10^{-3} lb.

26. 5.84×10^8 miles

27. Familiarize. From Example 3 we know that
1 light year = 5.88×10^{12} mi. Let $y =$ the number of light years from the earth to Sirius.

Translate. The distance from the earth to Sirius is y light years or $(5.88 \times 10^{12})y$ mi. It is also given by 4.704×10^{13} mi. We write an equation:

$$(5.88 \times 10^{12})y = 4.704 \times 10^{13}$$

Carry out. We solve the equation.

$$(5.88 \times 10^{12})y = 4.704 \times 10^{13}$$

$$y = \frac{4.704 \times 10^{13}}{5.88 \times 10^{12}}$$

$$y = \frac{4.704}{5.88} \times \frac{10^{13}}{10^{12}}$$

$$y = 0.8 \times 10$$

$$y = 8$$

Check. Since light travels 5.88×10^{12} mi in one year, in 8.00 yr it will travel $8.00 \times 5.88 \times 10^{12} = 4.704 \times 10^{13}$ mi, the distance from the earth to Sirius. The answer checks.

State. It is 8 light years from the earth to Sirius.

3. 1×10^5 light years

9. Familiarize. First we will find d, the number of drops in a pound. Then we will find b, the number of bacteria in a drop of U.S. mud.

Translate. To find d we convert 1 pound to drops:

$$d = 1 \text{ lb} \cdot \frac{16 \text{oz}}{1 \text{ lb}} \cdot \frac{6 \text{ tsp}}{1 \text{ oz}} \cdot \frac{60 \text{ drops}}{1 \text{ tsp}}.$$

Then we divide to find b:

$$b = \frac{4.55 \times 10^{11}}{d}.$$

Carry out. We do the calculations.

$$d = 1 \text{ lb} \cdot \frac{16 \text{oz}}{1 \text{ lb}} \cdot \frac{6 \text{ tsp}}{1 \text{ oz}} \cdot \frac{60 \text{ drops}}{1 \text{ tsp}}$$

$$= 5760 \text{ drops}$$

Now we find b.

$$b = \frac{4.55 \times 10^{11}}{5760} = \frac{4.55 \times 10^{11}}{5.760 \times 10^3} \approx 0.790 \times 10^8 \approx$$

$$(7.90 \times 10^{-1}) \times 10^8 = 7.90 \times 10^7$$

Check. If there are about 7.90×10^7 bacteria in a drop of U.S. mud, then in a pound there are about

$$\frac{7.90 \times 10^7}{1 \text{ drop}} \cdot \frac{60 \text{ drops}}{1 \text{ tsp}} \cdot \frac{6 \text{ tsp}}{1 \text{ oz}} \cdot \frac{16 \text{ oz}}{1 \text{ lb}} =$$

$$\frac{45,504 \times 10^7}{1 \text{ lb}} \approx 4.55 \times 10^{11} \text{ bacteria per pound. The answer checks.}$$

State. About 7.90×10^7 bacteria live in a drop of U.S. mud.

0. 1.475×10^{12} mi

1. Familiarize. First we will find the distance C around Jupiter at the equator, in km. Then we will use the formula Speed × Time = Distance to find the speed s at which Jupiter's equator is spinning.

Translate. We will use the formula for the circumference of a circle to find the distance around Jupiter at the equator:

$$C = \pi d = \pi(1.43 \times 10^5).$$

Then we find the speed s at which Jupiter's equator is spinning:

$$\underbrace{\text{Speed}}_{s} \times \underbrace{\text{Time}}_{10} = \underbrace{\text{Distance}}_{C}$$

Carry out. First we find C.

$$C = \pi(1.43 \times 10^5) \approx 4.49 \times 10^5$$

Then we find s.

$$s \times 10 = C$$

$$s \times 10 = 4.49 \times 10^5$$

$$s = \frac{4.49 \times 10^5}{10}$$

$$s = 4.49 \times 10^4$$

Check. At 4.49×10^4 km/h, in 10 hr, Jupiter's equator travels $4.49 \times 10^4 \times 10$, or 4.49×10^5 km. A circle with circumference 4.49×10^5 km has a diameter of $\frac{4.49 \times 10^5}{\pi} \approx 1.43 \times 10^5$ km. The answer checks.

State. Jupiter's equator spins at a speed of about 4.49×10^4 km/h.

32. $\$6.7 \times 10^4$

33.
$$d = rt$$

$$\frac{1}{t} \cdot d = \frac{1}{t} \cdot rt \quad \text{Multiplying both sides by } \frac{1}{t}$$

$$\frac{d}{t} = r \quad \text{Simplifying}$$

34. $t = \frac{d}{r}$

35.
$$F = ma$$

$$\frac{1}{m} \cdot F = \frac{1}{m} \cdot ma \quad \text{Multiplying both sides by } \frac{1}{m}$$

$$\frac{F}{m} = a \quad \text{Simplifying}$$

36. $w = \frac{A}{l}$

37.
$$W = EI$$

$$\frac{1}{E} \cdot W = \frac{1}{E} \cdot EI \quad \text{Multiplying both sides by } \frac{1}{E}$$

$$\frac{W}{E} = I \quad \text{Simplifying}$$

38. $E = \frac{W}{I}$

39.
$$V = lwh$$

$$\frac{1}{lw} \cdot V = \frac{1}{lw} \cdot lwh \quad \text{Multiplying both sides by } \frac{1}{lw}$$

$$\frac{V}{lw} = h \quad \text{Simplifying}$$

40. $r = \dfrac{I}{Pt}$

41. $\qquad L = \dfrac{k}{d^2}$

$\qquad d^2 \cdot L = d^2 \cdot \dfrac{k}{d^2} \quad$ Multiplying both sides by d^2

$\qquad d^2 L = k \qquad$ Simplifying

42. $m = \dfrac{Fr}{v^2}$

43. $\qquad G = w + 150n$

$\qquad G - w = 150n \qquad$ Subtracting w from both sides

$\qquad \dfrac{1}{150}(G - w) = \dfrac{1}{150} \cdot 150n \quad$ Multiplying both sides by $\dfrac{1}{150}$

$\qquad \dfrac{G - w}{150} = n \qquad$ Simplifying

44. $t = \dfrac{P - b}{0.5}$

45. $\qquad 2w + 2h + l = p$

$\qquad l = p - 2w - 2h \quad$ Adding $-2w - 2h$ to both sides

46. $w = \dfrac{p - 2h - l}{2}$

47. $\qquad Ax + By = C$

$\qquad By = C - Ax \qquad$ Subtracting Ax from both sides

$\qquad \dfrac{1}{B} \cdot By = \dfrac{1}{B}(C - Ax) \quad$ Multiplying both sides by $\dfrac{1}{B}$

$\qquad y = \dfrac{C - Ax}{B} \qquad$ Simplifying

48. $l = \dfrac{P - 2w}{2}$, or $l = \dfrac{P}{2} - w$

49. $\qquad C = \dfrac{5}{9}(F - 32)$

$\qquad \dfrac{9}{5} \cdot C = \dfrac{9}{5} \cdot \dfrac{5}{9}(F - 32) \quad$ Multiplying both sides by $\dfrac{9}{5}$

$\qquad \dfrac{9}{5}C = F - 32 \qquad$ Simplifying

$\qquad \dfrac{9}{5}C + 32 = F \qquad$ Adding 32 to both sides

50. $I = \dfrac{10}{3}T + 12,000$

51. $\qquad A = \dfrac{h}{2}(b_1 + b_2)$

$\qquad \dfrac{2}{h} \cdot A = \dfrac{2}{h} \cdot \dfrac{h}{2}(b_1 + b_2) \quad$ Multiplying both sides by $\dfrac{2}{h}$

$\qquad \dfrac{2A}{h} = b_1 + b_2 \qquad$ Simplifying

$\qquad \dfrac{2A}{h} - b_1 = b_2 \qquad$ Subtracting b_1 from both sides

52. $h = \dfrac{2A}{b_1 + b_2}$

53. $\qquad v = \dfrac{d_2 - d_1}{t}$

$\qquad t \cdot v = t \cdot \dfrac{d_2 - d_1}{t} \qquad$ Clearing the fraction

$\qquad tv = d_2 - d_1$

$\qquad tv \cdot \dfrac{1}{v} = (d_2 - d_1) \cdot \dfrac{1}{v} \quad$ Multiplying both sides by $\dfrac{1}{v}$

$\qquad t = \dfrac{d_2 - d_1}{v}$

54. $m = \dfrac{s_2 - s_1}{v}$

55. $\qquad v = \dfrac{d_2 - d_1}{t}$

$\qquad t \cdot v = t \cdot \dfrac{d_2 - d_1}{t} \quad$ Clearing the fraction

$\qquad tv = d_2 - d_1$

$\qquad tv - d_2 = -d_1 \qquad$ Subtracting d_2 from both sides

$\qquad -1 \cdot (tv - d_2) = -1 \cdot (-d_1) \quad$ Multiplying both sides by -1

$\qquad -tv + d_2 = d_1,$

$\qquad$ or $d_2 - tv = d_1$

56. $s_1 = s_2 - vm$

57. $\qquad r = m + mnp$

$\qquad r = m(1 + np) \qquad$ Factoring

$\qquad r \cdot \dfrac{1}{1 + np} = m(1 + np) \cdot \dfrac{1}{1 + np}$

$\qquad \dfrac{r}{1 + np} = m$

58. $x = \dfrac{p}{1 - yz}$

59. $\qquad y = ab - ac^2$

$\qquad y = a(b - c^2) \qquad$ Factoring

$\qquad y \cdot \dfrac{1}{b - c^2} = a(b - c^2) \cdot \dfrac{1}{b - c^2}$

$\qquad \dfrac{y}{b - c^2} = a$

). $m = \dfrac{d}{n - p^3}$

1. $3x + 6y = 9$

$\quad\quad 6y = -3x + 9 \quad$ Adding $-3x$ to both sides

$\quad\quad\ y = \dfrac{-3x + 9}{6} \quad$ Dividing both sides by 6

$\quad\quad\ y = \dfrac{-x + 3}{2} \quad$ Simplifying

2. $y = \dfrac{4x - 6}{7}$

3. $\quad\ x = y - 7$

$\quad x + 7 = y \quad\quad$ Adding 7 to both sides

4. $y = 4x - 2$

5. $y - 3(x + 2) = 4 + 2y$

$\quad y - 3x - 6 = 4 + 2y \quad$ Using the distributive law

$\quad\ -3x - 6 = 4 + y \quad$ Adding $-y$ to both sides

$\quad -3x - 10 = y \quad\quad$ Subtracting 4 from both sides

6. $y = \dfrac{2x}{3}$

7. $4y + x^2 = x + 1$

$\quad 4y = -x^2 + x + 1 \quad$ Adding $-x^2$ to both
$\quad\quad\quad\quad\quad\quad\quad\quad$ sides

$\quad\ y = \dfrac{-x^2 + x + 1}{4} \quad$ Dividing both sides by
$\quad\quad\quad\quad\quad\quad\quad\quad$ 4

8. $y = 2x^2 + 3x$

9. **Familiarize.** In an algebra book or a business math book we can find a formula for simple interest, $I = Prt$, when I is the interest, P is the principal, r is the interest rate, and t is the time, in years.

Translate. We want to find the interest rate, so we solve the formula for r.

$$I = Prt$$
$$\frac{1}{Pt} \cdot I = \frac{1}{Pt} \cdot Prt$$
$$\frac{I}{Pt} = r$$

Carry out. The model $r = \dfrac{I}{Pt}$ can be used to find the rate of interest at which an amount (the principal) must be invested in order to earn a given amount. We substitute $\$2600$ for P, $\dfrac{1}{2}$ for t $\left(6 \text{ months} = \dfrac{1}{2} \text{ yr}\right)$, and $\$156$ for I.

$$\frac{I}{Pt} = r$$
$$\frac{\$156}{\$2600\left(\dfrac{1}{2}\right)} = r$$
$$\frac{156}{1300} = r$$
$$0.12 = r$$
$$12\% = r$$

Check. Since $\$2600(0.12)\left(\dfrac{1}{2}\right) = \$156$, the answer checks.

State. The interest rate must be 12%.

70. $1571.43

71. **Familiarize.** In a geometry book or an algebra book we can find a formula for the area of a parallelogram, $A = bh$, where b is the base and h is the height.

Translate. We solve the formula for h.

$$A = bh$$
$$\frac{1}{b} \cdot A = \frac{1}{b} \cdot bh$$
$$\frac{A}{b} = h$$

Carry out. The model $h = \dfrac{A}{b}$ can be used to find the height of any parallelogram for which the area and base are known. We substitute 78 for A and 13 for b.

$$h = \frac{A}{b}$$
$$h = \frac{78}{13}$$
$$h = 6$$

Check. We repeat the calculation. The answer checks.

State. The height is 6 cm.

72. 12 cm

73. **Familiarize.** We will use Thurnau's model, $P = 9.337da - 299$.

Translate. Since we want to find the diameter of the fetus' head, we solve for d.

$$P = 9.337da - 299$$
$$P + 299 = 9.337da$$
$$\frac{P + 299}{9.337a} = d$$

Carry out. Substitute 1614 for P and 24.1 for a in the formula and calculate:

$$\frac{1614 + 299}{9.337(24.1)} = d$$
$$8.5 \approx d$$

Check. We repeat the calculation. The answer checks.

State. The diameter of the fetus' head at 29 weeks is about 8.5 cm.

74. 27.6 cm

75. Familiarize. The formula for the area of a trapezoid is $A = \frac{1}{2}h(b_1 + b_2)$, where A is the area, h is the height, and b_1 and b_2 are the bases.

Translate. The unknown dimension is the height, so we solve the formula for h. This was done in Exercise 22. We have

$$h = \frac{2A}{b_1 + b_2}.$$

Carry out. We substitute.

$$h = \frac{2A}{b_1 + b_2}$$

$$h = \frac{2 \cdot 90}{8 + 12}$$

$$h = \frac{180}{20}$$

$$h = 9$$

Check. We repeat the calculation. The answer checks.

State. The unknown dimension, the height of the trapezoid, is 9 ft.

76. 25 ft

77. Observe that 9% of $1000 is $90, so $90 is the amount of simple interest that would be earned in 1 yr. Thus, it will take 1 yr for the investment to be worth $1090.

78. 6 years

79. Familiarize. We will use Goiten's model, $I = 1.08(T/N)$. Note that 8 hr $= 8 \times 1$ hr $= 8 \times 60$ min $= 480$ min.

Translate. We solve the formula for N.

$$I = 1.08\left(\frac{T}{N}\right)$$

$$N \cdot I = N(1.08)\left(\frac{T}{N}\right)$$

$$NI = 1.08T$$

$$N = \frac{1.08T}{I}$$

Carry out. We substitute 480 for T and 15 for I.

$$N = \frac{1.08T}{I}$$

$$N = \frac{1.08(480)}{15}$$

$$N = 34.56$$

$$N \approx 34 \qquad \text{Rounding down}$$

Check. We repeat the calculations. The answer checks.

State. Dr. Cruz should schedule 34 appointments in on day.

80. 463 min, or 7.7 hr

81. *Writing Exercise*

82. *Writing Exercise*

83. $y = -x$

To find an ordered pair, we choose any number for x an then determine y. For example, if $x = -3$, then y $-(-3) = 3$. We find several ordered pairs, plot them, an draw the line.

x	y	(x, y)
-3	3	$(-3, 3)$
-1	1	$(-1, 1)$
0	0	$(0, 0)$
4	-4	$(4, -4)$

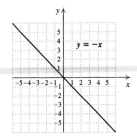

84.

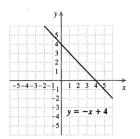

85. $x + y = 5$

It will be helpful to solve for y first. Subtracting x c both sides of the equation, we have $y = -x + 5$. Nov to find an ordered pair, we choose any number for x an then determine y. For example, if $x = -1$, then y $-(-1) + 5 = 1 + 5 = 6$. We find several ordered pairs, plot them, and draw the graph.

x	y	(x, y)
-1	6	$(-1, 6)$
0	5	$(0, 5)$
1	4	$(1, 4)$
4	1	$(4, 1)$

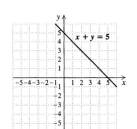

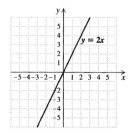

7. $y = |x + 1|$, or $y = \text{abs}(x + 1)$

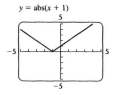

8. $y = -2x^2$

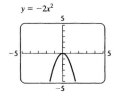

9. *Writing Exercise*

9. *Writing Exercise*

1. *Familiarize*. First we find the volume of the ring. Note that the inner diameter is 2 cm, so the inner radius is 2/2 or 1 cm. Then the volume of the ring is the volume of a right circular cylinder with height 0.5 cm and radius 1 + 0.15, or 1.15 cm, less the volume of a right circular cylinder with height 0.5 cm and radius 1 cm. Recall that the formula for the volume of a right circular cylinder is $V = \pi r^2 h$. Then the volume of the ring is

$$\pi(1.15)^2(0.5) - \pi(1)^2(0.5) = 0.16125\pi \text{ cm}^3.$$

Translate. To find the weight of the ring we will use the formula $D = \dfrac{m}{V}$. Solving for m, we get

$$D = \frac{m}{V}$$

$$V \cdot D = V \cdot \frac{m}{V}$$

$$V \cdot D = m$$

Carry out. We substitute in the formula $m = V \cdot D$.

$$m = 0.16125\pi(21.5)$$

$$m \approx 10.9$$

Check. We repeat the calculations. The answer checks.

State. The ring will weigh about 10.9 g.

92. About 26 g

93. *Familiarize*. We will use the formulas for density and for the volume of a right circular cylinder. Note that the radius of the penny is $\dfrac{1.85 \text{ cm}}{2}$, or 0.925 cm.

Translate. Solving the formula $D = \dfrac{m}{V}$ for V, we get $V = \dfrac{m}{D}$. Also, the volume of a right circular cylinder with radius r and height h is given by $V = \pi r^2 h$, so we have $\pi r^2 h = \dfrac{m}{D}$. Solve for h by multiplying by $\dfrac{1}{\pi r^2}$ on both sides of the equation:

$$h = \frac{m}{\pi r^2 D}$$

Carry out. Substitute 8.93 for D, 177.6 for m, and 0.925 for r.

$$h = \frac{m}{\pi r^2 D}$$

$$h = \frac{177.6}{\pi(0.925)^2(8.93)}$$

$$h \approx 7.4$$

Check. We repeat the calculation. The answer checks.

State. A role of pennies is about 7.4 cm tall.

94. About 610 cm

95. *Familiarize*. The average score on the first four tests is $\dfrac{83 + 91 + 78 + 81}{4}$, or 83.25. Let $x =$ the number of points above this average that Tico scores on the next test. Then the score on the fifth test is $83.25 + x$.

Translate.

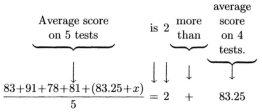

Carry out. Carry out some algebraic manipulation.

$$\frac{83 + 91 + 78 + 81 + (83.25 + x)}{5} = 2 + 83.25$$

$$\frac{416.25 + x}{5} = 85.25$$

$$416.25 + x = 426.25$$

$$x = 10$$

Check. If Tico scores 10 points more than the average of the first four tests on the fifth test, his score will be $83.25 + 10$, or 93.25. Then the five-test average will be $\dfrac{83 + 91 + 78 + 81 + 93.25}{5}$, or 85.25. This is 2 points above the four-test average, so the answer checks.

State. Tico must score 10 points above the four-test average in order to raise the average 2 points.

96. 90 in^2

97. Familiarize. Let p = the price of the house in 1994. In 1995 real estate prices increased 6%, so the house was worth $p + 0.06p$, or $1.06p$. In 1996 prices increased 2%, so the house was then worth $1.06p + 0.02(1.06p)$, or $1.02(1.06p)$. In 1997 prices dropped 1%, so the value of the house became $1.02(1.06p) - 0.01(1.02)(1.06p)$, or $0.99(1.02)(1.06p)$.

Translate.

$$\underbrace{\text{The price of the house in 1997}} \quad \text{was} \quad \$117,743.$$

$$0.99(1.02)(1.06p) \quad = \quad 117,743$$

Carry out. We carry out some algebraic manipulation.

$$0.99(1.02)(1.06p) = 117,743$$

$$p = \frac{117,743}{0.99(1.02)(1.06)}$$

$$p \approx 110,000$$

Check. If the price of the house in 1994 was \$110,000, then in 1995 it was worth 1.06(\$110,000), or \$116,600. In 1996 it was worth 1.02(\$116,600), or \$118,932, and in 1997 it was worth 0.99(\$118,932), or \$117,743. Our answer checks.

State. The house was worth \$110,000 in 1994.

98. $\dfrac{100}{100 - n}$

99.
$$s = v_i t + \frac{1}{2}at^2$$

$$s - v_i t = \frac{1}{2}at^2$$

$$2(s - v_i t) = at^2 \quad \text{Multiplying both sides by 2}$$

$$\frac{2(s - v_i t)}{t^2} = a,$$

$$\text{or } \frac{2s - 2v_i t}{t^2} = a$$

100. $l = \dfrac{A - w^2}{4w}$

101.
$$\frac{P_1 V_1}{T_1} = \frac{P_2 V_2}{T_2}$$

$$P_1 V_1 T_2 = P_2 V_2 T_1 \quad \text{Multiplying both sides by } T_1 T_2$$

$$T_2 = \frac{P_2 V_2 T_1}{P_1 V_1} \quad \text{Multiplying both sides by } \frac{1}{P_1 V_1}$$

102. $T_1 = \dfrac{P_1 V_1 T_2}{P_2 V_2}$

103.
$$\frac{b}{a - b} = c$$

$$b = c(a - b) \quad \text{Multiplying both sides by } a - b$$

$$b = ac - bc$$

$$b + bc = ac \quad \text{Adding } bc \text{ to both sides}$$

$$b(1 + c) = ac \quad \text{Factoring}$$

$$b = \frac{ac}{1 + c} \quad \text{Multiplying both sides by } \frac{1}{1 + c}$$

104. $d = \dfrac{me^2}{f}$

105.
$$\frac{a}{a + b} = c$$

$$a = c(a + b)$$

$$a = ac + bc$$

$$a - ac = bc$$

$$a(1 - c) = bc$$

$$a = \frac{bc}{1 - c}$$

106. $t = \dfrac{1}{s}$

107. *Writing Exercise*

108. *Writing Exercise*

109. *Writing Exercise*

110.

$y = (2x^3 + 7)/5$

111. First we solve the equation for y.
$$y - 2(x^2 + y) = 0$$
$$y - 2x^2 - 2y = 0$$
$$-2x^2 - y = 0$$
$$-2x^2 = y$$

Now enter $y = -2x^2$ on the equation-editor screen and graph the equation.

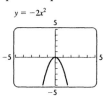

$y = -2x^2$

2.

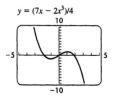

$y = (7x - 2x^3)/4$

3. First we solve the equation for y.

$$|x + 2| - 3y = 4y$$
$$|x + 2| = 7y$$
$$\frac{|x + 2|}{7} = y$$

Now enter $y = \text{abs}(x + 2)/7$ on the equation-editor screen and graph the equation.

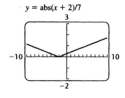

$y = \text{abs}(x + 2)/7$

Exercise Set 2.4

1. Graph: $f(x) = 2x - 7$.

We make a table of values. Then we plot the corresponding points and connect them.

x	$f(x)$
1	-5
2	-3
3	-1
5	3

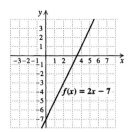

2.

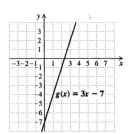

$g(x) = 3x - 7$

3. Graph: $g(x) = -\dfrac{1}{3}x + 2$.

We make a table of values. Then we plot the corresponding points and connect them.

x	$g(x)$
-3	3
0	2
3	1
6	0

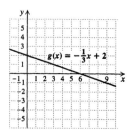

$g(x) = -\dfrac{1}{3}x + 2$

4.

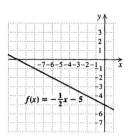

$f(x) = -\dfrac{1}{2}x - 5$

5. Graph: $h(x) = \dfrac{2}{5}x - 4$.

We make a table of values. Then we plot the corresponding points and connect them.

x	$h(x)$
-5	-6
0	-4
5	-2

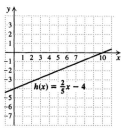

$h(x) = \dfrac{2}{5}x - 4$

6.

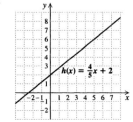

$h(x) = \dfrac{4}{5}x + 2$

7. $y = 5x + 7$

The y-intercept is $(0, 7)$, or simply 7.

8. $(0, -9)$

9. $f(x) = -2x - 6$

The y-intercept is $(0, -6)$, or simply -6.

10. $(0, 7)$

11. $y = -\dfrac{3}{8}x - 4.5$

The y-intercept is $(0, -4.5)$, or simply -4.5.

12. $(0, 2.2)$

13. $g(x) = 2.9x - 9$

The y-intercept is $(0, -9)$, or simply -9.

14. $(0, 5)$

15. $y = 37x + 204$

The y-intercept is $(0, 204)$, or simply 204.

16. $(0, 700)$

17. Slope $= \dfrac{\text{change in } y}{\text{change in } x} = \dfrac{5 - 9}{4 - 6} = \dfrac{-4}{-2} = 2$

18. $\dfrac{4}{3}$

19. Slope $= \dfrac{\text{change in } y}{\text{change in } x} = \dfrac{-4 - 8}{9 - 3} = \dfrac{-12}{6} = -2$

20. $\dfrac{3}{26}$

21. Slope $= \dfrac{\text{change in } y}{\text{change in } x} = \dfrac{8.7 - 12.4}{-5.2 - (-16.3)} = \dfrac{-3.7}{11.1} =$

$-\dfrac{37}{111} = -\dfrac{1}{3}$

22. $\dfrac{8}{269}$

23. Slope $= \dfrac{\text{change in } y}{\text{change in } x} = \dfrac{43.6 - 43.6}{4.5 - (-9.7)} = \dfrac{0}{14.2} = 0$

24. $\dfrac{1}{2}$

25. a) The graph of $y = 3x - 5$ has a positive slope, 3, and the y-intercept is $(0, -5)$. Thus, graph II matches this equation.

b) The graph of $y = 0.7x + 1$ has a positive slope, 0.7, and the y-intercept is $(0, 1)$. Thus graph IV matches this equation.

c) The graph of $y = -0.25x - 3$ has a negative slope, -0.25, and the y-intercept is $(0, -3)$. Thus graph III matches this equation.

d) The graph of $y = -4x + 2$ has a negative slope, -4, and the y-intercept is $(0, 2)$. Thus graph I matches this equation.

26. a) II

b) IV

c) I

d) III

27. $y = \dfrac{5}{2}x + 3$

Slope is $\dfrac{5}{2}$; y-intercept is $(0, 3)$.

From the y-intercept, we go *up* 5 units and to the *right* units. This gives us the point $(2, 8)$. We can now draw th graph.

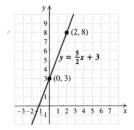

As a check, we can rename the slope and find anothe point.

$$\dfrac{5}{2} = \dfrac{5}{2} \cdot \dfrac{-1}{-1} = \dfrac{-5}{-2}$$

From the y-intercept, we go *down* 5 units and to the *left* units. This gives us the point $(-2, -2)$. Since $(-2, -2)$ on the line, we have a check.

28. Slope is $\dfrac{2}{5}$; y-intercept is $(0, 4)$.

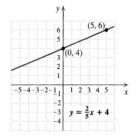

29. $f(x) = -\dfrac{5}{2}x + 1$

Slope is $-\dfrac{5}{2}$, or $\dfrac{-5}{2}$; y-intercept is $(0, 1)$.

From the y-intercept, we go *down* 5 units and to the *righ* 2 units. This gives us the point $(2, -4)$. We can now dra the graph.

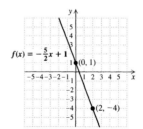

As a check, we can rename the slope and find anothe point.

$$\frac{-5}{2} = \frac{5}{-2}$$

From the y-intercept, we go *up* 5 units and to the *left* 2 units. This gives us the point $(-2, 6)$. Since $(-2, 6)$ is on the line, we have a check.

30. Slope is $-\dfrac{2}{5}$; y-intercept is $(0, 3)$.

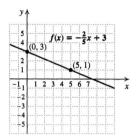

31. Convert to a slope-intercept equation.

$$2x - y = 5$$
$$-y = -2x + 5$$
$$y = 2x - 5$$

Slope is 2, or $\dfrac{2}{1}$; y-intercept is $(0, -5)$.

From the y-intercept, we go *up* 2 units and to the *right* 1 unit. This gives us the point $(1, -3)$. We can now draw the graph.

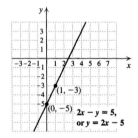

As a check, we can rename the slope and find another point.

$$2 = \frac{2}{1} \cdot \frac{3}{3} = \frac{6}{3}$$

From the y-intercept, we go *up* 6 units and to the *right* 3 units. This gives us the point $(3, 1)$. Since $(3, 1)$ is on the line, we have a check.

32. Slope is -2; y-intercept is $(0, 4)$.

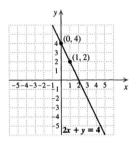

33. $f(x) = \dfrac{1}{3}x + 2$

Slope is $\dfrac{1}{3}$; y-intercept is $(0, 2)$.

From the y-intercept, we go *up* 1 unit and to the *right* 3 units. This gives us the point $(3, 3)$. We can now draw the graph.

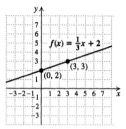

As a check, we can rename the slope and find another point.

$$\frac{1}{3} = \frac{1}{3} \cdot \frac{-1}{-1} = \frac{-1}{-3}$$

From the y-intercept, we go *down* 1 unit and to the *left* 3 units. This gives us the point $(-3, 1)$. Since $(-3, 1)$ is on the line, we have a check.

34. Slope is -3; y-intercept is $(0, 6)$.

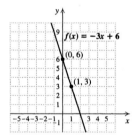

35. Convert to a slope intercept equation:

$$7y + 2x = 7$$
$$7y = -2x + 7$$
$$y = \frac{1}{7}(-2x + 7)$$
$$y = -\frac{2}{7}x + 1$$

Slope is $-\dfrac{2}{7}$ or $\dfrac{-2}{7}$; y-intercept is $(0, 1)$.

From the y-intercept we go *down* 2 units and to the *right* 7 units. This gives us the point $(7, -1)$. We can now draw the graph.

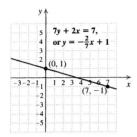

As a check we can rename the slope and find another point.

$$\frac{-2}{7} = \frac{-2}{7} \cdot \frac{-1}{-1} = \frac{2}{-7}$$

From the y-intercept, we go *up* 2 units and to the *left* 7 units. This gives us the point $(-7, 3)$. Since $(-7, 3)$ is on the line, we have a check.

36. Slope is $\frac{1}{4}$; y-intercept is $(0, -5)$.

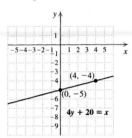

37. $f(x) = -0.25x$

Slope is -0.25, or $\frac{-1}{4}$; y-intercept is $(0, 0)$.

From the y-intercept, we go *down* 1 unit and to the *right* 4 units. This gives us the point $(4, -1)$. We can now draw the graph.

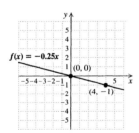

As a check, we can rename the slope and find another point.

$$\frac{-1}{4} = \frac{-1}{4} \cdot \frac{-1}{-1} = \frac{1}{-4}$$

From the y-intercept, we go *up* 1 unit and to the *left* 4 units. This gives us the point $(-4, 1)$. Since $(-4, 1)$ is on the line, we have a check.

38. Slope is 1.5, or $\frac{3}{2}$; y-intercept is $(0, -3)$.

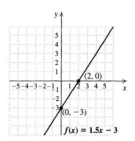

39. Convert to a slope-intercept equation.

$$4x - 5y = 10$$
$$-5y = 4x + 10$$
$$y = \frac{4}{5}x - 2$$

Slope is $\frac{4}{5}$; y-intercept is $(0, -2)$.

From the y-intercept, we go *up* 4 units and to the *right* units. This gives us the point $(5, 2)$. We can now draw th graph.

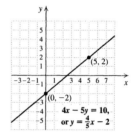

As a check, we choose some other value for x, say -5, an determine y:

$$y = \frac{4}{5}(-5) - 2 = -4 - 2 = -6$$

We plot the point $(-5, -6)$ and see that it *is* on the line.

40. Slope is $-\frac{5}{4}$; y-intercept is $(0, 1)$.

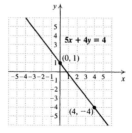

1. $f(x) = \dfrac{5}{4}x - 2$

Slope is $\dfrac{5}{4}$; y-intercept is $(0, -2)$.

From the y-intercept, we go *up* 5 units and to the *right* 4 units. This gives us the point $(4, 3)$. We can now draw the graph.

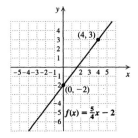

As a check, we choose some other value for x, say -2, and determine $f(x)$:

$$f(x) = \dfrac{5}{4}(-2) - 2 = -\dfrac{5}{2} - 2 = -\dfrac{9}{2}$$

We plot the point $\left(-2, -\dfrac{9}{2}\right)$ and see that it *is* on the line.

2. Slope is $\dfrac{4}{3}$; y-intercept is $(0, 2)$.

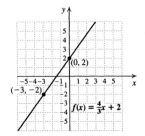

3. Convert to a slope-intercept equation:

$$12 - 4f(x) = 3x$$
$$-4f(x) = 3x - 12$$
$$f(x) = -\dfrac{1}{4}(3x - 12)$$
$$f(x) = -\dfrac{3}{4}x + 3$$

Slope is $-\dfrac{3}{4}$, or $\dfrac{-3}{4}$; y-intercept is $(0, 3)$.

From the y-intercept, we go *down* 3 units and to the *right* 4 units. This gives us the point $(4, 0)$. We can now draw the graph.

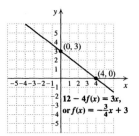

As a check, we choose some other value for x, say -4, and determine $f(x)$:

$$f(-4) = -\dfrac{3}{4}(-4) + 3 = 3 + 3 = 6$$

We plot the point $(-4, 6)$ and see that it *is* on the line.

44. Slope is $-\dfrac{2}{5}$; y-intercept is $(0, -3)$.

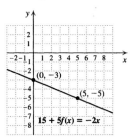

45. $g(x) = 2.5 = 0x + 2.5$

Slope is 0; y-intercept is $(0, 2.5)$.

From the y-intercept, we go up or down 0 units and any number of nonzero units to the left or right. Any point on the graph will lie on a horizontal line 2.5 units above the x-axis. We draw the graph.

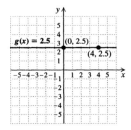

46. Slope is $\dfrac{3}{4}$; y-intercept is $(0, 0)$

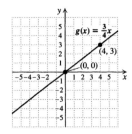

47. Use the slope-intercept equation, $f(x) = mx + b$, with $m = \frac{2}{3}$ and $b = -9$.
$$f(x) = mx + b$$
$$f(x) = \frac{2}{3}x + (-9)$$
$$f(x) = \frac{2}{3}x - 9$$

48. $f(x) = -\frac{3}{4}x + 12$

49. Use the slope-intercept equation, $f(x) = mx + b$, with $m = -5$ and $b = 2$.
$$f(x) = mx + b$$
$$f(x) = -5x + 2$$

50. $f(x) = 2x - 1$

51. Use the slope-intercept equation, $f(x) = mx + b$, with $m = -\frac{7}{9}$ and $b = 5$.
$$f(x) = mx + b$$
$$f(x) = -\frac{7}{9}x + 5$$

52. $f(x) = -\frac{4}{11}x + 9$

53. Use the slope-intercept equation, $f(x) = mx + b$, with $m = 5$ and $b = \frac{1}{2}$.
$$f(x) = mx + b$$
$$f(x) = 5x + \frac{1}{2}$$

54. $f(x) = 6x + \frac{2}{3}$

55. We can use the coordinates of any two points on the line. We'll use $(0, 30)$ and $(3, 3)$.
$$\text{Slope} = \frac{\text{change in } y}{\text{change in } x} = \frac{3 - 30}{3 - 0} = \frac{-27}{3} = -9$$
The value is decreasing at a rate of $900 per year.

56. The distance is decreasing at a rate of $6\frac{2}{3}$ m per second.

57. We can use the coordinates of any two points on the line. We'll use $(0, 0)$ and $(1, 3)$.
$$\text{Slope} = \frac{\text{change in } y}{\text{change in } x} = \frac{3 - 0}{1 - 0} = \frac{3}{1}, \text{ or } 3$$
The distance is increasing at a rate of 3 miles per hour.

58. The weight is increasing at a rate of $\frac{1}{2}$ pound per bag of feed.

59. We can use the coordinates of any two points on the line. We'll use $(0, 50)$ and $(2, 200)$.
$$\text{Slope} = \frac{\text{change in } y}{\text{change in } x} = \frac{200 - 50}{2 - 0} = 75$$
The number of pages read is increasing at a rate of 7 pages per day.

60. The distance is increasing at a rate of $\frac{1}{4}$ km per minute.

61. We can use the coordinates of any two points on the line. We'll use $(15, 470)$ and $(55, 510)$:
$$\text{Slope} = \frac{\text{change in } y}{\text{change in } x} = \frac{510 - 470}{55 - 15} = \frac{40}{40} = 1$$
The average SAT math score is increasing at a rate of 1 point per thousand dollars of family income.

62. The average SAT verbal score is increasing at a rate of 1 point per thousand dollars of family income.

63. The skier's speed is given by $\dfrac{\text{change in distance}}{\text{change in time}}$. Note that the skier reaches the 12-km mark 45 min after the 3-km mark was reached or after $15 + 45$, or 60 min. We will express time in hours: 15 min = 0.25 hr and 60 min = 1 hr. Then
$$\frac{\text{change in distance}}{\text{change in time}} = \frac{12 - 3}{1 - 0.25} = \frac{9}{0.75} = 12.$$
The speed is 12 km/h.

64. 10 mph

65. The rate at which the number of hits is increasing is given by $\dfrac{\text{change in number of hits}}{\text{change in time}}$.
$$\frac{\text{change in number of hits}}{\text{change in time}} = \frac{430,000 - 80,000}{2001 - 1999} =$$
$$\frac{350,000}{2} = 175,000$$
The number of hits is increasing at a rate of 175,000 hits/yr.

66. $\frac{5}{96}$ of the house per hour

67. The average rate of descent is given by $\dfrac{\text{change in altitude}}{\text{change in time}}$. We will express time in minutes:
$$1\frac{1}{2} \text{ hr} = \frac{3}{2} \text{ hr} \cdot \frac{60 \text{ min}}{1 \text{ hr}} = 90 \text{ min}$$
$$2 \text{ hr, } 10 \text{ min} = 2 \text{ hr} + 10 \text{ min} =$$
$$2 \text{ hr} \cdot \frac{60 \text{ min}}{1 \text{ hr}} + 10 \text{ min} = 120 \text{ min} + 10 \text{ min} = 130 \text{ min}$$
Then

$$\frac{\text{change in altitude}}{\text{change in time}} = \frac{0 - 12{,}000}{130 - 90} = \frac{-12{,}000}{40} =$$

$-300.$

The average rate of descent is 300 ft/min.

3. $\dfrac{71}{70}$ million/yr

9. a) Graph II indicated that 200 ml of fluid was dripped in the first 3 hr, a rate of $\dfrac{200}{3}$ ml/hr. It also indicates that 400 ml of fluid was dripped in the next 3 hr, a rate of $\dfrac{400}{3}$ ml/hr, and that this rate continues until the end of the time period shown. Since the rate of $\dfrac{400}{3}$ ml/hr is double the rate of $\dfrac{200}{3}$ ml/hr, this graph is appropriate for the given situation.

b) Graph IV indicates that 300 ml of fluid was dripped in the first 2 hr, a rate of 300/2, or 150 ml/hr. In the next 2 hr, 200 ml was dripped. This is a rate of 200/2, or 100 ml/hr. Then 100 ml was dripped in the next 3 hr, a rate of 100/3, or $33\dfrac{1}{3}$ ml/hr. Finally, in the remaining 2 hr, 0 ml of fluid was dripped, a rate of 0/2, or 0 ml/hr. Since the rate at which the fluid was given decreased as time progressed and eventually became 0, this graph is appropriate for the given situation.

c) Graph I is the only graph that shows a constant rate for 5 hours, in this case from 3 PM to 8 PM. Thus, it is appropriate for the given situation.

d) Graph III indicates that 100 ml of fluid was dripped in the first 4 hr, a rate of 100/4, or 25 ml/hr. In the next 3 hr, 200 ml was dripped. This is a rate of 200/3, or $66\dfrac{2}{3}$ ml/hr. Then 100 ml was dripped in the next hour, a rate of 100 ml/hr. In the last hour 200 ml was dripped, a rate of 200 ml/hr. Since the rate at which the fluid was given gradually increased, this graph is appropriate for the given situation.

0. a) III

b) IV

c) I

d) II

1. $C(x) = 25x + 75$

25 signifies that the cost per person is $25; 75 signifies that the setup cost for the party is $75.

2. 0.05 signifies that a salesperson earns a 5% commission on sales; 200 indicates that a saleperson's base salary is $200 per week.

73. $L(t) = \dfrac{1}{2}t + 1$

$\dfrac{1}{2}$ signifies that Tina's hair grows $\dfrac{1}{2}$ in. per month; 1 signifies that her hair is 1 in. long immediately after she gets it cut.

74. $\dfrac{1}{8}$ signifies that the grass grows $\dfrac{1}{8}$ in. per day; 2 signifies that the grass is 2 in. long immediately after it is cut.

75. $A(t) = \dfrac{3}{20}t + 72$ is of the form $y = mx + b$ with $m = \dfrac{3}{20}$ and $b = 72$.

$\dfrac{3}{20}$ signifies that the life expectancy of American women increases $\dfrac{3}{20}$ yr per year for years after 1950; 72 signifies that the life expectancy of American women in 1950 was 72 years.

76. $\dfrac{1}{5}$ signifies that the demand increases $\dfrac{1}{5}$ quadrillion joules per year for years after 1960; 20 signifies that the demand was 20 quadrillion joules in 1960.

77. $f(t) = 2.6t + 17.8$ is of the form $y = mx + b$ with $m = 2.6$ and $b = 17.8$.

2.6 signifies that sales increase $2.6 billion per year, for years after 1975; 17.8 signifies that sales in 1975 were $17.8 billion.

78. 0.1522 signifies that the price increases $0.1522 per year, for years since 1990; 4.29 signifies that the average cost of a movie ticket in 1990 was $4.29.

79. $C(d) = 0.75d + 2$ is of the form $y = mx + b$ with $m = 0.75$ and $b = 2$.

0.75 signifies that the cost per mile of a taxi ride is $0.75; 2 signifies that the minimum cost of a taxi ride is $2.

80. 0.3 signifies that the cost per mile of renting the truck is $0.30; 20 signifies that the minimum cost is $20.

81. $F(t) = -5000t + 90{,}000$

a) -5000 signifies that the truck's value depreciates $5000 per year; 90,000 signifies that the original value of the truck was $90,000.

b) We find the value of t for which $F(t) = 0$.

$$0 = -5000t + 90{,}000$$
$$5000t = 90{,}000$$
$$t = 18$$

It will take 18 yr for the truck to depreciate completely.

c) The truck's value goes from $90,000 when $t = 0$ to $0 when $t = 18$, so the domain of F is $\{x | 0 \le t \le 18\}$.

82. $V(t) = -2000t + 15,000$

    a) -2000 signifies that the color separator's value depreciates $2000 per year; 15,000 signifies that the original value of the separator was $15,000.

    b) 7.5 yr

    c) $\{t | 0 \le t \le 7.5\}$

83. $v(n) = -150n + 900$

    a) -150 signifies that the snowblower's value depreciates $150 per winter of use; 900 signifies that the original value of the snowblower was $900.

    b) We find the value of n for which $v(n) = 300$.
$$300 = -150n + 900$$
$$-600 = -150n$$
$$4 = n$$

    The snowblower's trade-in value will be $300 after 4 winters of use.

    c) First we find the value of n for which $v(n) = 0$.
$$0 = -150n + 900$$
$$-900 = -150n$$
$$6 = n$$

    The value of the snowblower goes from $900 when $n = 0$ to $0 when $n = 6$. Since the snowblower is used only in the winter we express the domain of v as $\{0, 1, 2, 3, 4, 5, 6\}$.

84. $T(x) = -300x + 2400$

    a) -300 signifies that the mower's value depreciates $300 per summer of use; 2400 signifies that the original value of the mower was $2400.

    b) After 4 summers of use

    c) $\{0, 1, 2, 3, 4, 5, 6, 7, 8\}$

85. *Writing Exercise*

86. *Writing Exercise*

87.
$$9\{2x - 3[5x + 2(-3x + y^0 - 2)]\}$$
$$= 9\{2x - 3[5x + 2(-3x + 1 - 2)]\} \quad (y^0 = 1)$$
$$= 9\{2x - 3[5x + 2(-3x - 1)]\}$$
$$= 9\{2x - 3[5x - 6x - 2]\}$$
$$= 9\{2x - 3[-x - 2]\}$$
$$= 9\{2x + 3x + 6\}$$
$$= 9\{5x + 6\}$$
$$= 45x + 54$$

88. $-125a^6b^9$

89. $(13m^2n^3)(-2m^5n) = 13(-2) \cdot m^2 \cdot m^5 \cdot n^3 \cdot n = -26m^{2+5}n^{3+1} = -26m^7n^4$

90. $8x - 3y - 12$

91. $(4x^2y^3)(-3x^5y)^0 = 4x^2y^3$

    (Any nonzero number raised to the 0 power is 1.)

92. $\dfrac{x^{10}y^{14}}{4}$

93. *Writing Exercise*

94. *Writing Exercise*

95. *Writing Exercise*

96. *Writing Exercise*

97. We first solve for y.
$$rx + py = s$$
$$py = -rx + s$$
$$y = -\frac{r}{p}x + \frac{s}{p}$$

    The slope is $-\dfrac{r}{p}$, and the y-intercept is $\left(0, \dfrac{s}{p}\right)$.

98. Slope: $-\dfrac{r}{r+p}$; y-intercept: $\left(0, \dfrac{s}{r+p}\right)$

99. See the answer section in the text.

100. False

101. Let $c = 2$ and $d = 3$. Then $f(cd) = f(2 \cdot 3) = f(6)$ $m \cdot 6 + b = 6m + b$, but $f(c)f(d) = f(2)f(3) = (m \cdot 2 + b)(m \cdot 3 + b) = 6m^2 + 5mb + b^2$. Thus, the given statement is false.

102. False

103. Let $c = 5$ and $d = 2$. Then $f(c - d) = f(5 - 2) = f(3)$ $m \cdot 3 + b = 3m + b$, but $f(c) - f(d) = f(5) - f(2)$ $(m \cdot 5 + b) - (m \cdot 2 + b) = 5m + b - 2m - b = 3m$. Thus the given statement is false.

104. $-\dfrac{31}{4}$

105. a) Graph III indicates that the first 2 mi and the last 3 mi were traveled in approximately the same length of time and at a fairly rapid rate. The mile following the first two miles was traveled at a much slower rate. This could indicate that the first two miles were driven, the next mile was swum and the last three miles were driven, so this graph is most appropriate for the given situation.

    b) The slope in Graph IV decreases at 2 mi and again at 3 mi. This could indicate that the first two miles were traveled by bicycle, the next mile was run, and the last 3 miles were walked, so this graph is most appropriate for the given situation.

c) The slope in Graph I decreases at 2 mi and then increases at 3 mi. This could indicate that the first two miles were traveled by bicycle, the next mile was hiked, and the last three miles were traveled by bus, so this graph is most appropriate for the given situation.

d) The slope in Graph II increases at 2 mi and again at 3 mi. This could indicate that the first two miles were hiked, the next mile was run, and the last three miles were traveled by bus, so this graph is most appropriate for the given situation.

. a) $-\dfrac{5c}{4b}$

b) Undefined

c) $\dfrac{a+d}{f}$

. $C(n) = 5n + 17.5$, for $1 \leq n \leq 10$,

$C(n) = 6n + 17.5$, for $11 \leq n \leq 20$,

$C(n) = 7n + 17.5$, for $21 \leq n \leq 30$,

$C(n) = 8n + 17.5$ for $n \geq 31$

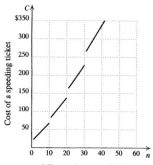

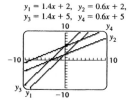

$y_1 = 1.4x + 2$, $y_2 = 0.6x + 2$,
$y_3 = 1.4x + 5$, $y_4 = 0.6x + 5$

---

ercise Set 2.5

. $y - 7 = 5$

$y = 12$

The graph of $y = 12$ is a horizontal line. Since $y - 7 = 5$ is equivalent to $y = 12$, the slope of the line $y - 7 = 5$ is 0.

. Undefined

3. $3x = 6$

$x = 2$

The graph of $x = 2$ is a vertical line. Since $3x = 6$ is equivalent to $x = 2$, the slope of the line $3x = 6$ is undefined.

4. 0

5. $4y = 20$

$y = 5$

The graph of $y = 5$ is a horizontal line. Since $4y = 20$ is equivalent to $y = 5$, the slope of the line $4y = 20$ is 0.

6. 0

7. $9 + x = 12$

$x = 3$

The graph of $x = 3$ is a vertical line. Since $9 + x = 12$ is equivalent to $x = 3$, the slope of the line $9 + x = 12$ is undefined.

8. Undefined

9. $2x - 4 = 3$

$2x = 7$

$x = \dfrac{7}{2}$

The graph of $x = \dfrac{7}{2}$ is a vertical line. Since $2x - 4 = 3$ is equivalent to $x = \dfrac{7}{2}$, the slope of the line $2x - 4 = 3$ is undefined.

10. 0

11. $5y - 4 = 35$

$5y = 39$

$y = \dfrac{39}{5}$

The graph of $y = \dfrac{39}{5}$ is a horizontal line. Since $5y - 4 = 35$ is equivalent to $y = \dfrac{39}{5}$, the slope of the line $5y - 4 = 35$ is 0.

12. Undefined

13. $3y + x = 3y + 2$

$x = 2$

The graph of $x = 2$ is a vertical line. Since $3y + x = 3y + 2$ is equivalent to $x = 2$, the slope of the line $3y + x = 3y + 2$ is undefined.

14. Undefined

15. $5x - 2 = 2x - 7$

$$5x = 2x - 5$$
$$3x = -5$$
$$x = -\frac{5}{3}$$

The graph of $x = -\frac{5}{3}$ is a vertical line. Since $5x-2 = 2x-7$ is equivalent to $x = -\frac{5}{3}$, the slope of the line $5x-2 = 2x-7$ is undefined.

16. 0

17. $y = -\frac{2}{3}x + 5$

The equation is written in slope-intercept form. We see that the slope is $-\frac{2}{3}$.

18. $-\frac{3}{2}$

19. Graph $y = 4$.

This is a horizontal line that crosses the y-axis at $(0, 4)$. If we find some ordered pairs, note that, for any x-value chosen, y must be 4.

x	y
-2	4
0	4
3	4

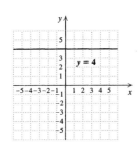

20.

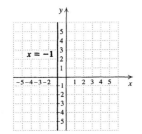

21. Graph $x = 2$.

This is a vertical line that crosses the x-axis at $(2, 0)$. If we find some ordered pairs, note that, for any y-value chosen, x must be -2.

x	y
-2	-1
-2	0
-2	2

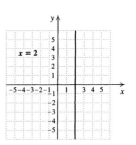

22.

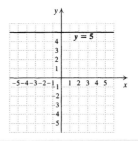

23. Graph $4 \cdot f(x) = 20$.

First solve for $f(x)$.

$$4 \cdot f(x) = 20$$
$$f(x) = 5$$

This is a horizontal line that crosses the vertical axis at $(0, 5)$.

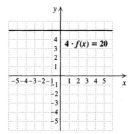

24.

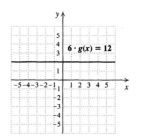

25. Graph $3x = -15$.

Since y does not appear, we solve for x.

$$3x = -15$$
$$x = -5$$

This is a vertical line that crosses the x-axis at $(-5, 0)$.

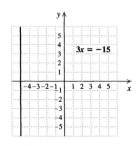

6.

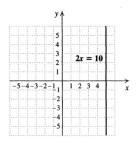

7. Graph $4 \cdot g(x) + 3x = 12 + 3x$.

First solve for $g(x)$.

$$4 \cdot g(x) + 3x = 12 + 3x$$
$$4 \cdot g(x) = 12 \qquad \text{Subtracting } 3x \text{ on}$$
$$\text{both sides}$$
$$g(x) = 3$$

This is a horizontal line that crosses the vertical axis at $(0, 3)$.

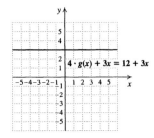

8.

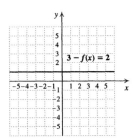

9. Graph $x + y = 5$.

To find the y-intercept, let $x = 0$ and solve for y.

$$0 + y = 5$$
$$y = 5$$

The y-intercept is $(0, 5)$.

To find the x-intercept, let $y = 0$ and solve for x.

$$x + 0 = 5$$
$$x = 5$$

The x-intercept is $(5, 0)$.

Plot these points and draw the line. A third point could be used as a check.

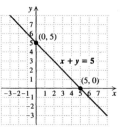

30.

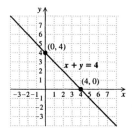

31. Graph $y = 2x + 8$.

To find the y-intercept, let $x = 0$ and solve for y.

$$y = 2 \cdot 0 + 8$$
$$y = 8$$

The y-intercept is $(0, 8)$.

To find the x-intercept, let $y = 0$ and solve for x.

$$0 = 2x + 8$$
$$-8 = 2x$$
$$-4 = x$$

The x-intercept is $(-4, 0)$.

Plot these points and draw the line. A third point could be used as a check.

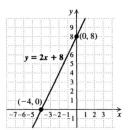

32.

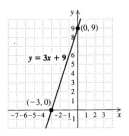

$$2x - 3 \cdot 0 = 18$$
$$2x = 18$$
$$x = 9$$

The x-intercept is $(9, 0)$.

Plot these points and draw the line. A third point coul
be used as a check.

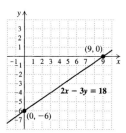

33. Graph $3x + 5y = 15$.

To find the y-intercept, let $x = 0$ and solve for y.

$$3 \cdot 0 + 5y = 15$$
$$5y = 15$$
$$y = 3$$

The y-intercept is $(0, 3)$.

To find the x-intercept, let $y = 0$ and solve for x.

$$3x + 5 \cdot 0 = 15$$
$$3x = 15$$
$$x = 5$$

The x-intercept is $(5, 0)$.

Plot these points and draw the line. A third point could
be used as a check.

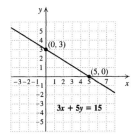

36.

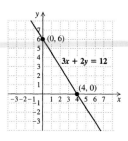

37. Graph $7x = 3y - 21$.

To find the y-intercept, let $x = 0$ and solve for y.

$$7 \cdot 0 = 3y - 21$$
$$0 = 3y - 21$$
$$21 = 3y$$
$$7 = y$$

The y-intercept is $(0, 7)$.

To find the x-intercept, let $y = 0$ and solve for x.

$$7x = 3 \cdot 0 - 21$$
$$7x = -21$$
$$x = -3$$

The x-intercept is $(-3, 0)$.

Plot these points and draw the line. A third point coul
be used as a check.

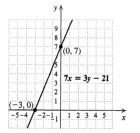

34.

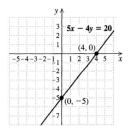

35. Graph $2x - 3y = 18$.

To find the y-intercept, let $x = 0$ and solve for y.

$$2 \cdot 0 - 3y = 18$$
$$-3y = 18$$
$$y = -6$$

The y-intercept is $(0, -6)$.

To find the x-intercept, let $y = 0$ and solve for x.

8.

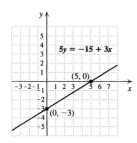

9. Graph $f(x) = 3x - 8$.

Because the function is in slope-intercept form, we know that the y-intercept is $(0, -8)$. To find the x-intercept, let $f(x) = 0$ and solve for x.

$$0 = 3x - 8$$
$$8 = 3x$$
$$\frac{8}{3} = x$$

The x-intercept is $\left(\frac{8}{3}, 0\right)$.

Plot these points and draw the line. A third point could be used as a check.

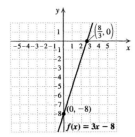

0.

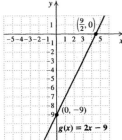

1. $1.4y - 3.5x = -9.8$

$14y - 35x = -98$ Multiplying by 10

$2y - 5x = -14$ Multiplying by $\frac{1}{7}$

Graph $2y - 5x = -14$.

To find the y-intercept, let $x = 0$.

$$2y - 5x = -14$$
$$2y - 5 \cdot 0 = -14$$
$$2y = -14$$
$$y = -7$$

$(0, -7)$ is the y-intercept.

To find the x-intercept, let $y = 0$.

$$2y - 5x = -14$$
$$2 \cdot 0 - 5x = -14$$
$$-5x = -14$$
$$x = 2.8$$

$(2.8, 0)$ is the x-intercept.

Plot these points and draw the line. A third point could be used as a check.

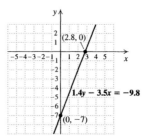

42.

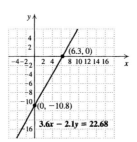

43. Graph $5x + 2g(x) = 7$

To find the y-intercept, let $x = 0$ and solve for $g(x)$.

$$5 \cdot 0 + 2g(x) = 7$$
$$2g(x) = 7$$
$$g(x) = \frac{7}{2}$$

$\left(0, \frac{7}{2}\right)$ is the y-intercept.

To find the x-intercept, let $g(x) = 0$ and solve for x.

$$5x + 2 \cdot 0 = 7$$
$$5x = 7$$
$$x = \frac{7}{5}$$

$\left(\frac{7}{5}, 0\right)$ is the x-intercept.

Plot these points and draw the line. A third point could be used as a check.

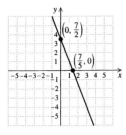

$$5x + 2g(x) = 7$$

44.

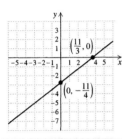

$$3x - 4f(x) = 11$$

45. $f(x) = 20 - 4x$, or $f(x) = -4x + 20$

From the equation we see that the y-intercept is $(0, 20)$. Next we find the x-intercept.

$$0 = 20 - 4x$$
$$4x = 20$$
$$x = 5$$

The x-intercept is $(5, 0)$. Thus, window (c) will show both intercepts.

46. (a)

47. $p(x) = -35x + 7000$

From the equation we see that the y-intercept is $(0, 7000)$. Next we find the x-intercept.

$$0 = -35x + 7000$$
$$35x = 7000$$
$$x = 200$$

The x-intercept is $(200, 0)$. Thus, window (d) will show both intercepts.

48. (b)

49. We first solve for y and determine the slope of each line.

$$x + 8 = y$$
$$y = x + 8 \quad \text{Reversing the order}$$

The slope of $y = x + 8$ is 1.

$$y - x = -5$$
$$y = x - 5$$

The slope of $y = x - 5$ is 1.

The slopes are the same; the lines are parallel.

50. Yes

51. We first solve for y and determine the slope of each line.

$$y + 9 = 3x$$
$$y = 3x - 9$$

The slope of $y = 3x - 9$ is 3.

$$3x - y = -2$$
$$3x + 2 = y$$
$$y = 3x + 2 \quad \text{Reversing the order}$$

The slope of $y = 3x + 2$ is 3.

The slopes are the same; the lines are parallel.

52. No

53. We determine the slope of each line.

The slope of $f(x) = 3x + 9$ is 3.

$$2y = 8x - 2$$
$$y = 4x - 1$$

The slope of $y = 4x - 1$ is 4.

The slopes are not the same; the lines are not parallel.

54. Yes

55. We determine the slope of each line.

The slope of $f(x) = 4x - 3$ is 4.

$$4y = 7 - x$$
$$4y = -x + 7$$
$$y = -\frac{1}{4}x + \frac{7}{4}$$

The slope of $4y = 7 - x$ is $-\frac{1}{4}$.

The product of their slopes is $4\left(-\frac{1}{4}\right)$, or -1; the line are perpendicular.

56. No

57. We determine the slope of each line.

$$x + 2y = 7$$
$$2y = -x + 7$$
$$y = -\frac{1}{2}x + \frac{7}{2}$$

The slope of $x + 2y = 7$ is $-\frac{1}{2}$.

$$2x + 4y = 4$$
$$4y = -2x + 4$$
$$y = -\frac{1}{2}x + 1.$$

The slope of $2x + 4y = 4$ is $-\frac{1}{2}$.

The product of their slopes is $\left(-\dfrac{1}{2}\right)\left(-\dfrac{1}{2}\right)$, or $\dfrac{1}{4}$; the lines are not perpendicular. For the lines to be perpendicular, the product must be -1.

68. Yes

69. The slope of the given line is 3. Therefore the slope of a line parallel to it is also 3. The y-intercept is $(0,9)$, so the equation of the desired function is $f(x) = 3x + 9$.

60. $f(x) = -5x - 2$

61. First we find the slope of the given line.
$$2x + y = 3$$
$$y = -2x + 3 \quad \text{Slope-intercept form}$$
The slope of the given line is -2, so the slope of a line parallel to it is also -2. The y-intercept is $(0,-5)$, so the equation of the desired function is $f(x) = -2x - 5$.

62. $f(x) = 3x + 1$.

63. First we find the slope of the given line.
$$2x + 5y = 8$$
$$5y = -2x + 8$$
$$\frac{1}{5} \cdot 5y = \frac{1}{5}(-2x + 8)$$
$$y = -\frac{2}{5}x + \frac{8}{5}$$
The slope of the given line is $-\dfrac{2}{5}$, so the slope of a line parallel to it is also $-\dfrac{2}{5}$.

The y-intercept is $-\dfrac{1}{3}$, so the equation of the desired function is $f(x) = -\dfrac{2}{5}x - \dfrac{1}{3}$.

64. $f(x) = \dfrac{1}{2}x + \dfrac{4}{5}$

65. $3y = 12$
$$y = 4, \text{ or } y = 0 \cdot x + 4$$
The slope is 0 and the y-intercept is $(0,-5)$. We have $f(x) = 0 \cdot x - 5$, or $f(x) = -5$.

66. $f(x) = 12$

67. The slope of the given line is 1. The slope of a line perpendicular to it is the opposite of the reciprocal of 1, or -1. The y-intercept is $(0,4)$, so we have $y = -1 \cdot x + 4$, or $y = -x + 4$.

68. $f(x) = -\dfrac{1}{2}x - 3$

69. First find the slope of the given line.
$$2x + 3y = 6$$
$$3y = -2x + 6$$
$$\frac{1}{3} \cdot 3y = \frac{1}{3}(-2x + 6)$$
$$y = -\frac{2}{3}x + 2$$
The slope of the given line is $-\dfrac{2}{3}$. The slope of a line perpendicular to it is the opposite of the reciprocal of $-\dfrac{2}{3}$, or $\dfrac{3}{2}$. The y-intercept is $(0,-4)$, so we have $f(x) = \dfrac{3}{2}x - 4$.

70. $f(x) = \dfrac{1}{2}x + 8$

71. First find the slope of the given line.
$$5x - y = 13$$
$$-y = -5x + 13$$
$$-1(-y) = -1(-5x + 13)$$
$$y = 5x - 13$$
The slope of the given line is 5. The slope of a line perpendicular to it is the opposite of the reciprocal of 5, or $-\dfrac{1}{5}$. The y-intercept is $\left(0, \dfrac{1}{5}\right)$, so we have $f(x) = -\dfrac{1}{5}x + \dfrac{1}{5}$.

72. $f(x) = -\dfrac{5}{2}x - \dfrac{1}{8}$

73. $5x - 3y = 15$

This equation is in the standard form for a linear equation, $Ax + By = C$, with $A = 5$, $B = -3$, and $C = 15$. Thus, it is a linear equation.

Solve for y to find the slope.
$$5x - 3y = 15$$
$$-3y = -5x + 15$$
$$y = \frac{5}{3}x - 5$$
The slope is $\dfrac{5}{3}$.

74. Linear; $-\dfrac{3}{5}$

75. $16 + 4y = 0$
$$4y = -16$$
This equation can be written in the standard form for a linear equation, $Ax + By = C$, with $A = 0$, $B = 4$, and $C = -16$. Thus, it is a linear equation.

Solve for y to find the slope.
$$4y = -16$$
$$y = -4$$

This is a horizontal line, so the slope is 0. (We can think of this as $y = 0 \cdot x - 4$.)

76. Linear; line is vertical

77. $3g(x) = 6x^2$

Replace $g(x)$ with y and attempt to write the equation in standard form.

$$3y = 6x^2$$
$$-6x^2 + 3y = 0$$

The equation is not linear, because it has an x^2-term.

78. Linear; $-\dfrac{1}{2}$

79.
$$3y = 7(2x - 4)$$
$$3y = 14x - 28$$
$$-14x + 3y = -28$$

The equation can be written in the standard form for a linear equation, $Ax + By = C$, with $A = -14$, $B = 3$, and $C = -28$. Thus, it is a linear equation. Solve for y to find the slope.

$$-14x + 3y = -28$$
$$3y = 14x - 28$$
$$y = \frac{14}{3}x - \frac{28}{3}$$

The slope is $\dfrac{14}{3}$.

80. Linear; $-\dfrac{6}{5}$

81. $g(x) - \dfrac{1}{x} = 0$

Replace $g(x)$ with y and attempt to write the equation in standard form.

$$y - \frac{1}{x} = 0$$
$$xy - 1 = 0 \quad \text{Multiplying by } x$$
$$xy = 1$$

The equation is not linear, because it has an xy-term.

82. Not linear

83. $\dfrac{f(x)}{5} = x^2$

Replace $f(x)$ with y and attempt to write the equation in standard form.

$$\frac{y}{5} = x^2$$
$$y = 5x^2$$
$$-5x^2 + y = 0$$

The equation is not linear, because it has an x^2-term.

84. Linear; 2

85. *Writing Exercise*

86. *Writing Exercise*

87. $-\dfrac{3}{7} \cdot \dfrac{7}{3} = -\dfrac{3 \cdot 7}{7 \cdot 3} = -1$

88. -1

89. $3(2x - y + 7) = 3 \cdot 2x - 3 \cdot y + 3 \cdot 7 = 6x - 3y + 21$

90. $-2x - 10y + 2$

91. $-5[x - (-3)] = -5[x + 3] = -5x - 15$

92. $-2x - 8$

93. $\dfrac{2}{3}\left[x - \left(-\dfrac{1}{2}\right)\right] - 1 = \dfrac{2}{3}\left[x + \dfrac{1}{2}\right] - 1 =$

$\dfrac{2}{3}x + \dfrac{1}{3} - 1 = \dfrac{2}{3}x - \dfrac{2}{3}$

94. $-\dfrac{3}{2}x - \dfrac{12}{5}$

95. *Writing Exercise*

96. *Writing Exercise*

97. The line contains the points $(5, 0)$ and $(0, -4)$. We use th points to find the slope.

$$\text{Slope} = \frac{-4 - 0}{0 - 5} = \frac{-4}{-5} = \frac{4}{5}$$

Then the slope-intercept equation is $y = \dfrac{4}{5}x - 4$. W rewrite this equation in standard form.

$$y = \frac{4}{5}x - 4$$
$$5y = 4x - 20 \quad \text{Multiplying by 5 on both sides}$$
$$-4x + 5y = -20 \quad \text{Standard form}$$

This equation can also be written as $4x - 5y = 20$.

98. $\left(-\dfrac{b}{m}, 0\right)$

99. $rx + 3y = p - s$

The equation is in standard form with $A = r$, $B = 3$, an $C = p - s$. It is linear.

100. Linear

101. Try to put the equation in standard form.

$$r^2x = py + 5$$
$$r^2x - py = 5$$

The equation is in standard form with $A = r^2$, $B = -$ and $C = 5$. It is linear.

2. Linear

3. Let equation A have intercepts $(a, 0)$ and $(0, b)$. Then equation B has intercepts $(2a, 0)$ and $(0, b)$.

Slope of $A = \dfrac{b - 0}{0 - a} = -\dfrac{b}{a}$

Slope of $B = \dfrac{b - 0}{0 - 2a} = -\dfrac{b}{2a} = \dfrac{1}{2}\left(-\dfrac{b}{a}\right)$

The slope of equation B is $\dfrac{1}{2}$ the slope of equation A.

4. $a = 5, b = 1$

5. First write the equation in standard form.

$$ax + 3y = 5x - by + 8$$

$$ax - 5x + 3y + by = 8 \qquad \text{Adding } -5x + by$$
$$\text{on both sides}$$

$$(a - 5)x + (3 + b)y = 8 \qquad \text{Factoring}$$

If the graph is a vertical line, then the coefficient of y is 0.

$$3 + b = 0$$
$$b = -3$$

Then we have $(a - 5)x = 8$.

If the line passes through $(4, 0)$, we have:

$$(a - 5)4 = 8 \quad \text{Substituting 4 for } x$$
$$a - 5 = 2$$
$$a = 7$$

6. a)

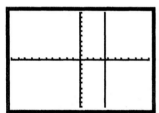

b)

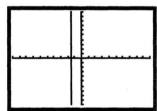

c)

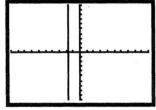

d)

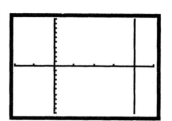

107. a) Solve each equation for y, enter each on the equation-editor screen, and then examine a table of values for the two functions. Since the difference between the y-values is the same for all x-values, the lines are parallel.

b) Solve each equation for y, enter each on the equation-editor screen, and then examine a table of values for the two functions. Since the difference between the y-values is not the same for all x-values, the lines are not parallel.

Exercise Set 2.6

1. $y - y_1 = m(x - x_1)$ Point-slope equation

$y - 3 = -2(x - 2)$ Substituting -2 for m, 2 for

$$x_1, \text{ and 3 for } y_1$$

To graph the equation, we count off a slope of $\dfrac{-2}{1}$, starting at $(2, 3)$, and draw the line.

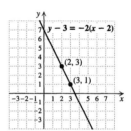

2. $y - 4 = 5(x - 7)$

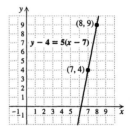

3. $y - y_1 = m(x - x_1)$ Point-slope equation

$y - 7 = 3(x - 4)$ Substituting 3 for m, 4 for x_1, and 7 for y_1

To graph the equation, we count off a slope of $\dfrac{3}{1}$, starting at $(4, 7)$ and draw the line.

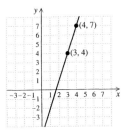

$y - 7 = 3(x - 4)$

4. $y - 3 = 2(x - 7)$

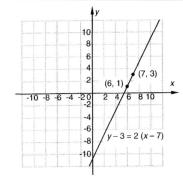

5. $y - y_1 = m(x - x_1)$ Point-slope equation

$y - (-4) = \dfrac{1}{2}[x - (-2)]$ Substituting $\dfrac{1}{2}$ for m, -2 for x_1, and -4 for y_1

To graph the equation, we count off a slope of $\dfrac{1}{2}$, starting at $(-2, -4)$, and draw the line.

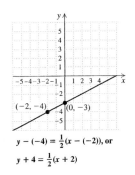

$y - (-4) = \frac{1}{2}(x - (-2))$, or

$y + 4 = \frac{1}{2}(x + 2)$

6. $y - (-7) = 1 \cdot [x - (-5)]$, or $y - (-7) = x - (-5)$

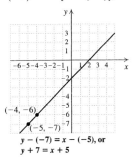

$y - (-7) = x - (-5)$, or
$y + 7 = x + 5$

7. $y - y_1 = m(x - x_1)$ Point-slope equation

$y - 0 = -1(x - 8)$ Substituting -1 for m, 8 for x_1, and 0 for y_1

To graph the equation, we count off a slope of $\dfrac{-1}{1}$, startin at $(8, 0)$ and draw the line.

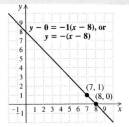

8. $y - 0 = -3[x - (-2)]$

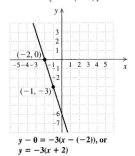

$y - 0 = -3(x - (-2))$, or
$y = -3(x + 2)$

9. $y - y_1 = m(x - x_1)$ Point-slope equation

$y - 8 = \dfrac{2}{5}[x - (-3)]$ Substituting $\dfrac{2}{5}$ for m, -3 for x_1, and 8 for y_1

We graph the equation by plotting $(-3, 8)$, counting off slope of $\dfrac{2}{5}$, and drawing the line.

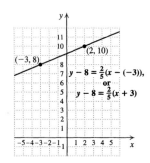

0. $y - (-5) = \frac{3}{4}(x - 1)$ or $y + 5 = \frac{3}{4}(x - 1)$

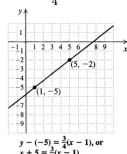

$y - (-5) = \frac{3}{4}(x - 1)$, or
$y + 5 = \frac{3}{4}(x - 1)$

1. $y - 4 = \frac{2}{7}(x - 1)$

$y - y_1 = m(x - x_1)$

$m = \frac{2}{7}$, $x_1 = 1$, and $y_1 = 4$, so the slope m is $\frac{2}{7}$ and a point (x_1, y_1) on the graph is $(1, 4)$.

2. 9; $(2, 3)$

3. $y + 2 = -5(x - 7)$

$y - (-2) = -5(x - 7)$

$y - y_1 = m(x - x_1)$

$m = -5$, $x_1 = 7$, and $y_1 = -2$, so the slope m is -5 and a point (x_1, y_1) on the graph is $(7, -2)$.

4. $-\frac{2}{9}$; $(-5, 1)$

5. $y - 1 = -\frac{5}{3}(x + 2)$

$y - 1 = -\frac{5}{3}[x - (-2)]$

$y - y_1 = m(x - x_1)$

$m = -\frac{5}{3}$, $x_1 = -2$, and $y_1 = 1$, so the slope m is $-\frac{5}{3}$ and a point (x_1, y_1) on the graph is $(-2, 1)$.

6. -4; $(9, -7)$

17. $y = \frac{4}{7}x$

The equation is of the form $y = mx$, so we know that its graph is a line through the origin with slope m. Thus, the slope is $\frac{4}{7}$ and a point on the graph is $(0, 0)$.

18. 3; $(0, 0)$

19. $y - y_1 = m(x - x_1)$ Point-slope equation

$y - (-3) = 4(x - 2)$ Substituting 4 for m, 2 for x_1, and -3 for y_1

$y + 3 = 4x - 8$ Simplifying

$y = 4x - 11$ Subtracting 3 from both sides

$f(x) = 4x - 11$ Using function notation

20. $f(x) = -4x + 1$

21. $y - y_1 = m(x - x_1)$ Point-slope equation

$y - (-7) = -\frac{3}{5}(x - 4)$ Substituting $-\frac{3}{5}$ for m, 4 for x_1, and -7 for y_1

$y + 7 = -\frac{3}{5}x + \frac{12}{5}$ Simplifying

$y = -\frac{3}{5}x - \frac{23}{5}$ Subtracting 7 from both sides

$f(x) = -\frac{3}{5}x - \frac{23}{5}$ Using function notation

22. $f(x) = -\frac{1}{5}x + \frac{3}{5}$

23. $y - y_1 = m(x - x_1)$ Point-slope equation

$y - (-4) = -0.6[x - (-3)]$ Substituting -0.6 for m, -3 for x_1, and -4 for y_1

$y + 4 = -0.6(x + 3)$

$y + 4 = -0.6x - 1.8$

$y = -0.6x - 5.8$

$f(x) = -0.6x - 5.8$ Using function notation

24. $f(x) = 2.3x - 14.2$

25. $m = \frac{2}{7}$; $(0, -5)$

Observe that the slope is $\frac{2}{7}$ and the y-intercept is $(0, -5)$. Thus, we have $f(x) = \frac{2}{7}x - 5$.

26. $f(x) = \frac{1}{4}x + 3$

27. First find the slope of the line:
$$m = \frac{6-4}{5-1} = \frac{2}{4} = \frac{1}{2}$$

Use the point-slope equation with $m = \frac{1}{2}$ and

$(1,4) = (x_1, y_1)$. (We could let $(5,6) = (x_1, y_1)$ instead and obtain an equivalent equation.)

$$y - 4 = \frac{1}{2}(x - 1)$$
$$y - 4 = \frac{1}{2}x - \frac{1}{2}$$
$$y = \frac{1}{2}x + \frac{7}{2}$$
$$f(x) = \frac{1}{2}x + \frac{7}{2} \quad \text{Using function notation}$$

28. $f(x) = -\frac{5}{2}x + 11$

29. First find the slope of the line:
$$m = \frac{3 - (-3)}{6.5 - 2.5} = \frac{3+3}{4} = \frac{6}{4} = 1.5$$

Use the point-slope equation with $m = 1.5$ and $(6.5, 3) = (x_1, y_1)$.

$$y - 3 = 1.5(x - 6.5)$$
$$y - 3 = 1.5x - 9.75$$
$$y = 1.5x - 6.75$$
$$f(x) = 1.5x - 6.75 \quad \text{Using function notation}$$

30. $f(x) = 0.6x - 2.5$

31. First find the slope of the line:
$$m = \frac{-2 - 3}{0 - 1} = \frac{-5}{-1} = 5$$

Observe that the y-intercept is $(0, -2)$. Then, using the slope-intercept equation we have $y = 5x - 2$.

We could also use the point-slope equation with $m = 5$ and $(1, 3) = (x_1, y_1)$.

$$y - 3 = 5(x - 1)$$
$$y - 3 = 5x - 5$$
$$y = 5x - 2$$
$$f(x) = 5x - 2$$

32. $f(x) = -\frac{4}{3}x - 4$

33. First find the slope of the line:
$$m = \frac{-6 - (-3)}{-4 - (-2)} = \frac{-6 + 3}{-4 + 2} = \frac{-3}{-2} = \frac{3}{2}$$

Use the point-slope equation with $m = \frac{3}{2}$ and $(-2, -3) = (x_1, y_1)$.

$$y - (-3) = \frac{3}{2}[x - (-2)]$$
$$y + 3 = \frac{3}{2}(x + 2)$$
$$y + 3 = \frac{3}{2}x + 3$$
$$y = \frac{3}{2}x$$
$$f(x) = \frac{3}{2}x \quad \text{Using function notation}$$

34. $f(x) = 3x + 5$

35. a) We form pairs of the type (t, R) where t is the number of years since 1930 and R is the record. We have two pairs, $(0, 46.8)$ and $(40, 43.8)$.

These are two points on the graph of the linear function we are seeking. We use the point-slope form to write an equation relating R and t:

$$m = \frac{43.8 - 46.8}{40 - 0} = \frac{-3}{40} = -0.075$$
$$R - 46.8 = -0.075(t - 0)$$
$$R - 46.8 = -0.075t$$
$$R = -0.075t + 46.8$$
$$R(t) = -0.075t + 46.8 \quad \text{Using function notation}$$

b) 2003 is 73 years since 1930, so to predict the record in 2003, we find $R(73)$:

$$R(73) = -0.075(73) + 46.8$$
$$= 41.325$$

The predicted record is 41.325 seconds in 2003.

2006 is 76 years since 1930, so to predict the record in 2006, we find $R(76)$:

$$R(76) = -0.075(76) + 46.8$$
$$= 41.1$$

The predicted record is 41.1 seconds in 2006.

c) Substitute 40 for $R(t)$ and solve for t:

$$40 = -0.075t + 46.8$$
$$-6.8 = -0.075t$$
$$91 \approx t$$

The record will be 40 seconds about 91 years after 1930, or in 2021.

36. a) $R(t) = -0.0075t + 3.85$

b) 3.31 min; 3.28 min

c) 2030

7. a) We form the pairs $(0, 178.6)$ and $(8, 243.1)$.

Use the point-slope form to write an equation relating A and t:
$$m = \frac{243.1 - 178.6}{8 - 0} = \frac{64.5}{8} = 8.0625$$
$$A - 178.6 = 8.0625(x - 0)$$
$$A - 178.6 = 8.0625x$$
$$A = 8.0625t + 178.6$$
$$A(t) = 8.0625t + 178.6 \qquad \text{Using function notation}$$

b) 2006 is 14 years since 1992, so we find $A(14)$:
$$A(14) = 8.0625(14) + 178.6$$
$$= 291.475$$

We predict that the amount of PAC contributions in 2006 will be \$291.475 million.

8. a) $A(p) = -2.5p + 26.5$

b) 11.5 million lb

9. a) We form the pairs $(0, 43.8)$ and $(5, 62.2)$. Use the point-slope form to write an equation relating N and t:
$$m = \frac{62.2 - 43.8}{5 - 0} = \frac{18.4}{5} = 3.68$$
$$N - 43.8 = 3.68(t - 0)$$
$$N - 43.8 = 3.68t$$
$$N = 3.68t + 43.8$$
$$N(t) = 3.68t + 43.8 \qquad \text{Using function notation}$$

b) 2005 is 12 years after 1993, so we find $N(12)$:
$$N(12) = 3.68(12) + 43.8$$
$$= 87.96$$

We predict that Americans will recycle about 87.96 million tons of garbage in 2005.

0. a) $A(p) = 2p - 11$

b) 1 million lb

1. a) We form the pairs $(0, 78.8)$ and $(10, 79.8)$.

Use the point-slope form to write an equation relating E and t:
$$m = \frac{79.8 - 78.8}{10 - 0} = \frac{1}{10}, \text{ or } 0.1$$
$$E - 78.8 = 0.1(t - 0)$$
$$E - 78.8 = 0.1t$$
$$E = 0.1t + 78.8$$
$$E(t) = 0.1t + 78.8 \qquad \text{Using function notation}$$

b) 2008 is 18 years after 1990, so we find $E(18)$:
$$E(18) = 0.1(18) + 78.8$$
$$= 80.6$$

We predict that the life expectancy of females in the United States in 2008 will be 80.6 years.

42. a) $E(t) = 0.26t + 71.8$

b) 76.22 years

43. a) We form the pairs $(0, 74.9)$ and $(4, 77.7)$.

Use the point-slope form to write an equation relating A and t:
$$m = \frac{77.7 - 74.9}{4 - 0} = \frac{2.8}{4} = 0.7$$
$$A - 74.9 = 0.7(t - 0)$$
$$A - 74.9 = 0.7t$$
$$A = 0.7t + 74.9$$
$$A(t) = 0.7t + 74.9 \quad \text{Using function notation}$$

b) 2006 is 12 years after 1994, so we find $A(12)$:
$$A(12) = 0.7(12) + 74.9$$
$$= 83.3$$

We predict that there will be 83.3 million acres of land in the national park system in 2006.

44. a) $P(d) = 0.03d + 1$

b) 21.7 atm

45. The points lie approximately on a straight line, so the data appear to be linear.

46. Linear

47. The points do not lie on a straight line, so the data are not linear.

48. Not linear

49. The points lie approximately on a straight line, so the data appear to be linear.

50. Not linear

51. a) We use the point-slope form to write an equation relating S and t, where S is in thousands and t is the number of years after 1964.
$$m = \frac{33 - 13}{24 - 8} = \frac{20}{16} = 1.25$$
$$S - 13 = 1.25(t - 8)$$
$$S - 13 = 1.25t - 10$$
$$S = 1.25t + 3$$
$$S(t) = 1.25t + 3 \qquad \text{Using function notation}$$

b) 2010 is 46 years after 1964, so we find $S(46)$.

$$S(46) = 1.25(46) + 3$$
$$= 60.5$$

We predict that there will be 60.5 thousand, or 60, 500, shopping centers in 2010.

52. a) $G(t) = 30.8t + 1550$

b) 2505 golf courses

53. a) Enter the data in a graphing calculator, letting x represent the number of years since 1900. Then use the linear regression feature to find the desired function. We have $W(x) = 0.183x + 62.13333333$.

b) $2008 - 1900 = 108$, so we use one of the methods discussed earlier in the text to find $W(108)$. We predict that the life expectancy in 2008 will be about 81.8 years. This estimate is higher than the one found in Exercise 41.

54. a) $M(x) = 0.1785714286x + 55.85714286$.

b) 74.96 yr; this estimate is lower

55. a) Enter the data in a graphing calculator, letting x represent the number of years after 1987. Then use the linear regression feature to find the desired function. We have $C(x) = -5.027678571x + 95.21892857$.

b) $2003 - 1987 = 16$, so we use one of the methods discussed earlier in the text to find $C(16)$. We predict that the average local monthly bill for a cellular phone will be about \$14.78 in 2003.

56. a) $E(x) = 75.93023256x + 8323.372093$

b) \$9083

57. *Writing Exercise*

58. *Writing Exercise*

59. $(3x^2 + 5x) + (2x - 4) = 3x^2 + (5 + 2)x - 4 = 3x^2 + 7x - 4$

60. $5t^2 - 6t - 3$

61. $\dfrac{2t - 6}{4t + 1} = \dfrac{2 \cdot 3 - 6}{4 \cdot 3 + 1} = \dfrac{6 - 6}{12 + 1} = \dfrac{0}{13} = 0$

62. 0

63. $2x - 5y = 2 \cdot 3 - 5(-1) = 6 + 5 = 11$

64. -34

65. *Writing Exercise*

66. *Writing Exercise*

67. *Familiarize*. The value C of the computer, in dollars after t months can be modeled by a line that contains the points $(6, 900)$ and $(8, 750)$.

Translate. We find an equation relating C and t.

$$m = \frac{750 - 900}{8 - 6} = \frac{-150}{2} = -75$$
$$C - 900 = -75(t - 6)$$
$$C - 900 = -75t + 450$$
$$C = -75t + 1350$$

Carry out. Using function notation we have $C(t) = -75t + 1350$. To find how much the computer cost we find $C(0)$:

$$C(0) = -75 \cdot 0 + 1350$$
$$= 1350$$

Check. We can repeat our calculations. We could also graph the function and determine that $(0, 1350)$ is on the graph.

State. The computer cost \$1350.

68. $21.1°C$

69. *Familiarize*. The total cost C of the phone, in dollars after t months, can be modeled by a line that contains the points $(5, 230)$ and $(9, 390)$.

Translate. We find an equation relating C and t.

$$m = \frac{390 - 230}{9 - 5} = \frac{160}{4} = 40$$
$$C - 230 = 40(t - 5)$$
$$C - 230 = 40t - 200$$
$$C = 40t + 30$$

Carry out. Using function notation we have $C(t) = 40t + 30$. To find the costs already incurred when the service began we find $C(0)$:

$$C(0) = 40 \cdot 0 + 30 = 30$$

Check. We can repeat the calculations. We could also graph the function and determine that $(0, 30)$ is on the graph.

State. Mel had already incurred \$30 in costs when his service just began.

70. \$11,000

71. *Familiarize*. The percentage of premiums paid out in benefits in 1993 was $103.6/124.7 \approx 83.1\%$. In 1996 the percentage was $113.8/137.1 \approx 83.0\%$. The percentage P of premiums, in dollars, paid out in benefits t years after 1993 can be modeled by a line that contains the points $(0, 83.1)$ and $(3, 83.0)$.

Translate. We find an equation relating P and t.

$$m = \frac{83.0 - 83.1}{3 - 0} = \frac{-0.1}{3} = -\frac{1}{30}$$

We know the slope and the y-intercept, so we use the slope-intercept equation.

$$P = -\frac{1}{30}t + 83.1$$

Carry out. Using function notation we have $P(t) = -\frac{1}{30}t + 83.1$. Since 2005 is 12 years after 1993, we find $P(12)$ to predict the percentage of premiums that will be paid out in benefits in 2005:

$$P(12) = -\frac{1}{30}(12) + 83.1 = 82.7$$

Check. We can repeat the calculations. We could also graph the function and determine that $(12, 82.7)$ is on the graph.

State. We estimate that 82.7% will be paid out in benefits in 2005. (Answers may vary slightly depending on when and how rounding occurred.)

72. $8.33 per pound

73. First solve the equation for y and determine the slope of the given line.

$$x + 2y = 6 \qquad \text{Given line}$$
$$2y = -x + 6$$
$$y = \frac{1}{2}x + 3 \quad m = -\frac{1}{2}$$

The slope of the given line is $-\frac{1}{2}$.

The slope of every line parallel to the given line must also be $-\frac{1}{2}$. We find the equation of the line with slope $-\frac{1}{2}$ and containing the point $(3, 7)$.

$$y - 7 = -\frac{1}{2}(x - 3)$$
$$y - 7 = -\frac{1}{2}x + \frac{3}{2}$$
$$y = -\frac{1}{2}x + \frac{17}{2}$$

74. $y = 3x + 7$

75. First solve the equation for y and determine the slope of the given line.

$$2x + y = -3 \qquad \text{Given line}$$
$$y = -2x - 3 \quad m = -2$$

The slope of the given line is -2.

The slope of a perpendicular line is given by the opposite of the reciprocal of -2, or $\frac{1}{2}$.

We find the equation of the line with slope $\frac{1}{2}$ containing the point $(2, 5)$.

$$y - 5 = \frac{1}{2}(x - 2) \quad \text{Substituting}$$
$$y - 5 = \frac{1}{2}x - 1$$
$$y = \frac{1}{2}x + 4$$

76. $y = -3x + 12$

77. The price must be a positive number, so we have $p > 0$. Furthermore, the amount of coffee sold must be a nonzero number. We have:

$$-2.5p + 26.5 \geq 0$$
$$-2.5p \geq -26.5$$
$$p \leq 10.6$$

Thus, the domain is $\{p | 0 < p \leq 10.6\}$.

78. $\{p | p > 5.5\}$

79. a) We have two pairs, $(3, -5)$ and $(7, -1)$. Use the point-slope form:

$$m = \frac{-1 - (-5)}{7 - 3} = \frac{-1 + 5}{4} = \frac{4}{4} = 1$$
$$y - (-5) = 1(x - 3)$$
$$y + 5 = x - 3$$
$$y = x - 8$$
$$g(x) = x - 8 \quad \text{Using function notation}$$

  b) $g(-2) = -2 - 8 = -10$

  c) $g(a) = a - 8$
   If $g(a) = 75$, we have
   $$a - 8 = 75$$
   $$a = 83.$$

Exercise Set 2.7

1. Since $f(2) = -3 \cdot 2 + 1 = -5$, and $g(2) = 2^2 + 2 = 6$, we have $f(2) + g(2) = -5 + 6 = 1$.

2. 7

3. Since $f(5) = -3 \cdot 5 + 1 = -14$ and $g(5) = 5^2 + 2 = 27$, we have $f(5) - g(5) = -14 - 27 = -41$.

4. -29

5. Since $f(-1) = -3(-1) + 1 = 4$ and $g(-1) = (-1)^2 + 2 = 3$, we have $f(-1) \cdot g(-1) = 4 \cdot 3 = 12$.

6. 42

7. Since $f(-4) = -3(-4) + 1 = 13$ and
$g(-4) = (-4)^2 + 2 = 18$, we have
$f(-4)/g(-4) = 13/18$.

8. $-\dfrac{8}{11}$

9. Since $g(1) = 1^2 + 2 = 3$ and
$f(1) = -3 \cdot 1 + 1 = -2$, we have
$g(1) - f(1) = 3 - (-2) = 3 + 2 = 5$.

10. $-\dfrac{6}{5}$

11. $(f+g)(x) = f(x)+g(x) = (-3x+1)+(x^2+2) = x^2-3x+3$

12. $x^2 + 3x + 1$

13. $(F + G)(x) = F(x) + G(x)$
$= x^2 - 2 + 5 - x$
$= x^2 - x + 3$

14. $a^2 - a + 3$

15. Using our work in Exercise 13, we have
$(F + G)(-4) = (-4)^2 - (-4) + 3$
$= 16 + 4 + 3$
$= 23.$

16. 33

17. $(F - G)(x) = F(x) - G(x)$
$= x^2 - 2 - (5 - x)$
$= x^2 - 2 - 5 + x$
$= x^2 + x - 7$
Then we have
$(F - G)(3) = 3^2 + 3 - 7$
$= 9 + 3 - 7$
$= 5.$

18. -1.

19. $(F \cdot G)(x) = F(x) \cdot G(x)$
$= (x^2 - 2)(5 - x)$
$= 5x^2 - x^3 - 10 + 2x$
Then we have
$(F \cdot G)(-3) = 5(-3)^2 - (-3)^3 - 10 + 2(-3)$
$= 5 \cdot 9 - (-27) - 10 - 6$
$= 45 + 27 - 10 - 6$
$= 56.$

20. 126

21. $(F/G)(x) = F(x)/G(x)$
$= \dfrac{x^2 - 2}{5 - x}, \ x \neq 5$

22. $-x^2 - x + 7$

23. Using our work in Exercise 21, we have
$(F/G)(-2) = \dfrac{(-2)^2 - 2}{5 - (-2)} = \dfrac{4 - 2}{5 + 2} = \dfrac{2}{7}.$

24. $-\dfrac{1}{6}$

25. $N(1980) = (R + W)(1980)$
$= R(1980) + W(1980)$
$\approx 0.75 + 2.5$
≈ 3.25
We estimate that 3.25 million U.S. women had children i꘡
1980.

26. $1.3 + 2.6 = 3.9$

27. The number of women under 30 who gave birth droppe꘡
from 1990 to 1998.

28. 30 and older

29. $(n + l)(98) = n(98) + l(98)$
From the middle line of the graph, we can see that
$n(98) + l(98) \approx 50$ million.
This represents the total number of passengers serviced b꘡
Newark and LaGuardia airports in 1998.

30. 41 million; the total number of passengers serviced k꘡
Kennedy and LaGuardia airports in 1998

31. $(k - l)(94) = k(94) - l(94)$
$\approx 29 - 21$
≈ 8 million
This represents how many more passengers used Kenne꘡
airport than LaGuardia airport in 1994.

32. 1 million; how many more passengers used Kennedy ai꘡
port than Newark airport in 1994.

33. $(n + l + k)(99) = n(99) + l(99) + k(99)$
From the top line of the graph, we can see that
$n(99) + l(99) + k(99) \approx 89$ million.
This represents the number of passengers serviced ꘡
Newark, LaGuardia, and Kennedy airports in 1999.

34. 69 million; the number of passengers serviced by Newar꘡
LaGuardia, and Kennedy airports in 1998

5. a) $f(x) = \dfrac{5}{x-3}$

Since $\dfrac{5}{x-3}$ cannot be computed when the denominator is 0, we find the x-value that causes $x-3$ to be 0:

$$x - 3 = 0$$

$$x = 3 \quad \text{Adding 3 to both sides}$$

Thus, 3 is not in the domain of f, while all other real numbers are. The domain of f is $\{x|x$ is a real number $and\ x \neq 3\}$.

b) $f(x) = \dfrac{7}{6-x}$

Since $\dfrac{7}{6-x}$ cannot be computed when the denominator is 0, we find the x-value that causes $6-x$ to be 0:

$$6 - x = 0$$

$$6 = x \quad \text{Adding } x \text{ on both sides}$$

Thus, 6 is not in the domain of f, while all other real numbers are. The domain of f is $\{x|x$ is a real number $and\ x \neq 6\}$.

c) $f(x) = 2x + 1$

Since we can compute $2x + 1$ for any real number x, the domain is the set of all real numbers.

d) $f(x) = x^2 + 3$

Since we can compute $x^2 + 3$ for any real number x, the domain is the set of all real numbers.

e) $f(x) = \dfrac{3}{2x-5}$

Since $\dfrac{3}{2x-5}$ cannot be computed when the denominator is 0, we find the x-value that causes $2x - 5$ to be 0:

$$2x - 5 = 0$$

$$2x = 5$$

$$x = \frac{5}{2}$$

Thus, $\dfrac{5}{2}$ is not in the domain of f, while all other real numbers are. The domain of f is $\left\{x|x \text{ is a real number } and\ x \neq \dfrac{5}{2}\right\}$.

f) $f(x) = |3x - 4|$

Since we can compute $|3x - 4|$ for any real number x, the domain is the set of all real numbers.

3. a) $\{x|x$ is a real number $and\ x \neq 1\}$

b) All real numbers

c) $\{x|x$ is a real number $and\ x \neq -3\}$

d) $\left\{x|x \text{ is a real number } and\ x \neq -\dfrac{4}{3}\right\}$

e) All real numbers

f) All real numbers

37. The domain of f and of g is all real numbers. Thus, Domain of $f + g =$ Domain of $f - g =$ Domain of $f \cdot g = \{x|x$ is a real number$\}$.

38. $\{x|x$ is a real number$\}$

39. Because division by 0 is undefined, we have

Domain of $f = \{x|x$ is a real number $and\ x \neq 3\}$,

and

Domain of $g = \{x|x$ is a real number$\}$.

Thus, Domain of $f + g =$ Domain of $f - g =$ Domain of $f \cdot g = \{x|x$ is a real number $and\ x \neq 3\}$.

40. $\{x|x$ is a real number $and\ x \neq 9\}$.

41. Because division by 0 is undefined, we have

Domain of $f = \{x|x$ is a real number $and\ x \neq 0\}$,

and

Domain of $g = \{x|x$ is a real number$\}$.

Thus, Domain of $f + g =$ Domain of $f - g =$ Domain of $f \cdot g = \{x|x$ is a real number $and\ x \neq 0\}$.

42. $\{x|x$ is a real number $and\ x \neq 0\}$.

43. Because division by 0 is undefined, we have

Domain of $f = \{x|x$ is a real number $and\ x \neq 1\}$,

and

Domain of $g = \{x|x$ is a real number$\}$.

Thus, Domain of $f + g =$ Domain of $f - g =$ Domain of $f \cdot g = \{x|x$ is a real number $and\ x \neq 1\}$.

44. $\{x|x$ is a real number $and\ x \neq 6\}$.

45. Because division by 0 is undefined, we have

Domain of $f = \{x|x$ is a real number $and\ x \neq 2\}$,

and

Domain of $g = \{x|x$ is a real number $and\ x \neq 4\}$.

Thus, Domain of $f + g =$ Domain of $f - g =$ Domain of $f \cdot g = \{x|x$ is a real number $and\ x \neq 2$ $and\ x \neq 4\}$.

46. $\{x|x$ is a real number $and\ x \neq 3$ $and\ x \neq 2\}$.

47. Domain of $f =$ Domain of $g = \{x|x$ is a real number$\}$.

Since $g(x) = 0$ when $x - 3 = 0$, we have $g(x) = 0$ when $x = 3$. We conclude that Domain of $f/g = \{x|x$ is a real number $and\ x \neq 3\}$.

48. $\{x | x$ is a real number *and* $x \neq 5\}$.

49. Domain of $f =$ Domain of $g =$
$\{x | x$ is a real number$\}$.

Since $g(x) = 0$ when $2x - 8 = 0$, we have $g(x) = 0$ when $x = 4$. We conclude that Domain of $f/g = \{x | x$ is a real number *and* $x \neq 4\}$.

50. $\{x | x$ is a real number *and* $x \neq 3\}$.

51. Domain of $f = \{x | x$ is a real number *and* $x \neq 4\}$.

Domain of $g = \{x | x$ is a real number$\}$.

Since $g(x) = 0$ when $5 - x = 0$, we have $g(x) = 0$ when $x = 5$. We conclude that Domain of $f/g = \{x | x$ is a real number *and* $x \neq 4$ *and* $x \neq 5\}$.

52. $\{x | x$ is a real number *and* $x \neq 2$ *and* $x \neq 7\}$.

53. Domain of $f = \{x | x$ is a real number *and* $x \neq -1\}$.

Domain of $g = \{x | x$ is a real number$\}$.

Since $g(x) = 0$ when $2x + 5 = 0$, we have $g(x) = 0$ when $x = -\dfrac{5}{2}$. We conclude that Domain of $f/g =$
$\left\{x \middle| x \text{ is a real number } and \ x \neq -1 \ and \ x \neq -\dfrac{5}{2}\right\}$.

54. $\left\{x \middle| x \text{ is a real number } and \ x \neq 2 \ and \ x \neq -\dfrac{7}{3}\right\}$.

55. $(F + G)(5) = F(5) + G(5) = 1 + 3 = 4$
$(F + G)(7) = F(7) + G(7) = -1 + 4 = 3$

56. $0; 2$

57. $(G - F)(7) = G(7) - F(7) = 4 - (-1) = 4 + 1 = 5$
$(G - F)(3) = G(3) - F(3) = 1 - 2 = -1$

58. $2; -\dfrac{1}{4}$

59. From the graph we see that Domain of
$F = \{x | 0 \leq x \leq 9\}$ and Domain of
$G = \{x | 3 \leq x \leq 10\}$. Then Domain of
$F + G = \{x | 3 \leq x \leq 9\}$. Since $G(x)$ is never 0, Domain of
$F/G = \{x | 3 \leq x \leq 9\}$.

60. Domain of $F - G = \{x | 3 \leq x \leq 9\}$;
Domain of $F \cdot G = \{x | 3 \leq x \leq 9\}$;
Domain of $G/F = \{x | 3 \leq x \leq 9$ *and* $x \neq 6$ *and* $x \neq 8\}$

61. We use $(F + G)(x) = F(x) + G(x)$.

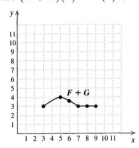

62.

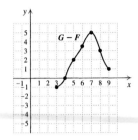

63. *Writing Exercise*

64. *Writing Exercise*

65. $3x^2 - 5y = 3(10)^2 - 5(6)$
$= 3(100) - 5(6)$
$= 300 - 30$
$= 270$

66. 16

67. $0.00000703 = \dfrac{0.00000703}{10^6} \cdot 10^6 = \dfrac{7.03}{10^6} =$
7.03×10^{-6}

68. 4.506×10^{12}

69. We move the decimal point 8 places to the right.
$4.3 \times 10^8 = 430,000,000$

70. 0.0006037

71. Let n represent the first integer; $x + (x + 1) = 145$.

72. Let x represent the number; $x - (-x) = 20$

73. *Writing Exercise*

74. *Writing Exercise*

5. Domain of $f = \left\{x \middle| x \text{ is a real number } and \ x \neq -\frac{5}{2}\right\}$; domain of $g = \{x | x \text{ is a real number } and \ x \neq -3\}$; $g(x) = 0$ when $x^4 - 1 = 0$, or when $x = 1$ or $x = -1$.

Then domain of $f/g = \Big\{x \Big| x \text{ is a real number } and$

$x \neq -\dfrac{5}{2} \text{ and } x \neq -3 \text{ and } x \neq 1 \text{ and } x \neq -1\Big\}.$

3. $\{x | x \text{ is a real number } and \ x \neq 4 \text{ and } x \neq 3 \text{ and } x \neq 2 \text{ and } x \neq -2\}.$

7. Answers may vary.

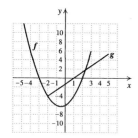

3. $\left\{x \middle| x \text{ is a real number } and \ -1 < x < 5 \text{ and } x \neq \dfrac{3}{2}\right\}$

9. The domain of each function is the set of first coordinates for that function.

Domain of $f = \{-2, -1, 0, 1, 2\}$ and

Domain of $g = \{-4, -3, -2, -1, 0, 1\}$.

Domain of $f + g =$ Domain of $f - g =$

Domain of $f \cdot g = \{-2, -1, 0, 1\}$.

Since $g(-1) = 0$, we conclude that Domain of $f/g = \{-2, 0, 1\}$.

. 5; 15; 2/3

. Answers may vary. $f(x) = \dfrac{1}{x+2}$, $g(x) = \dfrac{1}{x-5}$

. Left to the student

. Because $y_2 = 0$ when $x = 3$, the domain of $y_3 = \{x | x \text{ is a real number } and \ x \neq 3\}$. Since the graph produced using Connected mode contains the line $x = 3$, it does not represent y_3 accurately. The domain of the graph produced using Dot mode does not include 3, so it represents y_3 more accurately.

. Left to the student

85. Think of adding, subtracting, multiplying, or dividing the y-values for various x-values.

   a) IV

   b) I

   c) II

   d) III

Chapter 3

Systems of Linear Equations and Problem Solving

Exercise Set 3.1

---

1. We use alphabetical order for the variables. We replace x by 1 and y by 2.

$$\begin{array}{c|c} 4x - y = 2 \\ \hline 4 \cdot 1 - 2 \ ? \ 2 \\ 4 - 2 \\ \hline 2 & 2 \quad \text{TRUE} \end{array} \qquad \begin{array}{c|c} 10x - 3y = 4 \\ \hline 10 \cdot 1 - 3 \cdot 2 \ ? \ 4 \\ 10 - 6 \\ \hline 4 & 4 \quad \text{TRUE} \end{array}$$

The pair $(1, 2)$ makes both equations true, so it is a solution of the system.

2. Yes

3. We use alphabetical order for the variables. We replace x by 2 and y by 5.

$$\begin{array}{c|c} y = 3x - 1 \\ \hline 5 \ ? \ 3 \cdot 2 - 1 \\ 6 - 1 \\ \hline 5 & 5 \qquad \text{TRUE} \end{array} \qquad \begin{array}{c|c} 2x + y = 4 \\ \hline 2 \cdot 2 + 5 \ ? \ 4 \\ 4 + 5 \\ \hline 9 & 4 \quad \text{FALSE} \end{array}$$

The pair $(2, 5)$ is not a solution of $2x + y = 4$. Therefore, it is not a solution of the system of equations.

4. No

5. We replace x by 1 and y by 5.

$$\begin{array}{c|c} x + y = 6 \\ \hline 1 + 5 \ ? \ 6 \\ \hline 6 & 6 \quad \text{TRUE} \end{array} \qquad \begin{array}{c|c} y = 2x + 3 \\ \hline 5 \ ? \ 2 \cdot 1 + 3 \\ 2 + 3 \\ \hline 5 & 5 \qquad \text{TRUE} \end{array}$$

The pair $(1, 5)$ makes both equations true, so it is a solution of the system.

6. Yes

7. Observe that if we multiply both sides of the first equation by 2, we get the second equation. Thus, if we find that the given point makes the one equation true, we will also know that it makes the other equation true. We replace x by 3 and y by 1 in the first equation.

$$\begin{array}{c|c} 3x + 4y = 13 \\ \hline 3 \cdot 3 + 4 \cdot 1 \ ? \ 13 \\ 9 + 4 \\ \hline 13 & 13 \quad \text{TRUE} \end{array}$$

The pair $(3, 1)$ makes both equations true, so it is a solution of the system.

8. Yes

9. Graph both equations.

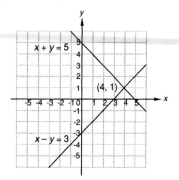

The solution (point of intersection) is apparently $(4, 1)$.

Check:

$$\begin{array}{c|c} x - y = 3 \\ \hline 4 - 1 \ ? \ 3 \\ \hline 3 & 3 \quad \text{TRUE} \end{array} \qquad \begin{array}{c|c} x + y = 5 \\ \hline 4 + 1 \ ? \ 5 \\ \hline 5 & 5 \quad \text{TRUE} \end{array}$$

The solution is $(4, 1)$.

10. $(3,1)$

11. Graph the equations.

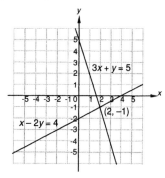

The solution (point of intersection) is apparently $(2, -1)$.

Check:

$$3x + y = 5$$

$$3 \cdot 2 + (-1) \ ? \ 5$$
$$6 - 1$$
$$5 \mid 5 \ \text{TRUE}$$

$$x - 2y = 4$$

$$2 - 2(-1) \ ? \ 4$$
$$2 + 2$$
$$4 \mid 4 \ \text{TRUE}$$

The solution is $(2, -1)$.

2. $(3,2)$

3. Graph both equations.

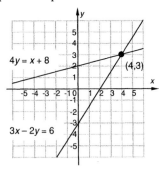

The solution (point of intersection) is apparently $(4, 3)$.

Check:

$$4y = x + 8$$

$$4 \cdot 3 \ ? \ 4 + 8$$
$$12 \mid 12 \ \ \ \text{TRUE}$$

$$3x - 2y = 6$$

$$3 \cdot 4 - 2 \cdot 3 \ ? \ 6$$
$$12 - 6$$
$$6 \mid 6 \ \text{TRUE}$$

The solution is $(4, 3)$.

4. $(1, -5)$

5. Graph both equations.

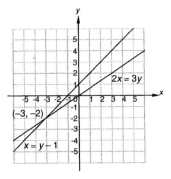

The solution (point of intersection) is apparently $(-3, -2)$.

Check:

$$x = y - 1$$

$$-3 \ ? \ -2 - 1$$
$$-3 \mid -3 \ \ \ \text{TRUE}$$

$$2x = 3y$$

$$2(-3) \ ? \ 3(-2)$$
$$-6 \mid -6 \ \ \ \text{TRUE}$$

The solution is $(-3, -2)$.

16. $(2,1)$

17. Graph both equations.

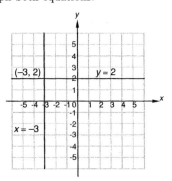

The ordered pair $(-3, 2)$ checks in both equations. It is the solution.

18. $(4, -5)$

19. Enter $y_1 = -5.43x + 10.89$ and $y_2 = 6.29x - 7.04$ on a graphing calculator and use the INTERSECT feature.

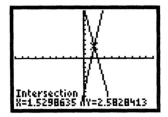

The solution is about $(1.53, 2.58)$.

20. Approximately $(-0.26, 57.06)$

21. Graph both equations.

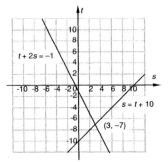

The solution (point of intersection) is apparently $(3, -7)$.

Check:

$$t + 2s = -1$$

$$-7 + 2 \cdot 3 \ ? \ -1$$
$$-7 + 6$$
$$-1 \mid -1 \ \text{TRUE}$$

$$s = t + 10$$

$$3 \ ? \ -7 + 10$$
$$3 \mid 3 \ \ \ \ \ \ \ \ \ \ \text{TRUE}$$

The solution is $(3, -7)$.

22. $(5, -8)$

23. Graph both equations.

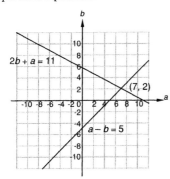

The solution (point of intersection) is apparently $(7, 2)$.

Check:

$$\begin{array}{c|c} 2b + a = 11 & a - b = 5 \\ \hline 2 \cdot 2 + 7 \ ? \ 11 & 7 - 2 \ ? \ 5 \\ 4 + 7 & 5 \quad \text{TRUE} \\ 11 \ | \ 11 \quad \text{TRUE} & \end{array}$$

The solution is $(7,2)$.

24. $(3, -2)$

25. Graph both equations.

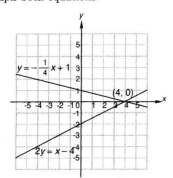

The solution (point of intersection) is apparently $(4, 0)$.

Check:

$$\begin{array}{c|c} y = -\dfrac{1}{4}x + 1 & 2y = x - 4 \\ \hline 0 \ ? \ -\dfrac{1}{4} \cdot 4 + 1 & 2 \cdot 0 \ ? \ 4 - 4 \\ -1 + 1 & 0 \ | \ 0 \quad \text{TRUE} \\ 0 \ | \ 0 \quad \text{TRUE} & \end{array}$$

The solution is $(4,0)$.

26. No solution

27. Solve each equation for y. We get
$$y = \frac{-2.18x + 13.78}{7.81} \text{ and } y = \frac{-5.79x + 8.94}{-3.45}. \text{ Graph thes}$$
equations on a graphing calculator and use the INTER
SECT feature.

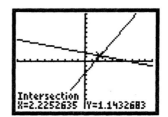

The solution is about $(2.23, 1.14)$.

28. No solution

29. Graph both equations.

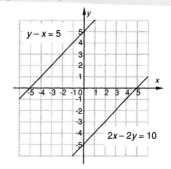

The lines are parallel. The system has no solution.

30. $(3, -4)$

31. Solve each equation for y. We get $y = \dfrac{45x + 33}{57}$ an
$y = \dfrac{30x + 22}{95}$. Graph these equations on a graphing cal
culator and use the INTERSECT feature.

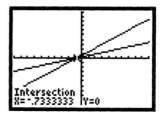

The solution is about $(-0.73, 0)$.

32. Approximately $(0.87, -0.32)$

3. Graph both equations.

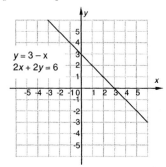

The graphs are the same. Any solution of one equation is a solution of the other. Each equation has infinitely many solutions. The solution set is the set of all pairs (x, y) for which $y = 3 - x$, or $\{(x, y) | y = 3 - x\}$. (In place of $y = 3 - x$ we could have used $2x + 2y = 6$ since the two equations are equivalent.)

4. $(-1, 2)$

5. Solve each equation for y. We get $y = \dfrac{1.9x - 1.7}{4.8}$ and $y = \dfrac{12.92x + 23.8}{32.64}$. Graph these equations on a graphing calculator and use the INTERSECT feature.

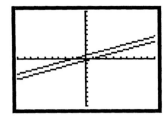

Note that the lines appear to be parallel. This is confirmed by the error message "NO SIGN CHNG" that is returned when we use the INTERSECT feature. The system of equations has no solution.

6. $\{(x, y) | 2x - 3y = 6\}$

7. A system of equations is consistent if it has at least one solution. Of the systems under consideration, only the ones in Exercises 29 and 35 have no solution. Therefore, all except the systems in Exercise 29 and 35 are consistent.

8. All except Exercises 26 and 28

9. A system of two equations in two variables is dependent if it has infinitely many solutions. Only the system in Exercise 33 is dependent.

10. Exercise 36

11. *Familiarize.* Let $x =$ the larger number and $y =$ the smaller number.

Translate.

The difference between two numbers is 11.

Rewording:

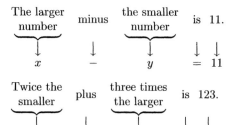

We have a system of equations:

$$x - y = 11,$$
$$3x + 2y = 123$$

42. Let $x =$ the first number and $y =$ the second number.

$$x + y = -42,$$
$$x - y = 52$$

43. *Familiarize.* Let $x =$ the number of less expensive brushes sold and $y =$ the number of more expensive brushes sold.

Translate. We organize the information in a table.

Kind of brush	Less expensive	More expensive	Total
Number sold	x	y	45
Price	\$8.50	\$9.75	
Amount taken in	$8.50x$	$9.75y$	398.75

The "Number sold" row of the table gives us one equation:

$$x + y = 45$$

The "Amount taken in" row gives us a second equation:

$$8.50x + 9.75y = 398.75$$

We have a system of equations:

$$x + y = 45,$$
$$8.50x + 9.75y = 398.75$$

We can multiply both sides of the second equation by 100 to clear the decimals:

$$x + y = 45,$$
$$850x + 975y = 39,875$$

44. Let $x =$ the number of polarfleece neckwarmers sold and $y =$ the number of wool neckwarmers sold.

$$x + y = 40,$$
$$9.9x + 12.75y = 421.65, \text{ or}$$

$$x + y = 40,$$
$$990x + 1275y = 42,165$$

miliarize. Let $x =$ the measure of one angle and $y =$ measure of the other angle.

anslate.

Two angles are supplementary.

wording: $\underbrace{\text{The sum of the measures}}$ is $180°$.

$$\underbrace{\qquad}_{x + y} \quad \underset{=}{\downarrow} \quad \underset{180}{\downarrow}$$

One angle is $3°$ less than twice the other.

wording: $\underbrace{\text{One angle}}$ is $\underbrace{\text{twice the other angle}}$ minus $3°$.

$$\underset{x}{\downarrow} \quad \underset{=}{\downarrow} \quad \underset{2y}{\downarrow} \quad \underset{-}{\downarrow} \quad \underset{3}{\downarrow}$$

have a system of equations:

$$x + y = 180,$$
$$x = 2y - 3$$

$x =$ the measure of the first angle and $y =$ the measure he second angle.

$$x + y = 90,$$
$$x + \frac{1}{2}y = 64$$

miliarize. Let $g =$ the number of two-point shots and the number of free throws made.

anslate. We organize the information in a table.

Kind of shot	Two-point	Free throw	Total
umber scored	g	t	64
Points per score	2	1	
Points scored	$2g$	t	100

m the "Number scored" row of the table we get one ation:

$$g + t = 64$$

"Points scored" row gives us another equation:

$$2g + t = 100$$

have a system of equations:

$$g + t = 64,$$
$$2g + t = 100$$

$x =$ the number of children's plates and $y =$ the num- of adult's plates served.

$$x + y = 250,$$
$$3.5x + 7y = 1347.5, \text{ or}$$

$$x + y = 250,$$
$$35x + 70y = 13,475$$

49. Familiarize. Let $h =$ the number of vials of Humulin Insulin sold and $n =$ the number of vials of Novolin Insulin sold.

Translate. We organize the information in a table.

Brand	Humulin	Novolin	Total
Number sold	h	n	50
Price	\$23.97	\$34.39	
Amount taken in	$23.97h$	$34.39n$	1406.90

The "Number sold" row of the table gives us one equation:

$$h + n = 50$$

The "Amount taken in" row gives us a second equation:

$$23.97h + 34.39n = 1406.90$$

We have a system of equations:

$$h + n = 50,$$
$$23.97h + 34.39n = 1406.90$$

We can multiply both sides of the second equation by 100 to clear the decimals:

$$h + n = 50,$$
$$2397h + 3439n = 140,690$$

50. Let $l =$ the length, in feet, and $w =$ the width, in feet.

$$2l + 2w = 288,$$
$$l = w + 44$$

51. Familiarize. The tennis court is a rectangle with perimeter 228 ft. Let $l =$ the length, in feet, and $w =$ width, in feet. Recall that for a rectangle with length l and width w, the perimeter P is given by $P = 2l + 2w$.

Translate. The formula for perimeter gives us one equation:

$$2l + 2w = 228$$

The statement relating width and length gives us another equation:

The width is 42 ft less than the length.

$$w \qquad = l - 42$$

We have a system of equations:

$$2l + 2w = 228,$$
$$w = l - 42$$

52. Let $x =$ the number of 2-pointers scored and $y =$ the number of 3-pointers scored.

$$x + y = 40,$$
$$2x + 3y = 89$$

3. *Familiarize*. Let $w =$ the number of wins and $t =$ the number of ties. Then the total number of points received from wins was $2w$ and the total number of points received from ties was t.

***Translate*.**

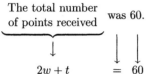

The total number of points received was 60.

$$2w + t = 60$$

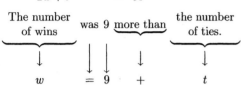

The number of wins was 9 more than the number of ties.

$$w = 9 + t$$

We have a system of equations:

$$2w + t = 60,$$
$$w = 9 + t$$

4. Let $x =$ the number of 30-sec commercials and $y =$ the number of 60-sec commercials.

$$x + y = 12,$$
$$30x + 60y = 600$$

5. *Familiarize*. Let $x =$ the number of ounces of lemon juice and $y =$ the number of ounces of linseed oil to be used.

***Translate*.**

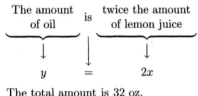

The amount of oil is twice the amount of lemon juice

$$y = 2x$$

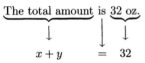

The total amount is 32 oz.

$$x + y = 32$$

We have a system of equations:

$$y = 2x,$$
$$x + y = 32$$

6. Let $l =$ the number of pallets of lumber produced and $p =$ the number of pallets of plywood produced.

$$l + p = 42,$$
$$25l + 40p = 1245$$

7. *Familiarize*. Let $x =$ the number of general-interest films rented and $y =$ the number of children's films rented. Then $3x$ is taken in from the general-interest rentals and $1.5y$ is taken in from the children's rentals.

***Translate*.**

The number of videos rented is 77.

$$x + y = 77$$

The amount taken in is \$213.

$$3x + 1.5y = 213$$

We have a system of equations:

$$x + y = 77,$$
$$3x + 1.5y = 213$$

Clearing decimals we have

$$x + y = 77,$$
$$30x + 15y = 2130$$

58. Let $c =$ the number of coach-class seats and $f =$ the number of first-class seats.

$$c + f = 152,$$
$$c = 5 + 6f$$

59. We will let x represent the number of years after 1970. First we will use the points $(10, 51.5)$ and $(28, 59.8)$ to find a linear equation that describes the percent of women in the work force.

$$m = \frac{59.8 - 51.5}{28 - 10} = \frac{8.3}{18} = \frac{8.3}{18} \cdot \frac{10}{10} = \frac{83}{180}$$

Now we use the point-slope form.

$$y - 51.5 = \frac{83}{180}(x - 10)$$

$$y - 51.5 = \frac{83}{180}x - \frac{83}{18}$$

$$y = \frac{83}{180}x + \frac{422}{9}$$

Next we will use the points $(10, 77.4)$ and $(28, 74.9)$ to find a linear equation that describes the percent of men in the work force.

$$m = \frac{74.9 - 77.4}{28 - 10} = \frac{-2.5}{18} = -\frac{2.5}{18} \cdot \frac{10}{10} = -\frac{25}{180} = -\frac{5}{36}$$

Now we use the point-slope form.

$$y - 77.4 = -\frac{5}{36}(x - 10)$$

$$y - 77.4 = -\frac{5}{36}x + \frac{25}{18}$$

$$y = -\frac{5}{36}x + \frac{7091}{90}$$

We have a system of equations.

$$y = \frac{83}{180}x + \frac{422}{9},$$

$$y = -\frac{5}{36}x + \frac{7091}{90},$$

where y is a percent and x is the number of years after 1970.

We graph these equations on a graphing calculator and use the Intersect feature to find the coordinates of the point of intersection, approximately $(53, 71.4)$. Thus, we estimate that there will be equal percentages of men and women in the work force about 53 years after 1970, or in 2023.

60. $y = -18.3125x + 6376,$

$y = 255.125x + 1634,$

where y represents vehicle production, in thousands, and x is in the number of years after 1980; 1997

61. We will let x represent the number of years after 1990. First we will use the points $(4, 80.8)$ and $(8, 84.1)$ to find a linear function that describes the amount of paper generated, in millions of tons.

$$m = \frac{84.1 - 80.8}{8 - 4} = 0.825$$

Now we use the point-slope form.

$$y - 80.8 = 0.825(x - 4)$$
$$y - 80.8 = 0.825x - 3.3$$
$$y = 0.825x + 77.5$$

Next we will use the points $(4, 36.5)$ and $(8, 41.6)$ to find a linear equation that describes the amount of paper recycled, in millions of tons.

$$m = \frac{41.6 - 36.5}{8 - 4} = \frac{5.1}{4} = 1.275$$

Now we use the point-slope form.

$$y - 36.5 = 1.275(x - 4)$$
$$y - 36.5 = 1.275x - 5.1$$
$$y = 1.275x + 31.4$$

We have a system of equations

$$y = 0.825x + 77.5,$$
$$y = 1.275x + 31.4,$$

where y is in millions of tons and x is the number of years after 1990.

We graph these equations on a graphing calculator and use the Intersect feature to find the coordinates of the point of intersection, approximately $(102, 162)$. Thus, we estimate that the amount of paper recycled will equal the amount generated about 102 years after 1990, or in 2092.

62. Let y represent the number of meals eaten;

$$y = -\frac{2}{3}x + 64,$$
$$y = \frac{4}{3}x + 55,$$

where x is the number of years after 1990; about 1994

63. We will let x represent the number of years after 1950. First we will use the points $(0, 38)$ and $(50, 24)$ to find a linear equation that describes the per capita consumption of milk, in gallons.

$$m = \frac{24 - 38}{50 - 0} = \frac{-14}{50} = -\frac{7}{25}$$

Using the slope-intercept equation we have

$$y = -\frac{7}{25}x + 38.$$

Next we will use the points $(0, 10)$ and $(50, 53)$ to find linear equation that describes the per capita consumption of soft drinks, in gallons.

$$m = \frac{53 - 10}{50 - 0} = \frac{43}{50}$$

Using the slope-intercept equation we have $y = \frac{43}{50}x + 10$

Then we have a system of equations

$$y = -\frac{7}{25}x + 38,$$
$$y = \frac{43}{50}x + 10,$$

where y is in gallons and x is in the number of years after 1950.

We graph these equations on a graphing calculator and use the Intersect feature to find the coordinates of the point of intersection, approximately $(25, 31)$. Thus, we estimate that per capita milk consumption equaled per capital soft drink consumption about 25 years after 1950, or in 1975

64. Let y represent the per capita consumption, in pounds;

$$y = -\frac{1}{20}x + 17.5,$$
$$y = \frac{1}{15}x + 7.1,$$

where x is the number of years after 1980; 2069

65. Enter the number of years after 1970 in L_1, the percent of women in the work force in L_2, and the percent of men in the work force in L_3. Then use the linear regression feature to fit a line to each set of data. For women we have $y_1 = 0.5660768453x + 44.59989889$ and for men we have $y_2 = -0.1678968655x + 79.48407482$, where x represents the number of years after 1970.

We graph these equations on a graphing calculator and use the Intersect feature to find the coordinates of the point of intersection, approximately $(48, 71.5)$. Thus, we predict that there will be equal percentages of women and men in the work force about 48 years after 1970, or in 2018.

66. For car production we have $y_1 = -8.073170732x + 6256.121951$ and for truck and bus production we have $y_2 = 266.6081301x + 1423.764228$, where y_1 and y_2 are in thousands and x represents the number of years after 1980; 1998

7. Enter the number of years after 1990 in L_1, the amount of paper waste generated in L_2, and the amount of paper recycled in L_3. Then use the linear regression feature to fit a line to each set of data. For paper waste generated we have $y_1 = 1.41x + 72.68$ and for paper recycled we have $y_2 = 1.81x + 28.86$, where y_1 and y_2 are in millions of tons and x represents the number of years after 1990.

We graph these equations on a graphing calculator and use the Intersect feature to find the coordinates of the point of intersection, approximately $(110, 227)$. Thus, we predict that the amount of paper waste generated will equal the amount recycled about 110 years after 1990, or in 2100.

8. For regular ice cream we have $y_1 = -0.0971857411x + 17.67298311$ and for lowfat ice cream we have $y_2 = 0.0628517824x + 6.896622889$, where y_1 and y_2 are in pounds and x represents the number of years after 1980; about 2047

9. *Writing Exercise*

0. *Writing Exercise*

1. $2(4x - 3) - 7x = 9$

$\qquad 8x - 6 - 7x = 9 \qquad$ Removing parentheses

$\qquad\quad x - 6 = 9 \qquad$ Collecting like terms

$\qquad\qquad x = 15 \qquad$ Adding 6 to both sides

The solution is 15.

2. $\dfrac{19}{12}$

3. $4x - 5x = 8x - 9 + 11x$

$\qquad -x = 19x - 9 \qquad$ Collecting like terms

$\qquad -20x = -9 \qquad$ Adding $-19x$ to both sides

$\qquad x = \dfrac{9}{20} \qquad$ Multiplying both sides by $-\dfrac{1}{20}$

The solution is $\dfrac{9}{20}$.

74. $\dfrac{13}{3}$

75. $3x + 4y = 7$

$\qquad 4y = -3y + 7 \qquad$ Adding $-3x$ to both sides

$\qquad y = \dfrac{1}{4}(-3x + 7) \qquad$ Multiplying both sides by $\dfrac{1}{4}$

$\qquad y = -\dfrac{3}{4}x + \dfrac{7}{4}$

76. $y = \dfrac{2}{5}x - \dfrac{9}{5}$

77. *Writing Exercise.*

78. *Writing Exercise*

79. The line representing the number of schools with CD-ROMs first lies above the line representing the number of schools with modems in 1994, so this is the year during which the number of schools with CD-ROMs first exceeded the number of schools with modems.

80. 1997; 15,000 schools per year

81. a) There are many correct answers. One can be found by expressing the sum and difference of the two numbers:

$$x + y = 6,$$
$$x - y = 4$$

b) There are many correct answers. For example, write an equation in two variables. Then write a second equation by multiplying the left side of the first equation by one nonzero constant and multiplying the right side by another nonzero constant.

$$x + y = 1,$$
$$2x + 2y = 3$$

c) There are many correct answers. One can be found by writing an equation in two variables and then writing a nonzero constant multiple of that equation:

$$x + y = 1,$$
$$2x + 2y = 2$$

82. a) Answers may vary. $(4, -5)$

b) Infinitely many

83. Substitute 4 for x and -5 for y in the first equation:

$$A(4) - 6(-5) = 13$$
$$4A + 30 = 13$$
$$4A = -17$$
$$A = -\dfrac{17}{4}$$

Substitute 4 for x and -5 for y in the second equation:

$$4 - B(-5) = -8$$
$$4 + 5B = -8$$
$$5B = -12$$
$$B = -\dfrac{12}{5}$$

We have $A = -\dfrac{17}{4}$, $B = -\dfrac{12}{5}$.

84. Let $x =$ Burl's age now and $y =$ his son's age now.

$$x = 2y,$$
$$x - 10 = 3(y - 10)$$

85. *Familiarize*. Let $x =$ the number of years Lou has taught and $y =$ the number of years Juanita has taught. Two years ago, Lou and Juanita had taught $x - 2$ and $y - 2$ years, respectively.

***Translate*.**

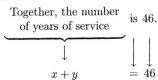

$\underbrace{\text{Together, the number of years of service}} \quad \underset{\downarrow}{\text{is 46.}}$

$$x + y \qquad = 46$$

Two years ago
$\underbrace{\text{Lou had taught 2.5 times as many years as Juanita.}}$

$$x - 2 = 2.5(y - 2)$$

We have a system of equations:

$$x + y = 46,$$
$$x - 2 = 2.5(y - 2)$$

86. Let $l =$ the original length, in inches, and $w =$ the original width, in inches.

$$2l + 2w = 156,$$
$$l = 4(w - 6)$$

87. *Familiarize*. Let $b =$ the number of ounces of baking soda and $v =$ the number of ounces of vinegar to be used. The amount of baking soda in the mixture will be four times the amount of vinegar.

***Translate*.**

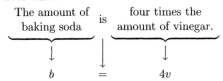

$\underbrace{\text{The amount of baking soda}} \quad \underset{\downarrow}{\text{is}} \quad \underbrace{\text{four times the amount of vinegar.}}$

$$b \qquad = \qquad 4v$$

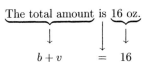

$\underbrace{\text{The total amount}} \text{ is } \underbrace{\text{16 oz.}}$

$$b + v \quad = \quad 16$$

We have a system of equations.

$$b = 4v,$$
$$b + v = 16$$

88. $(-5, 5), (3, 3)$

89. Graph both equations.

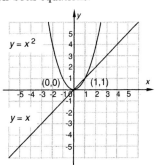

The solutions are apparently $(0, 0)$ and $(1, 1)$. Both pairs check.

90. (d)

91. The equations have the same slope and the same y-intercept. Thus their graphs are the same. Graph (c) matches this system.

92. (a)

93. The equations have the same slope and different y-intercepts. Thus their graphs are parallel lines. Graph (b) matches this system.

Exercise Set 3.2

1. $y = 5 - 4x, \qquad (1)$
   $2x - 3y = 13 \qquad (2)$

We substitute $5 - 4x$ for y in the second equation and solve for x.

$$2x - 3y = 13 \qquad (2)$$
$$2x - 3(5 - 4x) = 13 \quad \text{Substituting}$$
$$2x - 15 + 12x = 13$$
$$14x - 15 = 13$$
$$14x = 28$$
$$x = 2$$

Next we substitute 2 for x in either equation of the original system and solve for y.

$$y = 5 - 4x \qquad (1)$$
$$y = 5 - 4 \cdot 2 \quad \text{Substituting}$$
$$y = 5 - 8$$
$$y = -3$$

We check the ordered pair $(2, -3)$.

$$\begin{array}{c|c} y = 5 - 4x \\ \hline -3 \ ? \ 5 - 4 \cdot 2 \\ \quad\ \ 5 - 8 \\ \hline -3 \ \big|\ -3 \qquad \text{TRUE} \end{array}$$

$$\frac{2x - 3y = 13}{2 \cdot 2 - 3(-3) \ ? \ 13}$$
$$4 + 9 \ \Big|$$
$$13 \ \Big| \ 13 \quad \text{TRUE}$$

Since $(2, -3)$ checks, it is the solution.

2. $(-4, 3)$

3. $2y + x = 9,$ (1)
   $x = 3y - 3$ (2)

We substitute $3y - 3$ for x in the first equation and solve for y.

$$2y + x = 9 \qquad (1)$$
$$2y + (3y - 3) = 9 \qquad \text{Substituting}$$
$$5y - 3 = 9$$
$$5y = 12$$
$$y = \frac{12}{5}$$

Next we substitute $\frac{12}{5}$ for y in either equation of the original system and solve for x.

$$x = 3y - 3 \qquad (2)$$
$$x = 3 \cdot \frac{12}{5} - 3 = \frac{36}{5} - \frac{15}{5} = \frac{21}{5}$$

We check the ordered pair $\left(\frac{21}{5}, \frac{12}{5} \right)$.

$$\frac{2y + x = 9}{2 \cdot \frac{12}{5} + \frac{21}{15} \ ? \ 9}$$
$$\frac{24}{5} + \frac{21}{5} \ \Big|$$
$$\frac{45}{5} \ \Big|$$
$$9 \ \Big| \ 9 \quad \text{TRUE}$$

$$\frac{x = 3y - 3}{\frac{21}{5} \ ? \ 3 \cdot \frac{12}{5} - 3}$$
$$\Big| \ \frac{36}{5} - \frac{15}{5}$$
$$\frac{21}{5} \ \Big| \ \frac{21}{5} \qquad \text{TRUE}$$

Since $\left(\frac{21}{5}, \frac{12}{5} \right)$ checks, it is the solution.

4. $(-3, -15)$

5. $3s - 4t = 14,$ (1)
   $5s + t = 8$ (2)

We solve the second equation for t.

$$5s + t = 8 \qquad\qquad (2)$$
$$t = 8 - 5s \qquad (3)$$

We substitute $8 - 5s$ for t in the first equation and solve for s.

$$3s - 4t = 14 \qquad (1)$$
$$3s - 4(8 - 5s) = 14 \qquad \text{Substituting}$$
$$3s - 32 + 20s = 14$$
$$23s - 32 = 14$$
$$23s = 46$$
$$s = 2$$

Next we substitute 2 for s in Equation (1), (2), or (3). It is easiest to use Equation (3) since it is already solved for t.

$$t = 8 - 5 \cdot 2 = 8 - 10 = -2$$

We check the ordered pair $(2, -2)$.

$$\frac{3s - 4t = 14}{3 \cdot 2 - 4(-2) \ ? \ 14}$$
$$6 + 8 \ \Big|$$
$$14 \ \Big| \ 14 \quad \text{TRUE}$$

$$\frac{5s + t = 8}{5 \cdot 2 + (-2) \ ? \ 8}$$
$$10 - 2 \ \Big|$$
$$8 \ \Big| \ 8 \quad \text{TRUE}$$

Since $(2, -2)$ checks, it is the solution.

6. $(2, -7)$

7. $4x - 2y = 6,$ (1)
   $2x - 3 = y$ (2)

We substitute $2x - 3$ for y in the first equation and solve for x.

$$4x - 2y = 6 \quad (1)$$
$$4x - 2(2x - 3) = 6$$
$$4x - 4x + 6 = 6$$
$$6 = 6$$

We have an identity, or an equation that is always true. The equations are dependent and the solution set is infinite: $\{(x, y) | 2x - 3 = y\}$.

8. No solution

9. $-5s + t = 11$,   (1)

$4s + 12t = 4$ (2)

We solve the first equation for t.

$-5s + t = 11$ (1)

$t = 5s + 11$ (3)

We substitute $5s + 11$ for t in the second equation and solve for s.

$4s + 12t = 4$ (2)

$4s + 12(5s + 11) = 4$

$4s + 60s + 132 = 4$

$64s + 132 = 4$

$64s = -128$

$s = -2$

Next we substitute -2 for s in Equation (3).

$t = 5s + 11 = 5(-2) + 11 = -10 + 11 = 1$

We check the ordered pair $(-2, 1)$.

$$\frac{-5s + t = 11}{-5(-2) + 1 \ ? \ 11}$$
$$10 + 1 \ \Big| $$
$$11 \ \Big| \ 11 \quad \text{TRUE}$$

$$\frac{4s + 12t = 4}{4(-2) + 12 \cdot 1 \ ? \ 4}$$
$$-8 + 12 \ \Big|$$
$$4 \ \Big| \ 4 \quad \text{TRUE}$$

Since $(-2, 1)$ checks, it is the solution.

10. $(4, -1)$

11. $2x + 2y = 2$,   (1)

$3x - y = 1$ (2)

We solve the second equation for y.

$3x - y = 1$ (2)

$-y = -3x + 1$

$y = 3x - 1$ (3)

We substitute $3x - 1$ for y in the first equation and solve for x.

$2x + 2y = 2$ (1)

$2x + 2(3x - 1) = 2$

$2x + 6x - 2 = 2$

$8x - 2 = 2$

$8x = 4$

$x = \dfrac{1}{2}$

Next we substitute $\dfrac{1}{2}$ for x in Equation (3).

$y = 3x - 1 = 3 \cdot \dfrac{1}{2} - 1 = \dfrac{3}{2} - 1 = \dfrac{1}{2}$

The ordered pair $\left(\dfrac{1}{2}, \dfrac{1}{2}\right)$ checks in both equations. It is the solution.

12. $(3, -2)$

13. $3a - b = 7$,   (1)

$2a + 2b = 5$ (2)

We solve the first equation for b.

$3a - b = 7$ (1)

$-b = -3a + 7$

$b = 3a - 7$ (3)

We substitute $3a - 7$ for b in the second equation and solve for a.

$2a + 2b = 5$ (2)

$2a + 2(3a - 7) = 5$

$2a + 6a - 14 = 5$

$8a - 14 = 5$

$8a = 19$

$a = \dfrac{19}{8}$

We substitute $\dfrac{19}{8}$ for a in Equation (3).

$b = 3a - 7 = 3 \cdot \dfrac{19}{8} - 7 = \dfrac{57}{8} - \dfrac{56}{8} = \dfrac{1}{8}$

The ordered pair $\left(\dfrac{19}{8}, \dfrac{1}{8}\right)$ checks in both equations. It is the solution.

14. $\left(\dfrac{25}{23}, -\dfrac{11}{23}\right)$

15. $2x - 3 = y$ (1)

$y - 2x = 1$,   (2)

We substitute $2x - 3$ for y in the second equation and solve for x.

$y - 2x = 1$ (2)

$2x - 3 - 2x = 1$ Substituting

$-3 = 1$ Collecting like terms

We have a contradiction, or an equation that is always false. Therefore, there is no solution.

16. $\{(a, b) | a - 2b = 3\}$

17. $\quad x + 3y = \ 7$ (1)

$\dfrac{-x + 4y = \ 7}{0 + 7y = 14}$ (2)

$0 + 7y = 14$ Adding

$7y = 14$

$y = \ 2$

Substitute 2 for y in one of the original equations and solve for x.

$$x + 3y = 7 \quad (1)$$
$$x + 3 \cdot 2 = 7 \quad \text{Substituting}$$
$$x + 6 = 7$$
$$x = 1$$

Check:

$$\begin{array}{c|c}
x + 3y = 7 & \\
\hline
1 + 3 \cdot 2 \;?\; 7 & \\
1 + 6 & \\
7 \mid 7 \text{ TRUE}
\end{array}$$

$$\begin{array}{c|c}
-x + 4y = 7 & \\
\hline
-1 + 4 \cdot 2 \;?\; 7 & \\
-1 + 8 & \\
7 \mid 7 \text{ TRUE}
\end{array}$$

Since $(1, 2)$ checks, it is the solution.

18. $(2, 7)$

19.
$$2x + y = 6 \quad (1)$$
$$\underline{x - y = 3 \quad (2)}$$
$$3x + 0 = 9 \quad \text{Adding}$$
$$3x = 9$$
$$x = 3$$

Substitute 3 for x in one of the original equations and solve for y.

$$2x + y = 6 \quad (1)$$
$$2 \cdot 3 + y = 6 \quad \text{Substituting}$$
$$6 + y = 6$$
$$y = 0$$

We obtain $(3, 0)$. This checks, so it is the solution.

20. $(10, 2)$

21.
$$9x + 3y = -3 \quad (1)$$
$$\underline{2x - 3y = -8 \quad (2)}$$
$$11x + 0 = -11 \quad \text{Adding}$$
$$11x = -11$$
$$x = -1$$

Substitute -1 for x in Equation (1) and solve for y.

$$9x + 3y = -3$$
$$9(-1) + 3y = -3 \quad \text{Substituting}$$
$$-9 + 3y = -3$$
$$3y = 6$$
$$y = 2$$

We obtain $(-1, 2)$. This checks, so it is the solution.

22. $\left(\dfrac{1}{2}, -5\right)$

23.
$$5x + 3y = 19, \quad (1)$$
$$2x - 5y = 11 \quad (2)$$

We multiply twice to make two terms become opposites.

$$\text{From (1):} \quad 25x + 15y = 95 \quad \text{Multiplying by 5}$$
$$\text{From (2):} \quad \underline{6x - 15y = 33} \quad \text{Multiplying by 3}$$
$$31x + 0 = 128 \quad \text{Adding}$$
$$x = \frac{128}{31}$$

Substitute $\dfrac{128}{31}$ for x in Equation (1) and solve for y.

$$5x + 3y = 19$$
$$5 \cdot \frac{128}{31} + 3y = 19 \quad \text{Substituting}$$
$$\frac{640}{31} + 3y = \frac{589}{31}$$
$$3y = -\frac{51}{31}$$
$$\frac{1}{3} \cdot 3y = \frac{1}{3} \cdot \left(-\frac{51}{31}\right)$$
$$y = -\frac{17}{31}$$

We obtain $\left(\dfrac{128}{31}, -\dfrac{17}{31}\right)$. This checks, so it is the solution.

24. $\left(\dfrac{10}{21}, \dfrac{11}{14}\right)$

25.
$$5r - 3s = 24, \quad (1)$$
$$3r + 5s = 28 \quad (2)$$

We multiply twice to make two terms become additive inverses.

$$\text{From (1):} \quad 25r - 15s = 120 \quad \text{Multiplying by 5}$$
$$\text{From (2):} \quad \underline{9r + 15s = 84} \quad \text{Multiplying by 3}$$
$$34r + 0 = 204 \quad \text{Adding}$$
$$r = 6$$

Substitute 6 for r in Equation (2) and solve for s.

$$3r + 5s = 28$$
$$3 \cdot 6 + 5s = 28 \quad \text{Substituting}$$
$$18 + 5s = 28$$
$$5s = 10$$
$$s = 2$$

We obtain $(6, 2)$. This checks, so it is the solution.

26. $(1, 3)$

27. $6s + 9t = 12$, (1)

$4s + 6t = 5$ (2)

We multiply twice to make two terms become opposites.

From (1): $12s + 18t = 24$ Multiplying by 2

From (2): $\underline{-12s - 18t = -15}$ Multiplying by -3

$0 = 9$

We get a contradiction, or an equation that is always false. The system has no solution.

28. No solution

29. $\dfrac{1}{2}x - \dfrac{1}{6}y = 3$ (1)

$\dfrac{2}{5}x + \dfrac{1}{2}y = 2$, (2)

We first multiply each equation by the LCM of the denominators to clear fractions.

$3x - y = 18$ (3) Multiplying (1) by 6

$4x + 5y = 20$ (4) Multiplying (2) by 10

We multiply by 5 on both sides of Equation (3) and then add.

$15x - 5y = 90$ Multiplying (3) by 5

$\underline{4x + 5y = 20}$ (4)

$19x + 0 = 110$ Adding

$x = \dfrac{110}{19}$

Substitute $\dfrac{110}{19}$ for x in one of the equations in which the fractions were cleared and solve for y.

$3x - y = 18$ (3)

$3\left(\dfrac{110}{19}\right) - y = 18$ Substituting

$\dfrac{330}{19} - y = \dfrac{342}{19}$

$-y = \dfrac{12}{19}$

$y = -\dfrac{12}{19}$

We obtain $\left(\dfrac{110}{19}, -\dfrac{12}{19}\right)$. This checks, so it is the solution.

30. $(12, 15)$

31. $\dfrac{x}{2} + \dfrac{y}{3} = \dfrac{7}{6}$, (1)

$\dfrac{2x}{3} + \dfrac{3y}{4} = \dfrac{5}{4}$ (2)

We first multiply each equation by the LCM of the denominators to clear fractions.

$3x + 2y = 7$ (3) Multiplying (1) by 6

$8x + 9y = 15$ (4) Multiplying (2) by 12

We multiply twice to make two terms become opposites.

From (3): $27x + 18y = 63$ Multiplying by 9

From (4): $\underline{-16x - 18y = -30}$ Multiplying by -2

$11x = 33$ Adding

$x = 3$

Substitute 3 for x in one of the equations in which the fractions were cleared and solve for y.

$3x + 2y = 7$ (3)

$3 \cdot 3 + 2y = 7$ Substituting

$9 + 2y = 7$

$2y = -2$

$y = -1$

We obtain $(3, -1)$. This checks, so it is the solution.

32. $(-2, 3)$

33. $12x - 6y = -15$, (1)

$-4x + 2y = 5$ (2)

Observe that, if we multiply Equation (1) by $-\dfrac{1}{3}$, we obtain Equation (2). Thus, any pair that is a solution of Equation (1) is also a solution of Equation (2). The equations are dependent and the solution set is infinite $\{(x, y) | -4x + 2y = 5\}$.

34. $\{(s, t) | 6s + 9t = 12\}$

35. $0.2a + 0.3b = 1$,

$0.3a - 0.2b = 4$,

We first multiply each equation by 10 to clear decimals.

$2a + 3b = 10$ (1)

$3a - 2b = 40$ (2)

We multiply so that the b-terms can be eliminated.

From (1): $4a + 6b = 20$ Multiplying by 2

From (2): $\underline{9a - 6b = 120}$ Multiplying by 3

$13a + 0 = 140$ Adding

$a = \dfrac{140}{13}$

Substitute $\dfrac{140}{13}$ for a in Equation (1) and solve for b.

$$2a + 3b = 10$$

$$2 \cdot \frac{140}{13} + 3b = 10 \qquad \text{Substituting}$$

$$\frac{280}{13} + 3b = \frac{130}{13}$$

$$3b = -\frac{150}{13}$$

$$b = -\frac{50}{13}$$

We obtain $\left(\frac{140}{13}, -\frac{50}{13}\right)$. This checks, so it is the solution.

36. $(2, 3)$

37. $a - 2b = 16$, (1)

   $b + 3 = 3a$ (2)

We will use the substitution method. First solve Equation (1) for a.

$$a - 2b = 16$$

$$a = 2b + 16 \quad (3)$$

Now substitute $2b + 16$ for a in Equation (2) and solve for b.

$$b + 3 = 3a \qquad\qquad (2)$$

$$b + 3 = 3(2b + 16) \quad \text{Substituting}$$

$$b + 3 = 6b + 48$$

$$-45 = 5b$$

$$-9 = b$$

Substitute -9 for b in Equation (3).

$$a = 2(-9) + 16 = -2$$

We obtain $(-2, -9)$. This checks, so it is the solution.

38. $\left(\frac{1}{2}, -\frac{1}{2}\right)$

39. $10x + y = 306$, (1)

   $10y + x = 90$ (2)

We will use the substitution method. First solve Equation (1) for y.

$$10x + y = 306$$

$$y = -10x + 306 \quad (3)$$

Now substitute $-10x + 306$ for y in Equation (2) and solve for y.

$$10y + x = 90 \qquad\qquad (2)$$

$$10(-10x + 306) + x = 90 \quad \text{Substituting}$$

$$-100x + 3060 + x = 90$$

$$-99x + 3060 = 90$$

$$-99x = -2970$$

$$x = 30$$

Substitute 30 for x in Equation (3).

$$y = -10 \cdot 30 + 306 = 6$$

We obtain $(30, 6)$. This checks, so it is the solution.

40. $\left(-\frac{4}{3}, -\frac{19}{3}\right)$

41. $3y = x - 2$, (1)

   $x = 2 + 3y$ (2)

We will use the substitution method. Substitute $2 + 3y$ for x in the first equation and solve for y.

$$3y = x - 2 \qquad\qquad (1)$$

$$3y = 2 + 3y - 2 \quad \text{Substituting}$$

$$3y = 3y \qquad\qquad \text{Collecting like terms}$$

We get an identity. The system is dependent and the solution set is infinite: $\{(x, y) | x = 2 + 3y\}$.

42. No solution

43. $3s - 7t = 5$,

   $7t - 3s = 8$

First we rewrite the second equation with the variables in a different order. Then we use the elimination method.

$$3s - 7t = 5, \quad (1)$$

$$\underline{-3s + 7t = 8} \quad (2)$$

$$0 = 13$$

We get a contradiction, so the system has no solution.

44. $\{(s, t) | 2s - 13t = 120\}$

45. $0.05x + 0.25y = 22$, (1)

   $0.15x + 0.05y = 24$ (2)

We first multiply each equation by 100 to clear decimals.

$$5x + 25y = 2200$$

$$15x + 5y = 2400$$

We multiply by -5 on both sides of the second equation and add.

$$5x + 25y = \qquad 2200$$

$$\underline{-75x - 25y = -12{,}000} \quad \text{Multiplying (2) by } -5$$

$$-70x \qquad\quad = \quad -9800 \quad \text{Adding}$$

$$x = \frac{-9800}{-70}$$

$$x = 140$$

Substitute 140 for x in one of the equations in which the decimals were cleared and solve for y.

$$5x + 25y = 2200 \quad (1)$$
$$5 \cdot 140 + 25y = 2200 \quad \text{Substituting}$$
$$700 + 25y = 2200$$
$$25y = 1500$$
$$y = 60$$

We obtain $(140, 60)$. This checks, so it is the solution.

46. $(10, 5)$

47. $13a - 7b = 9, \quad (1)$
$2a - 8b = 6 \quad (2)$

We will use the elimination method. First we multiply so that the b-terms can be eliminated.

From (1): $104a - 56b = 72 \quad$ Multiplying by 8
From (2): $\underline{-14a + 56b = -42} \quad$ Multiplying by -7
$90a = 30 \quad$ Adding
$$a = \frac{1}{3}$$

Substitute $\frac{1}{3}$ for a in one of the equations and solve for b.

$$2a - 8b = 6 \quad (2)$$
$$2 \cdot \frac{1}{3} - 8b = 6$$
$$\frac{2}{3} - 8b = 6$$
$$-8b = \frac{16}{3}$$
$$b = -\frac{2}{3}$$

We obtain $\left(\frac{1}{3}, -\frac{2}{3}\right)$. This checks, so it is the solution.

48. $\left(-\dfrac{13}{45}, -\dfrac{37}{45}\right)$

49. The point of intersection is $(140, 60)$, so window (d) is the correct answer.

50. (a)

51. The point of intersection is $(30, 6)$, so window (b) is the correct answer.

52. (c)

53. *Writing Exercise*

54. *Writing Exercise*

55. *Familiarize.* Let $m =$ the number of $\frac{1}{4}$-mi units traveled after the first $\frac{1}{2}$ mi. The total distance traveled will be $\frac{1}{2}$ mi $+ m \cdot \frac{1}{4}$ mi.

Translate.

Fare for first $\frac{1}{2}$ mi plus fare for additional $\frac{1}{4}$-mi units is $5.20.

$$\underbrace{}_{\downarrow} \quad \underbrace{}_{\downarrow} \quad \underbrace{}_{\downarrow} \quad _{\downarrow} \quad _{\downarrow}$$
$$1 \quad + \quad 0.3m \quad = \quad 5.20$$

Carry out. We solve the equation.
$$1 + 0.3m = 5.20$$
$$0.3m = 4.20$$
$$m = 14$$

If the taxi travels the first $\frac{1}{2}$ mi plus 14 additional $\frac{1}{4}$-m units, then it travels a total of $\frac{1}{2} + 14 \cdot \frac{1}{4}$, or $\frac{1}{2} + \frac{7}{2}$, 4 mi.

Check. We have 4 mi $= \frac{1}{2}$ mi $+ \frac{7}{2}$ mi $= \frac{1}{2}$ mi $+ 14 \cdot \frac{1}{4}$ m The fare for traveling this distance is $1.00 + 0.30(14) $1.00 + $4.20 = 5.20. The answer checks.

State. It is 4 mi from Johnson Street to Elm Street.

56. 86

57. *Familiarize.* Let $a =$ the amount spent to remodel bat rooms, in billions of dollars. Then $2a =$ the amount spe to remodel kitchens. The sum of these two amounts is $ billion.

Translate.

Amount spent on bathrooms plus amount spent on kitchens is $35 billion.

$$\underbrace{}_{\downarrow} \quad _{\downarrow} \quad \underbrace{}_{\downarrow} \quad _{\downarrow} \quad _{\downarrow}$$
$$a \quad + \quad 2a \quad = \quad 35$$

Carry out. We solve the equation.
$$a + 2a = 35$$
$$3a = 35 \qquad \text{Combining like terms}$$
$$a = \frac{35}{3}, \text{ or } 11\frac{2}{3}$$

If $a = \frac{35}{3}$, then $2a = 2 \cdot \frac{35}{3} = \frac{70}{3} = 23\frac{1}{3}$.

Check. $\frac{70}{3}$ is twice $\frac{35}{3}$, and $\frac{35}{3} + \frac{70}{3} = \frac{105}{3} = 35$. T answer checks.

State. $11\frac{2}{3}$ billion was spent to remodel bathrooms, a $23\frac{1}{3}$ billion was spent to remodel kitchens.

58. 30 m, 90 m, 360 m

59. *Familiarize.* The total cost is the daily charge plus t mileage charge. The mileage charge is the cost per m times the number of miles driven. Let m = the number of miles that can be driven for \$80.

Translate. We reword the problem.

Daily rate	plus	Cost per mile	times	Number of miles driven	is	Amount.
↓	↓	↓	↓	↓	↓	↓
34.95	+	0.10	·	m	=	80

Carry out. We solve the equation.

$$34.95 + 0.10m = 80$$
$$100(34.95 + 0.10m) = 100(80) \quad \text{Clearing decimals}$$
$$3495 + 10m = 8000$$
$$10m = 4505$$
$$m = 450.5$$

Check. The mileage cost is found by multiplying 450.5 by \$0.10 obtaining \$45.05. Then we add \$45.05 to \$34.95, the daily rate, and get \$80.

State. The businessperson can drive 450.5 mi on the car-rental allotment.

0. 460.5 mi

1. *Writing Exercise*

2. *Writing Exercise*

3. First write $f(x) = mx + b$ as $y = mx + b$. Then substitute 1 for x and 2 for y to get one equation and also substitute -3 for x and 4 for y to get a second equation:

$$2 = m \cdot 1 + b$$
$$4 = m(-3) + b$$

Solve the resulting system of equations.

$$2 = m + b$$
$$4 = -3m + b$$

Multiply the second equation by -1 and add.

$$2 = m + b$$
$$\underline{-4 = 3m - b}$$
$$-2 = 4m$$
$$-\frac{1}{2} = m$$

Substitute $-\frac{1}{2}$ for m in the first equation and solve for b.

$$2 = -\frac{1}{2} + b$$
$$\frac{5}{2} = b$$

Thus, $m = -\frac{1}{2}$ and $b = \frac{5}{2}$.

64. $p = 2, q = -\dfrac{1}{3}$

65. Substitute -4 for x and -3 for y in both equations and solve for a and b.

$$-4a - 3b = -26, \quad (1)$$
$$-4b + 3a = 7 \qquad (2)$$

$$-12a - \quad 9b = -78 \quad \text{Multiplying (1) by 3}$$
$$\underline{12a - \quad 16b = \quad 28} \quad \text{Multiplying (2) by 4}$$
$$-25b = -50$$
$$b = \quad 2$$

Substitute 2 for b in Equation (2).

$$-4 \cdot 2 + 3a = 7$$
$$3a = 15$$
$$a = 5$$

Thus, $a = 5$ and $b = 2$.

66. $\left(\dfrac{a + 2b}{7}, \dfrac{a - 5b}{7}\right)$

67. $\dfrac{x + y}{2} - \dfrac{x - y}{5} = 1,$

$$\dfrac{x - y}{2} + \dfrac{x + y}{6} = -2$$

After clearing fractions we have:

$$3x + 7y = 10, \quad (1)$$
$$4x - 2y = -12 \quad (2)$$

$$6x + 14y = \quad 20 \quad \text{Multiplying (1) by 2}$$
$$\underline{28x - 14y = -84} \quad \text{Multiplying (2) by 7}$$
$$34x = -64$$
$$x = -\frac{32}{17}$$

Substitute $-\dfrac{32}{17}$ for x in Equation (1).

$$3\left(-\frac{32}{17}\right) + 7y = 10$$
$$7y = \frac{266}{17}$$
$$y = \frac{38}{17}$$

The solution is $\left(-\dfrac{32}{17}, \dfrac{38}{17}\right)$.

68. Approximately $(23.118879, -12.039964)$

69. $\dfrac{2}{x} + \dfrac{1}{y} = 0, \qquad 2 \cdot \dfrac{1}{x} + \dfrac{1}{y} = 0,$

$$\text{or}$$

$$\dfrac{5}{x} + \dfrac{2}{y} = -5 \qquad 5 \cdot \dfrac{1}{x} + 2 \cdot \dfrac{1}{y} = -5$$

Substitute u for $\frac{1}{x}$ and v for $\frac{1}{y}$.

$$2u + v = 0, \quad (1)$$
$$5u + 2v = -5 \quad (2)$$

$$-4u - 2v = 0 \quad \text{Multiplying (1) by } -2$$
$$\underline{5u + 2v = -5 \quad (2)}$$
$$u = -5$$

Substitute -5 for u in Equation (1).

$$2(-5) + v = 0$$
$$-10 + v = 0$$
$$v = 10$$

If $u = -5$, then $\frac{1}{x} = -5$. Thus $x = -\frac{1}{5}$.

If $v = 10$, then $\frac{1}{y} = 10$. Thus $y = \frac{1}{10}$.

The solution is $\left(-\frac{1}{5}, \frac{1}{10}\right)$.

70. $\left(-\frac{1}{4}, -\frac{1}{2}\right)$

Exercise Set 3.3

1. The Familiarize and Translate steps were done in Exercise 41 of Exercise Set 3.1

Carry out. We solve the system of equations

$$x - y = 11, \quad (1)$$
$$3x + 2y = 123 \quad (2)$$

where $x =$ the larger number and $y =$ the smaller number. We use elimination.

$$2x - 2y = 22 \quad \text{Multiplying (1) by 2}$$
$$\underline{3x + 2y = 123}$$
$$5x = 145$$
$$x = 29$$

Substitute 29 for x in (1) and solve for y.

$$29 - y = 11$$
$$-y = -18$$
$$y = 18$$

Check. The difference between the numbers is $29 - 18$, or 11. Also $2 \cdot 18 + 3 \cdot 29 = 36 + 87 = 123$. The numbers check.

State. The larger number is 29, and the smaller is 18.

2. 5 and -47

3. The Familiarize and Translate steps were done in Exercise 43 of Exercise Set 3.1

Carry out. We solve the system of equations

$$x + y = 45, \quad (1)$$
$$850x + 975y = 39,875 \quad (2)$$

where $x =$ the number of less expensive brushes sold and $y =$ the number of more expensive brushes sold. We use elimination. Begin by multiplying Equation (1) by -850.

$$-850x - 850y = -38,250 \quad \text{Multiplying (1)}$$
$$\underline{850x + 975y = 39,875}$$
$$125y = 1625$$
$$y = 13$$

Substitute 13 for y in (1) and solve for x.

$$x + 13 = 45$$
$$x = 32$$

Check. The number of brushes sold is $32 + 13$, or 45 The amount taken in was $\$8.50(32) + \$9.75(13) = \$272 + \$126.75 = \$398.75$. The answer checks.

State. 32 of the less expensive brushes were sold, and 1: of the more expensive brushes were sold.

4. 31 polarfleece, 9 wool

5. The Familiarize and Translate steps were done in Exercise 45 of Exercise Set 3.1

Carry out. We solve the system of equations

$$x + y = 180, \quad (1)$$
$$x = 2y - 3 \quad (2)$$

where $x =$ the measure of one angle and $y =$ the measur of the other angle. We use substitution.

Substitute $2y - 3$ for x in (1) and solve for y.

$$2y - 3 + y = 180$$
$$3y - 3 = 180$$
$$3y = 183$$
$$y = 61$$

Now substitute 61 for y in (2).

$$x = 2 \cdot 61 - 3 = 122 - 3 = 119$$

Check. The sum of the angle measures is $119° + 61°$, c $180°$, so the angles are supplementary. Also $2 \cdot 61° - 3° = 122° - 3° = 119°$. The answer checks.

State. The measures of the angles are $119°$ and $61°$.

6. $38°, 52°$

7. The Familiarize and Translate steps were done in Exer cise 47 of Exercise Set 3.1

Carry out. We solve the system of equations

$$g + t = 64, \quad (1)$$
$$2g + t = 100 \quad (2)$$

where g = the number of two-point shots and t = the number of free throws Chamberlain made. We use elimination.

$$-g - t = -64 \quad \text{Multiplying (1) by} -1$$
$$\underline{2g + t = 100}$$
$$g = 36$$

Substitute 36 for g in (1) and solve for t.

$$36 + t = 64$$
$$t = 28$$

Check. The total number of scores was 36+28, or 64. The total number of points was $2 \cdot 36 + 28 = 72 + 28 = 100$. The answer checks.

State. Chamberlain made 36 two-point shots and 28 free throws.

8. 115 children's plates, 135 adult's plates

9. The Familiarize and Translate steps were done in Exercise 49 of Exercise Set 3.1.

Carry out. We solve the system of equations

$$h + n = 50, \qquad (1)$$
$$2397h + 3439n = 140,690 \quad (2)$$

where h = the number of vials of Humulin Insulin sold and n = the number of vials of Novolin Insulin sold. We use elimination.

$$-2397h - 2397n = -119,850 \quad \text{Multiplying (1)}$$
$$\underline{2397h + 3439n = 140,690} \quad \text{by} -2397$$
$$1042n = 20,840$$
$$n = 20$$

Substitute 20 for n in (1) and solve for h.

$$h + 20 = 50$$
$$h = 30$$

Check. A total of $30 + 20$, or 50 vials, was sold. The amount collected was $\$23.97(30) + \$34.39(20) = \$719.10 + \$687.80 = \$1406.90$. The answer checks.

State. 30 vials of Humulin Insulin and 20 vials of Novolin Insulin were sold.

10. Length: 94 ft; width: 50 ft

11. The Familiarize and Translate steps were done in Exercise 51 of Exercise Set 3.1

Carry out. We solve the system of equations

$$2l + 2w = 228, \quad (1)$$
$$w = l - 42 \qquad (2)$$

where l = the length, in feet, and w = the width, in feet, of the tennis court. We use substitution.

Substitute $l - 42$ for w in (1) and solve for l.

$$2l + 2(l - 42)w = 228$$
$$2l + 2l - 84 = 228$$
$$4l - 84 = 228$$
$$4l = 312$$
$$l = 78$$

Now substitute 78 for l in (2).

$$w = 78 - 42 = 36$$

Check. The perimeter is $2 \cdot 78$ ft $+ 2 \cdot 36$ ft $= 156$ ft $+ 72$ ft $= 228$ ft. The width, 36 ft, is 42 ft less than the length, 78 ft. The answer checks.

State. The length of the tennis court is 78 ft, and the width is 36 ft.

12. 31 two-point field goal, 9 three-point field goals

13. The Familiarize and Translate steps were done in Exercise 53 of Exercise Set 3.1.

Carry out. We solve the system of equations

$$2w + t = 60, \quad (1)$$
$$w = 9 + t \qquad (2)$$

where w = the number of wins and t = the number of ties. We use substitution.

Substitute $9 + t$ for w in (1) and solve for t.

$$2(9 + t) + t = 60$$
$$18 + 2t + t = 60$$
$$18 + 3t = 60$$
$$3t = 42$$
$$t = 14$$

Now substitute 14 for t in (2).

$$w = 9 + 14 = 23$$

Check. The total number of points is $2 \cdot 23 + 14 = 46 + 14 = 60$. The number of wins, 23, is nine more than the number of ties, 14. The answer checks.

State. The Wildcats had 23 wins and 14 ties.

14. 4 30-sec commercials, and 8 60-sec commercials

15. The Familiarize and Translate steps were done in Exercise 55 of Exercise Set 3.1.

Carry out. We solve the system of equations

$$y = 2x, \qquad (1)$$
$$x + y = 32 \quad (2)$$

where x = the number of ounces of lemon juice and y = the number of ounces of linseed oil to be used. We use substitution.

Substitute $2x$ for y in (2) and solve for x.

$$x + 2x = 32$$
$$3x = 32$$
$$x = \frac{32}{3}, \text{ or } 10\frac{2}{3}$$

Now substitute $\frac{32}{3}$ for x in (1).

$$y = 2 \cdot \frac{32}{3} = \frac{64}{3}, \text{ or } 21\frac{1}{3}$$

Check. The amount of oil, $\frac{64}{3}$ oz, is twice the amount of lemon juice, $\frac{32}{3}$ oz. The mixture contains $\frac{32}{3}$ oz $+ \frac{64}{3}$ oz $= \frac{96}{3}$ oz $= 32$ oz. The answer checks.

State. $10\frac{2}{3}$ oz of lemon juice and $21\frac{1}{3}$ oz of linseed oil are needed.

16. 29 pallets of lumber, 13 pallets of plywood

17. The Familiarize and Translate steps were done in Exercise 57 of Exercise Set 3.1.

Carry out. We solve the system of equations
$$x + y = 77, \qquad (1)$$
$$30x + 15y = 2130 \quad (2)$$
where $x = $ the number of general-interest films rented and $y = $ the number of children's films rented. We use elimination.

$$
\begin{array}{ll}
-15x - 15y = -1155 & \text{Multiplying (1) by } -15 \\
\underline{30x + 15y = 2130} & \\
15x = 975 & \\
x = 65 &
\end{array}
$$

Substitute 65 for x in (1) and solve for y.
$$65 + y = 77$$
$$y = 12$$

Check. The total number of films rented is $65 + 12$, or 77. The total amount taken in was $\$3(65) + \$1.50(12) = \$195 + \$18 = \$213$. The answer checks.

State. 65 general-interest videos and 12 children's videos were rented.

18. 131 coach-class seats, 21 first-class seats

19. **Familiarize**. Let $f = $ the number of boxes of Flair pens sold and $u = $ the number of four-packs of Uniball pens sold.

Translate. We organize the information in a table.

	Flair boxes	Uniball four-packs	Total
Number sold	f	u	40
Price	$12	$8	
Total cost	$12f$	$8u$	372

We get one equation from the "Number sold" row of the table:
$$f + u = 40$$
The "Total cost" row yields a second equation:
$$12f + 8u = 372$$
We have translated to a system of equations:
$$f + u = 40, \quad (1)$$
$$12f + 8u = 372 \quad (2)$$

Carry out. We solve the system of equations using the elimination method.

$$
\begin{array}{ll}
-8f - 8u = -320 & \text{Multiplying (1) by } -8 \\
\underline{12f + 8u = 372} & \\
4f = 52 & \\
f = 13 &
\end{array}
$$

Now substitute 13 for f in (1) and solve for u.
$$13 + u = 40$$
$$u = 27$$

Check. The total number of boxes and four-packs sold is $13 + 27$, or 40. The total cost of these purchases $\$12 \cdot 13 + \$8 \cdot 27 = \$156 + \$216 = \$372$. The answer checks.

State. 13 boxes of Flair pens and 27 four-packs of Uniball pens were sold.

20. 18 graph-paper notebooks, 32 college-ruled notebooks

21. **Familiarize**. Let $k = $ the number of pounds of Kenyan French Roast coffee and $s = $ the number of pounds of Sumatran coffee to be used in the mixture. The value of the mixture will be $\$8.40(20)$, or $\$168$.

Translate. We organize the information in a table.

	Kenyan	Sumatran	Mixture
Number of pounds	k	s	20
Price per pound	$9	$8	$8.40
Value of coffee	$9k$	$8s$	168

The "Number of pounds" row of the table gives us one equation:
$$k + s = 20$$
The "Value of coffee" row yields a second equation:
$$9k + 8s = 168$$
We have translated to a system of equations:
$$k + s = 20, \quad (1)$$
$$9k + 8s = 168 \quad (2)$$

Carry out. We use the elimination method to solve the system of equations.

$$-8k - 8s = -160 \quad \text{Multiplying (1) by } -8$$
$$\underline{9k + 8s = 168}$$
$$k = 8$$

Substitute 8 for k in (1) and solve for s.

$$8 + s = 20$$
$$s = 12$$

Check. The total mixture contains 8 lb + 12 lb, or 20 lb. Its value is $\$9 \cdot 8 + \$8 \cdot 12 = \$72 + \$96 = \$168$. The answer checks.

State. 8 lb of Kenyan French Roast coffee and 12 lb of Sumatran coffee should be used.

22. 20 lb of cashews, 30 lb of Brazil nuts

23. Observe that the average of 40% and 10% is 25%:
$$\frac{40\% + 10\%}{2} = \frac{50\%}{2} = 25\%.$$ Thus, the caterer should use equal parts of the 40% and 10% mixtures. Since a 10-lb mixture is desired, the caterer should use 5 lb each of the 40% and the 10% mixture.

24. 150 lb of soybean meal, 200 lb of corn meal

25. *Familiarize*. Let x = the number of liters of 25% solution and y = the number of liters of 50% solution to be used. The mixture contains 40%(10 L), or 0.4(10 L) = 4 L of acid.

Translate. We organize the information in a table.

	25% solution	50% solution	Mixture
Number of liters	x	y	10
Percent of acid	25%	50%	40%
Amount of acid	$0.25x$	$0.5y$	4 L

We get one equation from the "Number of liters" row of the table.

$$x + y = 10$$

The last row of the table yields a second equation.

$$0.25x + 0.5y = 4$$

After clearing decimals, we have the problem translated to a system of equations:

$$x + y = 10, \quad (1)$$
$$25x + 50y = 400 \quad (2)$$

Carry out. We use the elimination method to solve the system of equations.

$$-25x - 25y = -250 \quad \text{Multiplying (1) by } -25$$
$$\underline{25x + 50y = 400}$$
$$25y = 150$$
$$y = 6$$

Substitute 6 for y in (1) and solve for x.

$$x + 6 = 10$$
$$x = 4$$

Check. The total amount of the mixture is 4 lb + 6 lb, or 10 lb. The amount of acid in the mixture is 0.25(4 L) + 0.5(6 L) = 1 L + 3 L = 4 L. The answer checks.

State. 4 L of the 25% solution and 6 L of the 50% solution should be mixed.

26. 12 lb of Deep Thought Granola, 8 lb of Oat Dream Granola

27. *Familiarize*. Let x = the amount of the 6% loan and y = the amount of the 9% loan. Recall that the formula for simple interest is

$$\text{Interest} = \text{Principal} \cdot \text{Rate} \cdot \text{Time}.$$

Translate. We organize the information in a table.

	6% loan	9% loan	Total
Principal	x	y	$12,000
Interest Rate	6%	9%	
Time	1 yr	1 yr	
Interest	$0.06x$	$0.09y$	$855

The "Principal" row of the table gives us one equation:

$$x + y = 12,000$$

The last row of the table yields another equation:

$$0.06x + 0.09y = 855$$

After clearing decimals, we have the problem translated to a system of equations:

$$x + y = 12,000 \quad (1)$$
$$6x + 9y = 85,500 \quad (2)$$

Carry out. We use the elimination method to solve the system of equations.

$$-6x - 6y = -72,000 \quad \text{Multiplying (1) by } -6$$
$$\underline{6x + 9y = 85,500}$$
$$3y = 13,500$$
$$y = 4500$$

Substitute 4500 for y in (1) and solve for x.

$$x + 4500 = 12,000$$
$$x = 7500$$

Check. The loans total $7500 + $4500, or $12,000. The total interest is $0.06(\$7500) + 0.09(\$4500) = \$450 + \$405 = \$855$. The answer checks.

State. The 6% loan was for $7500, and the 9% loan was for $4500.

28. $6800 at 9%, $8200 at 10%

29. **Familiarize**. Let x = the number of liters of Arctic Antifreeze and y = the number of liters of Frost-No-More in the mixture. The amount of alcohol in the mixture is $0.15(20\text{ L}) = 3$ L.

Translate. We organize the information in a table.

	18% solution	10% solution	Mixture
Number of liters	x	y	20
Percent of alcohol	18%	10%	15%
Amount of alcohol	$0.18x$	$0.1y$	3

We get one equation from the "Number of liters" row of the table:

$$x + y = 20$$

The last row of the table yields a second equation:

$$0.18x + 0.1y = 3$$

After clearing decimals we have the problem translated to a system of equations:

$$x + y = 20, \quad (1)$$
$$18x + 10y = 300 \quad (2)$$

Carry out. We use the elimination method to solve the system of equations.

$$-10x - 10y = -200 \quad \text{Multiplying (1) by } -10$$
$$\underline{18x + 10y = 300}$$
$$8x = 100$$
$$x = 12.5$$

Substitute 12.5 for x in (1) and solve for y.

$$12.5 + y = 20$$
$$y = 7.5$$

Check. The total amount of the mixture is 12.5 L + 7.5 L or 20 L. The amount of alcohol in the mixture is $0.18(12.5\text{ L}) + 0.1(7.5\text{ L}) = 2.25\text{ L} + 0.75\text{ L} = 3$ L. The answer checks.

State. 12.5 L of Arctic Antifreeze and 7.5 L of Frost-No-More should be used.

30. $169\dfrac{3}{13}$ lb of whole milk, $30\dfrac{10}{13}$ lb of cream

31. **Familiarize**. Let l = the length, in meters, and w = the width, in meters. Recall that the formula for the perimeter P of a rectangle with length l and width w is $P = 2l + 2w$.

Translate.

The perimeter is 190 m.
$$2l + 2w = 190$$

The width is one-fourth of the length.
$$w = \frac{1}{4} \cdot l$$

We have translated to a system of equations:

$$2l + 2w = 190, \quad (1)$$
$$w = \frac{1}{4}l$$

Carry out. We use the substitution method to solve the system of equations.

Substitute $\frac{1}{4}l$ for w in (1) and solve for l.

$$2l + 2\left(\frac{1}{4}l\right) = 190$$
$$2l + \frac{1}{2}l = 190$$
$$\frac{5}{2}l = 190$$
$$l = \frac{2}{5} \cdot 190 = 76$$

Now substitute 76 for l in (2).

$$l = \frac{1}{4} \cdot 76 = 19$$

Check. The perimeter is $2 \cdot 76\text{ m} + 2 \cdot 19\text{ m} = 152\text{ m} + 38\text{ m} = 190$ m. The width, 19 m, is one-fourth the length 76 m. The answer checks.

State. The length is 76 m, and the width is 19 m.

32. Length: 265 ft; width: 165 ft

33. **Familiarize**. The change from the $9.25 purchase is $20 - $9.25, or $10.75. Let x = the number of quarters and y = the number of fifty-cent pieces. The total value of the quarters, in dollars, is $0.25x$ and the total value of the fifty-cent pieces, in dollars, is $0.50y$.

Translate.

The total number of coins is 30.
$$x + y = 30$$

The total value of the coins is $10.75.
$$0.25x + 0.50y = 10.75$$

After clearing decimals we have the following system of equations:

$$x + \quad y = \quad 30, \quad (1)$$
$$25x + 50y = 1075 \quad (2)$$

Carry out. We use the elimination method to solve the system of equations.

$$-25x - 25y = -750 \quad \text{Multiplying (1) by } -25$$
$$\underline{25x + 50y = \ 1075}$$
$$25y = \ 325$$
$$y = \quad 13$$

Substitute 13 for y in (1) and solve for x.

$$x + 13 = 30$$
$$x = 17$$

Check. The total number of coins is $17 + 13$, or 30. The total value of the coins is $\$0.25(17) + \$0.50(13) = \$4.25 + \$6.50 = \$10.75$. The answer checks.

State. There were 17 quarters and 13 fifty-cent pieces.

4. 7 $5 bills, 15 $1 bills

5. **Familiarize**. We first make a drawing.

Slow train
d kilometers   75 km/h $(t + 2)$ hr

Fast train
d kilometers   125 km/h t hr

From the drawing we see that the distances are the same. Now complete the chart.

$$d \quad = \quad r \quad \cdot \quad t$$

	Distance	Rate	Time	
Slow train	d	75	$t + 2$	$\to d = 75(t+2)$
Fast train	d	125	t	$\to d = 125t$

Translate. Using $d = rt$ in each row of the table, we get a system of equations:

$$d = 75(t + 2),$$
$$d = 125t$$

Carry out. We solve the system of equations.

$$125t = 75(t + 2) \quad \text{Using substitution}$$
$$125t = 75t + 150$$
$$50t = 150$$
$$t = 3$$

Then $d = 125t = 125 \cdot 3 = 375$

Check. At 125 km/h, in 3 hr the fast train will travel $125 \cdot 3 = 375$ km. At 75 km/h, in $3 + 2$, or 5 hr the slow train will travel $75 \cdot 5 = 375$ km. The numbers check.

State. The trains will meet 375 km from the station.

36. 3 hr

37. **Familiarize**. We first make a drawing. Let $d =$ the distance and $r =$ the speed of the canoe in still water. Then when the canoe travels downstream its speed is $r + 6$, and its speed upstream is $r - 6$. From the drawing we see that the distances are the same.

Downstream, 6 mph current

d mi, $r + 6$, 4 hr

Upstream, 6 mph current

d mi, $r - 6$, 10 hr

Organize the information in a table.

	Distance	Rate	Time
With current	d	$r + 6$	4
Against current	d	$r - 6$	10

Translate. Using $d = rt$ in each row of the table, we get a system of equations:

$$d = 4(r + 6), \qquad d = 4r + 24,$$
$$\text{or}$$
$$d = 10(r - 6) \qquad d = 10r - 60$$

Carry out. Solve the system of equations.

$$4r + 24 = 10r - 60 \quad \text{Using substitution}$$
$$24 = 6r - 60$$
$$84 = 6r$$
$$14 = r$$

Check. When $r = 14$, then $r + 6 = 14 + 6 = 20$, and the distance traveled in 4 hr is $4 \cdot 20 = 80$ km. Also, $r - 6 = 14 - 6 = 8$, and the distance traveled in 10 hr is $8 \cdot 10 = 80$ km. The answer checks.

State. The speed of the canoe in still water is 14 km/h.

38. 24 mph

39. **Familiarize**. We make a drawing. Note that the plane's speed traveling toward London is $360 + 50$, or 410 mph, and the speed traveling toward New York City is $360 - 50$, or 310 mph. Also, when the plane is d mi from New York City, it is $3458 - d$ mi from London.

New York City   London
310 mph t hours t hours   410 mph

$\vdash$————————— 3458 mi————————$\dashv$

$\vdash$——— d ———$\vdash$——— 3458 mi $-d$ ———$\dashv$

Organize the information in a table.

	Distance	Rate	Time
Toward NYC	d	310	t
Toward London	$3458 - d$	410	t

Translate. Using $d = rt$ in each row of the table, we get a system of equations:

$$d = 310t, \quad (1)$$
$$3458 - d = 410t \quad (2)$$

Carry out. We solve the system of equations.

$$3458 - 310t = 410t \quad \text{Using substitution}$$
$$3458 = 720t$$
$$4.8028 \approx t$$

Substitute 4.8028 for t in (1).

$$d \approx 310(4.8028) \approx 1489$$

Check. If the plane is 1489 mi from New York City, it can return to New York City, flying at 310 mph, in $1489/310 \approx 4.8$ hr. If the plane is $3458 - 1489$, or 1969 mi from London, it can fly to London, traveling at 410 mph, in $1969/410 \approx 4.8$ hr. Since the times are the same, the answer checks.

State. The point of no return is about 1489 mi from New York City.

40. About 1524 mi

41. *Writing Exercise*

42. *Writing Exercise*

43. $2x - 3y + 12 = 2 \cdot 5 - 3 \cdot 2 + 12$
$$= 10 - 6 + 12$$
$$= 4 + 12$$
$$= 16$$

44. 11

45. $5a - 7b + 3c = 5(-2) - 7(3) + 3 \cdot 1$
$$= -10 - 21 + 3$$
$$= -31 + 3$$
$$= -28$$

46. -10

47. $4 - 2y + 3z = 4 - 2 \cdot \dfrac{1}{3} + 3 \cdot \dfrac{1}{4}$
$$= 4 - \frac{2}{3} + \frac{3}{4}$$
$$= \frac{48}{12} - \frac{8}{12} + \frac{9}{12}$$
$$= \frac{40}{12} + \frac{9}{12}$$
$$= \frac{49}{12}$$

48. $\dfrac{13}{10}$

49. *Writing Exercise*

50. *Writing Exercise*

51. The Familiarize and Translate steps were done in Exercise 84 of Exercise Set 3.1.

Carry out. We solve the system of equations

$$x = 2y, \quad (1)$$
$$x + 20 = 3y \quad (2)$$

where $x = $ Burl's age now and $y = $ his son's age now.

$$2y + 20 = 3y \quad \text{Substituting } 2y \text{ for } x \text{ in (2)}$$
$$20 = y$$

$$x = 2 \cdot 20 \quad \text{Substituting 20 for } y \text{ in (1)}$$
$$x = 40$$

Check. Burl's age now, 40, is twice his son's age now, 20. Ten years ago Burl was 30 and his son was 10, and $30 = 3 \cdot 10$. The numbers check.

State. Now Burl is 40 and his son is 20.

52. Lou: 32 years, Juanita: 14 years

53. The Familiarize and Translate steps were done in Exercise 86 of Exercise Set 3.1.

Carry out. We solve the system of equations

$$2l + 2w = 156, \quad (1)$$
$$l = 4(w - 6) \quad (2)$$

where $l = $ length, in inches, and $w = $ width, in inches.

$$2 \cdot 4(w - 6) + 2w = 156 \quad \text{Substituting } 4(w - 6) \text{ for } l \text{ in (1)}$$
$$8w - 48 + 2w = 156$$
$$10w - 48 = 156$$
$$10w = 204$$
$$w = \frac{204}{10}, \text{ or } \frac{102}{5}$$

$$l = 4\left(\frac{102}{5} - 6\right) \quad \text{Substituting } \frac{102}{5} \text{ for } w \text{ in (2)}$$
$$l = 4\left(\frac{102}{5} - \frac{30}{5}\right)$$
$$l = 4\left(\frac{72}{5}\right)$$
$$l = \frac{288}{5}$$

Check. The perimeter of a rectangle with width $\dfrac{102}{5}$ in. and length $\dfrac{288}{5}$ in. is

$$2\left(\frac{288}{5}\right) + 2\left(\frac{102}{5}\right) = \frac{576}{5} + \frac{204}{5} = \frac{780}{5} = 156 \text{ in.}$$

If 6 in. is cut off the width, the new width is

$\frac{102}{5} - 6 = \frac{102}{5} - \frac{30}{5} = \frac{72}{5}$. The length, $\frac{288}{5}$, is

$4\left(\frac{72}{5}\right)$. The numbers check.

State. The original piece of posterboard had width $\frac{102}{5}$ in. and length $\frac{288}{5}$ in.

54. $\frac{64}{5}$ oz of baking soda, $\frac{16}{5}$ oz of vinegar

55. Familiarize. Let k = the number of pounds of Kona coffee that must be added to the Mexican coffee, and m = the number of pounds of coffee in the mixture.

Translate. We organize the information in a table.

	Mexican	Kona	Mixture
Number of pounds	40	k	m
Percent of Kona	0%	100%	30%
Amount of Kona	0	k	$0.3m$

We get one equation from the "Number of pounds" row of the table:

$40 + k = m$

The last row of the table gives us a second equation:

$k = 0.3m$

After clearing the decimal we have the problem translated to a system of equations:

$$40 + k = m, \quad (1)$$
$$10k = 3m \quad (2)$$

Carry out. We use substitution to solve the system of equations. First we substitute $40 + k$ for m in (2).

$$10k = 3m \quad (2)$$
$$10k = 3(40 + k) \quad \text{Substituting}$$
$$10k = 120 + 3k$$
$$7k = 120$$
$$k = \frac{120}{7}$$

Although the problem asks only for k, the amount of Kona coffee that should be used, we will also find m in order to check the answer.

$$40 + k = m \quad (1)$$
$$40 + \frac{120}{7} = m \quad \text{Substituting } \frac{120}{7} \text{ for } k$$
$$\frac{280}{7} + \frac{120}{7} = m$$
$$\frac{400}{7} = m$$

Check. If $\frac{400}{7}$ lb of coffee contain $\frac{120}{7}$ lb of Kona coffee, then the percent of Kona beans in the mixture is $\frac{120/7}{400/7} = \frac{120}{7} \cdot \frac{7}{400} = \frac{3}{10}$, or 30%. The answer checks.

State. $\frac{120}{7}$ lb of Kona coffee should be added to the Mexican coffee.

56. 1.8 L

57. Familiarize. Let d = the distance, in km, that Natalie jogs in a trip to school, and let t = the time, in hr, that she jogs. We organize the information in a table.

	Distance	Rate	Time
Jogging	d	8	t
Walking	$6 - d$	4	$1 - t$

Translate. Using $d = rt$ in each row of the table we get a system of equations:

$$d = 8t, \quad (1)$$
$$6 - d = 4(1 - t) \quad (2)$$

Carry out. We use substitution to solve the system of equations.

$$6 - 8t = 4(1 - t) \quad \text{Substituting } 8t \text{ for } d \text{ in (2)}$$
$$6 - 8t = 4 - 4t$$
$$2 - 8t = -4t$$
$$2 = 4t$$
$$\frac{1}{2} = t$$

Substitute $\frac{1}{2}$ for t in (1).

$$d = 8 \cdot \frac{1}{2} = 4$$

Check. If Natalie jogs 4 km in $\frac{1}{2}$ hr, then she walks $6 - 4$ or 2 km, in $1 - \frac{1}{2}$, or $\frac{1}{2}$ hr. At a rate of 8 km/h, in $\frac{1}{2}$ hr she can jog $8 \cdot \frac{1}{2}$, or 4 km. At a rate of 4 km/h, in $\frac{1}{2}$ hr she can walk $4 \cdot \frac{1}{2}$, or 2 km. Then the total time is $\frac{1}{2}$ hr + $\frac{1}{2}$ hr, or 1 hr, and the total distance is 4 km + 2 km, or 6 km. The answer checks.

State. Natalie jogs 4 km in a trip to school.

58. 180 members

59. Familiarize. Let x = the ten's digit and y = the unit's digit. Then the number is $10x + y$. If the digits are interchanged, the new number is $10y + x$.

Translate.

Ten's digit is 2 more than 3 times unit's digit.

$$x \quad = 2 \quad + \quad 3 \quad \cdot \quad y$$

If the digits are interchanged,

$\underbrace{\text{new number}}$ is half of $\underbrace{\text{given number}}$ minus 13.

$$10y + x = \frac{1}{2} \cdot (10x + y) - 13$$

The system of equations is

$$x = 2 + 3y, \qquad (1)$$

$$10y + x = \frac{1}{2}(10x + y) - 13 \quad (2)$$

Carry out. We use the substitution method. Substitute $2 + 3y$ for x in (2).

$$10y + (2 + 3y) = \frac{1}{2}[10(2 + 3y) + y] - 13$$

$$13y + 2 = \frac{1}{2}[20 + 30y + y] - 13$$

$$13y + 2 = \frac{1}{2}[20 + 31y] - 13$$

$$13y + 2 = 10 + \frac{31}{2}y - 13$$

$$13y + 2 = \frac{31}{2}y - 3$$

$$5 = \frac{5}{2}y$$

$$2 = y$$

$$x = 2 + 3 \cdot 2 \quad \text{Substituting 2 for } y \text{ in (1)}$$

$$x = 2 + 6$$

$$x = 8$$

Check. If $x = 8$ and $y = 2$, the given number is 82 and the new number is 28. In the given number the ten's digit, 8, is two more than three times the unit's digit, 2. The new number is 13 less than one-half the given number: $28 = \frac{1}{2}(82) - 13$. The values check.

State. The given integer is 82.

60. First train: 36 km/h; second train: 54 km/h

61. Familiarize. Let $x =$ the number of gallons of pure brown and $y =$ the number of gallons of neutral stain that should be added to the original 0.5 gal. Note that a total of 1 gal of stain needs to be added to bring the amount of stain up to 1.5 gal. The original 0.5 gal of stain contains 20%(0.5 gal), or 0.2(0.5 gal) = 0.1 gal of brown stain. The final solution contains 60%(1.5 gal), or 0.6(1.5 gal) = 0.9 gal of brown stain. This is composed of the original 0.1 gal and the x gal that are added.

Translate.

$\underbrace{\text{The amount of stain added}}$ was 1 gal.

$$x + y = 1$$

$\underbrace{\begin{array}{c}\text{The amount of brown stain}\\\text{in the final solution}\end{array}}$ is 0.9 gal.

$$0.1 + x = 0.9$$

We have a system of equations.

$$x + y = 1, \qquad (1)$$

$$0.1 + x = 0.9 \quad (2)$$

Carry out. First we solve (2) for x.

$$0.1 + x = 0.9$$

$$x = 0.8$$

Then substitute 0.8 for x in (1) and solve for y.

$$0.8 + y = 1$$

$$y = 0.2$$

Check. Total amount of stain: $0.5 + 0.8 + 0.2 = 1.5$ gal

Total amount of brown stain: $0.1 + 0.8 = 0.9$ gal

Total amount of neutral stain: $0.8(0.5) + 0.2 = 0.4 + 0.2 = 0.6$ gal $= 0.4(1.5$ gal)

The answer checks.

State. 0.8 gal of pure brown and 0.2 gal of neutral stain should be added.

62. City: 261 miles, highway: 204 miles

63. Observe that if 100% acetone is added to water to create a 10% acetone solution, then the ratio of acetone to water is 10% to 90%, or 10 to 90, or 1 to 9. Thus, for each liter of acetone, 9 liters of water are required. If 5 extra liters of acetone are added to the vat, then $9 \cdot 5$, or 45 L of additional water must be added to bring the concentration down to 10%.

64. 4 boys, 3 girls

65. The 1.5 gal mixture contains $0.1 + x$ gal of pure brown stain. (See Exercise 61.). Thus, the function $P(x) = \dfrac{0.1 + x}{1.5}$ gives the percentage of brown in the mixture as a decimal quantity. Using the Intersect feature, we confirm that when $x = 0.8$, then $P(x) = 0.6$ or 60%.

Exercise Set 3.4

1. Substitute $(2, -1, -2)$ into the three equations, using alphabetical order.

$$\begin{array}{c|c}\underline{x + y - 2z = 5}\\ 2 + (-1) - 2(-2) ~ ? ~ 5\\ 2 - 1 + 4\\ 5 ~\big|~ 5 \quad \text{TRUE}\end{array}$$

$$\overline{\qquad 2x - y - z = 7 \qquad}$$
$$\frac{2 \cdot 2 - (-1) - (-2) \;?\; 7}{4 + 1 + 2 \;\Big|}$$
$$7 \;\Big|\; 7 \quad \text{TRUE}$$

$$\overline{\qquad -x - 2y + 3z = 6 \qquad}$$
$$\frac{-2 - 2(-1) + 3(-2) \;?\; 6}{-2 + 2 - 6 \;\Big|}$$
$$-6 \;\Big|\; 6 \quad \text{FALSE}$$

The triple $(2, -1, -2)$ does not make the third equation true, so it is not a solution of the system.

2. Yes

3. $x + y + z = 6, \quad (1)$
$2x - y + 3z = 9, \quad (2)$
$-x + 2y + 2z = 9 \quad (3)$

1., 2. The equations are already in standard form with no fractions or decimals.

3. Add Equations (1) and (2) to eliminate y:
$$x + y + z = 6 \quad (1)$$
$$\underline{2x - y + 3z = 9 \quad (2)}$$
$$3x + 4z = 15 \quad (4) \quad \text{Adding}$$

4. Use a different pair of equations and eliminate y:
$$4x - 2y + 6z = 18 \quad \text{Multiplying (2) by 2}$$
$$\underline{-x + 2y + 2z = 9 \quad (3)}$$
$$3x + 8z = 27 \quad (5)$$

5. Now solve the system of Equations (4) and (5).
$$3x + 4z = 15 \quad (4)$$
$$3x + 8z = 27 \quad (5)$$

$$-3x - 4z = -15 \quad \text{Multiplying (4) by } -1$$
$$\underline{3x + 8z = 27}$$
$$4z = 12$$
$$z = 3$$

$$3x + 4 \cdot 3 = 15 \quad \text{Substituting 3 for } z \text{ in (4)}$$
$$3x + 12 = 15$$
$$3x = 3$$
$$x = 1$$

6. Substitute in one of the original equations to find y.

$$1 + y + 3 = 6 \quad \text{Substituting 1 for } x \text{ and 3}$$
$$\phantom{1 + y + 3 = 6 \quad \text{Substit}} \text{for } z \text{ in (1)}$$
$$y + 4 = 6$$
$$y = 2$$

We obtain $(1, 2, 3)$. This checks, so it is the solution.

4. $(4, 0, 2)$

5. $2x - y - 3z = -1, \quad (1)$
$2x - y + z = -9, \quad (2)$
$x + 2y - 4z = 17 \quad (3)$

1., 2. The equations are already in standard form with no fractions or decimals.

3., 4. We eliminate z from two different pairs of equations.
$$2x - y - 3z = -1 \quad (1)$$
$$\underline{6x - 3y + 3z = -27 \quad \text{Multiplying (2) by 3}}$$
$$8x - 4y = -28 \quad (4) \quad \text{Adding}$$

$$8x - 4y + 4z = -36 \quad \text{Multiplying (2) by 4}$$
$$\underline{x + 2y - 4z = 17 \quad (3)}$$
$$9x - 2y = -19 \quad (5) \quad \text{Adding}$$

5. Now solve the system of Equations (4) and (5).
$$8x - 4y = -28 \quad (4)$$
$$9x - 2y = -19 \quad (5)$$

$$8x - 4y = -28 \quad (4)$$
$$\underline{-18x + 4y = 38 \quad \text{Multiplying (5) by } -2}$$
$$-10x = 10 \quad \text{Adding}$$
$$x = -1$$

$$8(-1) - 4y = -28 \quad \text{Substituting } -1 \text{ for } x \text{ in}$$
$$\phantom{8(-1) - 4y = -28 \quad \text{Substit}} (4)$$
$$-8 - 4y = -28$$
$$-4y = -20$$
$$y = 5$$

6. Substitute in one of the original equations to find z.
$$2(-1) - 5 + z = -9 \quad \text{Substituting } -1 \text{ for } x$$
$$\phantom{2(-1) - 5 + z = -9 \quad \text{Substit}} \text{and 5 for } y \text{ in (2)}$$
$$-2 - 5 + z = -9$$
$$-7 + z = -9$$
$$z = -2$$

We obtain $(-1, 5, -2)$. This checks, so it is the solution.

6. $(2, -2, 2)$

7. $2x - 3y + z = 5,$ (1)

$\quad x + 3y + 8z = 22,$ (2)

$\quad 3x - y + 2z = 12$ (3)

1., 2. The equations are already in standard form with no fractions or decimals.

3., 4. We eliminate y from two different pairs of equations.

$$
\begin{array}{l}
2x - 3y + z = 5 \quad (1) \\
\underline{x + 3y + 8z = 22 \quad (2)} \\
3x \quad\quad + 9z = 27 \quad (4) \quad \text{Adding}
\end{array}
$$

$$
\begin{array}{l}
x + 3y + 8z = 22 \quad (2) \\
\underline{9x - 3y + 6z = 36} \quad \text{Multiplying (3) by 3} \\
10x \quad\quad + 14z = 58 \quad (5) \quad \text{Adding}
\end{array}
$$

5. Solve the system of Equations (4) and (5).

$$
\begin{array}{l}
3x + 9z = 27 \quad (4) \\
10x + 14z = 58 \quad (5)
\end{array}
$$

$$
\begin{array}{l}
30x + 90z = 270 \quad \text{Multiplying (4) by 10} \\
\underline{-30x - 42z = -174} \quad \text{Multiplying (5) by} -3 \\
48z = 96 \quad \text{Adding} \\
z = 2
\end{array}
$$

$3x + 9 \cdot 2 = 27$ Substituting 2 for z in (4)

$3x + 18 = 27$

$3x = 9$

$x = 3$

6. Substitute in one of the original equations to find y.

$2 \cdot 3 - 3y + 2 = 5$ Substituting 3 for x and 2 for z in (1)

$-3y + 8 = 5$

$-3y = -3$

$y = 1$

We obtain $(3, 1, 2)$. This checks, so it is the solution.

8. $(3, -2, 1)$

9. $3a - 2b + 7c = 13,$ (1)

$\quad a + 8b - 6c = -47,$ (2)

$\quad 7a - 9b - 9c = -3$ (3)

1., 2. The equations are already in standard form with no fractions or decimals.

3., 4. We eliminate a from two different pairs of equations.

$$
\begin{array}{l}
3a - 2b + 7c = 13 \quad (1) \\
\underline{-3a - 24b + 18c = 141} \quad \text{Multiplying (2) by} -3 \\
- 26b + 25c = 154 \quad (4) \quad \text{Adding}
\end{array}
$$

$$
\begin{array}{l}
-7a - 56b + 42c = 329 \quad \text{Multiplying (2) by} -7 \\
\underline{7a - 9b - 9c = -3} \quad (3) \\
- 65b + 33c = 326 \quad (5) \quad \text{Adding}
\end{array}
$$

5. Now solve the system of Equations (4) and (5).

$$
\begin{array}{l}
-26b + 25c = 154 \quad (4) \\
-65b + 33c = 326 \quad (5)
\end{array}
$$

$$
\begin{array}{l}
-130b + 125c = 770 \quad \text{Multiplying (4) by 5} \\
\underline{130b - 66c = -652} \quad \text{Multiplying (5) by} -2 \\
59c = 118 \\
c = 2
\end{array}
$$

$-26b + 25 \cdot 2 = 154$ Substituting 2 for c in (4)

$-26b + 50 = 154$

$-26b = 104$

$b = -4$

6. Substitute in one of the original equations to find a.

$a + 8(-4) - 6(2) = -47$ Substituting -4 for b and 2 for c in (2)

$a - 32 - 12 = -47$

$a - 44 = -47$

$a = -3$

We obtain $(-3, -4, 2)$. This checks, so it is the solution.

10. $(7, -3, -4)$

11. $2x + 3y + z = 17,$ (1)

$\quad x - 3y + 2z = -8,$ (2)

$\quad 5x - 2y + 3z = 5$ (3)

1., 2. The equations are already in standard form with no fractions or decimals.

3., 4. We eliminate y from two different pairs of equations.

$$
\begin{array}{l}
2x + 3y + z = 17 \quad (1) \\
\underline{x - 3y + 2z = -8} \quad (2) \\
3x \quad\quad + 3z = 9 \quad (4) \quad \text{Adding}
\end{array}
$$

$$
\begin{array}{l}
4x + 6y + 2z = 34 \quad \text{Multiplying (1) by 2} \\
\underline{15x - 6y + 9z = 15} \quad \text{Multiplying (3) by 3} \\
19x \quad\quad + 11z = 49 \quad (5) \quad \text{Adding}
\end{array}
$$

5. Now solve the system of Equations (4) and (5).

$$3x + 3z = 9 \quad (4)$$
$$19x + 11z = 49 \quad (5)$$

$$\begin{array}{ll} 33x + 33z = 99 & \text{Multiplying (4) by 11} \\ -57x - 33z = -147 & \text{Multiplying (5) by } -3 \\ \hline -24x = -48 \\ x = 2 \end{array}$$

$$3 \cdot 2 + 3z = 9 \quad \text{Substituting 2 for } x \text{ in (4)}$$
$$6 + 3z = 9$$
$$3z = 3$$
$$z = 1$$

6. Substitute in one of the original equations to find y.

$$2 \cdot 2 + 3y + 1 = 17 \quad \text{Substituting 2 for } x \text{ and}$$
$$ 1 \text{ for } z \text{ in (1)}$$
$$3y + 5 = 17$$
$$3y = 12$$
$$y = 4$$

We obtain $(2, 4, 1)$. This checks, so it is the solution.

2. $(2, 1, 3)$

3.
$$2x + y + z = -2, \quad (1)$$
$$2x - y + 3z = 6, \quad (2)$$
$$3x - 5y + 4z = 7 \quad (3)$$

1., 2. The equations are already in standard form with no fractions or decimals.

3., 4. We eliminate y from two different pairs of equations.

$$\begin{array}{ll} 2x + y + z = -2 & (1) \\ 2x - y + 3z = 6 & (2) \\ \hline 4x + 4z = 4 & (4) \quad \text{Adding} \end{array}$$

$$\begin{array}{ll} 10x + 5y + 5z = -10 & \text{Multiplying (1) by 5} \\ 3x - 5y + 4z = 7 & (3) \\ \hline 13x + 9z = -3 & (5) \quad \text{Adding} \end{array}$$

5. Now solve the system of Equations (4) and (5).

$$4x + 4z = 4 \quad (4)$$
$$13x + 9z = -3 \quad (5)$$

$$\begin{array}{ll} 36x + 36z = 36 & \text{Multiplying (4) by 9} \\ -52x - 36z = 12 & \text{Multiplying (5) by } -4 \\ \hline -16x = 48 & \text{Adding} \\ x = -3 \end{array}$$

$$4(-3) + 4z = 4 \quad \text{Substituting } -3 \text{ for } x \text{ in (4)}$$
$$-12 + 4z = 4$$
$$4z = 16$$
$$z = 4$$

6. Substitute in one of the original equations to find y.

$$2(-3) + y + 4 = -2 \quad \text{Substituting } -3 \text{ for}$$
$$ x \text{ and 4 for } z \text{ in (1)}$$
$$y - 2 = -2$$
$$y = 0$$

We obtain $(-3, 0, 4)$. This checks, so it is the solution.

14. $(2, -5, 6)$

15.
$$x - y + z = 4, \quad (1)$$
$$5x + 2y - 3z = 2, \quad (2)$$
$$4x + 3y - 4z = -2 \quad (3)$$

1., 2. The equations are already in standard form with no fractions or decimals.

3., 4. We eliminate z from two different pairs of equations.

$$\begin{array}{ll} 3x - 3y + 3z = 12 & \text{Multiplying (1) by 3} \\ 5x + 2y - 3z = 2 & (2) \\ \hline 8x - y = 14 & (4) \quad \text{Adding} \end{array}$$

$$\begin{array}{ll} 4x - 4y + 4z = 16 & \text{Multiplying (1) by 4} \\ 4x + 3y - 4z = -2 & (3) \\ \hline 8x - y = 14 & (5) \quad \text{Adding} \end{array}$$

5. Now solve the system of Equations (4) and (5).

$$8x - y = 14 \quad (4)$$
$$8x - y = 14 \quad (5)$$

$$\begin{array}{ll} 8x - y = 14 & (4) \\ -8x + y = -14 & \text{Multiplying (5) by } -1 \\ \hline 0 = 0 & (6) \end{array}$$

Equation (6) indicates Equations (1), (2), and (3) are dependent. (Note that if Equation (1) is subtracted from Equation (2), the result is Equation (3).) We could also have concluded that the equations are dependent by observing that Equations (4) and (5) are identical.

16. The equations are dependent.

17.
$$a + 2b + c = 1, \quad (1)$$
$$7a + 3b - c = -2, \quad (2)$$
$$a + 5b + 3c = 2 \quad (3)$$

1., 2. The equations are already in standard form with no fractions or decimals.

3., 4. We eliminate c from two different pairs of equations.

$$a + 2b + c = 1 \quad (1)$$
$$\underline{7a + 3b - c = -2} \quad (2)$$
$$8a + 5b = -1 \quad (4)$$

$$21a + 9b - 3c = -6 \quad \text{Multiplying (2) by 3}$$
$$\underline{a + 5b + 3c = 2}$$
$$22a + 14b = -4 \quad (5)$$

5. Now solve the system of Equations (4) and (5).

$$8a + 5b = -1 \quad (4)$$
$$22a + 14b = -4 \quad (5)$$

$$112a + 70b = -14 \quad \text{Multiplying (4) by 14}$$
$$\underline{-110a - 70b = 20} \quad \text{Multiplying (5) by } -5$$
$$2a = 6$$
$$a = 3$$

$$8 \cdot 3 + 5b = -1 \quad \text{Substituting in (4)}$$
$$24 + 5b = -1$$
$$5b = -25$$
$$b = -5$$

6. Substitute in one of the original equations to find c.

$$3 + 2(-5) + c = 1 \quad \text{Substituting in (1)}$$
$$-7 + c = 1$$
$$c = 8$$

We obtain $(3, -5, 8)$. This checks, so it is the solution.

18. $\left(\dfrac{1}{2}, 4, -6\right)$

19. $5x + 3y + \dfrac{1}{2}z = \dfrac{7}{2},$
$0.5x - 0.9y - 0.2z = 0.3,$
$3x - 2.4y + 0.4z = -1$

1. All equations are already in standard form.

2. Multiply the first equation by 2 to clear the fractions. Also, multiply the second and third equations by 10 to clear the decimals.

$$10x + 6y + z = 7, (1)$$
$$5x - 9y - 2z = 3, (2)$$
$$30x - 24y + 4z = -10 \quad (3)$$

3., 4. We eliminate z from two different pairs of equations.

$$20x + 12y + 2z = 14 \quad \text{Multiplying (1) by 2}$$
$$\underline{5x - 9y - 2z = 3} \quad (2)$$
$$25x + 3y = 17 \quad (4)$$

$$10x - 18y - 4z = 6 \quad \text{Multiplying(2) by 2}$$
$$\underline{30x - 24y + 4z = -10} \quad (3)$$
$$40x - 42y = -4 \quad (5)$$

5. Now solve the system of Equations (4) and (5).

$$25x + 3y = 17 \quad (4)$$
$$40x - 42y = -4 \quad (5)$$

$$350x + 42y = 238 \quad \text{Multiplying (4) by 14}$$
$$\underline{40x - 42y = -4} \quad (5)$$
$$390x = 234$$
$$x = \dfrac{3}{5}$$

$$25\left(\dfrac{3}{5}\right) + 3y = 17 \quad \text{Substituting in (4)}$$
$$15 + 3y = 17$$
$$3y = 2$$
$$y = \dfrac{2}{3}$$

6. Substitute in one of the original equations to find z.

$$10\left(\dfrac{3}{5}\right) + 6\left(\dfrac{2}{3}\right) + z = 7 \quad \text{Substituting in (1)}$$
$$6 + 4 + z = 7$$
$$10 + z = 7$$
$$z = -3$$

We obtain $\left(\dfrac{3}{5}, \dfrac{2}{3}, -3\right)$. This checks, so it is the solution.

20. $\left(\dfrac{1}{2}, \dfrac{1}{3}, \dfrac{1}{6}\right)$

21. $3p + 2r = 11, \quad (1)$
$ q - 7r = 4, \quad (2)$
$p - 6q = 1 \quad (3)$

1., 2. The equations are already in standard form with no fractions or decimals.

3., 4. Note that there is no q in Equation (1). We will use Equations (2) and (3) to obtain another equation with no q-term.

$$6q - 42r = 24 \quad \text{Multiplying (2) by 6}$$
$$\underline{p - 6q = 1} \quad (3)$$
$$p - 42r = 25 \quad (4)$$

5. Solve the system of Equations (1) and (4).

$$3p + 2r = 11 \quad (1)$$
$$p - 42r = 25 \quad (4)$$

$3p +\quad 2r =\quad 11\quad (1)$

$\underline{-3p + 126r = -75}$ Multiplying (4) by -3

$\qquad 128r = -64$

$\qquad\qquad r = -\dfrac{1}{2}$

$3p + 2\left(-\dfrac{1}{2}\right) = 11$ Substituting in (1)

$\qquad\quad 3p - 1 = 11$

$\qquad\qquad 3p = 12$

$\qquad\qquad\ p = 4$

6. Substitute in Equation (2) or (3) to find q.

$q - 7\left(-\dfrac{1}{2}\right) = 4$ Substituting in (2)

$\qquad\ q + \dfrac{7}{2} = 4$

$\qquad\qquad q = \dfrac{1}{2}$

We obtain $\left(4, \dfrac{1}{2}, -\dfrac{1}{2}\right)$. This checks, so it is the solution.

2. $\left(\dfrac{1}{2}, \dfrac{2}{3}, -\dfrac{5}{6}\right)$

3. $\quad x +\quad y + z = 105,\quad (1)$

$\qquad\quad 10y - z = 11,\quad (2)$

$\ 2x -\quad 3y \quad = 7 \qquad (3)$

1., 2. The equations are already in standard form with no fractions or decimals.

3., 4. Note that there is no z in Equation (3). We will use Equations (1) and (2) to obtain another equation with no z-term.

$x +\quad y + z = 105\quad (1)$

$\underline{\quad 10y - z =\ 11\quad (2)}$

$x + 11y \qquad = 116\quad (4)$

5. Now solve the system of Equations (3) and (4).

$2x - 3y = 7 \qquad (3)$

$x + 11y = 116\quad (4)$

$2x -\quad 3y =\qquad 7\ \ (3)$

$\underline{-2x - 22y = -232}$ Multiplying (4) by -2

$\quad - 25y = -225$

$\qquad\quad y =\qquad 9$

$x + 11 \cdot 9 = 116$ Substituting in (4)

$\quad x + 99 = 116$

$\qquad\quad x = 17$

6. Substitute in Equation (1) or (2) to find z.

$17 + 9 + z = 105$ Substituting in (1)

$\qquad 26 + z = 105$

$\qquad\qquad z = 79$

We obtain $(17, 9, 79)$. This checks, so it is the solution.

24. $(15, 33, 9)$

25. $\quad 2a -\quad 3b \qquad = 2,\quad (1)$

$\qquad 7a \qquad + 4c = \dfrac{3}{4},\quad (2)$

$\qquad\quad -3b + 2c = 1\quad (3)$

1. The equations are already in standard form.

2. Multiply Equation (2) by 4 to clear the fraction. The resulting system is

$2a -\quad 3b \qquad = 2,\quad (1)$

$28a \qquad + 16c = 3,\quad (4)$

$\qquad -3b +\ 2c = 1\quad (3)$

3. Note that there is no b in Equation (2). We will use Equations (1) and (3) to obtain another equation with no b-term.

$2a - 3b \qquad = \quad 2\ \ (1)$

$\underline{\qquad 3b - 2c = -1}$ Multiplying (3) by -1

$2a \qquad - 2c =\quad 1\ \ (5)$

5. Now solve the system of Equations (4) and (5).

$28a + 16c = 3\quad (4)$

$2a -\quad 2c = 1\quad (5)$

$28a + 16c =\quad 3\quad (4)$

$\underline{16a - 16c =\quad 8}$ Multiplying (5) by 8

$44a \qquad = 11$

$\qquad\ a = \dfrac{1}{4}$

$2 \cdot \dfrac{1}{4} - 2c = 1$ Substituting $\dfrac{1}{4}$ for a in (5)

$\qquad \dfrac{1}{2} - 2c = 1$

$\qquad\quad -2c = \dfrac{1}{2}$

$\qquad\qquad c = -\dfrac{1}{4}$

6. Substitute in Equation (1) or (2) to find b.

$$2\left(\frac{1}{4}\right) - 3b = 2 \quad \text{Substituting } \frac{1}{4} \text{ for } a \text{ in (1)}$$

$$\frac{1}{2} - 3b = 2$$

$$-3b = \frac{3}{2}$$

$$b = -\frac{1}{2}$$

We obtain $\left(\frac{1}{4}, -\frac{1}{2}, -\frac{1}{4}\right)$. This checks, so it is the solution.

26. $(3, 4, -1)$

27. $x + y + z = 182, \quad (1)$
$ y = 2 + 3x, \quad (2)$
$ z = 80 + x \quad (3)$

Observe, from Equations (2) and (3), that we can substitute $2 + 3x$ for y and $80 + x$ for z in Equation (1) and solve for x.

$$x + y + x = 182$$
$$x + (2 + 3x) + (80 + x) = 182$$
$$5x + 82 = 182$$
$$5x = 100$$
$$x = 20$$

Now substitute 20 for x in Equation (2).

$$y = 2 + 3x = 2 + 3 \cdot 20 = 2 + 60 = 62$$

Finally, substitute 20 for x in Equation (3).

$$z = 80 + x = 80 + 20 = 100.$$

We obtain $(20, 62, 100)$. This checks, so it is the solution.

28. $(2, 5, -3)$

29. $x + y = 0, \quad (1)$
$x + z = 1, \quad (2)$
$2x + y + z = 2 \quad (3)$

1., 2. The equations are already in standard form with no fractions or decimals.

3., 4. Note that there is no z in Equation (1). We will use Equations (2) and (3) to obtain another equation with no z-term.

$$\begin{array}{l} -x - z = -1 \quad \text{Multiplying (2) by } -1 \\ \underline{2x + y + z = 2 \quad (3)} \\ x + y = 1 \quad (4) \end{array}$$

5. Now solve the system of Equations (1) and (4).

$$x + y = 0 \quad (1)$$
$$x + y = 1 \quad (4)$$

$$\begin{array}{l} x + y = 0 \quad (1) \\ \underline{-x - y = -1 \quad \text{Multiplying (4) by } -1} \\ 0 = -1 \quad \text{Adding} \end{array}$$

We get a false equation, or contradiction. There is ▮ solution.

30. No solution

31. $y + z = 1, \quad (1)$
$x + y + z = 1, \quad (2)$
$x + 2y + 2z = 2 \quad (3)$

1., 2. The equations are already in standard form with no fractions or decimals.

3., 4. Note that there is no x in Equation (1). We will use Equations (2) and (3) to obtain another equation with no x-term.

$$\begin{array}{l} -x - y - z = -1 \quad \text{Multiplying (2)} \\ \text{by } -1 \\ \underline{x + 2y + 2z = 2 \quad (3)} \\ y + z = 1 \quad (4) \end{array}$$

Equations (1) and (4) are identical. This means that Equations (1), (2), and (3) are dependent. (We have seen tha if Equation (2) is multiplied by -1 and added to Equatic (3), the result is Equation (1).)

32. The equations are dependent.

33. *Writing Exercise*

34. *Writing Exercise*

35. Let x represent the larger number and y represent th smaller number. Then we have $x = 2y$.

36. Let x represent the first number and y represent the secon number; $x + y = 3x$

37. Let x, $x + 1$, and $x + 2$ represent the numbers. Then w have $x + (x + 1) + (x + 2) = 45$.

38. Let x and y represent the numbers; $x + 2y = 17$

39. Let x and y represent the first two numbers and let represent the third number. Then we have $x + y = 5z$.

40. Let x and y represent the numbers; $xy = 2(x + y)$

41. *Writing Exercise*

42. *Writing Exercise*

3. $\dfrac{x+2}{3} - \dfrac{y+4}{2} + \dfrac{z+1}{6} = 0,$

$\dfrac{x-4}{3} + \dfrac{y+1}{4} - \dfrac{z-2}{2} = -1,$

$\dfrac{x+1}{2} + \dfrac{y}{2} + \dfrac{z-1}{4} = \dfrac{3}{4}$

1., 2. We clear fractions and write each equation in standard form.

To clear fractions, we multiply both sides of each equation by the LCM of its denominators. The LCM's are 6, 12, and 4, respectively.

$$6\left(\dfrac{x+2}{3} - \dfrac{y+4}{2} + \dfrac{z+1}{6}\right) = 6\cdot 0$$

$$2(x+2) - 3(y+4) + (z+1) = 0$$

$$2x + 4 - 3y - 12 + z + 1 = 0$$

$$2x - 3y + z = 7$$

$$12\left(\dfrac{x-4}{3} + \dfrac{y+1}{4} - \dfrac{z-2}{2}\right) = 12\cdot(-1)$$

$$4(x-4) + 3(y+1) - 6(z-2) = -12$$

$$4x - 16 + 3y + 3 - 6z + 12 = -12$$

$$4x + 3y - 6z = -11$$

$$4\left(\dfrac{x+1}{2} + \dfrac{y}{2} + \dfrac{z-1}{4}\right) = 4\cdot\dfrac{3}{4}$$

$$2(x+1) + 2(y) + (z-1) = 3$$

$$2x + 2 + 2y + z - 1 = 3$$

$$2x + 2y + z = 2$$

The resulting system is

$2x - 3y + z = 7,$ (1)

$4x + 3y - 6z = -11,$ (2)

$2x + 2y + z = 2$ (3)

3., 4. We eliminate z from two different pairs of equations.

$12x - 18y + 6z = 42$ Multiplying (1) by 6

$\underline{4x + 3y - 6z = -11}$ (2)

$16x - 15y = 31$ (4)   Adding

$2x - 3y + z = 7$ (1)

$\underline{-2x - 2y - z = -2}$ Multiplying (3) by -1

$-5y = 5$ (5)   Adding

5. Solve (5) for y: $-5y = 5$

$y = -1$

Substitute -1 for y in (4):

$16x - 15(-1) = 31$

$16x + 15 = 31$

$16x = 16$

$x = 1$

6. Substitute 1 for x and -1 for y in (1):

$2\cdot 1 - 3(-1) + z = 7$

$5 + z = 7$

$z = 2$

We obtain $(1, -1, 2)$. This checks, so it is the solution.

44. $(1, -2, 4, -1)$

45. $w + x - y + z = 0,$ (1)

$w - 2x - 2y - z = -5,$ (2)

$w - 3x - y + z = 4,$ (3)

$2w - x - y + 3z = 7$ (4)

The equations are already in standard form with no fractions or decimals.

Start by eliminating z from three different pairs of equations.

$w + x - y + z = 0$ (1)

$\underline{w - 2x - 2y - z = -5}$ (2)

$2w - x - 3y = -5$ (5)   Adding

$w - 2x - 2y - z = -5$ (2)

$\underline{w - 3x - y + z = 4}$ (3)

$2w - 5x - 3y = -1$ (6)   Adding

$3w - 6x - 6y - 3z = -15$ Multiplying (2) by 3

$\underline{2w - x - y + 3z = 7}$ (4)

$5w - 7x - 7y = -8$ (7)   Adding

Now solve the system of equations (5), (6), and (7).

$2w - x - 3y = -5,$ (5)

$2w - 5x - 3y = -1,$ (6)

$5w - 7x - 7y = -8.$ (7)

$2w - x - 3y = -5$ (5)

$\underline{-2w + 5x + 3y = 1}$ Multiplying (6) by -1

$4x = -4$

$x = -1$

Substituting -1 for x in (5) and (7) and simplifying, we have

$2w - 3y = -6,$ (8)

$5w - 7y = -15.$ (9)

Now solve the system of Equations (8) and (9).

$10w - 15y = -30$ Multiplying (8) by 5

$\underline{-10w + 14y = 30}$ Multiplying (9) by -2

$-y = 0$

$y = 0$

Substitute 0 for y in Equation (8) or (9) and solve for w.

$2w - 3 \cdot 0 = -6$ Substituting in (8)

$2w = -6$

$w = -3$

Substitute in one of the original equations to find z.

$-3 - 1 - 0 + z = 0$ Substituting in (1)

$-4 + z = 0$

$z = 4$

We obtain $(-3, -1, 0, 4)$. This checks, so it is the solution.

46. $\left(-1, \dfrac{1}{5}, -\dfrac{1}{2}\right)$

47.
$$\dfrac{2}{x} + \dfrac{2}{y} - \dfrac{3}{z} = 3,$$
$$\dfrac{1}{x} - \dfrac{2}{y} - \dfrac{3}{z} = 9,$$
$$\dfrac{7}{x} - \dfrac{2}{y} + \dfrac{9}{z} = -39$$

Let u represent $\dfrac{1}{x}$, v represent $\dfrac{1}{y}$, and w represent $\dfrac{1}{z}$. Substituting, we have

$2u + 2v - 3w = 3$ (1)

$u - 2v - 3w = 9$ (2)

$7u - 2v + 9w = -39$ (3)

1., 2. The equations in u, v, and w are in standard form with no fractions or decimals.

3., 4. We eliminate v from two different pairs of equations.

$\begin{array}{ll} 2u + 2v - 3w = 3 & (1) \\ \underline{u - 2v - 3w = 9} & (2) \\ 3u \quad\quad - 6w = 12 & (4) \text{ Adding} \end{array}$

$\begin{array}{ll} 2u + 2v - 3w = 3 & (1) \\ \underline{7u - 2v + 9w = -39} & (3) \\ 9u \quad\quad + 6w = -36 & (5) \text{ Adding} \end{array}$

5. Now solve the system of Equations (4) and (5).

$\begin{array}{ll} 3u - 6w = 12, & (4) \\ \underline{9u + 6w = -36} & (5) \\ 12u \quad\quad = -24 \\ \quad u = -2 \end{array}$

$3(-2) - 6w = 12$ Substituting in (4)

$-6 - 6w = 12$

$-6w = 18$

$w = -3$

6. Substitute in Equation (1), (2), or (3) to find v.

$2(-2) + 2v - 3(-3) = 3$ Substituting in (1)

$2v + 5 = 3$

$2v = -2$

$v = -1$

Solve for x, y, and z. We substitute -2 for u, -1 for v, and -3 for w.

$$u = \dfrac{1}{x} \quad\quad v = \dfrac{1}{y} \quad\quad w = \dfrac{1}{z}$$

$$-2 = \dfrac{1}{x} \quad\quad -1 = \dfrac{1}{y} \quad\quad -3 = \dfrac{1}{z}$$

$$x = \dfrac{1}{2} \quad\quad y = -1 \quad\quad z = -\dfrac{1}{3}$$

We obtain $\left(-\dfrac{1}{2}, -1, -\dfrac{1}{3}\right)$. This checks, so it is the solution.

48. 12

49.
$5x - 6y + kz = -5,$ (1)

$x + 3y - 2z = 2,$ (2)

$2x - y + 4z = -1$ (3)

Eliminate y from two different pairs of equations.

$\begin{array}{ll} 5x - 6y + \quad kz = -5 & (1) \\ \underline{2x + 6y - \quad 4z = 4} & \text{Multiplying (2) by 2} \\ 7x \quad\quad + (k-4)z = -1 & (4) \end{array}$

$\begin{array}{ll} x + 3y - \quad 2z = 2 & (2) \\ \underline{6x - 3y + 12z = -3} & \text{Multiplying (3) by 3} \\ 7x \quad\quad + 10z = -1 & (5) \end{array}$

Solve the system of Equations (4) and (5).

$7x + (k-4)z = -1$ (4)

$7x + \quad\quad 10z = -1$ (5)

$\begin{array}{ll} -7x - \quad (k-4)z = 1 & \text{Multiplying (4) by } -1 \\ \underline{7x + \quad\quad 10z = -1} & (5) \\ (-k + 14)z = 0 & (6) \end{array}$

The system is dependent for the value of k that makes Equation (6) true. This occurs when $-k + 14$ is 0. We solve for k:

$-k + 14 = 0$

$14 = k$

50. $3x + 4y + 2z = 12$

51. $z = b - mx - ny$

Three solutions are $(1, 1, 2)$, $(3, 2, -6)$, and $\left(\dfrac{3}{2}, 1, 1\right)$. We substitute for x, y, and z and then solve for b, m, and n.

$$2 = b - m - n,$$
$$-6 = b - 3m - 2n,$$
$$1 = b - \frac{3}{2}m - n$$

1., 2. Write the equations in standard form. Also, clear the fraction in the last equation.

$$b - m - n = 2, \quad (1)$$
$$b - 3m - 2n = -6, \quad (2)$$
$$2b - 3m - 2n = 2 \quad (3)$$

3., 4. Eliminate b from two different pairs of equations.

$$\begin{array}{ll} b - m - n = 2 & (1) \\ -b + 3m + 2n = 6 & \text{Multiplying (2) by } -1 \\ \hline 2m + n = 8 & (4) \quad \text{Adding} \end{array}$$

$$\begin{array}{ll} -2b + 2m + 2n = -4 & \text{Multiplying (1) by } -2 \\ 2b - 3m - 2n = 2 & (3) \\ \hline -m = -2 & (5) \quad \text{Adding} \end{array}$$

5. We solve Equation (5) for m:

$$-m = -2$$
$$m = 2$$

Substitute in Equation (4) and solve for n.

$$2 \cdot 2 + n = 8$$
$$4 + n = 8$$
$$n = 4$$

6. Substitute in one of the original equations to find b.

$$\begin{array}{ll} b - 2 - 4 = 2 & \text{Substituting 2 for } m \\ & \text{and 4 for } n \text{ in (1)} \\ b - 6 = 2 \\ b = 8 \end{array}$$

The solution is $(8, 2, 4)$, so the equation is $z = 8 - 2x - 4y$.

2. Answers may vary.

$$x + y + z = 1,$$
$$2x + 2y + 2z = 2,$$
$$x + y + z = 3$$

Exercise Set 3.5

1. **Familiarize**. Let $x =$ the first number, $y =$ the second number, and $z =$ the third number.

Translate.

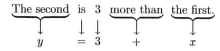

The second is 3 more than the first.

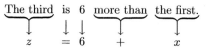

The third is 6 more than the first.

We now have a system of equations.

$$\begin{array}{lll} x + y + z = 57, & \text{or} & x + y + z = 57, \\ y = 3 + x & & -x + y = 3, \\ z = 6 + x & & -x + z = 6 \end{array}$$

Carry out. Solving the system we get $(16, 19, 22)$.

Check. The sum of the three numbers is $16 + 19 + 22$, or 57. The second number, 19, is three more than the first number, 16. The third number, 22, is 6 more than the first number, 16. The numbers check.

State. The numbers are 16, 19, and 22.

2. 4, 2, -1

3. **Familiarize**. Let $x =$ the first number, $y =$ the second number, and $z =$ the third number.

Translate.

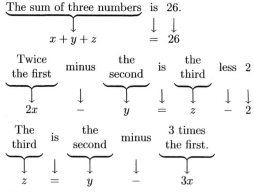

We now have a system of equations.

$$\begin{array}{lll} x + y + z = 26, & \text{or} & x + y + z = 26, \\ 2x - y = z - 2, & & 2x - y - z = -2, \\ z = y - 3x & & 3x - y + z = 0 \end{array}$$

Carry out. Solving the system we get $(8, 21, -3)$.

Check. The sum of the numbers is $8 + 21 - 3$, or 26. Twice the first minus the second is $2 \cdot 8 - 21$, or -5, which is 2 less than the third. The second minus three times the first is $21 - 3 \cdot 8$, or -3, which is the third. The numbers check.

State. The numbers are 8, 21, and -3.

4. 17, 9, 79

5. Familiarize. We first make a drawing.

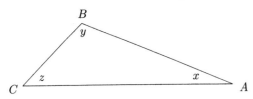

We let x, y, and z represent the measures of angles A, B, and C, respectively. The measures of the angles of a triangle add up to $180°$.

Translate.

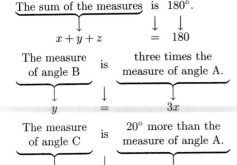

The sum of the measures is $180°$.

$$x + y + z = 180$$

The measure of angle B is three times the measure of angle A.

$$y = 3x$$

The measure of angle C is $20°$ more than the measure of angle A.

$$z = x + 20$$

We now have a system of equations.

$$x + y + z = 180,$$
$$y = 3x,$$
$$z = x + 20$$

Carry out. Solving the system we get $(32, 96, 52)$.

Check. The sum of the measures is $32° + 96° + 52°$, or $180°$. Three times the measure of angle A is $3 \cdot 32°$, or $96°$, the measure of angle B. $20°$ more than the measure of angle A is $32° + 20°$, or $52°$, the measure of angle C. The numbers check.

State. The measures of angles A, B, and C are $32°$, $96°$, and $52°$, respectively.

6. $25°$, $50°$, $105°$

7. Familiarize. Let $x =$ the cost of automatic transmission, $y =$ the cost of power door locks, and $z =$ the cost of air conditioning. The prices of the options are added to the basic price of $\$12{,}685$.

Translate.

The basic model plus automatic transmission plus

$$12{,}685 + x +$$

power door locks was $\$14{,}070$.

$$y = 14{,}070$$

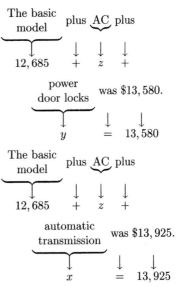

The basic model plus AC plus

$$12{,}685 + z +$$

power door locks was $\$13{,}580$.

$$y = 13{,}580$$

The basic model plus AC plus

$$12{,}685 + z +$$

automatic transmission was $\$13{,}925$.

$$x = 13{,}925$$

We now have a system of equations.

$$12{,}685 + x + y = 14{,}070,$$
$$12{,}685 + z + y = 13{,}580,$$
$$12{,}685 + z + x = 13{,}925$$

Carry out. Solving the system we get $(865, 520, 375)$.

Check. The basic model with automatic transmission an power door locks costs $\$12{,}685 + \$865 + \$520$, or $\$14{,}07$ The basic model with AC and power door locks cost $\$12{,}685 + \$375 + \$520$, or $\$13{,}580$. The basic model wit AC and automatic transmission costs $\$12{,}685 + \$375$ $\$865$, or $\$13{,}925$. The numbers check.

State. Automatic transmission costs $\$865$, power doc locks cost $\$520$, and AC costs $\$375$.

8. A: 1500; B: 1900; C: 2300

9. We know that Elrod, Dot, and Wendy can weld 74 line feet per hour when working together. We also know tha Elrod and Dot together can weld 44 linear feet per hou which leads to the conclusion that Wendy can weld $74 - 4$ or 30 linear feet per hour alone. We also know that Elr and Wendy together can weld 50 linear feet per hour. Th along with the earlier conclusion that Wendy can weld 3 linear feet per hour alone, leads to two conclusions: Elr can weld $50 - 30$, or 20 linear feet per hour alone and D can weld $74 - 50$, or 24 linear feet per hour alone.

10. Sven: 220; Tillie: 250; Isaiah: 270

11. Familiarize. Let $x =$ the number of 10-oz cups, $y =$ t number of 14-oz cups, and $z =$ the number of 20-oz cu that Kyle filled. Note that five 96-oz pots contain $5 \cdot 96$ o or 480 oz of coffee. Also, x 10-oz cups contain a total $10x$ oz of coffee and bring in $\$1.05x$, y 14-oz cups conta

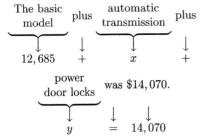

$14y$ oz and bring in $\$1.35y$, and z 20-oz cups contain $20z$ oz and bring in $\$1.65z$.

Translate.

The total number of coffees served   was   34.

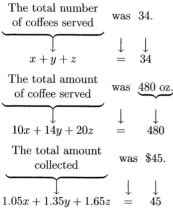

$$x + y + z = 34$$

The total amount of coffee served   was   480 oz.

$$10x + 14y + 20z = 480$$

The total amount collected   was   \$45.

$$1.05x + 1.35y + 1.65z = 45$$

Now we have a system of equations.

$$x + y + z = 34,$$
$$10x + 14y + 20z = 480,$$
$$1.05x + 1.35y + 1.65z = 45$$

Carry out. Solving the system we get $(11, 15, 8)$.

Check. The total number of coffees served was $11+15+8$, or 34, The total amount of coffee served was $10 \cdot 11 + 14 \cdot 15 + 20 \cdot 8 = 110 + 210 + 160 = 480$ oz. The total amount collected was $\$1.05(11) + \$1.35(15) + \$1.65(8) = \$11.55 + \$20.25 + \$13.20 = \$45$. The numbers check.

State. Kyle filled 11 10-oz cups, 15 14-oz cups, and 8 20-oz cups.

2. \$36 billion on television ads, \$7.7 billion on radio ads

3. Familiarize. Let $x =$ the amount invested in the first fund, $y =$ the amount invested in the second fund, and $z =$ the amount invested in the third fund. Then the earnings from the investments were $0.1x$, $0.06y$, and $0.15z$.

Translate.

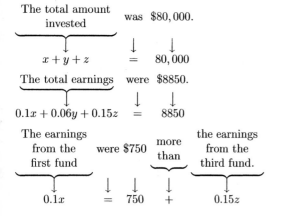

The total amount invested   was   \$80,000.

$$x + y + z = 80,000$$

The total earnings   were   \$8850.

$$0.1x + 0.06y + 0.15z = 8850$$

The earnings from the first fund   were \$750   more than   the earnings from the third fund.

$$0.1x = 750 + 0.15z$$

Now we have a system of equations.

$$x + y + z = 80,000$$
$$0.1x + 0.06y + 0.15z = 8850,$$
$$0.1x = 750 + 0.15z$$

Carry out. Solving the system we get $(45,000, 10,000, 25,000)$.

Check. The total investment was $\$45,000+ \$10,000 + \$25,000$, or \$80,000. The total earnings were $0.1(\$45,000) + 0.06(10,000) + 0.15(25,000) = \$4500 + \$600 + \$3750 = \$8850$. The earnings from the first fund, \$4500, were \$750 more than the earnings from the second fund, \$3750.

State. \$45,000 was invested in the first fund, \$10,000 in the second fund, and \$25,000 in the third fund.

14. 10 small drinks, 25 medium drinks, 5 large drinks

15. Familiarize. Let $r =$ the number of servings of roast beef, $p =$ the number of baked potatoes, and $b =$ the number of servings of broccoli. Then r servings of roast beef contain $300r$ Calories, $20r$ g of protein, and no vitamin C. In p baked potatoes there are $100p$ Calories, $5p$ g of protein, and $20p$ mg of vitamin C. And b servings of broccoli contain $50b$ Calories, $5b$ g of protein, and $100b$ mg of vitamin C. The patient requires 800 Calories, 55 g of protein, and 220 mg of vitamin C.

Translate. Write equations for the total number of calories, the total amount of protein, and the total amount of vitamin C.

$$300r + 100p + 50b = 800 \quad \text{(Calories)}$$
$$20r + 5p + 5b = 55 \quad \text{(protein)}$$
$$20p + 100b = 220 \quad \text{(vitamin C)}$$

We now have a system of equations.

Carry out. Solving the system we get $(2, 1, 2)$.

Check. Two servings of roast beef provide 600 Calories, 40 g of protein, and no vitamin C. One baked potato provides 100 Calories, 5 g of protein, and 20 mg of vitamin C. And 2 servings of broccoli provide 100 Calories, 10 g of protein, and 200 mg of vitamin C. Together, then, they provide 800 Calories, 55 g of protein, and 220 mg of vitamin C. The values check.

State. The dietician should prepare 2 servings of roast beef, 1 baked potato, and 2 servings of broccoli.

16. $1\frac{1}{8}$ servings of roast beef, $2\frac{3}{4}$ baked potatoes, $3\frac{3}{4}$ servings of asparagus

17. Let x, y, and z represent the average number of times a man, a woman, and a one-year-old child cry each month, respectively.

Translate.

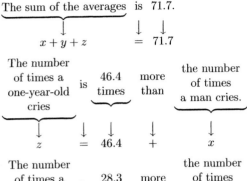

The sum of the averages is 71.7.

$$x + y + z = 71.7$$

The number of times a one-year-old cries is 46.4 times more than the number of times a man cries.

$$z = 46.4 + x$$

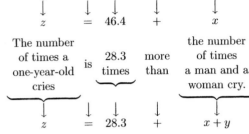

The number of times a one-year-old cries is 28.3 times more than the number of times a man and a woman cry.

$$z = 28.3 + x + y$$

Now we have a system of equations.

$$x + y + z = 71.7,$$
$$z = 46.4 + x,$$
$$z = 28.3 + x + y$$

Carry out. Solving the system, we get $(3.6, 18.1, 50)$.

Check. The sum of the average number times a man, a woman, and a one-year-old child cry each month is $3.6 + 18.1 + 50 = 71.7$. The number of times a one-year-old child cries, 50, is 46.4 more than 3.6, the average number of times a man cries each month and is 28.3 more than $3.6 + 18.1$, or 21.7, the average number of times a man and a woman cry. These numbers check.

State. In a month, a man cries an average of 3.6 times, a woman cries 18.1 times, and a one-year old child cries 50 times.

18. 385 for Asian-Americans, 200 for African-Americans, 154 for Caucasians

19. Familiarize. Let x, y, and z represent the number of 2-point field goals, 3-point field goals, and 1-point foul shots made, respectively. The total number of points scored from each of these types of goals is $2x$, $3y$, and z.

Translate.

The total number of points was 92.

$$2x + 3y + z = 92$$

The total number of baskets was 50.

$$x + y + z = 50$$

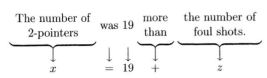

The number of 2-pointers was 19 more than the number of foul shots.

$$x = 19 + z$$

Now we have a system of equations.

$$2x + 3y + z = 92,$$
$$x + y + z = 50,$$
$$x = 19 + z$$

Carry out. Solving the system we get $(32, 5, 13)$.

Check. The total number of points was $2 \cdot 32 + 3 \cdot 5 + 13 = 64 + 15 + 13 = 92$. The number of baskets was $32 + 5 + 13$ or 50. The number of 2-pointers, 32, was 19 more than the number of foul shots, 13. The numbers check.

State. The Knicks made 32 two-point field goals, 5 three-point field goals, and 13 foul shots.

20. 1869

21. *Writing Exercise*

22. *Writing Exercise*

23. $5(-3) + 7 = -15 + 7 = -8$

24. 33

25. $-6(8) + (-7) = -48 + (-7) = -55$

26. -71

27. $-7(2x - 3y + 5z) = -7 \cdot 2x - 7(-3y) - 7(5z)$
$$= -14x + 21y - 35z$$

28. $-24a - 42b + 54c$

29. $\quad -4(2a + 5b) + 3a + 20b$
$$= -8a - 20b + 3a + 20b$$
$$= -8a + 3a - 20b + 20b$$
$$= -5a$$

30. $11x$

31. *Writing Exercise*

32. *Writing Exercise*

33. Familiarize. Let x = the one's digit, y = the ten's digit and z = the hundred's digit. Then the number is represented by $100z + 10y + x$. When the digits are reversed the resulting number is represented by $100x + 10y + z$.

Translate.

The sum of the digits is 14.

$$x + y + z = 14$$

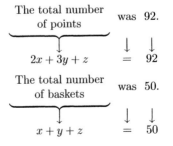

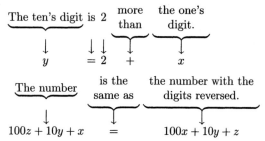

$$100z + 10y + x = 100x + 10y + z$$

Now we have a system of equations.

$$x + y + z = 14,$$
$$y = 2 + x,$$
$$100z + 10y + x = 100x + 10y + z$$

Carry out. Solving the system we get $(4, 6, 4)$.

Check. If the number is 464, then the sum of the digits is $4 + 6 + 4$, or 14. The ten's digit, 6, is 2 more than the one's digit, 4. If the digits are reversed the number is unchanged The result checks.

State. The number is 464.

4. 20 yr

5. Familiarize. Let $x =$ the number of adults, $y =$ the number of students, and $z =$ the number of children in attendance.

Translate. The given information gives rise to two equations.

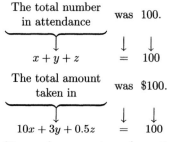

Now we have a system of equations.

$$x + y + z = 100,$$
$$10x + 3y + 0.5z = 100$$

Multiply the second equation by 2 to clear the decimal:

$$x + y + z = 100, \quad (1)$$
$$20x + 6y + z = 200. \quad (2)$$

Carry out. We use the elimination method.

$$
\begin{aligned}
-x - y - z &= -100 \quad \text{Multiplying (1) by } -1 \\
20x + 6y + z &= 200 \quad (2) \\
\hline
19x + 5y &= 100 \quad (3)
\end{aligned}
$$

In (3), note that 5 is a factor of both $5y$ and 100. Therefore, 5 must also be a factor of $19x$, and hence of x, since 5 is not a factor of 19. Then for some positive integer n, $x = 5n$. (We require $n > 0$, since the number of adults

clearly cannot be negative and must also be nonzero since the exercise states that the audience consists of *adults*, students, and children.) We have

$$19 \cdot 5n + 5y = 100, \text{ or}$$
$$19n + y = 20. \quad \text{Dividing by 5 on both sides}$$

Since n and y must both be positive, $n = 1$. Otherwise, $19n + y$ would be greater than 20. Then $x = 5 \cdot 1$, or 5.

$$19 \cdot 5 + 5y = 100 \quad \text{Substituting in (3)}$$
$$95 + 5y = 100$$
$$5y = 5$$
$$y = 1$$

$$5 + 1 + z = 100 \quad \text{Substituting in (1)}$$
$$6 + z = 100$$
$$z = 94$$

Check. The number of people in attendance was $5 + 1 + 94$, or 100. The amount of money taken in was $\$10 \cdot 5 + \$3 \cdot 1 + \$0.50(94) = \$50 + \$3 + \$47 = \$100$. The numbers check.

State. There were 5 adults, 1 student, and 94 children.

36. 35 tickets

37. Familiarize. We first make a drawing with additional labels.

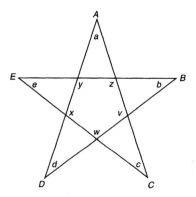

We let a, b, c, d, and e represent the angle measures at the tips of the star. We also label the interior angles of the pentagon v, w, x, y, and z. We recall the following geometric fact:

The sum of the measures of the interior angles of a polygon of n sides is given by $(n - 2)180°$.

Using this fact we know:

1. The sum of the angle measures of a triangle is $(3 - 2)180°$, or $180°$.

2. The sum of the angle measures of a pentagon is $(5 - 2)180°$, or $3(180°)$.

Translate. Using fact (1) listed above we obtain a system of 5 equations.

$$a + v + d = 180$$
$$b + w + e = 180$$
$$c + x + a = 180$$
$$d + y + b = 180$$
$$e + z + c = 180$$

Carry out. Adding we obtain

$$2a + 2b + 2c + 2d + 2e + v + w + x + y + z =$$
$$5(180)$$
$$2(a + b + c + d + e) + (v + w + x + y + z) =$$
$$5(180)$$

Using fact (2) listed above we substitute $3(180)$ for $(v + w + x + y + z)$ and solve for $(a + b + c + d + e)$.

$$2(a + b + c + d + e) + 3(180) = 5(180)$$
$$2(a + b + c + d + e) = 2(180)$$
$$a + b + c + d + e = 180$$

Check. We should repeat the above calculations.

State. The sum of the angle measures at the tips of the star is $180°$.

Exercise Set 3.6

1. $9x - 2y = 5,$

$3x - 3y = 11$

Write a matrix using only the constants.

$$\begin{bmatrix} 9 & -2 & \vdots & 5 \\ 3 & -3 & \vdots & 11 \end{bmatrix}$$

Multiply row 2 by 3 to make the first number in row 2 a multiple of 9.

$$\begin{bmatrix} 9 & -2 & \vdots & 5 \\ 9 & -9 & \vdots & 33 \end{bmatrix} \quad \text{New Row 2} = 3(\text{Row 2})$$

Multiply row 1 by -1 and add it to row 2.

$$\begin{bmatrix} 9 & -2 & \vdots & 5 \\ 0 & -7 & \vdots & 28 \end{bmatrix} \quad \begin{array}{l} \text{New Row 2} = -1(\text{Row 1}) + \\ \text{Row 2} \end{array}$$

Reinserting the variables, we have

$$9x - 2y = 5, \quad (1)$$
$$-7y = 28. \quad (2)$$

Solve Equation (2) for y.

$$-7y = 28$$
$$y = -4$$

Substitute -4 for y in Equation (1) and solve for x.

$$9x - 2y = 5$$
$$9x - 2(-4) = 5$$
$$9x + 8 = 5$$
$$9x = -3$$
$$x = -\frac{1}{3}$$

The solution is $\left(-\frac{1}{3}, -4 \right)$.

2. $(2, -1)$

3. $x + 4y = 8,$

$3x + 5y = 3$

We first write a matrix using only the constants.

$$\begin{bmatrix} 1 & 4 & \vdots & 8 \\ 3 & 5 & \vdots & 3 \end{bmatrix}$$

Multiply the first row by -3 and add it to the second row

$$\begin{bmatrix} 1 & 4 & \vdots & 8 \\ 0 & -7 & \vdots & -21 \end{bmatrix} \begin{array}{l} \text{New Row 2} = -3(\text{Row 1}) + \\ \text{Row 2} \end{array}$$

Reinserting the variables, we have

$$x + 4y = 8, \quad (1)$$
$$-7y = -21. \quad (2)$$

Solve Equation (2) for y.

$$-7y = -21$$
$$y = 3$$

Substitute 3 for y in Equation (1) and solve for x.

$$x + 4 \cdot 3 = 8$$
$$x + 12 = 8$$
$$x = -4$$

The solution is $(-4, 3)$.

4. $(-3, 2)$

5. $6x - 2y = 4,$

$7x + y = 13$

Write a matrix using only the constants.

$$\begin{bmatrix} 6 & -2 & \vdots & 4 \\ 7 & 1 & \vdots & 13 \end{bmatrix}$$

Multiply the second row by 6 to make the first number in row 2 a multiple of 6.

$$\begin{bmatrix} 6 & -2 & \vdots & 4 \\ 42 & 6 & \vdots & 78 \end{bmatrix} \quad \text{New Row 2} = 6(\text{Row 2})$$

Now multiply the first row by -7 and add it to the second row.

$$\begin{bmatrix} 6 & -2 & \vdots & 4 \\ 0 & 20 & \vdots & 50 \end{bmatrix} \begin{array}{l} \text{New Row 2} = -7(\text{Row 1}) + \\ \text{Row 2} \end{array}$$

Reinserting the variables, we have

$$6x - 2y = 4, \quad (1)$$
$$20y = 50. \quad (2)$$

Solve Equation (2) for y.

$$20y = 50$$
$$y = \frac{5}{2}$$

Substitute $\frac{5}{2}$ for y in Equation (1) and solve for x.

$$6x - 2y = 4$$
$$6x - 2\left(\frac{5}{2}\right) = 4$$
$$6x - 5 = 4$$
$$6x = 9$$
$$x = \frac{3}{2}$$

The solution is $\left(\frac{3}{2}, \frac{5}{2}\right)$.

6. $\left(-1, \frac{5}{2}\right)$

7. $3x + 2y + 2z = 3,$
   $\quad x + 2y - z = 5,$
   $\quad 2x - 4y + z = 0$

We first write a matrix using only the constants.

$$\begin{bmatrix} 3 & 2 & 2 & | & 3 \\ 1 & 2 & -1 & | & 5 \\ 2 & -4 & 1 & | & 0 \end{bmatrix}$$

First interchange rows 1 and 2 so that each number below the first number in the first row is a multiple of that number.

$$\begin{bmatrix} 1 & 2 & -1 & | & 5 \\ 3 & 2 & 2 & | & 3 \\ 2 & -4 & 1 & | & 0 \end{bmatrix}$$

Multiply row 1 by -3 and add it to row 2.

Multiply row 1 by -2 and add it to row 3.

$$\begin{bmatrix} 1 & 2 & -1 & | & 5 \\ 0 & -4 & 5 & | & -12 \\ 0 & -8 & 3 & | & -10 \end{bmatrix}$$

Multiply row 2 by -2 and add it to row 3.

$$\begin{bmatrix} 1 & 2 & -1 & | & 5 \\ 0 & -4 & 5 & | & -12 \\ 0 & 0 & -7 & | & 14 \end{bmatrix}$$

Reinserting the variables, we have

$$x + 2y - z = 5, \quad (1)$$
$$-4y + 5z = -12, \quad (2)$$
$$-7z = 14. \quad (3)$$

Solve (3) for z.

$$-7z = 14$$
$$z = -2$$

Substitute -2 for z in (2) and solve for y.

$$-4y + 5(-2) = -12$$
$$-4y - 10 = -12$$
$$-4y = -2$$
$$y = \frac{1}{2}$$

Substitute $\frac{1}{2}$ for y and -2 for z in (1) and solve for x.

$$x + 2 \cdot \frac{1}{2} - (-2) = 5$$
$$x + 1 + 2 = 5$$
$$x + 3 = 5$$
$$x = 2$$

The solution is $\left(2, \frac{1}{2}, -2\right)$.

8. $\left(\frac{3}{2}, -4, -3\right)$

9. $p - 2q - 3r = 3,$
   $\quad 2p - q - 2r = 4,$
   $\quad 4p + 5q + 6r = 4$

We first write a matrix using only the constants.

$$\begin{bmatrix} 1 & -2 & -3 & | & 3 \\ 2 & -1 & -2 & | & 4 \\ 4 & 5 & 6 & | & 4 \end{bmatrix}$$

$$\begin{bmatrix} 1 & -2 & -3 & | & 3 \\ 0 & 3 & 4 & | & -2 \\ 0 & 13 & 18 & | & -8 \end{bmatrix}$$

New Row 2 = -2(Row 1) + Row 2
New Row 3 = -4(Row 1) + Row 3

$$\begin{bmatrix} 1 & -2 & -3 & | & 3 \\ 0 & 3 & 4 & | & -2 \\ 0 & 39 & 54 & | & -24 \end{bmatrix}$$

New Row 3 = 3(Row 3)

$$\begin{bmatrix} 1 & -2 & -3 & | & 3 \\ 0 & 3 & 4 & | & -2 \\ 0 & 0 & 2 & | & 2 \end{bmatrix}$$

New Row 3 = -13(Row 2)+ Row 3

Reinserting the variables, we have

$$p - 2q - 3r = 3, \quad (1)$$
$$3q + 4r = -2, \quad (2)$$
$$2r = 2 \quad (3)$$

Solve (3) for r.

$$2r = 2$$
$$r = 1$$

Substitute 1 for r in (2) and solve for q.

$$3q + 4 \cdot 1 = -2$$
$$3q + 4 = -2$$
$$3q = -6$$
$$q = -2$$

Substitute -2 for q and 1 for r in (1) and solve for p.

$$p - 2(-2) - 3 \cdot 1 = 3$$
$$p + 4 - 3 = 3$$
$$p + 1 = 3$$
$$p = 2$$

The solution is $(2, -2, 1)$.

10. $(-1, 2, -2)$

11.
$$3p \qquad + 2r = 11,$$
$$q - 7r = 4,$$
$$p - 6q \qquad = 1$$

We first write a matrix using only the constants.

$$\begin{bmatrix} 3 & 0 & 2 & | & 11 \\ 0 & 1 & -7 & | & 4 \\ 1 & -6 & 0 & | & 1 \end{bmatrix}$$

$$\begin{bmatrix} 1 & -6 & 0 & | & 1 \\ 0 & 1 & -7 & | & 4 \\ 3 & 0 & 2 & | & 11 \end{bmatrix}$$ Interchange Row 1 and Row 3

$$\begin{bmatrix} 1 & -6 & 0 & | & 1 \\ 0 & 1 & -7 & | & 4 \\ 0 & 18 & 2 & | & 8 \end{bmatrix}$$ New Row 3 $= -3$(Row 1) $+$ Row 3

$$\begin{bmatrix} 1 & -6 & 0 & | & 1 \\ 0 & 1 & -7 & | & 4 \\ 0 & 0 & 128 & | & -64 \end{bmatrix}$$ New Row 3 $= -18$(Row 2) $+$ Row 3

Reinserting the variables, we have

$$p - 6q \qquad = 1, \qquad (1)$$
$$q - 7r = 4, \qquad (2)$$
$$128r = -64. \qquad (3)$$

Solve (3) for r.

$$128r = -64$$
$$r = -\frac{1}{2}$$

Substitute $-\frac{1}{2}$ for r in (2) and solve for q.

$$q - 7r = 4$$
$$q - 7\left(-\frac{1}{2}\right) = 4$$
$$q + \frac{7}{2} = 4$$
$$q = \frac{1}{2}$$

Substitute $\frac{1}{2}$ for q in (1) and solve for p.

$$p - 6 \cdot \frac{1}{2} = 1$$
$$p - 3 = 1$$
$$p = 4$$

The solution is $\left(4, \frac{1}{2}, -\frac{1}{2}\right)$.

12. $\left(\frac{1}{2}, \frac{2}{3}, -\frac{5}{6}\right)$

13.
$$3x + y = 8,$$
$$4x + 5y - 3z = 4,$$
$$7x + 2y - 9z = 1$$

The coefficient matrix is:

$$\begin{bmatrix} 3 & 1 & 0 & 8 \\ 4 & 5 & -3 & 4 \\ 7 & 2 & -9 & 1 \end{bmatrix}$$

We enter this on a graphing calculator and use the "rref(" command along with Frac to find the reduced row-echelon form of the matrix with the elements expressed in fractional form.

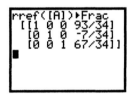

We see that $x = \dfrac{93}{34}$, $y = -\dfrac{7}{34}$, and $z = \dfrac{67}{34}$. The solution is $\left(\dfrac{93}{34}, -\dfrac{7}{34}, \dfrac{67}{34}\right)$.

14. $\left(\frac{3}{5}, \frac{5}{2}, -2\right)$

15.
$$-0.01x + 0.7y = -0.9,$$
$$0.5x - 0.3y + 0.18z = 0.01,$$
$$50x + 6y - 75z = 12$$

The coefficient matrix is:

$$\begin{bmatrix} -0.01 & 0.7 & 0 & -0.9 \\ 0.5 & -0.3 & 0.18 & 0.01 \\ 50 & 6 & -75 & 12 \end{bmatrix}$$

We enter this on a graphing calculator and use the " rref(" command to find the row-echelon form of the matrix.

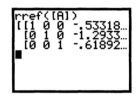

We see that $x \approx -0.5332$, $y \approx -1.2933$, and $z \approx -0.6189$. The solution is $(-0.5332, -1.2933, -0.6189)$.

6. $\left(-\dfrac{167}{68}, -\dfrac{227}{68}, \dfrac{7}{2}\right)$

7. We will rewrite the equations with the variables in alphabetical order:
$$-2w + 2x + 2y - 2z = -10,$$
$$w + x + y + z = -5,$$
$$3w + x - y + 4z = -2,$$
$$w + 3x - 2y + 2z = -6$$

Write a matrix using only the constants.

$$\begin{bmatrix} -2 & 2 & 2 & -2 & | & -10 \\ 1 & 1 & 1 & 1 & | & -5 \\ 3 & 1 & -1 & 4 & | & -2 \\ 1 & 3 & -2 & 2 & | & -6 \end{bmatrix}$$

$$\begin{bmatrix} -1 & 1 & 1 & -1 & | & -5 \\ 1 & 1 & 1 & 1 & | & -5 \\ 3 & 1 & -1 & 4 & | & -2 \\ 1 & 3 & -2 & 2 & | & -6 \end{bmatrix}$$ New Row 1 = $\frac{1}{2}$(Row 1)

$$\begin{bmatrix} -1 & 1 & 1 & -1 & | & -5 \\ 0 & 2 & 2 & 0 & | & -10 \\ 0 & 4 & 2 & 1 & | & -17 \\ 0 & 4 & -1 & 1 & | & -11 \end{bmatrix}$$

New Row 2 = Row 1 + Row 2
New Row 3 = 3(Row 1) + Row 3
New Row 4 = Row 1 + Row 4

$$\begin{bmatrix} -1 & 1 & 1 & -1 & | & -5 \\ 0 & 2 & 2 & 0 & | & -10 \\ 0 & 0 & -2 & 1 & | & 3 \\ 0 & 0 & -5 & 1 & | & 9 \end{bmatrix}$$

New Row 3 = −2(Row 2) + Row 3
New Row 4 = −2(Row 2) + Row 4

$$\begin{bmatrix} -1 & 1 & 1 & -1 & | & -5 \\ 0 & 2 & 2 & 0 & | & -10 \\ 0 & 0 & -2 & 1 & | & 3 \\ 0 & 0 & -10 & 2 & | & 18 \end{bmatrix}$$ New Row 4 = 2(Row 4)

$$\begin{bmatrix} -1 & 1 & 1 & -1 & | & -5 \\ 0 & 2 & 2 & 0 & | & -10 \\ 0 & 0 & -2 & 1 & | & 3 \\ 0 & 0 & 0 & -3 & | & 3 \end{bmatrix}$$ New Row 4 = −5(Row 3) + Row 4

Reinserting the variables, we have
$$-w + x + y - z = -5, \quad (1)$$
$$2x + 2y = -10, \quad (2)$$
$$-2y + z = 3, \quad (3)$$
$$-3z = 3. \quad (4)$$

Solve (4) for z.
$$-3z = 3$$
$$z = -1$$

Substitute -1 for z in (3) and solve for y.
$$-2y + (-1) = 3$$
$$-2y = 4$$
$$y = -2$$

Substitute -2 for y in (2) and solve for x.
$$2x + 2(-2) = -10$$
$$2x - 4 = -10$$
$$2x = -6$$
$$x = -3$$

Substitute -3 for x, -2 for y, and -1 for z in (1) and solve for w.
$$-w + (-3) + (-2) - (-1) = -5$$
$$-w - 3 - 2 + 1 = -5$$
$$-w - 4 = -5$$
$$-w = -1$$
$$w = 1$$

The solution is $(1, -3, -2, -1)$.

18. $(7, 4, 5, 6)$

19. *Familiarize.* Let d = the number of dimes and q = the number of quarters. The value of d dimes is \$$0.10d$, and the value of q quarters is \$$0.25q$.

Translate.

$\underbrace{\text{Total number of coins}}_{d + q}$ is 43.

$= 43$

Total value of coins is $7.60.

$$0.10d + 0.25q = 7.60$$

After clearing decimals, we have this system.

$$d + q = 43,$$
$$10d + 25q = 760$$

Carry out. Solve using matrices.

$$\begin{bmatrix} 1 & 1 & \vdots & 43 \\ 10 & 25 & \vdots & 760 \end{bmatrix}$$

$$\begin{bmatrix} 1 & 1 & \vdots & 43 \\ 0 & 15 & \vdots & 330 \end{bmatrix} \text{ New Row 2} = -10(\text{Row 1}) + \text{Row 2}$$

Reinserting the variables, we have

$$d + q = 43, \quad (1)$$
$$15q = 330. \quad (2)$$

Solve (2) for q.

$$15q = 330$$
$$q = 22$$

$$d + 22 = 43 \quad \text{Substituting in (2)}$$
$$d = 21$$

Check. The sum of the two numbers is 43. The total value is $0.10(21) + $0.25(22) = $2.10 + $5.50 = 7.60. The answer checks.

State. There are 21 dimes and 22 quarters.

20. 4 dimes, 30 nickels

21. Familiarize. We let x represent the number of pounds of the $4.05 kind and y represent the number of pounds of the $2.70 kind of granola. We organize the information in a table.

Granola	Number of pounds	Price per pound	Value
$4.05 kind	x	$4.05	$4.05x$
$2.70 kind	y	$2.70	$2.70y$
Mixture	15	$3.15	$3.15 × 15 or $47.25

Translate.

Total number of pounds is 15.

$$x + y = 15$$

Total value of mixture is $47.25.

$$4.05x + 2.70y = 47.25$$

After clearing decimals, we have this system:

$$x + y = 15,$$
$$405x + 270y = 4725$$

Carry out. Solve using matrices.

$$\begin{bmatrix} 1 & 1 & \vdots & 15 \\ 405 & 270 & \vdots & 4725 \end{bmatrix}$$

$$\begin{bmatrix} 1 & 1 & \vdots & 15 \\ 0 & -135 & \vdots & -1350 \end{bmatrix} \text{ New Row 2} = -405(\text{Row 1}) + \text{Row 2}$$

Reinserting the variables, we have

$$x + y = 15, \quad (1)$$
$$-135y = -1350 \quad (2)$$

Solve (2) for y.

$$-135y = -1350$$
$$y = 10$$

Substitute 10 for y in (1) and solve for x.

$$x + 10 = 15$$
$$x = 5$$

Check. The sum of the numbers is 15. The total value is $4.05(5) + $2.70(10), or $20.25 + $27.00, or $47.25. The numbers check.

State. 5 pounds of the $4.05 per lb granola and 10 pounds of the $2.70 per lb granola should be used.

22. 14 pounds of nuts, 6 pounds of oats

23. Familiarize. We let x, y, and z represent the amounts invested at 7%, 8%, and 9%, respectively. Recall the formula for simple interest:

$$\text{Interest} = \text{Principal} \times \text{Rate} \times \text{Time}$$

Translate. We organize the information in a table.

	First Investment	Second Investment	Third Investment	Total
P	x	y	z	$2500
R	7%	8%	9%	
T	1 yr	1 yr	1 yr	
I	$0.07x$	$0.08y$	$0.09z$	$212

The first row gives us one equation:

$$x + y + z = 2500$$

The last row gives a second equation:

$$0.07x + 0.08y + 0.09z = 212$$

Amount invested at 9% is $1100 more than amount invested at 8%.

$$z = $1100 + y$$

After clearing decimals, we have this system:

$$x + y + z = 2500,$$
$$7x + 8y + 9z = 21,200,$$
$$-y + z = 1100$$

Carry out. Solve using matrices.

$$\begin{bmatrix} 1 & 1 & 1 & \vdots & 2500 \\ 7 & 8 & 9 & \vdots & 21,200 \\ 0 & -1 & 1 & \vdots & 1100 \end{bmatrix}$$

$$\begin{bmatrix} 1 & 1 & 1 & \vdots & 2500 \\ 0 & 1 & 2 & \vdots & 3700 \\ 0 & -1 & 1 & \vdots & 1100 \end{bmatrix} \quad \begin{array}{l} \text{New Row 2} = \\ -7(\text{Row 1}) + \text{Row 2} \end{array}$$

$$\begin{bmatrix} 1 & 1 & 1 & \vdots & 2500 \\ 0 & 1 & 2 & \vdots & 3700 \\ 0 & 0 & 3 & \vdots & 4800 \end{bmatrix} \quad \begin{array}{l} \text{New Row 3} = \\ \text{Row 2} + \text{Row 3} \end{array}$$

Reinserting the variables, we have

$$x + y + z = 2500, \quad (1)$$
$$y + 2z = 3700, \quad (2)$$
$$3z = 4800 \quad (3)$$

Solve (3) for z.

$$3z = 4800$$
$$z = 1600$$

Substitute 1600 for z in (2) and solve for y.

$$y + 2 \cdot 1600 = 3700$$
$$y + 3200 = 3700$$
$$y = 500$$

Substitute 500 for y and 1600 for z in (1) and solve for x.

$$x + 500 + 1600 = 2500$$
$$x + 2100 = 2500$$
$$x = 400$$

Check. The total investment is \$400 + \$500 + \$1600, or \$2500. The total interest is 0.07(\$400) + 0.08(\$500) + 0.09(\$1600) = \$28 + \$40 + \$144 = \$212. The amount invested at 9%, \$1600, is \$1100 more than the amount invested at 8%, \$500. The numbers check.

State. \$400 is invested at 7%, \$500 is invested at 8%, and \$1600 is invested at 9%.

34. \$500 at 8%, \$400 at 9%, \$2300 at 10%

35. *Writing Exercise*

36. *Writing Exercise*

37. $5(-3) - (-7)4 = -15 - (-28) = -15 + 28 = 13$

28. -22

29.
$$-2(5 \cdot 3 - 4 \cdot 6) - 3(2 \cdot 7 - 15) + 4(3 \cdot 8 - 5 \cdot 4)$$
$$= -2(15 - 24) - 3(14 - 15) + 4(24 - 20)$$
$$= -2(-9) - 3(-1) + 4(4)$$
$$= 18 + 3 + 16$$
$$= 21 + 16$$
$$= 37$$

30. 422

31. *Writing Exercise*

32. *Writing Exercise*

33. **Familiarize.** Let w, x, y, and z represent the thousand's, hundred's, ten's, and one's digits, respectively.

Translate.

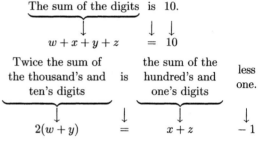

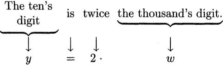

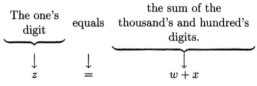

We have a system of equations which can be written as

$$w + x + y + z = 10,$$
$$2w - x + 2y - z = -1,$$
$$-2w + y = 0,$$
$$w + x - z = 0.$$

Carry out. We can use matrices to solve the system. We get $(1, 3, 2, 4)$.

Check. The sum of the digits is 10. Twice the sum of 1 and 2 is 6. This is one less than the sum of 3 and 4. The ten's digit, 2, is twice the thousand's digit, 1. The one's digit, 4, equals $1 + 3$. The numbers check.

State. The number is 1324.

34. $x = \dfrac{ce - bf}{ae - bd}$, $y = \dfrac{af - cd}{ae - bd}$

Exercise Set 3.7

1. $C(x) = 45x + 300,000$ $R(x) = 65x$

a) $P(x) = R(x) - C(x)$

$\qquad = 65x - (45x + 300,000)$

$\qquad = 65x - 45x - 300,000$

$\qquad = 20x - 300,000$

b) To find the break-even point we solve the system

$\qquad R(x) = 65x,$

$\qquad C(x) = 45x + 300,000.$

Since $R(x) = C(x)$ at the break-even point, we can rewrite the system:

$\qquad R(x) = 65x,$ (1)

$\qquad R(x) = 45x + 300,000$ (2)

We solve using substitution.

$\quad 65x = 45x + 300,000$ Substituting $65x$ for $R(x)$ in (2)

$\quad 20x = 300,000$

$\quad\ \ x = 15,000$

Thus, 15,000 units must be produced and sold in order to break even.

2. a) $P(x) = 45x - 270,000$

b) 6000 units

3. $C(x) = 10x + 120,000$ $R(x) = 60x$

a) $P(x) = R(x) - C(x)$

$\qquad = 60x - (10x + 120,000)$

$\qquad = 60x - 10x - 120,000$

$\qquad = 50x - 120,000$

b) Solve the system

$\qquad R(x) = 60x,$

$\qquad C(x) = 10x + 120,000.$

Since both $R(x)$ and $C(x)$ are in dollars and they are equal at the break-even point, we can rewrite the system:

$\qquad d = 60x,$ (1)

$\qquad d = 10x + 120,000$ (2)

We solve using substitution.

$\quad 60x = 10x + 120,000$ Substituting $60x$ for d in (2)

$\quad 50x = 120,000$

$\quad\ \ x = 2400$

Thus, 2400 units must be produced and sold in order to break even.

4. a)$P(x) = 55x - 49,500$

b) 900 units

5. $C(x) = 40x + 22,500$ $R(x) = 85x$

a) $P(x) = R(x) - C(x)$

$\qquad = 85x - (40x + 22,500)$

$\qquad = 85x - 40x - 22,500$

$\qquad = 45x - 22,500$

b) Solve the system

$\qquad R(x) = 85x,$

$\qquad C(x) = 40x + 22,500.$

Since both $R(x)$ and $C(x)$ are in dollars and they are equal at the break-even point, we can rewrite the system:

$\qquad d = 85x,$ (1)

$\qquad d = 40x + 22,500$ (2)

We solve using substitution.

$\quad 85x = 40x + 22,500$ Substituting $85x$ for d in (2)

$\quad 45x = 22,500$

$\quad\ \ x = 500$

Thus, 500 units must be produced and sold in order to break even.

6. a) $P(x) = 80x - 10,000$

b) 125 units

7. $C(x) = 22x + 16,000$ $R(x) = 40x$

a) $P(x) = R(x) - C(x)$

$\qquad = 40x - (22x + 16,000)$

$\qquad = 40x - 22x - 16,000$

$\qquad = 18x - 16,000$

b) Solve the system

$\qquad R(x) = 40x,$

$\qquad C(x) = 22x + 16,000.$

Since both $R(x)$ and $C(x)$ are in dollars and they are equal at the break-even point, we can rewrite the system:

$\qquad d = 40x,$ (1)

$\qquad d = 22x + 16,000$ (2)

We solve using substitution.

$\quad 40x = 22x + 16,000$ Substituting $40x$ for d in (2)

$\quad 18x = 16,000$

$\quad\ \ x \approx 889$ units

Thus, 889 units must be produced and sold in order to break even.

8. a) $P(x) = 40x - 75,000$

b) 1875 units

9. $C(x) = 75x + 100,000 \qquad R(x) = 125x$

a) $P(x) = R(x) - C(x)$
$$= 125x - (75x + 100,000)$$
$$= 125x - 75x - 100,000$$
$$= 50x - 100,000$$

b) Solve the system
$$R(x) = 125x,$$
$$C(x) = 75x + 100,000.$$

Since $R(x) = C(x)$ at the break-even point, we can rewrite the system:
$$R(x) = 125x, \qquad (1)$$
$$R(x) = 75x + 100,000 \quad (2)$$

We solve using substitution.

$125x = 75x + 100,000$ Substituting $125x$
for $R(x)$ in (2)
$$50x = 100,000$$
$$x = 2000$$

To break even 2000 units must be produced and sold.

0. a) $P(x) = 30x - 120,000$

b) 4000 units

1. $D(p) = 1000 - 10p,$
$$S(p) = 230 + p$$

Since both demand and supply are quantities, the system can be rewritten:
$$q = 1000 - 10p, \quad (1)$$
$$q = 230 + p \qquad (2)$$

Substitute $1000 - 10p$ for q in (2) and solve.
$$1000 - 10p = 230 + p$$
$$770 = 11p$$
$$70 = p$$

The equilibrium price is $70 per unit. To find the equilibrium quantity we substitute $70 into either $D(p)$ or $S(p)$.
$$D(70) = 1000 - 10 \cdot 70 = 1000 - 700 = 300$$

The equilibrium quantity is 300 units.

The equilibrium point is ($70, 300).

2. ($10, 1400)

3. $D(p) = 760 - 13p,$
$$S(p) = 430 + 2p$$

Rewrite the system:
$$q = 760 - 13p, \quad (1)$$
$$q = 430 + 2p \qquad (2)$$

Substitute $760 - 13p$ for q in (2) and solve.
$$760 - 13p = 430 + 2p$$
$$330 = 15p$$
$$22 = p$$

The equilibrium price is $22 per unit.

To find the equilibrium quantity we substitute $22 into either $D(p)$ or $S(p)$.
$$S(22) = 430 + 2(22) = 430 + 44 = 474$$

The equilibrium quantity is 474 units.

The equilibrium point is ($22, 474).

14. ($10, 370)

15. $D(p) = 7500 - 25p,$
$$S(p) = 6000 + 5p$$

Rewrite the system:
$$q = 7500 - 25p, \quad (1)$$
$$q = 6000 + 5p \qquad (2)$$

Substitute $7500 - 25p$ for q in (2) and solve.
$$7500 - 25p = 6000 + 5p$$
$$1500 = 30p$$
$$50 = p$$

The equilibrium price is $50 per unit.

To find the equilibrium quantity we substitute $50 into either $D(p)$ or $S(p)$.
$$D(50) = 7500 - 25(50) = 7500 - 1250 = 6250$$

The equilibrium quantity is 6250 units.

The equilibrium point is ($50, 6250).

16. ($40, 7600)

17. $D(p) = 1600 - 53p,$
$$S(p) = 320 + 75p$$

Rewrite the system:
$$q = 1600 - 53p, \quad (1)$$
$$q = 320 + 75p \qquad (2)$$

Substitute $1600 - 53p$ for q in (2) and solve.
$$1600 - 53p = 320 + 75p$$
$$1280 = 128p$$
$$10 = p$$

The equilibrium price is $10 per unit.

To find the equilibrium quantity we substitute $10 into either $D(p)$ or $S(p)$.
$$S(10) = 320 + 75(10) = 320 + 750 = 1070$$

The equilibrium quantity is 1070 units.

The equilibrium point is ($10, 1070).

18. ($36, 4060)

19. a) $C(x) =$ Fixed costs + Variable costs

$C(x) = 125,300 + 450x,$

where x is the number of computers produced.

b) Each computer sells for $800. The total revenue is 800 times the number of computers sold. We assume that all computers produced are sold.

$R(x) = 800x$

c) $P(x) = R(x) - C(x)$

$P(x) = 800x - (125,300 + 450x)$

$\quad = 800x - 125,300 - 450x$

$\quad = 350x - 125,300$

d) $\quad P(x) = 350x - 125,300$

$P(100) = 350(100) - 125,300$

$\quad = 35,000 - 125,300$

$\quad = -90,300$

The company will realize a $90,300 loss when 100 computers are produced and sold.

$P(400) = 350(400) - 125,300$

$\quad = 140,000 - 125,300$

$\quad = 14,700$

The company will realize a profit of $14,700 from the production and sale of 400 computers.

e) Solve the system

$R(x) = 800x,$

$C(x) = 125,300 + 450x.$

Since both $R(x)$ and $C(x)$ are in dollars and they are equal at the break-even point, we can rewrite the system:

$d = 800x,$ (1)

$d = 125,300 + 450x$ (2)

We solve using substitution.

$800x = 125,300 + 450x$ Substituting $800x$ for d in (2)

$350x = 125,300$

$x \approx 358$ Rounding up

The firm will break even if it produces and sells 358 computers and takes in a total of $R(358) = 800 \cdot 358 = \$286,400$ in revenue. Thus, the break-even point is (358 computers, $286,400).

20. a) $C(x) = 22,500 + 40x$

b) $R(x) = 85x$

c) $P(x) = 45x - 22,500$

d) $112,500 profit; $4500 loss

e) (500 lamps, $42,500)

21. a) $C(x) =$ Fixed costs + Variable costs

$C(x) = 16,404 + 6x,$

where x is the number of caps produced, in dozens.

b) Each dozen caps sell for $18. The total revenue is 18 times the number of caps sold, in dozens. We assume that all caps produced are sold.

$R(x) = 18x$

c) $P(x) = R(x) - C(x)$

$P(x) = 18x - (16,404 + 6x)$

$\quad = 18x - 16,404 - 6x$

$\quad = 12x - 16,404$

d) $P(3000) = 12(3000) - 16,404$

$\quad = 36,000 - 16,404$

$\quad = 19,596$

The company will realize a profit of $19,596 when 3000 dozen caps are produced and sold.

$P(1000) = 12(1000) - 16,404$

$\quad = 12,000 - 16,404$

$\quad = -4404$

The company will realize a $4404 loss when 1000 dozen caps are produced and sold.

e) Solve the system

$R(x) = 18x,$

$C(x) = 16,404 + 6x.$

Since both $R(x)$ and $C(x)$ are in dollars and they are equal at the break-even point, we can rewrite the system:

$d = 18x,$ (1)

$d = 16,404 + 6x$ (2)

We solve using substitution.

$18x = 16,404 + 6x$ Substituting $18x$ for d in (2)

$12x = 16,404$

$x = 1367$

The firm will break even if it produces and sells 1367 dozen caps and takes in a total of $R(1367) = 18 \cdot 1367 = \$24,606$ in revenue. Thus, the break-even point is (1367 dozen caps, $24,606).

22. a) $C(x) = 10,000 + 30x$

b) $R(x) = 80x$

c) $P(x) = 50x - 10,000$

d) $90,000 profit; $7500 loss

e) (200 sport coats, $16,000)

3. a) $D(p) = -14.97p + 987.35$,

     $S(p) = 98.55p - 5.13$

   Rewrite the system:

     $q = -14.97p + 987.35$,   (1)

     $q = 98.55p - 5.13$ (2)

   Substitute $-14.97p + 987.35$ for q in (2) and solve.

    $-14.97p + 987.35 = 98.55p - 5.13$

             $992.48 = 113.52p$

              $8.74 \approx p$

   The equilibrium price is $8.74 per unit. A price of $8.74 per unit should be charged in order to have equilibrium between supply and demand.

  b) $R(x) = 8.74x$,

     $C(x) = 2.10x + 5265$

   Rewrite the system:

     $d = 8.74x$,   (1)

     $d = 2.10x + 5265$ (2)

   We solve using substitution.

    $8.74x = 2.10x + 5265$ Substituting $8.74x$ for d

                        in (2)

     $6.64x = 5265$

        $x \approx 793$

   Thus 793 units must be sold in order to break even.

4. a) (4526 units, $4,390,220)

  b) $870

5. *Writing Exercise*

6. *Writing Exercise*

7. $3x - 9 = 27$

       $3x = 36$ Adding 9 to both sides

        $x = 12$ Dividing both sides by 3

   The solution is 12.

8. 15

9. $4x - 5 = 7x - 13$

      $-5 = 3x - 13$ Subtracting $4x$ from both sides

       $8 = 3x$ Adding 13 to both sides

     $\dfrac{8}{3} = x$ Dividing both sides by 3

   The solution is $\dfrac{8}{3}$.

10. 4

31. $7 - 2(x - 8) = 14$

     $7 - 2x + 16 = 14$ Removing parentheses

        $-2x + 23 = 14$ Collecting like terms

             $-2x = -9$ Subtracting 23 from both sides

               $x = \dfrac{9}{2}$ Dividing both sides by -2

   The solution is $\dfrac{9}{2}$.

32. $\dfrac{1}{3}$

33. *Writing Exercise*

34. *Writing Exercise*

35. The supply function contains the points ($2, 100) and ($8, 500). We find its equation:

$$m = \frac{500 - 100}{8 - 2} = \frac{400}{6} = \frac{200}{3}$$

$$y - y_1 = m(x - x_1) \quad \text{Point-slope form}$$

$$y - 100 = \frac{200}{3}(x - 2)$$

$$y - 100 = \frac{200}{3}x - \frac{400}{3}$$

$$y = \frac{200}{3}x - \frac{100}{3}$$

We can equivalently express supply S as a function of price p:

$$S(p) = \frac{200}{3}p - \frac{100}{3}$$

The demand function contains the points ($1, 500) and ($9, 100). We find its equation:

$$m = \frac{100 - 500}{9 - 1} = \frac{-400}{8} = -50$$

$$y - y_1 = m(x - x_1)$$

$$y - 500 = -50(x - 1)$$

$$y - 500 = -50x + 50$$

$$y = -50x + 550$$

We can equivalently express demand D as a function of price p:

$$D(p) = -50p + 550$$

We have a system of equations

$$S(p) = \frac{200}{3}p - \frac{100}{3},$$

$$D(p) = -50p + 550.$$

Rewrite the system:

$$q = \frac{200}{3}p - \frac{100}{3}, \quad (1)$$

$$q = -50p + 550 \quad (2)$$

Substitute $\dfrac{200}{3}p - \dfrac{100}{3}$ for q in (2) and solve.

$$\dfrac{200}{3}p - \dfrac{100}{3} = -50p + 550$$

$200p - 100 = -150p + 1650$ Multiplying by 3
                               to clear fractions

$$350p - 100 = 1650$$
$$350p = 1750$$
$$p = 5$$

The equilibrium price is \$5 per unit.

To find the equilibrium quantity, we substitute \$5 into either $S(p)$ or $D(p)$.

$$D(5) = -50(5) + 550 = -250 + 550 = 300$$

The equilibrium quantity is 300 units.

The equilibrium point is ($5, 300$).

36. 308 pairs

37. a) Enter the data and use the linear regression feature to get $S(p) = 15.97p - 1.05$.

    b) Enter the data and use the linear regression feature to get $D(p) = -11.26p + 41.16$.

    c) Find the point of intersection of the graphs of the functions found in parts (a) and (b).

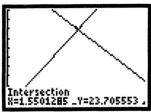

    We see that the equilibrium point is ($1.55, 23.7 million jars).

38. a) $S(p) = 3.8p - 1.82$

    b) $D(p) = -1.44p + 7.64$

    c) ($1.81, 5.0 thousand)

Chapter 4

Inequalities and Problem Solving

Exercise Set 4.1

1. $x - 1 \geq 7$

-4: We substitute and get $-4 - 1 \geq 7$, or
$-5 \geq 7$, a false sentence. Therefore, -4 is not
a solution.

0: We substitute and get $0 - 1 \geq 7$, or
$-1 \geq 7$, a false sentence. Therefore, 0 is not
a solution.

8: We substitute and get $8 - 1 \geq 7$, or $7 \geq 7$,
a true sentence. Therefore, 8 is a solution.

13: We substitute and get $13 - 1 \geq 7$, or $12 \geq 7$,
a true sentence. Therefore, 13 is a solution.

2. Yes; yes; no; no

3. $t - 6 > 2t - 1$

0: We substitute and get $0 - 6 > 2 \cdot 0 - 1$, or
$-6 > -1$, a false sentence. Therefore, 0 is
not a solution.

-8: We substitute and get $-8 - 6 > 2(-8) - 1$,
or $-14 > -17$, a true sentence. Therefore,
-8 is a solution.

-9: We substitute and get $-9 - 6 > 2(-9) - 1$,
or $-15 > -19$, a true sentence. Therefore,
-9 is a solution.

-3: We substitute and get $-3 - 6 > 2(-3) - 1$,
or $-9 > -7$, a false sentence. Therefore, -3
is not a solution.

4. No; yes; yes; no

5. $y < 6$

Graph: The solutions consist of all real numbers less than
6, so we shade all numbers to the left of 5 and use an open
circle at 6 to indicate that it is not a solution.

Set builder notation: $\{y|y < 6\}$

Interval notation: $(-\infty, 6)$

6.

Set builder notation: $\{x|x > 4\}$

Interval notation: $(4, \infty)$

7. $x \geq -4$

Graph: We shade all numbers to the right of -4 and use a
solid endpoint at -4 to indicate that it is also a solution.

Set builder notation: $\{x|x \geq -4\}$

Interval notation: $[-4, \infty)$

8.

Set builder notation: $\{t|t \leq 6\}$

Interval notation: $(-\infty, 6]$

9. $t > -3$

Graph: We shade all numbers to the right of -3 and use
an open circle at -3 to indicate that it is not a solution.

Set builder notation: $\{t|t > -3\}$

Interval notation: $(-3, \infty)$

10.

Set builder notation: $\{y|y < -3\}$

Interval notation: $(-\infty, -3)$

11. $x \leq -7$

Graph: We shade all numbers to the left of -7 and use a
solid endpoint at -7 to indicate that it is also a solution.

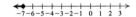

Set builder notation: $\{x|x \leq -7\}$

Interval notation: $(-\infty, -7]$

12.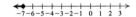

Set builder notation: $\{x|x \geq -6\}$

Interval notation: $[-6, \infty)$

13.
$$x + 9 > 4$$
$$x + 9 + (-9) > 4 + (-9) \qquad \text{Adding } -9$$
$$x > -5$$

The solution set is $\{x|x > -5\}$, or $(-5, \infty)$.

4. $\{x|x > -3\}$, or $(-3, \infty)$

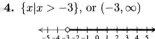

5.
$$a + 7 \leq -13$$
$$a + 7 + (-7) \leq -13 + (-7) \quad \text{Adding } -7$$
$$a \leq -20$$
The solution set is $\{a|a \leq -20\}$, or $(-\infty, -20]$.

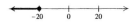

6. $\{a|a \leq -21\}$, or $(-\infty, -21]$

7.
$$x - 5 \leq 7$$
$$x - 5 + 5 \leq 7 + 5 \quad \text{Adding } 5$$
$$x \leq 12$$
The solution set is $\{x|x \leq 12\}$, or $(-\infty, 12]$.

8. $\{t|t \geq -5\}$, or $[-5, \infty)$

9.
$$y - 9 > -18$$
$$y - 9 + 9 > -18 + 9 \quad \text{Adding } 9$$
$$y > -9$$
The solution set is $\{y|y > -9\}$, or $(-9, \infty)$.

0. $\{y|y > -6\}$, or $(-6, \infty)$

1.
$$y - 20 \leq -6$$
$$y - 20 + 20 \leq -6 + 20 \quad \text{Adding } 20$$
$$y \leq 14$$
The solution set is $\{y|y \leq 14\}$, or $(-\infty, 14]$.

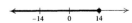

2. $\{x|x \leq 9\}$, or $(-\infty, 9]$

3.
$$9t < -81$$
$$\frac{1}{9} \cdot 9t < \frac{1}{9}(-81) \quad \text{Multiplying by } \frac{1}{9}$$
$$t < -9$$

The solution set is $\{t|t < -9\}$, or $(-\infty, -9)$.

24. $\{x|x \geq 3\}$, or $[3, \infty)$

25.
$$0.5x < 25$$
$$\frac{1}{0.5}(0.5x) < \frac{1}{0.5}(25) \quad \text{Multiplying by } \frac{1}{0.5}$$
$$x < \frac{25}{0.5}$$
$$x < 50$$
The solution set is $\{x|x < 50\}$, or $(-\infty, 50)$.

26. $\{x|x < -60\}$, or $(-\infty, -60)$

27.
$$-8y \leq 3.2$$
$$-\frac{1}{8}(-8y) \geq -\frac{1}{8}(3.2) \quad \text{Multiplying by } -\frac{1}{8} \text{ and reversing the inequality symbol}$$
$$y \geq -0.4$$
The solution set is $\{y|y \geq -0.4\}$, or $[-0.4, \infty)$.

28. $\{x|x \leq 0.9\}$, or $(-\infty, 0.9]$

29.
$$-\frac{5}{6}y \leq -\frac{3}{4}$$
$$-\frac{6}{5}\left(-\frac{5}{6}y\right) \geq -\frac{6}{5}\left(-\frac{3}{4}\right) \quad \text{Multiplying by } -\frac{6}{5} \text{ and reversing the inequality symbol}$$
$$y \geq \frac{9}{10}$$
The solution set is $\left\{y|y \geq \frac{9}{10}\right\}$, or $\left[\frac{9}{10}, \infty\right)$.

30. $\left\{x|x \leq \frac{5}{6}\right\}$, or $\left(-\infty, \frac{5}{6}\right]$

31.
$$5y + 13 > 28$$
$$5y + 13 + (-13) > 28 + (-13) \quad \text{Adding } -13$$
$$5y > 15$$
$$\frac{1}{5} \cdot 5y > \frac{1}{5} \cdot 15 \quad \text{Multiplying by } \frac{1}{5}$$
$$y > 3$$

The solution set is $\{y | y > 3\}$, or $(3, \infty)$.

32. $\{x | x < 6\}$, or $(-\infty, 6)$

33.
$$-9x + 3x \geq -24$$
$$-6x \geq -24 \quad \text{Combining like terms}$$
$$-\frac{1}{6}(-6x) \leq -\frac{1}{6}(-24) \quad \text{Multiplying by } -\frac{1}{6}$$
$$\text{and reversing the}$$
$$\text{inequality symbol}$$
$$x \leq 4$$

The solution set is $\{x | x \leq 4\}$, or $(-\infty, 4]$.

34. $\{y | y \leq -3\}$, or $(-\infty, -3]$

35. $f(x) = 8x - 9$, $g(x) = 3x - 11$
$$f(x) < g(x)$$
$$8x - 9 < 3x - 11$$
$$5x - 9 < -11 \quad \text{Adding } -3x$$
$$5x < -2 \quad \text{Adding } 9$$
$$x < -\frac{2}{5} \quad \text{Multiplying by } \frac{1}{5}$$

The solution set is $\left\{x \middle| x < -\frac{2}{5}\right\}$, or $\left(-\infty, -\frac{2}{5}\right)$.

36. $\left\{x \middle| x > \frac{2}{3}\right\}$, or $\left(\frac{2}{3}, \infty\right)$

37. $f(x) = 0.4x + 5$, $g(x) = 1.2x - 4$
$$g(x) \geq f(x)$$
$$1.2x - 4 \geq 0.4x + 5$$
$$0.8x - 4 \geq 5 \quad \text{Adding } -0.4x$$
$$0.8x \geq 9 \quad \text{Adding } 4$$
$$x \geq 11.25 \quad \text{Multiplying by } \frac{1}{0.8}$$

The solution set is $\{x | x \geq 11.25\}$, or $[11.25, \infty)$.

38. $\left\{x \middle| x \geq \frac{1}{2}\right\}$, or $\left[\frac{1}{2}, \infty\right)$

39. We see that the graph of $f(x)$ lies on or above the graph of $g(x)$ for values of x that are 2 or greater. Thus, the solution set is $\{x | x \geq 2\}$, or $[2, \infty)$.

40. $\{x | x < -1\}$, or $(-\infty, -1)$

41. We see that the graph of y_1 lies below the graph of y_2 for values of x less than 3. Thus, the solution set is $\{x | x < 3\}$ or $(-\infty, 3)$.

42. $\{x | x \leq 6\}$, or $(-\infty, 6]$

43. a) The graph of y_1 lies above the graph of y_2 for x-values to the left of the point of intersection. The solution set is $\{x | x < 4\}$, or $(-\infty, 4)$.

b) The graph of y_2 lies on or below the graph of y_3 for x-values at and to the right of the point of intersection. The solution set is $\{x | x \geq 2\}$, or $[2, \infty)$.

c) The graph of y_3 lies on or above the graph of y_1 at and to the right of the point of intersection. The solution set is $\{x | x \geq 3.2\}$, or $[3.2, \infty)$.

44. a) $\{x | x \leq -2\}$, or $(-\infty, -2]$

b) $\left\{x \middle| x > \frac{8}{3}\right\}$, or $\left(\frac{8}{3}, \infty\right)$

c) $\left\{x \middle| x > \frac{4}{5}\right\}$, or $\left(\frac{4}{5}, \infty\right)$

45.
$$4(3y - 2) \geq 9(2y + 5)$$
$$12y - 8 \geq 18y + 45$$
$$-6y - 8 \geq 45$$
$$-6y \geq 53$$
$$y \leq -\frac{53}{6}$$

The solution set is $\left\{y \middle| y \leq -\frac{53}{6}\right\}$, or $\left(-\infty, -\frac{53}{6}\right]$.

46. $\left\{ m \middle| m \le \dfrac{49}{10} \right\}$, or $\left(-\infty, \dfrac{49}{10} \right]$

47. $5(t-3) + 4t < 2(7 + 2t)$

$5t - 15 + 4t < 14 + 4t$

$9t - 15 < 14 + 4t$

$5t - 15 < 14$

$5t < 29$

$t < \dfrac{29}{5}$

The solution set is $\left\{ t \middle| t < \dfrac{29}{5} \right\}$, or $\left(-\infty, \dfrac{29}{5} \right)$.

48. $\left\{ x \middle| x > -\dfrac{2}{17} \right\}$, or $\left(-\dfrac{2}{17}, \infty \right)$

49. $5[3m - (m+4)] > -2(m-4)$

$5(3m - m - 4) > -2(m-4)$

$5(2m - 4) > -2(m-4)$

$10m - 20 > -2m + 8$

$12m - 20 > 8$

$12m > 28$

$m > \dfrac{28}{12}$

$m > \dfrac{7}{3}$

The solution set is $\left\{ m \middle| m > \dfrac{7}{3} \right\}$, or $\left(\dfrac{7}{3}, \infty \right)$.

50. $\left\{ x \middle| x \le -\dfrac{23}{2} \right\}$, or $\left(-\infty, -\dfrac{23}{2} \right]$

51. $19 - (2x+3) \le 2(x+3) + x$

$19 - 2x - 3 \le 2x + 6 + x$

$16 - 2x \le 3x + 6$

$16 - 5x \le 6$

$-5x \le -10$

$x \ge 2$

The solution set is $\{ x | x \ge 2 \}$, or $[2, \infty)$.

52. $\{ c | c \le 1 \}$, or $(-\infty, 1]$

53. $\dfrac{1}{4}(8y + 4) - 17 \ < \ -\dfrac{1}{2}(4y - 8)$

$2y + 1 - 17 \ < \ -2y + 4$

$2y - 16 \ < \ -2y + 4$

$4y - 16 < 4$

$4y \ < \ 20$

$y \ < \ 5$

The solution set is $\{ y | y < 5 \}$, or $(-\infty, 5)$.

54. $\{ x | x > 6 \}$, or $(6, \infty)$

55. $2[8 - 4(3 - x)] - 2 \ge 8[2(4x - 3) + 7] - 50$

$2[8 - 12 + 4x] - 2 \ge 8[8x - 6 + 7] - 50$

$2[-4 + 4x] - 2 \ge 8[8x + 1] - 50$

$-8 + 8x - 2 \ge 64x + 8 - 50$

$8x - 10 \ge 64x - 42$

$-56x - 10 \ge -42$

$-56x \ge -32$

$x \le \dfrac{32}{56}$

$x \le \dfrac{4}{7}$

The solution set is $\left\{ x \middle| x \le \dfrac{4}{7} \right\}$, or $\left(-\infty, \dfrac{4}{7} \right]$.

56. $\left\{ t \middle| t \ge -\dfrac{27}{19} \right\}$, or $\left[-\dfrac{27}{19}, \infty \right)$

57. *Familiarize.* Let $m =$ the mileage. Then the mileage charge is $\$0.20m$ and the total cost of the rental is $\$45 + \$0.20m$.

Translate. We write an inequality stating that the cost of the rental is at most $\$75$.

$45 + 0.20m \le 75$

Carry out.

$45 + 0.20m \le 75$

$0.20m \le 30$

$m \le 150$

Check. We can do a partial check by substituting a value for m greater than 150. When $m = 151$, the rental cost is $\$45 + \$0.20(151) = \$75.20$. This is more than the budget amount, $\$75$. We cannot check all possible values for m, so we stop here.

State. The budget will not be exceeded for mileages less than or equal to 150 mi.

58. Mileages greater than 65 mi

59. *Familiarize.* Let $v =$ the blue book value of the car. Since the car was not replaced, we know that $\$9200$ does not exceed 80% of the blue book value.

Translate. We write an inequality stating that $\$9200$ does not exceed 80% of the blue book value.

$9200 \le 0.8v$

Carry out.

$9200 \le 0.8v$

$11,500 \le v$ Multiplying by $\dfrac{1}{0.8}$

Check. We can do a partial check by substituting a value for v greater than 11,500. When $v = 11,525$, then 80% of

v is $0.8(11,525)$, or 9220. This is greater than 9200; that is, 9200 does not exceed this amount. We cannot check all possible values for v, so we stop here.

State. The blue book value of the car is $11,500$ or more.

60. Calls shorter than 3.5 minutes

61. Familiarize. Let $m = $ the number of peak local minutes used. Then the charge for the minutes used is $0.022m$ and the total monthly charge is $13.55 + 0.022m$.

Translate. We write an inequality stating that the monthly charge is at least 39.40.

$$13.55 + 0.022m \geq 39.40$$

Carry out.

$$13.55 + 0.022m \geq 39.40$$
$$0.022m \geq 25.85$$
$$m \geq 1175$$

Check. We can do a partial check by substituting a value for m less than 1175. When $m = 1174$, the monthly charge is $13.55 + 0.022(1174) \approx 39.38$. This is less than the maximum charge of 39.40. We cannot check all possible values for m, so we stop here.

State. A customer must speak on the phone for 1175 local peak minutes or more if the maximum charge is to apply.

62. For 5170 local off-peak minutes or more

63. Familiarize. Let $c = $ the number of checks per month. Then the Anywhere plan will cost $0.20c$ per month and the Acu-checking plan will cost $2 + 0.12c$ per month.

Translate. We write an inequality stating that the Acu-checking plan costs less than the Anywhere plan.

$$2 + 0.12c < 0.20c$$

Carry out.

$$2 + 0.12c < 0.20c$$
$$2 < 0.08c$$
$$25 < c$$

Check. We can do a partial check by substituting a value for c less than 25 and a value for c greater than 25. When $c = 24$, the Acu-checking plan costs $2 + 0.12(24)$, or 4.88, and the Anywhere plan costs $0.20(24)$, or 4.80, so the Anywhere plan is less expensive. When $c = 26$, the Acu-checking plan costs $2 + 0.12(26)$, or 5.12, and the Anywhere plan costs $0.20(26)$, or 5.20, so Acu-checking is less expensive. We cannot check all possible values for c, so we stop here.

State. The Acu-checking plan costs less for more than 25 checks per month.

64. More than 4.25 hr

65. Familiarize. We list the given information in a table.

Plan A: Monthly Income	Plan B: Monthly Income
$400 salary	$610
8% of sales	5% of sales
Total: 400 + 8% of sales	Total: 610 + 5% of sales

Suppose Toni had gross sales of $5000 one month. Then under plan A she would earn

$$\$400 + 0.08(\$5000), \text{ or } \$800.$$

Under plan B she would earn

$$\$610 + 0.05(\$5000), \text{ or } \$860.$$

This shows that, for gross sales of $5000, plan B is better.

If Toni had gross sales of $10,000 one month, then under plan A she would earn

$$\$400 + 0.08(\$10,000), \text{ or } \$1200.$$

Under plan B she would earn

$$\$610 + 0.05(\$10,000), \text{ or } \$1110.$$

This shows that, for gross sales of $10,000, plan A is better. To determine all values for which plan A is better we solve an inequality.

Translate.

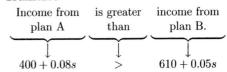

$$400 + 0.08s \qquad > \qquad 610 + 0.05s$$

Carry out.

$$400 + 0.08s > 610 + 0.05s$$
$$400 + 0.03s > 610$$
$$0.03s > 210$$
$$s > 7000$$

Check. For $s = \$7000$, the income from plan A is

$$\$400 + 0.08(\$7000), \text{ or } \$960$$

and the income from plan B is

$$\$610 + 0.05(\$7000), \text{ or } \$960.$$

This shows that for sales of $7000 Toni's income is the same from each plan. In the Familiarize step we showed that, for a value less than $7000, plan B is better and for a value greater than $7000, plan A is better. Since we cannot check all possible values, we stop here.

State. Toni should select plan A for gross sales greater than $7000.

66. Values of n greater than $85\dfrac{5}{7}$

67. *Familiarize.* Let m = the amount of the medical bills. Then under plan A Giselle would pay $50 + 0.2(m - \$50)$. Under plan B she would pay $250 + 0.1(m - \$250)$.

Translate. We write an inequality stating than the cost of plan B is less than the cost of plan A.

$$250 + 0.1(m - 250) < 50 + 0.2(m - 50)$$

Carry out.

$$250 + 0.1(m - 250) < 50 + 0.2(m - 50)$$
$$250 + 0.1m - 25 < 50 + 0.2m - 10$$
$$225 + 0.1m < 40 + 0.2m$$
$$185 + 0.1m < 0.2m$$
$$185 < 0.1m$$
$$1850 < m$$

Check. We can do a partial check by substituting a value for m less than $1850 and a value for m greater than $1850. When $m = \$1840$, plan A costs $50 + 0.2(\$1840 - \$50)$, or $408, and plan B costs $250 + 0.1(\$1840 - \$250)$, or $409. When $m = \$1860$, plan A costs $50 + 0.2(\$1860 - \$50)$, or $412, and plan B costs $250 + 0.1(\$1860 - \$250)$, or $411, so plan B will save Giselle money. We cannot check all possible values for m, so we stop here.

State. Plan B will save Giselle money for medical bills greater than $1850.

68. Parties of more than 80

69. *Familiarize.* Let n = the number of people who attend. Then the total receipts are $\$6 \cdot n$, and the amount of receipts over $750 is $\$6 \cdot n - \$750$. The band will receive $750 plus 15% of $\$6 \cdot n - \$750$, or $750 + 0.15(\$6 \cdot n - \$750)$.

Translate. We write an inequality stating that the amount the band receives is at least $1200.

$$750 + 0.15(6n - 750) \geq 1200$$

Carry out.

$$750 + 0.15(6n - 750) \geq 1200$$
$$750 + 0.9n - 112.5 \geq 1200$$
$$0.9n + 637.5 \geq 1200$$
$$0.9n \geq 562.5$$
$$n \geq 625$$

Check. When $n = 625$, the band receives $\$75 + 0.15(\$6 \cdot 625 - \$750)$, or $750 + 0.15(\$3000)$, or $750 + \$450$, or $1200. When $n = 626$, the band receives $750 + 0.15(\$6 \cdot 626 - \$750)$, or $750 + 0.15(\$3006)$, or $750 + \$450.90$, or $1200.90. Since the band receives exactly $1200 when 625 people attend and more than $1200 when 626 people attend, we have performed a partial check. We cannot check all possible solutions, so we stop here.

State. At least 625 people must attend in order for the band to receive at least $1200.

70. a) Fahrenheit temperatures less than $1945.4°$

b) Fahrenheit temperatures less than $1761.44°$

71. a) *Familiarize.* Find the values of x for which $R(x) < C(x)$.

Translate.

$$26x < 90,000 + 15x$$

Carry out.

$$11x < 90,000$$
$$x < 8181\frac{9}{11}$$

Check. $R\left(8181\frac{9}{11}\right) = \$212,727.27 = C\left(8181\frac{9}{11}\right)$.

Calculate $R(x)$ and $C(x)$ for some x greater than $8181\frac{9}{11}$ and for some x less than $8181\frac{9}{11}$.

Suppose $x = 8200$:

$$R(x) = 26(8200) = 213,200 \quad \text{and}$$
$$C(x) = 90,000 + 15(8200) = 213,000.$$

In this case $R(x) > C(x)$.

Suppose $x = 8000$:

$$R(x) = 26(8000) = 208,000 \quad \text{and}$$
$$C(x) = 90,000 + 15(8000) = 210,000.$$

In this case $R(x) < C(x)$.

Then for $x < 8181\frac{9}{11}$, $R(x) < C(x)$.

State. We will state the result in terms of integers, since the company cannot sell a fraction of a radio. For 8181 or fewer radios the company loses money.

b) Our check in part a) shows that for $x > 8181\frac{9}{11}$, $R(x) > C(x)$ and the company makes a profit. Again, we will state the result in terms of an integer. For more than 8181 radios the company makes money.

72. a) $\{p | p < 10\}$

b) $\{p | p > 10\}$

73. *Familiarize.* 1992 and 1997 are 2 and 7 years after 1990, respectively, so we have the data points $(2, 39,000)$ and $(7, 82,722)$.

Translate. First we find the slope of the line.

$$m = \frac{82,722 - 39,000}{7 - 2} = \frac{43,722}{5} = 8744.4$$

Using the point-slope equation, we have:

$$y - y_1 = m(x - x_1)$$
$$y - 39,000 = 8744.4(x - 2)$$
$$y - 39,000 = 8744.4x - 17,488.8$$
$$y = 8744.4x + 21,511.2$$

Letting t represent the number of trampoline injuries x years after 1990, we have

$$t(x) = 8744.4x + 21,511.2.$$

To predict the years in which the number of trampoline injuries will exceed 100,000 we solve the inequality $t(x) > 100,000$, or

$$8744.4x + 21,511.2 > 100,000.$$

Carry out. We solve the inequality.

$$8744.4x + 21,511.2 > 100,000$$
$$8744.4x > 78,488.8$$
$$x > 8.98$$

Check. As a partial check we can substitute a value for x greater than 8.98.

$$t(9) = 8744.4(9) + 21,511.2 = 100,210.8$$

Since $t(9) > 100,000$, we have a partial check.

State. There will be more than 100,000 trampoline injuries more than 8.98 years after 1990, or in years after 1998.

74. $b(x) = 9.75x + 135.75$; years after 2017

75. Familiarize. 1990 and 1998 are 5 and 13 years after 1985, respectively. For people who swim we have the data points $(5, 67.5)$ and $(13, 58.2)$. For people who exercise with equipment we have $(5, 35.3)$ and $(13, 46.1)$.

Translate. First we find a function describing the number of people, in millions, who swim x years after 1985. The slope of the line is:

$$m = \frac{58.2 - 67.5}{13 - 5} = \frac{-9.3}{8} = -1.1625$$

Using the point-slope equation, we have:

$$y - y_1 = m(x - x_1)$$
$$y - 67.5 = -1.1625(x - 5)$$
$$y - 67.5 = -1.1625x + 5.8125$$
$$y = -1.1625x + 73.3125$$
$$f(x) = -1.1625x + 73.3125$$

Now we find a function describing the number of people, in millions, who exercise with equipment x years after 1985. The slope of the line is:

$$m = \frac{46.1 - 35.3}{13 - 5} = \frac{10.8}{8} = 1.35$$

Using the point-slope equation, we have:

$$y - y_1 = m(x - x_1)$$
$$y - 35.3 = 1.35(x - 5)$$
$$y - 35.3 = 1.35x - 6.75$$
$$y = 1.35x + 28.55$$
$$g(x) = 1.35x + 28.55$$

To estimate the years in which the number of people who exercise with equipment will be greater than the number of people who swim, solve the inequality $g(x) > f(x)$, or

$$1.35x + 28.55 > -1.1625x + 73.3125.$$

Carry out. We solve the inequality.

$$1.35x + 28.55 > -1.1625x + 73.3125$$
$$1.35x > -1.1625x + 44.7625$$
$$2.5125x > 44.7625$$
$$x > 17.8$$

Check. As a partial check we can substitute a value greater than 17.8 in each function.

$$f(18) = -1.1625(18) + 73.3125 = 52.3875$$
$$g(18) = 1.35(18) + 28.55 = 52.85$$

Since $g(18) > f(18)$, we have a partial check.

State. We estimate that the number of people who exercise with equipment will be greater than the number of people who swim more than 17.8 years after 1985, or in years after 2002.

76. Public: $f(x) = 0.12625x - 0.0605$;

Private: $g(x) = 0.015375x + 2.02325$;

Years after 1998

77. Enter the data in a graphing calculator, letting x represent the number of years after 1900. Then use the linear regression feature to find the function $f(x) = 0.0056862962x + 1.891866531$. To predict the years in which the world record in the high jump will exceed 2.5 meters, graph $y_1 = f(x)$ along with $y_2 = 2.5$ and find the point of intersection. We use the window $[0, 150, 0, 3]$, Xscl $= 30$.

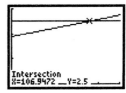

We see that the record will be 2.5 meters about 106.9 years after 1900, so we predict that the record will exceed 2.5 meters more than 106.9 years after 1900, or in years after 2006.

78. $f(x) = -0.0058087885x + 3.962605066$; years after 2033

79. Enter the data in a graphing calculator, letting x represent the number of years after 1985. Then use the linear regression feature to find the desired functions. For swimmers we have $f(x) = -1.174729242x + 73.30288805$ and for people who exercise with equipment we have $g(x) = 1.273826715x + 31.20415162$. To estimate the year in which the number of people who exercise with equipment will be greater than the number of people who swim, graph $y_1 = f(x)$ and $y_2 = g(x)$ and find the point of intersection. We use the window $[0, 30, 0, 100]$, Xscl $= 3$, Yscl $= 10$.

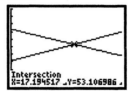

We see that the numbers will be equal about 17.2 years after 1985, so we predict that the number of people who exercise with equipment will be greater than the number who swim more than 17.2 years after 1985, or in years after 2002.

0. Public: $f(x) = 0.0922664165x + 0.4818424015$;

Private: $g(x) = 0.0588968105x + 1.402990619$;

Years after 2007

1. *Writing Exercise*

2. *Writing Exercise*

3. $f(x) = \dfrac{3}{x - 2}$

Since $\dfrac{3}{x - 2}$ cannot be computed when $x - 2$ is 0, we solve an equation:

$$x - 2 = 0$$
$$x = 2$$

The domain is $\{x | x \text{ is a real number } and\ x \neq 2\}$.

4. $\{x | x \text{ is a real number } and\ x \neq -3\}$

5. $f(x) = \dfrac{5x}{7 - 2x}$

Since $\dfrac{5x}{7 - 2x}$ cannot be computed when $7 - 2x$ is 0, we solve an equation:

$$7 - 2x = 0$$
$$7 = 2x$$
$$\dfrac{7}{2} = x$$

The domain is $\left\{x \middle| x \text{ is a real number } and\ x \neq \dfrac{7}{2}\right\}$.

6. $\left\{x \middle| x \text{ is a real number } and\ x \neq \dfrac{9}{4}\right\}$

7. $9x - 2(x - 5) = 9x - 2x + 10 = 7x + 10$

8. $22x - 7$

9. *Writing Exercise*

90. *Writing Exercise*

91. $3ax + 2x \geq 5ax - 4$

$$2x - 2ax \geq -4$$
$$2x(1 - a) \geq -4$$
$$x(1 - a) \geq -2$$
$$x \leq -\dfrac{2}{1 - a}, \text{ or } \dfrac{2}{a - 1}$$

We reversed the inequality symbol when we divided because when $a > 1$, then $1 - a < 0$.

The solution set is $\left\{x \middle| x \leq \dfrac{2}{a - 1}\right\}$.

92. $\left\{y \middle| y \geq -\dfrac{10}{b + 4}\right\}$

93. $a(by - 2) \geq b(2y + 5)$

$$aby - 2a \geq 2by + 5b$$
$$aby - 2by \geq 2a + 5b$$
$$y(ab - 2b) \geq 2a + 5b$$
$$y \geq \dfrac{2a + 5b}{ab - 2b}, \text{ or } \dfrac{2a + 5b}{b(a - 2)}$$

The inequality symbol remained unchanged when we divided because when $a > 2$ and $b > 0$, then $ab - 2b > 0$.

The solution set is $\left\{y \middle| y \geq \dfrac{2a + 5b}{b(a - 2)}\right\}$.

94. $\left\{x \middle| x < \dfrac{4c + 3d}{6c - 2d}\right\}$

95. $c(2 - 5x) + dx > m(4 + 2x)$

$$2c - 5cx + dx > 4m + 2mx$$
$$-5cx + dx - 2mx > 4m - 2c$$
$$x(-5c + d - 2m) > 4m - 2c$$
$$x[d - (5c + 2m)] > 4m - 2c$$
$$x > \dfrac{4m - 2c}{d - (5c + 2m)}$$

The inequality symbol remained unchanged when we divided because when $5c + 2m < d$, then $d - (5c + 2m) > 0$.

The solution set is $\left\{x \middle| x > \dfrac{4m - 2c}{d - (5c + 2m)}\right\}$.

96. $\left\{x \middle| x < \dfrac{-3a + 2d}{c - (4a + 5d)}\right\}$

97. False. If $a = 2$, $b = 3$, $c = 4$, and $d = 5$, then $2 < 3$ and $4 < 5$ but $2 - 4 = 3 - 5$.

98. False; $-3 < -2$, but $9 > 4$.

99. *Writing Exercise*

100. *Writing Exercise*

101. $x + 5 \le 5 + x$

$\qquad 5 \le 5 \qquad$ Subtracting x

We get an inequality that is true for all real numbers x. Thus the solution set is all real numbers.

102. $\emptyset$

103. $0^2 = 0$, $x^2 > 0$ for $x \ne 0$

The solution is $\{x | x$ is a real number and $x \ne 0\}$.

Exercise Set 4.2

1. $\{7, 9, 11\} \cap \{9, 11, 13\}$

The numbers 9 and 11 are common to both sets, so the intersection is $\{9, 11\}$.

2. $\{2, 4, 8, 9, 10\}$

3. $\{1, 5, 10, 15\} \cup \{5, 15, 20\}$

The numbers in either or both sets are 1, 5, 10, 15, and 20, so the union is $\{1, 5, 10, 15, 20\}$.

4. $\{5\}$

5. $\{a, b, c, d, e, f\} \cap \{b, d, f\}$

The letters b, d, and f are common to both sets, so the intersection is $\{b, d, f\}$.

6. $\{a, b, c\}$

7. $\{r, s, t\} \cup \{r, u, t, s, v\}$

The letters in either or both sets are r, s, t, u, and v, so the union is $\{r, s, t, u, v\}$.

8. $\{m, o, p\}$

9. $\{3, 6, 9, 12\} \cap \{5, 10, 15\}$

There are no numbers common to both sets, so the solution set has no members. It is $\emptyset$.

10. $\{1, 4, 5, 6, 8, 9\}$

11. $\{3, 5, 7\} \cup \emptyset$

The numbers in either or both sets are 3, 5, and 7, so the union is $\{3, 5, 7\}$.

12. $\emptyset$

13. $3 < x < 8$

This inequality is an abbreviation for the conjunction $3 < x \ and \ x < 8$. The graph is the intersection of two separate solution sets: $\{x | 3 < x\} \cap \{x | x < 8\} = \{x | 3 < x < 8\}$.

Interval notation: $(3, 8)$

14.

$[0, 4]$

15. $-6 \le y \le -2$

This inequality is an abbreviation for the conjunction $-6 \le y \ and \ y \le -2$.

Interval notation: $[-6, -2]$

16.

$[-9, -5)$

17. $x < -2 \ or \ x > 3$

The graph of this disjunction is the union of the graphs of the individual solution sets $\{x | x < -2\}$ and $\{x | x > 3\}$.

Interval notation: $(-\infty, -2) \cup (3, \infty)$

18.

$(-\infty, -5) \cup (1, \infty)$

19. $x \le -1 \ or \ x > 5$

Interval notation: $(-\infty, -1] \cup (5, \infty)$

20.

$(-\infty, -5] \cup (2, \infty)$

21. $-4 \le -x < 2$

$\qquad 4 \ge x > -2 \qquad$ Multiplying by -1 and reversing the inequality symbols

$-2 < x \le 4 \qquad$ Rewriting

Interval notation: $(-2, 4]$

32.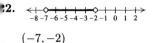

$(-7, -2)$

33. $t < 2$ *or* $t < 5$

Observe that every number that is less than 2 is also less than 5. Then $t < 2$ *or* $t < 5$ is equivalent to $t < 5$ and the graph of this disjunction is the set $\{t | t < 5\}$.

Interval notation: $(-\infty, 5)$

33. $x > -2$ *and* $x < 4$

This conjunction can be abbreviated as $-2 < x < 4$.

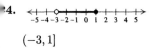

Interval notation: $(-2, 4)$

34.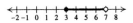

$(-1, \infty)$

34.

$(-3, 1]$

35. $x > -1$ *or* $x \leq 3$

The graph of this disjunction is the union of the graphs of the individual solution sets:

$\{x | x > -1\} \cup \{x | x \leq 3\}$ = the set of all real numbers.

Interval notation: $(-\infty, \infty)$

35. $5 > a$ *or* $a > 7$

Interval notation: $(-\infty, 5) \cup (7, \infty)$

36.

$(-\infty, -3) \cup [2, \infty)$

36.

$(-\infty, \infty)$

37. $x \geq 5$ *or* $-x \geq 4$

Multiplying the second inequality by -1 and reversing the inequality symbols, we get $x \geq 5$ *or* $x \leq -4$.

Interval notation: $(-\infty - 4] \cup [5, \infty)$

37. $x \geq 5$ *and* $x > 7$

The graph of this conjunction is the intersection of two separate solution sets: $\{x | x \geq 5\} \cap \{x | x > 7\} = \{x | x > 7\}$.

Interval notation: $(7, \infty)$

38.

$(-\infty, -6) \cup (-3, \infty)$

38.

$(-\infty, -4]$

39. $4 > y$ *and* $y \geq -6$

This conjunction can be abbreviated as $-6 \leq x < 4$.

Interval notation: $[-6, 4)$

39.

$$-1 < t + 2 < 7$$
$$-1 - 2 < t < 7 - 2$$
$$-3 < t < 5$$

The solution set is $\{t | -3 < t < 5\}$, or $(-3, 5)$.

40.

$(-6, 0]$

40. $\{t | -4 < t \leq 4\}$, or $(-4, 4]$

41. $x < 7$ *and* $x \geq 3$

This conjunction can be abbreviated as $3 \leq x < 7$.

Interval notation: $[3, 7)$

41.

$$2 < x + 3 \quad and \quad x + 1 \leq 5$$
$$-1 < x \quad and \quad x \leq 4$$

We can abbreviate the answer as $-1 < x \leq 4$. The solution set is $\{x | -1 < x \leq 4\}$, or $(-1, 4]$.

42.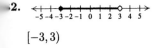

$[-3, 3)$

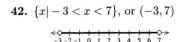

42. $\{x| - 3 < x < 7\}$, or $(-3, 7)$

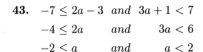

43. $-7 \le 2a - 3 \quad and \quad 3a + 1 < 7$

$\quad\quad -4 \le 2a \quad\quad and \quad\quad 3a < 6$

$\quad\quad -2 \le a \quad\quad\; and \quad\quad\quad a < 2$

We can abbreviate the answer as $-2 \le a < 2$. The solution set is $\{a| - 2 \le a < 2\}$, or $[-2, 2)$.

44. $\{n| - 2 \le n \le 4\}$, or $[-2, 4]$

45. $x + 7 \le -2$ or $x + 7 \ge -3$

Observe that any real number is either less than or equal to -2 or greater than or equal to -3. Then the solution set is $\{x|x$ is a real number$\}$, or $(-\infty, \infty)$.

46. $\{x|x < -8$ or $x \ge -1\}$, or $(-\infty, -8) \cup [-1, \infty)$

47. $2 \le 3x - 1 \le 8$

$\quad 3 \le 3x \le 9$

$\quad 1 \le x \le 3$

The solution set is $\{x|1 \le x \le 3\}$, or $[1, 3]$.

48. $\{x|1 \le x \le 4\}$, or $[1, 4]$

49. $-21 \le -2x - 7 < 0$

$\quad -14 \le -2x < 7$

$\quad 7 \ge x > -\dfrac{7}{2}$, or

$\quad -\dfrac{7}{2} < x \le 7$

The solution set is $\left\{x\middle| -\dfrac{7}{2} < x \le 7\right\}$, or $\left(-\dfrac{7}{2}, 7\right]$.

50. $\left\{t\middle| - 4 < t \le -\dfrac{10}{3}\right\}$, or $\left(-4, -\dfrac{10}{3}\right]$

51. $3x - 1 \le 2$ or $3x - 1 \ge 8$

$\quad\quad 3x \le 3 \quad or \quad\quad 3x \ge 9$

$\quad\quad\; x \le 1 \quad or \quad\quad\; x \ge 3$

The solution set is $\{x|x \le 1$ or $x \ge 3\}$, or $(-\infty, 1] \cup [3, \infty)$.

52. $\{x|x \le 1$ or $x \ge 5\}$, or $(-\infty, 1] \cup [5, \infty)$

53. $2x - 7 < -1$ or $2x - 7 > 1$

$\quad\quad 2x < 6 \quad or \quad\quad 2x > 8$

$\quad\quad\; x < 3 \quad or \quad\quad\; x > 4$

The solution set is $\{x|x < 3$ or $x > 4\}$, or $(-\infty, 3) \cup (4, \infty)$.

54. $\left\{x\middle| x < -4$ or $x > \dfrac{2}{3}\right\}$, or $(-\infty, -4) \cup \left(\dfrac{2}{3}, \infty\right)$

55. $6 > 2a - 1$ or $-4 \le -3a + 2$

$\quad\quad 7 > 2a \quad\quad or \quad -6 \le -3a$

$\quad\quad \dfrac{7}{2} > a \quad\quad or \quad\quad 2 \ge a$

The solution set is $\left\{a\middle|\dfrac{7}{2} > a\right\} \cup \{a|2 \ge a\} =$

$\left\{a\middle|\dfrac{7}{2} > a\right\}$, or $\left\{a\middle|a < \dfrac{7}{2}\right\}$, or $\left(-\infty, \dfrac{7}{2}\right)$.

56. All real numbers, or $(-\infty, \infty)$

57. $a + 4 < -1$ and $3a - 5 < 7$

$\quad\quad a < -5 \quad and \quad\quad 3a < 12$

$\quad\quad a < -5 \quad and \quad\quad\; a < 4$

The solution set is $\{a|a < -5\} \cap \{a|a < 4\} = \{a|a < -5\}$, or $(-\infty, -5)$.

68. $\{a|a > 4\}$, or $(4, \infty)$

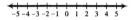

69. $3x + 2 < 2$ _or_ $4 - 2x < 14$

$\qquad 3x < 0$ _or_ $\quad -2x < 10$

$\qquad x < 0$ _or_ $\qquad x > -5$

The solution set is $\{x|x < 0\} \cup \{x|x > -5\}$ = the set of all real numbers, or $(-\infty, \infty)$.

70. $\{x|x \le -2 \text{ or } x > 3\}$, or $(-\infty, -2] \cup (3, \infty)$

71. $2t - 7 \le 5$ _or_ $5 - 2t > 3$

$\quad 2t \le 12$ _or_ $\quad -2t > -2$

$\qquad t \le 6$ _or_ $\qquad t < 1$

The solution set is $\{t|t \le 6\} \cup \{t|t < 1\} = \{t|t \le 6\}$, or $(-\infty, 6]$.

72. $\{a|a \ge -1\}$, or $[-1, \infty)$

73. From the graph we observe that the values of x for which $2x - 5 > -7$ _and_ $2x - 5 < 7$ are $\{x| -1 < x < 6\}$, or $(-1, 6)$.

74. $\{x|x < -3 \text{ or } x > 6\}$, or $(-\infty, -3) \cup (6, \infty)$

75. $f(x) = \dfrac{9}{x + 7}$

$f(x)$ cannot be computed when the denominator is 0. Since $x + 7 = 0$ is equivalent to $x = -7$, we have Domain of $f = \{x|x$ is a real number _and_ $x \ne -7\} = (-\infty, -7) \cup (-7, \infty)$.

76. $(-\infty, -3) \cup (-3, \infty)$

77. $f(x) = \sqrt{x - 6}$

The expression $\sqrt{x - 6}$ is not a real number when $x - 6$ is negative. Thus, the domain of f is the set of all x-values for which $x - 6 \ge 0$. Since $x - 6 \ge 0$ is equivalent to $x \ge 6$, we have Domain of $f = [6, \infty)$.

78. $[2, \infty)$

79. $f(x) = \dfrac{x + 3}{2x - 5}$

$f(x)$ cannot be computed when the denominator is 0. Since $2x - 5 = 0$ is equivalent to $x = \dfrac{5}{2}$, we have

Domain of $f = \left\{x \middle| x \text{ is a real number } and \ x \ne \dfrac{5}{2}\right\}$, or $\left(-\infty, \dfrac{5}{2}\right) \cup \left(\dfrac{5}{2}, \infty\right)$.

70. $\left(-\infty, -\dfrac{4}{3}\right) \cup \left(-\dfrac{4}{3}, \infty\right)$

71. $f(x) = \sqrt{2x + 8}$

The expression $\sqrt{2x + 8}$ is not a real number when $2x + 8$ is negative. Thus, the domain of f is the set of all x-values for which $2x + 8 \ge 0$. Since $2x + 8 \ge 0$ is equivalent to $x \ge -4$, we have Domain of $f = [-4, \infty)$.

72. $(-\infty, 2]$

73. $f(x) = \sqrt{8 - 2x}$

The expression $\sqrt{8 - 2x}$ is not a real number when $8 - 2x$ is negative. Thus, the domain of f is the set of all x-values for which $8 - 2x \ge 0$. Since $8 - 2x \ge 0$ is equivalent to $x \le 4$, we have Domain of $f = (-\infty, 4]$.

74. $(-\infty, 5]$

75. $f(x) = \sqrt{x + 3}$, $g(x) = \sqrt{4 - x}$

The domain of f is the set of all x-values for which $x + 3 \ge 0$, or $[-3, \infty)$. The domain of g is the set of all x-values for which $4 - x \ge 0$, or $(-\infty, 4]$. The intersection of the domains is $[-3, 4]$.

76. $\left[-\dfrac{1}{2}, \dfrac{3}{5}\right]$

77. $f(x) = \sqrt{4x - 3}$, $g(x) = \sqrt{2x + 9}$

The domain of f is the set of all x-values for which $4x - 3 \ge 0$, or $\left[\dfrac{3}{4}, \infty\right)$. The domain of g is the set of all x-values for which $2x + 9 \ge 0$, or $\left[-\dfrac{9}{2}, \infty\right)$. The intersection of the domains is $\left[\dfrac{3}{4}, \infty\right)$.

78. $(-\infty, 1]$

79. $f(x) = \dfrac{x}{x - 5}$, $g(x) = \sqrt{3x + 2}$

The domain of f is $\{x|x$ is a real number _and_ $x \ne 5\}$. The domain of g is $\left[-\dfrac{2}{3}, \infty\right)$. The intersection of the domains is $\left[-\dfrac{2}{3}, 5\right) \cup (5, \infty)$.

80. $\left(-\infty, -\dfrac{1}{2}\right) \cup \left(-\dfrac{1}{2}, \dfrac{4}{5}\right]$

81. _Writing Exercise_

82. *Writing Exercise*

83. Graph: $y = 5$

The graph of any constant function $y = c$ is a horizontal line that crosses the vertical axis at $(0, c)$. Thus, the graph of $y = 5$ is a horizontal line that crosses the vertical axis at $(0, 5)$.

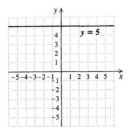

84.

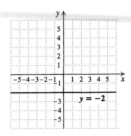

85. Graph $f(x) = |x|$

We make a table of values, plot points, and draw the graph.

x	$f(x)$
-5	5
-2	2
0	0
1	1
4	4

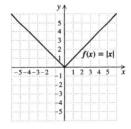

86.

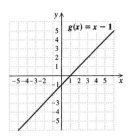

87. Graph both equations.

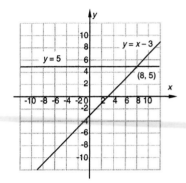

The solution (point of intersection) is apparently $(8, 5)$.

$$\frac{y = x - 3}{5 \ ? \ 8 - 3}$$
$$5 \ | \ 5 \qquad \text{TRUE}$$

$$\frac{y = 5}{5 \ ? \ 5 \quad \text{TRUE}}$$

The solution is $(8, 5)$.

88. $(-5, -3)$

89. *Writing Exercise*

90. *Writing Exercise*

91. **Familiarize.** Let $c =$ the number of crossings per year. Then at the $3 per crossing rate, the total cost of c crossings is $3c$. Two six-month passes cost $2 \cdot \$15$, or $\$30$. The additional $0.50 per crossing toll brings the total cost of c crossings to $\$30 + \$0.50c$. A one-year pass costs $\$150$ regardless of the number of crossings.

Translate. We write an inequality that states that the cost of c crossings per year using the six-month passes is less than the cost using the $3 per crossing toll and is less than the cost using the one-year pass.

$$30 + 0.50c < 3c \text{ and } 30 + 0.50c < 150$$

Carry out. We solve the inequality.

$$30 + 0.50c < 3c \quad \text{and} \quad 30 + 0.50c < 150$$
$$30 < 2.5c \quad \text{and} \qquad 0.50c < 120$$
$$12 < c \qquad \text{and} \qquad c < 240$$

This result can be written as $12 < c < 240$.

Check. When we substitute values of c less than 12, between 12 and 240, and greater than 240, we find that the result checks. Since we cannot check every possible value of c, we stop here.

State. For more than 12 crossings but less than 240 crossings per year the six-month passes are the most economical choice.

92. $0 \text{ ft} \le d \le 198 \text{ ft}$

93. Solve $32 < f(x) < 46$, or $32 < 2(x + 10) < 46$.

$$32 < 2(x + 10) < 46$$
$$32 < 2x + 20 < 46$$
$$12 < 2x < 26$$
$$6 < x < 13$$

For U.S. dress sizes between 6 and 13, dress sizes in Italy will be between 32 and 46.

94. From 2011 through 2036

95. a) Substitute $\frac{5}{9}(F - 32)$ for C in the given inequality.

$$1063 \le \frac{5}{9}(F - 32) < 2660$$
$$9 \cdot 1063 \le 9 \cdot \frac{5}{9}(F - 32) < 9 \cdot 2660$$
$$9567 \le 5(F - 32) < 23{,}940$$
$$9567 \le 5F - 160 < 23{,}940$$
$$9727 \le 5F < 24{,}100$$
$$1945.4 \le F < 4820$$

The inequality for Fahrenheit temperatures is $1945.4° \le F < 4820°$.

b) Substitute $\frac{5}{9}(F - 32)$ for C in the given inequality.

$$960.8 \le \frac{5}{9}(F - 32) < 2180$$
$$9(960.8) \le 9 \cdot \frac{5}{9}(F - 32) < 9 \cdot 2180$$
$$8647.2 \le 5(F - 32) < 19{,}620$$
$$8647.2 \le 5F - 160 < 19{,}620$$
$$8807.2 \le 5F < 19{,}780$$
$$1761.44 \le F < 3956$$

The inequality for Fahrenheit temperatures is $1761.44° \le F < 3956°$.

96. $1965 \le t \le 1981$

97. $4a - 2 \le a + 1 \le 3a + 4$

$$4a - 2 \le a + 1 \quad and \quad a + 1 \le 3a + 4$$
$$3a \le 3 \qquad and \qquad -3 \le 2a$$
$$a \le 1 \qquad and \qquad -\frac{3}{2} \le a$$

The solution set is $\left\{a \middle| -\frac{3}{2} \le a \le 1\right\}$, or $\left[-\frac{3}{2}, 1\right]$.

98. $\left\{m \middle| m < \frac{6}{5}\right\}$, or $\left(-\infty, \frac{6}{5}\right)$

99. $x - 10 < 5x + 6 \le x + 10$

$$-10 < 4x + 6 \le 10$$
$$-16 < 4x \le 4$$
$$-4 < x \le 1$$

The solution set is $\{x | -4 < x \le 1\}$, or $(-4, 1]$.

100. $\left\{x \middle| -\frac{1}{8} < x < \frac{1}{2}\right\}$, or $\left(-\frac{1}{8}, \frac{1}{2}\right)$

101. If $-b < -a$, then $-1(-b) > -1(-a)$, or $b > a$, or $a < b$. The statement is true.

102. False

103. Let $a = 5$, $c = 12$, and $b = 2$. Then $a < c$ and $b < c$, but $a \not< b$. The given statement is false.

104. True

105. $f(x) = \dfrac{\sqrt{5 + 2x}}{x - 1}$

The expression $\sqrt{5 + 2x}$ is not a real number when $5 + 2x$ is negative. Then for $5 + 2x \ge 0$, or for $x \ge -\frac{5}{2}$, the numerator of $f(x)$ is a real number. In addition, $f(x)$ cannot be computed when the denominator is 0. Since $x - 1 = 0$ is equivalent to $x = 1$, we have Domain of $f = \left\{x \middle| x \ge -\frac{5}{2} \text{ and } x \ne 1\right\}$, or $\left[-\frac{5}{2}, 1\right) \cup (1, \infty)$.

106. $\left\{x \middle| x \le \frac{3}{4} \text{ and } x \ne -7\right\}$, or $(-\infty, -7) \cup \left(-7, \frac{3}{4}\right]$

107. Left to the student

Exercise Set 4.3

1. The solutions of $|x + 2| = 3$ are the first coordinates of the points of intersection of $y_1 = \text{abs}(x + 2)$ and $y_2 = 3$. They are -5 and 1, so the solution set is $\{-5, 1\}$.

2. $[-5, 1]$

3. The graph of $y_1 = \text{abs}(x + 2)$ lies below the graph of $y_2 = 3$ for $\{x| -5 < x < 1\}$, or on $(-5, 1)$.

4. $(-\infty, -5) \cup (1, \infty)$

5. The graph of $y_1 = \text{abs}(x + 2)$ lies on or above the graph of $y_2 = 3$ for $\{x| x \le -5 \text{ or } x \ge 1\}$, or on $(\infty, -5] \cup [1, \infty)$.

6. $\emptyset$

7. $\quad |x| = 4$

$\quad\quad x = -4 \text{ or } x = 4 \quad$ Using the absolute-value principle

The solution set is $\{-4, 4\}$.

8. $\{-9, 9\}$

9. $|x| = -5$

The absolute value of a number is always nonnegative. Therefore, the solution set is $\emptyset$.

10. $\emptyset$

11. $\quad |y| = 7.3$

$\quad\quad y = -7.3 \text{ or } y = 7.3 \quad$ Using the absolute-value principle

The solution set is $\{-7.3, 7.3\}$.

12. $\{0\}$

13. $\quad |m| = 0$

$\quad\quad m = 0$

$\quad\quad \{0\}$

The only number whose absolute value is 0 is 0. The solution set is $\{0\}$.

14. $\{-5.5, 5.5\}$

15. $|5x + 2| = 7$

$5x + 2 = -7 \text{ or } 5x + 2 = 7 \quad$ Absolute-value principle

$\quad 5x = -9 \text{ or } \quad\quad 5x = 5$

$\quad x = -\dfrac{9}{5} \text{ or } \quad\quad x = 1$

The solution set is $\left\{-\dfrac{9}{5}, 1\right\}$.

16. $\left\{-\dfrac{1}{2}, \dfrac{7}{2}\right\}$

17. $|7x - 2| = -9$

Absolute value is always nonnegative, so the equation has no solution. The solution set is $\emptyset$.

18. $\emptyset$

19. $|x - 3| = 8$

$\quad x - 3 = -8 \text{ or } x - 3 = 8 \quad$ Absolute value principle

$\quad\quad x = -5 \text{ or } \quad\quad x = 11$

The solution set is $\{-5, 11\}$.

20. $\{-4, 8\}$

21. $|x - 6| = 1$

$\quad x - 6 = -1 \text{ or } x - 6 = 1$

$\quad\quad x = 5 \quad\text{ or } \quad\quad x = 7$

The solution set is $\{5, 7\}$.

22. $\{2, 8\}$

23. $|x - 4| = 5$

$\quad x - 4 = -5 \text{ or } x - 4 = 5$

$\quad\quad x = -1 \text{ or } \quad\quad x = 9$

The solution set is $\{-1, 9\}$.

24. $\{-2, 16\}$

25. $\quad |2y| - 5 = 13$

$\quad\quad |2y| = 18 \quad$ Adding 5

$\quad 2y = -18 \text{ or } 2y = 18$

$\quad y = -9 \quad\text{ or } \quad y = 9$

The solution set is $\{-9, 9\}$.

26. $\{-8, 8\}$

27. $\quad 7|z| + 2 = 16 \quad$ Adding -2

$\quad\quad 7|z| = 14 \quad$ Multiplying by $\dfrac{1}{7}$

$\quad\quad |z| = 2$

$\quad z = -2 \text{ or } z = 2$

The solution set is $\{-2, 2\}$.

28. $\left\{-\dfrac{11}{5}, \dfrac{11}{5}\right\}$

. $\left|\dfrac{4-5x}{6}\right| = 3$

$\dfrac{4-5x}{6} = -3 \quad or \quad \dfrac{4-5x}{6} = 3$

$4-5x = -18 \quad or \quad 4-5x = 18$

$-5x = -22 \quad or \quad -5x = 14$

$x = \dfrac{22}{5} \quad or \quad x = -\dfrac{14}{5}$

The solution set is $\left\{-\dfrac{14}{5}, \dfrac{22}{5}\right\}$.

. $\{-7, 8\}$

. $|t-7| + 1 = 4 \quad$ Adding -1

$|t-7| = 3$

$t-7 = -3 \quad or \quad t-7 = 3$

$t = 4 \quad or \quad t = 10$

The solution set is $\{4, 10\}$.

2. $\{-12, 2\}$

3. $3|2x-5| - 7 = -1$

$3|2x-5| = 6$

$|2x-5| = 2$

$2x-5 = -2 \quad or \quad 2x-5 = 2$

$2x = 3 \quad or \quad 2x = 7$

$x = \dfrac{3}{2} \quad or \quad x = \dfrac{7}{2}$

The solution set is $\left\{\dfrac{3}{2}, \dfrac{7}{2}\right\}$.

4. $\left\{-\dfrac{1}{3}, 3\right\}$

5. $|3x-4| = 8$

$3x-4 = -8 \quad or \quad 3x-4 = 8$

$3x = -4 \quad or \quad 3x = 12$

$x = -\dfrac{4}{3} \quad or \quad x = 4$

The solution set is $\left\{-\dfrac{4}{3}, 4\right\}$.

6. $\left\{-\dfrac{3}{2}, \dfrac{17}{2}\right\}$

7. $|x| - 2 = 6.3$

$|x| = 8.3$

$x = -8.3 \quad or \quad x = 8.3$

The solution set is $\{-8.3, 8.3\}$.

8. $\{-11, 11\}$

39. $\left|\dfrac{3x-2}{5}\right| = 2$

$\dfrac{3x-2}{5} = -2 \quad or \quad \dfrac{3x-2}{5} = 2$

$3x-2 = -10 \quad or \quad 3x-2 = 10$

$3x = -8 \quad or \quad 3x = 12$

$x = -\dfrac{8}{3} \quad or \quad x = 4$

The solution set is $\left\{-\dfrac{8}{3}, 4\right\}$.

40. $\{-1, 2\}$

41. $|x+4| = |2x-7|$

$x+4 = 2x-7 \quad or \quad x+4 = -(2x-7)$

$4 = x-7 \quad or \quad x+4 = -2x+7$

$11 = x \quad or \quad 3x+4 = 7$

$3x = 3$

$x = 1$

The solution set is $\{1, 11\}$.

42. $\left\{-\dfrac{11}{2}, \dfrac{1}{4}\right\}$

43. $|x-9| = |x+6|$

$x-9 = x+6 \quad or \quad x-9 = -(x+6)$

$-9 = 6 \quad or \quad x-9 = -x-6$

False – $\quad\quad\quad\quad 2x-9 = -6$

yields no $\quad\quad\quad\quad 2x = 3$

solution $\quad\quad\quad\quad x = \dfrac{3}{2}$

The solution set is $\left\{\dfrac{3}{2}\right\}$.

44. $\left\{-\dfrac{1}{2}\right\}$

45. $|5t+7| = |4t+3|$

$5t+7 = 4t+3 \quad or \quad 5t+7 = -(4t+3)$

$t+7 = 3 \quad or \quad 5t+7 = -4t-3$

$t = -4 \quad or \quad 9t+7 = -3$

$9t = -10$

$t = -\dfrac{10}{9}$

The solution set is $\left\{-4, -\dfrac{10}{9}\right\}$.

46. $\left\{-\dfrac{3}{5}, 5\right\}$

47. $|n - 3| = |3 - n|$

$n - 3 = 3 - n$ or $n - 3 = -(3 - n)$

$2n - 3 = 3$ or $n - 3 = -3 + n$

$2n = 6$ or $-3 = -3$

$n = 3$ True for all real values of n

The solution set is the set of all real numbers.

48. All real numbers

49. $|7 - a| = |a + 5|$

$7 - a = a + 5$ or $7 - a = -(a + 5)$

$7 = 2a + 5$ or $7 - a = -a - 5$

$2 = 2a$ or $7 = -5$

$1 = a$ False

The solution set is $\{1\}$.

50. $\left\{ -\dfrac{1}{2} \right\}$

51. $\left| \dfrac{1}{2}x - 5 \right| = \left| \dfrac{1}{4}x + 3 \right|$

$\dfrac{1}{2}x - 5 = \dfrac{1}{4}x + 3$ or $\dfrac{1}{2}x - 5 = -\left(\dfrac{1}{4}x + 3 \right)$

$\dfrac{1}{4}x - 5 = 3$ or $\dfrac{1}{2}x - 5 = -\dfrac{1}{4}x - 3$

$\dfrac{1}{4}x = 8$ or $\dfrac{3}{4}x - 5 = -3$

$x = 32$ or $\dfrac{3}{4}x = 2$

$x = \dfrac{8}{3}$

The solution set is $\left\{ 32, \dfrac{8}{3} \right\}$.

52. $\left\{ -\dfrac{48}{37}, -\dfrac{144}{5} \right\}$

53. $|a| \le 7$

$-7 \le a \le 7$ Part (b)

The solution set is $\{a| -7 \le a \le 7\}$, or $[-7, 7]$.

54. $\{x| -2 < x < 2\}$, or $(-2, 2)$

55. $|x| > 8$

$x < -8$ or $8 < x$ Part (c)

The solution set is $\{x|x < -8$ or $x > 8\}$, or $(-\infty, -8) \cup (8, \infty)$.

56. $\{a|a \le -3$ or $a \ge 3\}$, or $(-\infty, -3] \cup [3, \infty)$

57. $|t| > 0$

$t < 0$ or $0 < t$ Part (c)

The solution set is $\{t|t < 0$ or $t > 0\}$, or $\{t|t \ne 0\}$, or $(-\infty, 0) \cup (0, \infty)$.

58. $\{t|t \le -1.7$ or $t \ge 1.7\}$, or $(-\infty, -1.7] \cup [1.7, \infty)$

59. $|x - 3| < 5$

$-5 < x - 3 < 5$ Part (b)

$-2 < x < 8$

The solution set is $\{x| -2 < x < 8\}$, or $(-2, 8)$.

60. $\{x| -2 < x < 4\}$, or $(-2, 4)$

61. $|x + 2| \le 6$

$-6 \le x + 2 \le 6$ Part (b)

$-8 \le x \le 4$ Adding -2

The solution set is $\{x| -8 \le x \le 4\}$, or $[-8, 4]$.

62. $\{x| -5 \le x \le -3\}$, or $[-5, -3]$

63. $|x - 3| + 2 > 7$

$|x - 3| > 5$ Adding -2

$x - 3 < -5$ or $5 < x - 3$ Part (c)

$x < -2$ or $8 < x$

The solution set is $\{x|x < -2$ or $x > 8\}$, or $(-\infty, -2) \cup (8, \infty)$.

64. All real numbers, or $(-\infty, \infty)$

65. $|2y - 7| > -5$

Since absolute value is never negative, any value of $2y - 7$, and hence any value of y, will satisfy the inequality. The solution set is the set of all real numbers, or $(-\infty, \infty)$.

66. $\left\{ y \middle| y < -\dfrac{4}{3} \ or \ y > 4 \right\}$, or $\left(-\infty, -\dfrac{4}{3} \right) \cup (4, \infty)$

67. $|3a - 4| + 2 \geq 8$

$|3a - 4| \geq 6$ Adding -2

$3a - 4 \leq -6 \ or \ 6 \leq 3a - 4$ Part (c)

$3a \leq -2 \ or \ 10 \leq 3a$

$a \leq -\dfrac{2}{3} \ or \ \dfrac{10}{3} \leq a$

The solution set is $\left\{ a \middle| a \leq -\dfrac{2}{3} \ or \ a \geq \dfrac{10}{3} \right\}$, or

$\left(-\infty, -\dfrac{2}{3} \right] \cup \left[\dfrac{10}{3}, \infty \right)$.

68. $\left\{ a \middle| a \leq -\dfrac{3}{2} \ or \ a \geq \dfrac{13}{2} \right\}$, or $\left(-\infty, -\dfrac{3}{2} \right] \cup \left[\dfrac{13}{2}, \infty \right)$

69. $|y - 3| < 12$

$-12 < y - 3 < 12$ Part (b)

$-9 < y < 15$ Adding 3

The solution set is $\{y | -9 < y < 15\}$, or $(-9, 15)$.

70. $\{p | -1 < p < 5\}$ or $(-1, 5)$

71. $9 - |x + 4| \leq 5$

$-|x + 4| \leq -4$

$|x + 4| \geq 4$ Multiplying by -1

$x + 4 \leq -4 \ or \ 4 \leq x + 4$ Part (c)

$x \leq -8 \ or \ 0 \leq x$

The solution set is $\{x | x \leq -8 \ or \ x \geq 0\}$, or $(-\infty, -8] \cup [0, \infty)$.

72. $\{x | x \leq 2 \ or \ x \geq 8\}$, or $(-\infty, 2] \cup [8, \infty)$

73. $|4 - 3y| > 8$

$4 - 3y < -8 \ or \ 8 < 4 - 3y$ Part (c)

$-3y < -12 \ or \ 4 < -3y$ Adding -4

$y > 4 \ or \ -\dfrac{4}{3} > y$ Multiplying by $-\dfrac{1}{3}$

The solution set is $\left\{ y \middle| y < -\dfrac{4}{3} \ or \ y > 4 \right\}$, or

$\left(-\infty, -\dfrac{4}{3} \right) \cup (4, \infty)$.

74. $\emptyset$

75. $|3 - 4x| < -5$

Absolute value is always nonnegative, so the inequality has no solution. The solution set is $\emptyset$.

76. $\left\{ a \middle| -\dfrac{7}{2} \leq a \leq 6 \right\}$, or $\left[-\dfrac{7}{2}, 6 \right]$

77. $\left| \dfrac{2 - 5x}{4} \right| \geq \dfrac{2}{3}$

$\dfrac{2 - 5x}{4} \leq -\dfrac{2}{3} \ or \ \dfrac{2}{3} \leq \dfrac{2 - 5x}{4}$ Part (c)

$2 - 5x \leq -\dfrac{8}{3} \ or \ \dfrac{8}{3} \leq 2 - 5x$ Multiplying by 4

$-5x \leq -\dfrac{14}{3} \ or \ \dfrac{2}{3} \leq -5x$ Adding -2

$x \geq \dfrac{14}{15} \ or \ -\dfrac{2}{15} \geq x$ Multiplying by $-\dfrac{1}{5}$

The solution set is $\left\{ x \middle| x \leq -\dfrac{2}{15} \ or \ x \geq \dfrac{14}{15} \right\}$, or

$\left(-\infty, -\dfrac{2}{15} \right] \cup \left[\dfrac{14}{15}, \infty \right)$.

78. $\left\{ x \middle| x < -\dfrac{43}{24} \ or \ x > \dfrac{9}{8} \right\}$, or $\left(-\infty, -\dfrac{43}{24} \right) \cup \left(\dfrac{9}{8}, \infty \right)$

79. $|m + 5| + 9 \leq 16$

$|m + 5| \leq 7$ Adding -9

$-7 \leq m + 5 \leq 7$

$-12 \leq m \leq 2$

The solution set is $\{m| - 12 \leq m \leq 2\}$, or $[-12, 2]$.

80. $\{t|t \leq 6 \ \ or \ \ t \geq 8\}$, or $(-\infty, 6] \cup [8, \infty)$

81. $25 - 2|a + 3| > 19$

$-2|a + 3| > -6$

$|a + 3| < 3$ Multiplying by $-\dfrac{1}{2}$

$-3 < a + 3 < 3$ Part (b)

$-6 < a < 0$

The solution set is $\{a| - 6 < a < 0\}$, or $(-6, 0)$.

82. $\left\{a\middle| - \dfrac{13}{2} < a < \dfrac{5}{2}\right\}$, or $\left(-\dfrac{13}{2}, \dfrac{5}{2}\right)$

83. $|2x - 3| \leq 4$

$-4 \leq 2x - 3 \leq 4$ Part (b)

$-1 \leq 2x \leq 7$ Adding 3

$-\dfrac{1}{2} \leq x \leq \dfrac{7}{2}$ Multiplying by $\dfrac{1}{2}$

The solution set is $\left\{x\middle| - \dfrac{1}{2} \leq x \leq \dfrac{7}{2}\right\}$, or $\left[-\dfrac{1}{2}, \dfrac{7}{2}\right]$.

84. $\left\{x\middle| - 1 \leq x \leq \dfrac{1}{5}\right\}$, or $\left[-1, \dfrac{1}{5}\right]$

85. $2 + |3x - 4| \geq 13$

$|3x - 4| \geq 11$

$3x - 4 \leq -11 \ \ or \ \ 11 \leq 3x - 4$ Part (c)

$3x \leq -7 \ \ or \ \ 15 \leq 3x$

$x \leq -\dfrac{7}{3} \ \ or \ \ 5 \leq x$

The solution set is $\left\{x\middle|x \leq -\dfrac{7}{3} \ \ or \ \ x \geq 5\right\}$, or

$\left(-\infty, -\dfrac{7}{3}\right] \cup [5, \infty)$.

86. $\left\{x\middle|x \leq -\dfrac{23}{9} \ \ or \ \ x \geq 3\right\}$, or $\left(-\infty, -\dfrac{23}{9}\right] \cup [3, \infty)$

87. $7 + |2x - 1| < 16$

$|2x - 1| < 9$

$-9 < 2x - 1 < 9$ Part (b)

$-8 < 2x < 10$

$-4 < x < 5$

The solution set is $\{x| - 4 < x < 5\}$, or $(-4, 5)$.

88. $\left\{x\middle| - \dfrac{16}{3} < x < 4\right\}$, or $\left(-\dfrac{16}{3}, 4\right)$

89. *Writing Exercise*

90. *Writing Exercise*

91. $2x - 3y = 7$,   (1)

$3x + 2y = -10$ (2)

We will use the elimination method. First, multiply equation (1) by 2 and equation (2) by 3 and add to eliminate a variable.

$4x - 6y = \ \ \ 14$

$\underline{9x + 6y = -30}$

$13x \qquad = -16$

$x = -\dfrac{16}{13}$

Now substitute $-\dfrac{16}{13}$ for x in either of the original equations and solve for y.

$$3x + 2y = -10 \qquad (2)$$

$$3\left(-\frac{16}{13}\right) + 2y = -10$$

$$-\frac{48}{13} + 2y = -10$$

$$2y = -\frac{130}{13} + \frac{48}{13}$$

$$2y = -\frac{82}{13}$$

$$y = -\frac{41}{13}$$

The solution is $\left(-\dfrac{16}{13}, -\dfrac{41}{13}\right)$.

92. $(-2, -3)$

93. $x = -2 + 3y, \quad (1)$

$x - 2y = 2 \qquad (2)$

We will use the substitution method. We substitute $-2 + 3y$ for x in equation (2).

$$x - 2y = 2 \quad (2)$$
$$(-2 + 3y) - 2y = 2 \quad \text{Substituting}$$
$$-2 + y = 2$$
$$y = 4$$

Now substitute 4 for y in equation (1) and find x.

$$x = -2 + 3 \cdot 4 = -2 + 12 = 10$$

The solution is $(10, 4)$.

94. $(-1, 7)$

95. Graph both equations.

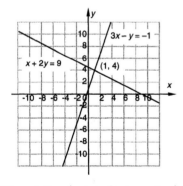

The solution (point of intersection) is apparently $(1, 4)$.

Check:

$x + 2y = 9$		$3x - y = -1$	
$1 + 2 \cdot 4 \;?\; 9$		$3 \cdot 1 - 4 \;?\; -1$	
$1 + 8$		$3 - 4$	
9	9 TRUE	-1	-1 TRUE

The solution is $(1, 4)$.

96. $(24, -41)$

97. *Writing Exercise*

98. *Writing Exercise*

99. From the definition of absolute value, $|3t - 5| = 3t - 5$ only when $3t - 5 \geq 0$. Solve $3t - 5 \geq 0$.

$$3t - 5 \geq 0$$
$$3t \geq 5$$
$$t \geq \frac{5}{3}$$

The solution set is $\left\{t \middle| t \geq \dfrac{5}{3}\right\}$, or $\left[\dfrac{5}{3}, \infty\right)$.

100. $\{x | x \text{ is a real number}\}$, or $(-\infty, \infty)$

101. $2 \leq |x - 1| \leq 5$

$2 \leq |x - 1| \; and \; |x - 1| \leq 5$.

For $2 \leq |x - 1|$:

$$x - 1 \leq -2 \;\; or \;\; 2 \leq x - 1$$
$$x \leq -1 \;\; or \;\; 3 \leq x$$

The solution set of $2 \leq |x-1|$ is $\{x | x \leq -1 \; or \; x \geq 3\}$.

For $|x - 1| \leq 5$:

$$-5 \leq x - 1 \leq 5$$
$$-4 \leq x \leq 6$$

The solution set of $|x - 1| \leq 5$ is $\{x | -4 \leq x \leq 6\}$.

The solution set of $2 \leq |x - 1| \leq 5$ is

$$\{x | x \leq -1 \; or \; x \geq 3\} \cap \{x | -4 \leq x \leq 6\}$$
$$= \{x | -4 \leq x \leq -1 \; or \; 3 \leq x \leq 6\}, \; or$$
$$[-4, -1] \cup [3, 6].$$

102. $\left\{-\dfrac{1}{7}, \dfrac{7}{3}\right\}$

103. $t - 2 \leq |t - 3|$

$$t - 3 \leq -(t - 2) \;\; or \;\; t - 2 \leq t - 3$$
$$t - 3 \leq -t + 2 \;\;\; or \;\;\; -2 \leq -3$$
$$2t - 3 \leq 2 \qquad\qquad \text{False}$$
$$2t \leq 5$$
$$t \leq \frac{5}{2}$$

The solution set is $\left\{t \middle| t \leq \dfrac{5}{2}\right\}$, or $\left(-\infty, \dfrac{5}{2}\right]$.

104. $|x| < 3$

105. Using part (b), we find that $-5 \leq y \leq 5$ is equivalent to $|y| \leq 5$.

106. $|x| \geq 6$

107. $x < -4 \ or \ 4 < x$

$\qquad |x| > 4 \qquad$ Using part (c)

108. $|x + 3| > 5$

109. $-5 < x < 1$

$\qquad -3 < x + 2 < 3 \qquad$ Adding 2

$\qquad |x + 2| < 3 \qquad$ Using part (b)

110. $|x - 7| < 2$, or $|7 - x| < 2$.

111. The distance from x to 5 is $|x - 5|$ or $|5 - x|$, so we have $|x - 5| < 1$, or $|5 - x| < 1$.

112. $|x - 3| \leq 4$

113. The length of the segment from -4 to 8 is $|-4 - 8| = |-12| = 12$ units. The midpoint of the segment is $\dfrac{-4 + 8}{2} = \dfrac{4}{2} = 2$. Thus, the interval extends 12/2, or 6, units on each side of 2. An inequality for which the open interval is the solution set is $|x - 2| < 6$.

114. $|x + 4| < 3$

115. The length of the segment from 2 to 12 is $|2 - 12| = |-10| = 10$ units. The midpoint of the segment is $\dfrac{2 + 12}{2} = \dfrac{14}{2} = 7$. Thus, the interval extends 10/2, or 5, units on each side of 7. An inequality for which the closed interval is the solution set is $|x - 7| \leq 5$.

116. $\left\{d \Big| 5\dfrac{1}{2} \text{ ft} \leq d \leq 6\dfrac{1}{2} \text{ ft}\right\}$, or $\left[5\dfrac{1}{2} \text{ ft}, \ 6\dfrac{1}{2} \text{ ft}\right]$

Exercise Set 4.4

1. We replace x with -4 and y with 2.

$$\dfrac{2x + 3y < -1}{2(-4) + 3 \cdot 2 \ ? \ -1}$$

$\qquad -8 + 6 \ \Big|$

$\qquad\qquad -2 \ \Big| \ -1 \qquad$ TRUE

Since $-2 < -1$ is true, $(-4, 2)$ is a solution.

2. No

3. We replace x with 8 and y with 14.

$$\dfrac{2y - 3x \geq 9}{2 \cdot 14 - 3 \cdot 8 \ ? \ 9}$$

$\qquad 28 - 24 \ \Big|$

$\qquad\qquad 4 \ \Big| \ 9 \qquad$ FALSE

Since $4 > 9$ is false, $(8, 14)$ is not a solution.

4. Yes

5. Graph: $y > \dfrac{1}{2}x$

We first graph the line $y = \dfrac{1}{2}x$. We draw the line dashed since the inequality symbol is $>$. To determine which half-plane to shade, test a point not on the line. We try $(0, 1)$:

$$\dfrac{y > \dfrac{1}{2}x}{1 \ ? \ \dfrac{1}{2} \cdot 0}$$

$\qquad 1 \ \Big| \ 0 \qquad$ TRUE

Since $1 > 0$ is true, $(0, 1)$ is a solution as are all of the points in the half-plane containing $(0, 1)$. We shade that half-plane and obtain the graph.

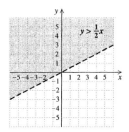

6.

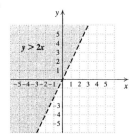

7. Graph: $y \geq x - 3$

First graph the line $y = x - 3$. Draw it solid since the inequality symbol is $\geq$. Test the point $(0, 0)$ to determine if it is a solution.

$$\dfrac{y \geq x - 3}{0 \ ? \ 0 - 3}$$

$\qquad 0 \ \Big| \ -3 \qquad$ TRUE

Since $0 \geq -3$ is true, we shade the half-plane that contains $(0,0)$ and obtain the graph.

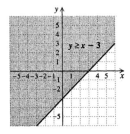

8.

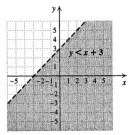

9. Graph: $y \leq x + 4$

First graph the line $y = x + 4$. Draw it solid since the inequality symbol is $\leq$. Test the point $(0,0)$ to determine if it is a solution.

$$\frac{y \leq x + 4}{0\ ?\ 0 + 4}$$
$$0 \mid 4 \qquad \text{TRUE}$$

Since $0 \leq 4$ is true, we shade the half-plane that contains $(0,0)$ and obtain the graph.

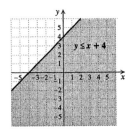

10.

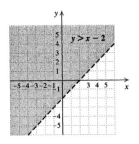

11. Graph: $x - y \leq 5$

First graph the line $x - y = 5$. Draw a solid line since the inequality symbol is $\leq$. Test the point $(0,0)$ to determine if it is a solution.

$$\frac{x - y \leq 5}{0 - 0\ ?\ 5}$$
$$0 \mid 5 \qquad \text{TRUE}$$

Since $0 \leq 5$ is true, we shade the half-plane that contains $(0,0)$ and obtain the graph.

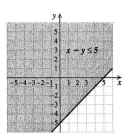

12.

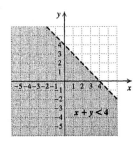

13. Graph: $2x + 3y < 6$

First graph $2x + 3y = 6$. Draw the line dashed since the inequality symbol is $<$. Test the point $(0,0)$ to determine if it is a solution.

$$\frac{2x + 3y < 6}{2 \cdot 0 + 3 \cdot 0\ ?\ 6}$$
$$0 \mid 6 \qquad \text{TRUE}$$

Since $0 < 6$ is true, we shade the half-plane containing $(0,0)$ and obtain the graph.

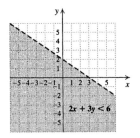

14.

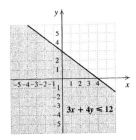

15. Graph: $2x - y \leq 4$

We first graph $2x - y = 4$. Draw the line solid since the inequality symbol is $\leq$. Test the point $(0,0)$ to determine if it is a solution.

$$\frac{2x - y \leq 4}{2 \cdot 0 - 0 \;?\; 4}$$
$$0 \;\big|\; 4 \qquad \text{TRUE}$$

Since $0 \leq 4$ is true, we shade the half-plane containing $(0,0)$ and obtain the graph.

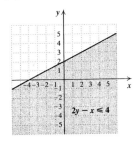

16.

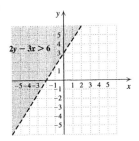

17. Graph: $2x - 2y \geq 8 + 2y$

$$2x - 4y \geq 8$$

First graph $2x - 4y = 8$. Draw the line solid since the inequality symbol is $\geq$. Test the point $(0,0)$ to determine if it is a solution.

$$\frac{2x - 4y \geq 8}{2 \cdot 0 - 4 \cdot 0 \;?\; 8}$$
$$0 \;\big|\; 8 \qquad \text{FALSE}$$

Since $0 \geq 8$ is false, we shade the half-plane that does not contain $(0,0)$ and obtain the graph.

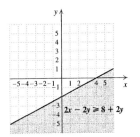

18.

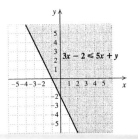

19. Graph: $y \geq 2$

We first graph $y = 2$. Draw the line solid since the inequality symbol is $\geq$. Test the point $(0,0)$ to determine if it is a solution.

$$\frac{y \geq 2}{0 \;?\; 2} \qquad \text{FALSE}$$

Since $0 \geq 2$ is false, we shade the half-plane that does not contain $(0,0)$ and obtain the graph.

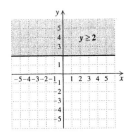

20.

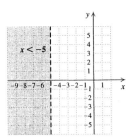

21. Graph: $x \leq 7$

We first graph $x = 7$. We draw the line solid since the inequality symbol is $\leq$. Test the point $(0,0)$ to determine if it is a solution.

$$x \leq 7$$

0 ? 7   TRUE

Since $0 \leq 7$ is true, we shade the half-plane containing $(0,0)$ and obtain the graph.

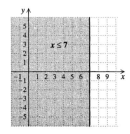

22.

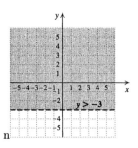

n

23. Graph: $-2 < y < 6$

This is a system of inequalities:
$$-2 < y,$$
$$y < 6$$

We graph the equation $-2 = y$ and see that the graph of $-2 < y$ is the half-plane above the line $-2 = y$. We also graph $y = 6$ and see that the graph of $y < 6$ is the half-plane below the line $y = 6$.

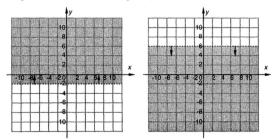

Finally, we shade the intersection of these graphs.

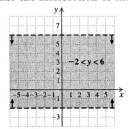

24.

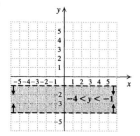

25. Graph: $-4 \leq x \leq 5$

This is a system of inequalities:
$$-4 \leq x,$$
$$x \leq 5$$

Graph $-4 \leq x$ and $x \leq 5$.

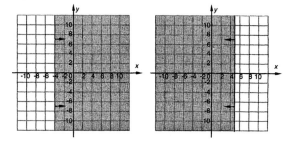

Then we shade the intersection of these graphs.

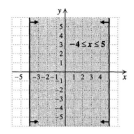

26.

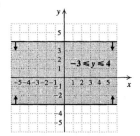

27. Graph: $0 \leq y \leq 3$

This is a system of inequalities:
$$0 \leq y,$$
$$y \leq 3$$

Graph $0 \leq y$ and $y \leq 3$.

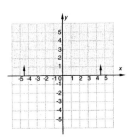 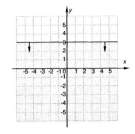

Then we shade the intersection of these graphs.

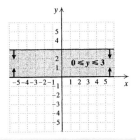

28.

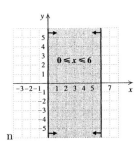

29.

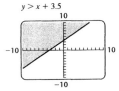

30.

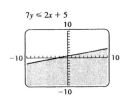

31. First get y alone on one side of the inequality.

$$8x - 2y < 11$$
$$-2y < -8x + 11$$
$$y > \frac{-8x + 11}{-2}$$

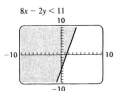

32.

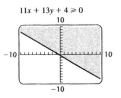

33. Graph: $y > x,$

$\qquad\qquad y < -x + 2$

We graph the lines $y = x$ and $y = -x + 2$, using dashed lines. We indicate the region for each inequality by the arrows at the ends of the lines. Note where the regions overlap and shade the region of solutions.

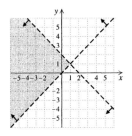

34.

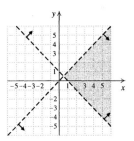

35. Graph: $y \geq x,$

$\qquad\qquad y \leq 2x - 4$

Graph $y = x$ and $y = 2x - 4$, using solid lines. Indicate the region for each inequality by arrows, and shade the region where they overlap.

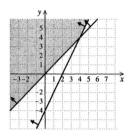

36.

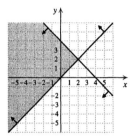

37. Graph: $y \le -3,$

   $x \ge -1$

Graph $y = -3$ and $x = -1$ using solid lines. Indicate the region for each inequality by arrows, and shade the region where they overlap.

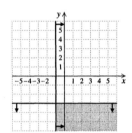

38.

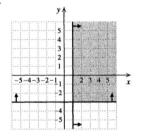

39. Graph: $x > -4,$

   $y < -2x + 3$

Graph the lines $x = -4$ and $y = -2x + 3$, using dashed lines. Indicate the region for each inequality by arrows, and shade the region where they overlap.

40.

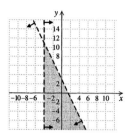

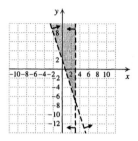

41. Graph: $y \le 3,$

   $y \ge -x + 2$

Graph the lines $y = 3$ and $y = -x + 2$, using solid lines. Indicate the region for each inequality by arrows, and shade the region where they overlap.

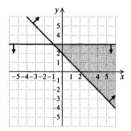

42.

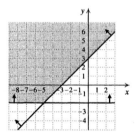

43. Graph: $x + y \le 6,$

   $x - y \le 4$

Graph the lines $x + y = 6$ and $x - y = 4$, using solid lines. Indicate the region for each inequality by arrows, and shade the region where they overlap.

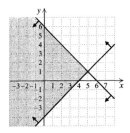

44.

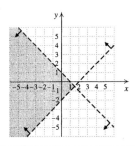

45. Graph: $y + 3x > 0$,

$\qquad\qquad y + 3x < 2$

Graph the lines $y + 3x = 0$ and $y + 3x = 2$, using dashed lines. Indicate the region for each inequality by arrows, and shade the region where they overlap.

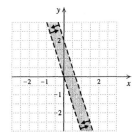

46.

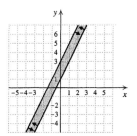

47. Graph: $y \le 2x - 1$,   (1)

$\qquad\qquad y \ge -2x + 1$,   (2)

$\qquad\qquad x \le 3$ (3)

Graph the lines $y = 2x - 1$, $y = -2x + 1$, and $x = 3$ using solid lines. Indicate the region for each inequality by arrows, and shade the region where they overlap.

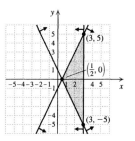

To find the vertex we solve three different systems of related equations.

From (1) and (2) we have $y = 2x - 1$,

$\qquad\qquad\qquad\qquad\qquad y = -2x + 1$.

Solving, we obtain the vertex $\left(\dfrac{1}{2}, 0\right)$.

From (1) and (3) we have $y = 2x - 1$,

$\qquad\qquad\qquad\qquad\qquad x = 3$.

Solving, we obtain the vertex $(3, 5)$.

From (2) and (3) we have $y = -2x + 1$,

$\qquad\qquad\qquad\qquad\qquad x = 3$.

Solving, we obtain the vertex $(3, -5)$.

48.

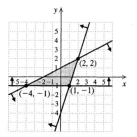

49. Graph: $x + 2y \le 12$,   (1)

$\qquad\qquad 2x + y \le 12$ (2)

$\qquad\qquad\quad x \ge 0$,   (3)

$\qquad\qquad\quad y \ge 0$ (4)

Graph the lines $x + 2y = 12$, $2x + y = 12$, $x = 0$, and $y = 0$ using solid lines. Indicate the region for each inequality by arrows, and shade the region where they overlap.

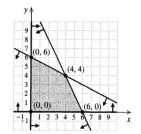

To find the vertices we solve four different systems of equations.

From (1) and (2) we have $x + 2y = 12,$
$$2x + y = 12.$$
Solving, we obtain the vertex $(4, 4)$.

From (1) and (3) we have $x + 2y = 12,$
$$x = 0.$$
Solving, we obtain the vertex $(0, 6)$.

From (2) and (4) we have $2x + y = 12,$
$$y = 0.$$
Solving, we obtain the vertex $(6, 0)$.

From (3) and (4) we have $x = 0,$
$$y = 0.$$
Solving, we obtain the vertex $(0, 0)$.

0.

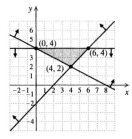

1.   Graph: $8x + 5y \leq 40,$ (1)
$$x + 2y \leq 8$$ (2)
$$x \geq 0,$$ (3)
$$y \geq 0$$ (4)

Graph the lines $8x + 5y = 40$, $x + 2y = 8$, $x = 0$, and $y = 0$ using solid lines. Indicate the region for each inequality by arrows, and shade the region where they overlap.

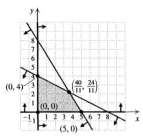

To find the vertices we solve four different systems of equations.

From (1) and (2) we have $8x + 5y = 40,$
$$x + 2y = 8.$$
Solving, we obtain the vertex $\left(\dfrac{40}{11}, \dfrac{24}{11} \right)$.

From (1) and (4) we have $8x + 5y = 40,$
$$y = 0.$$
Solving, we obtain the vertex $(5, 0)$.

From (2) and (3) we have $x + 2y = 8,$
$$x = 0.$$
Solving, we obtain the vertex $(0, 4)$.

From (3) and (4) we have $x = 0,$
$$y = 0.$$
Solving, we obtain the vertex $(0, 0)$.

52.

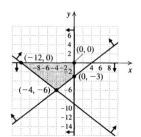

53.   Graph: $y - x \geq 1,$ (1)
$$y - x \leq 3,$$ (2)
$$2 \leq x \leq 5$$ (3)

Think of (3) as two inequalities:
$$2 \leq x,$$ (4)
$$x \leq 5$$ (5)

Graph the lines $y - x = 1$, $y - x = 3$, $x = 2$, and $x = 5$, using solid lines. Indicate the region for each inequality by arrows, and shade the region where they overlap.

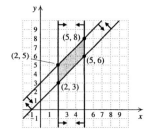

To find the vertices we solve four different systems of equations.

From (1) and (4) we have $y - x = 1,$
$$x = 2.$$
Solving, we obtain the vertex $(2, 3)$.

From (1) and (5) we have $y - x = 1,$
$$x = 5.$$
Solving, we obtain the vertex $(5, 6)$.

From (2) and (4) we have $y - x = 3,$
$$x = 2.$$
Solving, we obtain the vertex $(2, 5)$.
From (2) and (5) we have $y - x = 3,$
$$x = 5.$$
Solving, we obtain the vertex $(5, 8)$.

54.

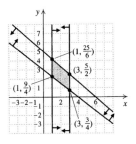

55. *Writing Exercise*

56. *Writing Exercise*

57. **Familiarize**. We let x and y represent the number of pounds of peanuts and fancy nuts in the mixture, respectively. We organize the given information in a table.

Type of nuts	Peanuts	Fancy	Mixture
Amount	x	y	10
Price per pound	$2.50	$7	
Value	$2.5x$	$7y$	40

Translate. We get a system of equations from the first and third rows of the table.
$$x + \ y = 10,$$
$$2.5x + 7y = 40$$
Clearing decimals we have
$$x + \ y = 10, \quad (1)$$
$$25x + 70y = 400. \quad (2)$$
Carry out. We use the elimination method. Multiply Equation (1) by -25 and add.

$$-25x - 25y = -250$$
$$\underline{25x + 70y = 400}$$
$$45y = 150$$
$$y = \frac{10}{3}, \quad \text{or } 3\frac{1}{3}$$

Substitute $\frac{10}{3}$ for y in Equation (1) and solve for x.
$$x + y = 10$$
$$x + \frac{10}{3} = 10$$
$$x = \frac{20}{3}, \quad \text{or } 6\frac{2}{3}$$

Check. The sum of $6\frac{2}{3}$ and $3\frac{1}{3}$ is 10. The value of the mixture is $2.5\left(\dfrac{20}{3}\right) + 7\left(\dfrac{10}{3}\right)$, or $\dfrac{50}{3} + \dfrac{70}{3}$, or \$40. These numbers check.

State. $6\frac{2}{3}$ lb of peanuts and $3\frac{1}{3}$ lb of fancy nuts should be used.

58. Hendersons: 10 bags; Savickis: 4 bags

59. **Familiarize**. Let $x =$ the number of cardholders tickets that were sold and $y =$ the number of non-cardholders tickets. We arrange the information in a table.

	Card-holders	Non-card-holders	Total
Price	\$1.25	\$2	
Number sold	x	y	203
Money taken in	$1.25x$	$2y$	\$310

Translate. The last two rows of the table give us two equations. The total number of tickets sold was 203, so we have
$$x + y = 203.$$
The total amount of money collected was \$310, so we have
$$1.25x + 2y = 310.$$
We can multiply the second equation on both sides by 100 to clear decimals. The resulting system is
$$x + y = 203, \quad (1)$$
$$125x + 200y = 31,000. \quad (2)$$
Carry out. We use the elimination method. We multiply on both sides of Equation (1) by -125 and then add.

$$-125x - 125y = -25,375 \quad \text{Multiplying by } -125$$
$$\underline{125x + 200y = \ \ 31,000}$$
$$75y = \ \ \ \ 5625$$
$$y = \ \ \ \ \ \ 75$$

We go back to Equation (1) and substitute 75 for y.
$$x + y = 203$$
$$x + 75 = 203$$
$$x = 128$$

Check. The number of tickets sold was $128 + 75$, or 203. The money collected was $\$1.25(128) + \$2(75)$, or $\$160 + \$150$, or \$310. These numbers check.

State. 128 cardholders tickets and 75 non-cardholder tickets were sold.

60. 70 student tickets, 130 adult tickets

61. *Familiarize*. The formula for the area of a triangle with base b and height h is $A = \frac{1}{2}bh$.

Translate. Substitute 200 for A and 16 for b in the formula.

$$A = \frac{1}{2}bh$$

$$200 = \frac{1}{2} \cdot 16 \cdot h$$

Carry out. We solve the equation.

$$200 = \frac{1}{2} \cdot 16 \cdot h$$

$200 = 8h$ Multiplying

$25 = h$ Dividing by 8 on both sides

Check. The area of a triangle with base 16 ft and height 25 ft is $\frac{1}{2} \cdot 16 \cdot 25$, or 200 ft^2. The answer checks.

State. The seed can fill a triangle that is 25 ft tall.

62. 11%

63. *Writing Exercise*

64. *Writing Exercise*

65. Graph: $x + y > 8$,

           $x + y \le -2$

Graph the line $x + y = 8$ using a dashed line and graph $x + y = -2$, using a solid line. Indicate the region for each inequality by arrows. The regions do not overlap (the solution set is $\emptyset$), so we do not shade any portion of the graph.

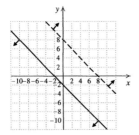

66.

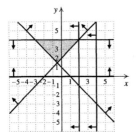

67. Graph: $x - 2y \le 0$,

                $-2x + y \le 2$,

                $x \le 2$,

                $y \le 2$,

                $x + y \le 4$

Graph the five inequalities above, and shade the region where they overlap.

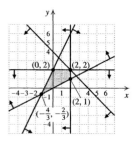

68. $x \ge -2$,

     $y \le 2$,

     $x \le 0$,

     $y \ge 0$;

     $x \ge 0$,

     $y \le 2$,

     $x \le 2$,

     $y \ge 0$;

     $x \ge 0$,

     $y \le 0$,

     $x \le 2$,

     $y \ge -2$;

     $x \ge -2$,

     $y \le 0$,

     $x \le 0$,

     $y \ge -2$

69. Both the width and the height must be positive, but they must be less than 62 in. in order to be checked as luggage, so we have:

$$0 < w \le 62,$$

$$0 < h \le 62$$

The girth is represented by $2w + 2h$ and the length is 62 in. In order to meet postal regulations the sum of the girth and the length cannot exceed 108 in., so we have:

$$62 + 2w + 2h \le 108, \text{ or}$$

$$2w + 2h \le 46, \text{ or}$$

$$w + h \le 23$$

Thus, have a system of inequalities:

$$0 < w \leq 62,$$
$$0 < h \leq 62,$$
$$w + h \leq 23$$

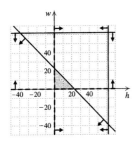

70. $2w + t \geq 60,$
$$w \geq 0,$$
$$t \geq 0$$

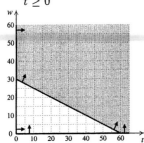

71. Graph: $35c + 75a > 1000,$
$$c \geq 0,$$
$$a \geq 0$$

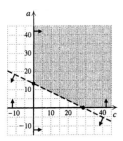

72. $0 < L \leq 94,$
$$0 < W \leq 50$$

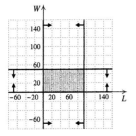

73. The shaded region lies below the graphs of $y = x$ and $y = 2$ and both lines are solid. Thus, we have
$$y \leq x,$$
$$y \leq 2.$$

74. $y \leq x + 1,$
$$x \leq 3,$$
$$y \geq -2$$

75. The shaded region lies below the graphs of $y = x + 2$ and $y = -x + 4$ and above $y = 0$, and all of the lines are solid. Thus, we have
$$y \leq x + 2,$$
$$y \leq -x + 4,$$
$$y \geq 0.$$

76. $y \leq x + 2,$
$$y \geq x - 3.$$

Chapter 5

Polynomials and Polynomial Functions

Exercise Set 5.1

1. $-6x^5 - 8x^3 + x^2 + 3x - 4$

Term	$-6x^5$	$-8x^3$	x^2	$3x$	-4
Degree	5	3	2	1	0
Degree of polynomial	5				

2. 3, 2, 1, 0; 3

3. $y^3 + 2y^7 + x^2y^4 - 8$

Term	y^3	$2y^7$	x^2y^4	-8
Degree	3	7	6	0
Degree of polynomial	7			

4. 2, 5, 7, 0; 7

5. $a^5 + 4a^2b^4 + 6ab + 4a - 3$

Term	a^5	$4a^2b^4$	$6ab$	$4a$	-3
Degree	5	6	2	1	0
Degree of polynomial	6				

6. 6, 8, 4, 2, 0; 8

7. $-4y^3 - 6y^2 + 7y + 19$; $-4y^3$; -4

8. $-18y^4 + 11y^3 + 6y^2 - 5y + 3$; $-18y^4$; -18

9. $3x^7 + 5x^2 - x + 12$; $3x^7$; 3

10. $-10x^4 + 7x^2 - 3x + 9$; $-10x^4$; -10

11. $-a^7 + 8a^5 + 5a^3 - 19a^2 + a$; $-a^7$; -1

12. $a^9 + 11a^4 + a^3 - 5a^2 - 7$; a^9; 1

13. $-9 + 6x - 5x^2 + 3x^4$

14. $9 + 4x - x^3 - 3x^4$

15. $3xy^3 + x^2y^2 + 7x^3y - 5x^4$

16. $-9xy + 5x^2y^2 + 8x^3y^2 - 5x^4$

17. $-7ab + 4ax - 7ax^2 + 4x^6$

18. $-12a + 5xy^8 + 4ax^3 - 3ax^5 + 5x^5$

19. $P(x) = 3x^2 - 2x + 7$

$P(4) = 3 \cdot 4^2 - 2 \cdot 4 + 7$

$\qquad = 48 - 8 + 7$

$\qquad = 47$

$P(0) = 3 \cdot 0^2 - 2 \cdot 0 + 7$

$\qquad = 0 - 0 + 7$

$\qquad = 7$

20. $-51; 5$

21. $\quad P(y) = 8y^3 - 12y - 5$

$P(-2) = 8(-2)^3 - 12(-2) - 5$

$\qquad = -64 + 24 - 5$

$\qquad = -45$

$P\left(\frac{1}{3}\right) = 8\left(\frac{1}{3}\right)^3 - 12 \cdot \frac{1}{3} - 5$

$\qquad = 8 \cdot \frac{1}{27} - 4 - 5$

$\qquad = \frac{8}{27} - 9$

$\qquad = \frac{8}{27} - \frac{243}{27}$

$\qquad = -\frac{235}{27}, \text{ or } -8\frac{19}{27}$

22. $282; -9$

23. $-7x + 5 = -7 \cdot 4 + 5 = -28 + 5 = -23$

24. 3

25. $x^3 - 5x^2 + x = 4^3 - 5 \cdot 4^2 + 4 = 64 - 5 \cdot 16 + 4 = 64 - 80 + 4 = -12$

26. 51

27. $\quad f(x) = -5x^3 + 3x^2 - 4x - 3$

$f(-1) = -5(-1)^3 + 3(-1)^2 - 4(-1) - 3$

$\qquad = 5 + 3 + 4 - 3$

$\qquad = 9$

28. -6

29. $p(n) = n^3 - 3n^2 + 2n$

$$p(20) = 20^3 - 3 \cdot 20^2 + 2 \cdot 20$$
$$= 8000 - 1200 + 40$$
$$= 6840$$

A president, vice president, and treasurer can be elected in 6840 ways.

30.   1320

31. $s(t) = 16t^2$

$$s(3) = 16 \cdot 3^2 = 16 \cdot 9 = 144$$

The scaffold is 144 ft high.

32.   400 ft

33.   Evaluate the polynomial function for $x = 75$:

$$R(x) = 280x - 0.4x^2$$
$$R(75) = 280 \cdot 75 - 0.4(75)^2$$
$$= 21,000 - 0.4(5625)$$
$$= 21,000 - 2250 = 18,750$$

The total revenue is \$18,750.

34.   \$24,000

35.   Evaluate the polynomial function for $x = 75$:

$$C(x) = 5000 + 0.6x^2$$
$$C(75) = 5000 + 0.6(75)^2$$
$$= 5000 + 0.6(5625)$$
$$= 5000 + 3375$$
$$= 8375$$

The total cost is \$8375.

36.   \$11,000

37. $P(a) = 0.4a^2 - 40a + 1039$

$$P(20) = 0.4(20)^2 - 40(20) + 1039$$
$$= 0.4(400) - 40(20) + 1039$$
$$= 160 - 800 + 1039$$
$$= 399$$

There are approximately 399 accidents daily involving an 18-year-old driver.

38.   289 accidents

39.   1995 is 6 years since 1989. Locate 6 on the horizontal axis. From there move vertically to the graph and then horizontally to the vertical axis. This locates a value of about 5.2. Thus, the attendance at NASCAR races in 1995 was about 5.2 million.

40.   8 million

41.   Locate 8 on the horizontal axis. From there move vertically to the graph and then horizontally to the vertical axis. This locates a value of about 6.4. Thus, $A(8) \approx 6.4$ million.

42.   10.1 million

43.   Locate 2 on the horizontal axis. From there move vertically to the graph and then horizontally to the $M(t)$-axis. This locates a value of about 340. Thus, about 340 mg of ibuprofen is in the the bloodstream 2 hr after 400 mg have been swallowed.

44.   About 185 mg

45. M has a minimum value of 0 and a maximum value of about 345, so the range is $[0, 345]$.

46. $[0, 6]$

47.   We evaluate the polynomial for $h = 6.3$ and $r = 1.2$:

$$2\pi rh + 2\pi r^2 = 2\pi(1.2)(6.3) + 2\pi(1.2)^2 \approx 56.5$$

The surface area is about 56.5 in^2.

48.   44.5 in^2

49.   The function has a maximum value of 3 and no minimum is indicated, so the range is $(-\infty, 3]$.

50. $(-\infty, \infty)$

51.   There is no maximum or minimum value indicated by the graph, so the range is $(-\infty, \infty)$.

52. $[-4, \infty)$

53.   The function has a minimum value of -4 and no maximum is indicated, so the range is $[-4, \infty)$.

54. $(-\infty, \infty)$

55.   The function has a minimum value of -65 and no maximum is indicated, so the range is $[-65, \infty)$.

56. $(-\infty, \infty)$

57.   We graph $y = x^2 + 2x + 1$ in the standard viewing window.

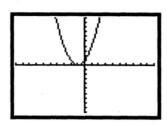

The range appears to be $[0, \infty)$.

58. $[-6.25, \infty)$

59. We graph $y = -2x^2 + 5$ in the standard viewing window.

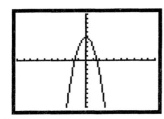

The range appears to be $(-\infty, 5]$.

60. $(-\infty, 1]$

61. We graph $y = -2x^3 + x + 5$ in the standard viewing window.

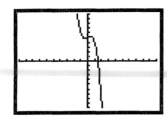

The range appears to be $(-\infty, \infty)$.

62. $(-\infty, -8.3]$

63. We graph $y = x^4 + 2x^3 - 5$ in the standard viewing window.

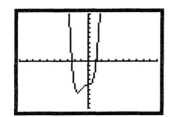

Using TRACE and ZOOM, we estimate that the range is $[-6.7, \infty)$.

64. $(-\infty, \infty)$

65. $5a + 6 - 4 + 2a^3 - 6a + 2$
$= 2a^3 + (5 - 6)a + (6 - 4 + 2)$
$= 2a^3 - a + 4$

66. $5x^2 - x + 15$

67. $3a^2b + 4b^2 - 9a^2b - 7b^2$
$= (3 - 9)a^2b + (4 - 7)b^2$
$= -6a^2b - 3b^2$

68. $-8x^3 - 3x^2y^2$

69. $9x^2 - 3xy + 12y^2 + x^2 - y^2 + 5xy + 4y^2$
$= (9 + 1)x^2 + (-3 + 5)xy + (12 - 1 + 4)y^2$
$= 10x^2 + 2xy + 15y^2$

70. $11a^2 + 3ab - 3b^2$

71. $(8a + 6b - 3c) + (4a - 2b + 2c)$
$= (8 + 4)a + (6 - 2)b + (-3 + 2)c$
$= 12a + 4b - c$

72. $16x + 7y - 5z$

73. $(a^2 - 3b^2 + 4c^2) + (-5a^2 + 2b^2 - c^2)$
$= (1 - 5)a^2 + (-3 + 2)b^2 + (4 - 1)c^2$
$= -4a^2 - b^2 + 3c^2$

74. $-5x^2 + 4y^2 - 11z^2$

75. $(x^2 + 2x - 3xy - 7) + (-3x^2 - x + 2xy + 6)$
$= (1 - 3)x^2 + (2 - 1)x + (-3 + 2)xy + (-7 + 6)$
$= -2x^2 + x - xy - 1$

76. $2a^2 + 3b - 4ab + 4$

77. $(8x^2y - 3xy^2 + 4xy) + (-2x^2y - xy^2 + xy)$
$= (8 - 2)x^2y + (-3 - 1)xy^2 + (4 + 1)xy$
$= 6x^2y - 4xy^2 + 5xy$

78. $22ab - 18ac - 3bc$

79. $(2r^2 + 12r - 11) + (6r^2 - 2r + 4) + (r^2 - r - 2)$
$= (2 + 6 + 1)r^2 + (12 - 2 - 1)r + (-11 + 4 - 2)$
$= 9r^2 + 9r - 9$

80. $-3x^2 - x - 3$

81. $\left(\dfrac{1}{8}xy - \dfrac{3}{5}x^3y^2 + 4.3y^3\right) +$
    $\left(-\dfrac{1}{3}xy - \dfrac{3}{4}x^3y^2 - 2.9y^3\right)$
$= \left(\dfrac{1}{8} - \dfrac{1}{3}\right)xy + \left(-\dfrac{3}{5} - \dfrac{3}{4}\right)x^3y^2 + (4.3 - 2.9)y^3$
$= \left(\dfrac{3}{24} - \dfrac{8}{24}\right)xy + \left(-\dfrac{12}{20} - \dfrac{15}{20}\right)x^3y^2 + 1.4y^3$
$= -\dfrac{5}{24}xy - \dfrac{27}{20}x^3y^2 + 1.4y^3$

82. $-\dfrac{2}{15}xy + \dfrac{19}{12}xy^2 + 1.7x^2y$

83. $5x^3 - 7x^2 + 3x - 9$
   a) $-(5x^3 - 7x^2 + 3x - 9)$ Writing the opposite
                                    of P as $-P$
   b) $-5x^3 + 7x^2 - 3x + 9$ Changing the sign
                                    of every term

84. $-(-8y^4 - 18y^3 + 4y - 7)$, $8y^4 + 18y^3 - 4y + 7$

85. $-12y^5 + 4ay^4 - 7by^2$
　a) $-(-12y^5 + 4ay^4 - 7by^2)$
　b) $12y^5 - 4ay^4 + 7by^2$

86. $-(7ax^3y^2 - 8by^4 - 7abx - 12ay)$,
　$-7ax^3y^2 + 8by^4 + 7abx + 12ay$

87. $\quad (7x - 5) - (-3x + 4)$
　$= (7x - 5) + (3x - 4)$
　$= 10x - 9$

88. $14y + 7$

89. $\quad (-3x^2 + 2x + 9) - (x^2 + 5x - 4)$
　$= (-3x^2 + 2x + 9) + (-x^2 - 5x + 4)$
　$= -4x^2 - 3x + 13$

90. $-13y^2 + 2y + 11$

91. $\quad (6a - 2b + c) - (3a + 2b - 2c)$
　$= (6a - 2b + c) + (-3a - 2b + 2c)$
　$= 3a - 4b + 3c$

92. $3x - 10y + 4z$

93. $\quad (3x^2 - 2x - x^3) - (5x^2 - 8x - x^3)$
　$= (3x^2 - 2x - x^3) + (-5x^2 + 8x + x^3)$
　$= -2x^2 + 6x$

94. $5y^2 + 6y + 3y^3$

95. $\quad (5a^2 + 4ab - 3b^2) - (9a^2 - 4ab + 2b^2)$
　$= (5a^2 + 4ab - 3b^2) + (-9a^2 + 4ab - 2b^2)$
　$= -4a^2 + 8ab - 5b^2$

96. $-5y^2 - 6yz - 12z^2$

97. $\quad (6ab - 4a^2b + 6ab^2) - (3ab^2 - 10ab - 12a^2b)$
　$= (6ab - 4a^2b + 6ab^2) + (-3ab^2 + 10ab + 12a^2b)$
　$= 8a^2b + 16ab + 3ab^2$

98. $17xy + 5x^2y^2 - 7y^3$

99. $\quad \left(\dfrac{5}{8}x^4 - \dfrac{1}{4}x^2 - \dfrac{1}{2}\right) - \left(-\dfrac{3}{8}x^4 + \dfrac{3}{4}x^2 + \dfrac{1}{2}\right)$
　$= \left(\dfrac{5}{8}x^4 - \dfrac{1}{4}x^2 - \dfrac{1}{2}\right) + \left(\dfrac{3}{8}x^4 - \dfrac{3}{4}x^2 - \dfrac{1}{2}\right)$
　$= x^4 - x^2 - 1$

100. $\dfrac{29}{24}y^4 - \dfrac{5}{4}y^2 - 11.2y + \dfrac{8}{15}$

101. $\quad P(x) = R(x) - C(x)$
　$P(x) = (280x - 0.4x^2) - (5000 + 0.6x^2)$
　$P(x) = (280x - 0.4x^2) + (-5000 - 0.6x^2)$
　$P(x) = 280x - x^2 - 5000$
　$P(70) = 280 \cdot 70 - 70^2 - 5000$
　$\quad\quad = 19,600 - 4900 - 5000 = 9700$
The profit is $9700.

102. $8000

103. Since $(5x^3 + 2x + 3) + (3x^3 - 1) = 8x^3 + 2x + 2$, statements
　(a) and (c) are true.

104. (a) and (c)

105. Since $(x^2 + 8x + 1) - (-x^2 + 3x + 5) \neq 5x - 4$, statements
　(b) and (d) are true.

106. (b) and (d)

107. *Writing Exercise*

108. *Writing Exercise*

109. $2(x + 3) + 5(x + 2) = 2x + 6 + 5x + 10 = 7x + 16$

110. $10a + 59$

111. $a(a - 1) + 4(a - 1) = a^2 - a + 4a - 4 = a^2 + 3a - 4$

112. $x^2 - x - 2$

113. $x^5 \cdot x^4 = x^{5+4} = x^9$

114. a^8

115. *Writing Exercise*

116. *Writing Exercise*

117. $\quad 2[P(x)]$
　$= 2(13x^5 - 22x^4 - 36x^3 + 40x^2 - 16x + 75)$
　$= 26x^5 - 44x^4 - 72x^3 + 80x^2 - 32x + 150$
Use columns to add:
　$26x^5 - 44x^4 - 72x^3 + 80x^2 - 32x + 150$
　$\underline{42x^5 - 37x^4 + 50x^3 - 28x^2 + 34x + 100}$
　$68x^5 - 81x^4 - 22x^3 + 52x^2 + \ \ 2x + 250$

118. $-3x^5 - 29x^4 - 158x^3 + 148x^2 - 82x + 125$

119. $\quad 2[Q(x)]$
　$= 2(42x^5 - 37x^4 + 50x^3 - 28x^2 + 34x + 100)$
　$= 84x^5 - 74x^4 + 100x^3 - 56x^2 + 68x + 200$

$3[P(x)]$
$= 3(13x^5 - 22x^4 - 36x^3 + 40x^2 - 16x + 75)$
$= 39x^5 - 66x^4 - 108x^3 + 120x^2 - 48x + 225$

Use columns to subtract, adding the opposite of $3[P(x)]$:

$$\begin{array}{r} 84x^5 - 74x^4 + 100x^3 - 56x^2 + 68x + 200 \\ -39x^5 + 66x^4 + 108x^3 - 120x^2 + 48x - 225 \\ \hline 45x^5 - 8x^4 + 208x^3 - 176x^2 + 116x - 25 \end{array}$$

120. $178x^5 - 199x^4 + 6x^3 + 76x^2 + 38x + 600$

121. First we find the number of truffles in the display.

$$N(x) = \frac{1}{6}x^3 + \frac{1}{2}x^2 + \frac{1}{3}x$$

$$N(5) = \frac{1}{6} \cdot 5^3 + \frac{1}{2} \cdot 5^2 + \frac{1}{3} \cdot 5$$

$$= \frac{1}{6} \cdot 125 + \frac{1}{2} \cdot 25 + \frac{5}{3}$$

$$= \frac{125}{6} + \frac{25}{2} + \frac{5}{3}$$

$$= \frac{125}{6} + \frac{75}{6} + \frac{10}{6}$$

$$= \frac{210}{6} = 35$$

There are 35 truffles in the display. Now find the volume of one truffle. Each truffle's diameter is 3 cm, so the radius is $\frac{3}{2}$, or 1.5 cm.

$$V(r) = \frac{4}{3}\pi r^3$$

$$V(1.5) \approx \frac{4}{3}(3.14)(1.5)^3 \approx 14.13 \text{ cm}^3$$

Finally, multiply the number of truffles and the volume of a truffle to find the total volume of chocolate.

$$35(14.13 \text{ cm}^3) = 494.55 \text{ cm}^3$$

The display contains about 494.55 cm^3 of chocolate.

122. 9.8 cm

123. The area of the base is $x \cdot x$, or x^2.
The area of each side is $x \cdot (x - 2)$.
The total area of all four sides is $4x(x - 2)$.

The surface area of this box can be expressed as a polynomial function.

$$S(x) = x^2 + 4x(x - 2)$$
$$= x^2 + 4x^2 - 8x$$
$$= 5x^2 - 8x$$

124. $200\pi rh + 20,000\pi r^2$ cm^2; $0.02\pi rh + 2\pi r^2$ m^2

125. $(2x^{2a} + 4x^a + 3) + (6x^{2a} + 3x^a + 4)$
$= (2 + 6)x^{2a} + (4 + 3)x^a + (3 + 4)$
$= 8x^{2a} + 7x^a + 7$

126. $x^{6a} - 5x^{5a} - 4x^{4a} + x^{3a} - 2x^{2a} + 8$

127. $(2x^{5b} + 4x^{4b} + 3x^{3b} + 8) -$
$(x^{5b} + 2x^{3b} + 6x^{2b} + 9x^b + 8)$
$= (2 - 1)x^{5b} + 4x^{4b} + (3 - 2)x^{3b} - 6x^{2b} - 9x^b + (8 - 8)$
$= x^{5b} + 4x^{4b} + x^{3b} - 6x^{2b} - 9x^b$

Exercise Set 5.2

1. $8a^2 \cdot 4a = (8 \cdot 4)(a^2 \cdot a) = 32a^3$

2. $-10x^4$

3. $5x(-4x^2y) = 5(-4)(x \cdot x^2)y = -20x^3y$

4. $-6a^3b^4$

5. $2x^3y^2(-5x^2y^4) = 2(-5)(x^3 \cdot x^2)(y^2 \cdot y^4) = -10x^5y^6$

6. $-56a^3b^4c^6$

7. $7x(3 - x) = 7x \cdot 3 - 7x \cdot x$
$= 21x - 7x^2$

8. $3a^3 - 12a^2$

9. $5cd(4c^2d - 5cd^2)$
$= 5cd \cdot 4c^2d - 5cd \cdot 5cd^2$
$= 20c^3d^2 - 25c^2d^3$

10. $2a^4 - 5a^5$

11. $(2x + 5)(3x - 4)$
$= 6x^2 - 8x + 15x - 20$ \quad FOIL
$= 6x^2 + 7x - 20$

12. $8a^2 + 10ab - 3b^2$

13. $(m + 2n)(m - 3n)$
$= m^2 - 3mn + 2mn - 6n^2$ \quad FOIL
$= m^2 - mn - 6n^2$

14. $m^2 - 25$

15. $(3y + 8x)(y - 7x)$
$= 3y^2 - 21xy + 8xy - 56x^2$ \quad FOIL
$= 3y^2 - 13xy - 56x^2$

16. $x^2 - xy - 2y^2$

17. $(a^2 - 2b^2)(a^2 - 3b^2)$
$= a^4 - 3a^2b^2 - 2a^2b^2 + 6b^4$ \quad FOIL
$= a^4 - 5a^2b^2 + 6b^4$

18. $6m^4 - 13m^2n^2 + 5n^4$

19.
$$(x - 4)(x^2 + 4x + 16)$$
$$= (x - 4)(x^2) + (x - 4)(4x) + (x - 4)(16)$$
$$\qquad\qquad\qquad \text{Distributive law}$$
$$= x(x^2) - 4(x^2) + x(4x) - 4(4x) + x(16) - 4(16)$$
$$\qquad\qquad\qquad \text{Distributive law}$$
$$= x^3 - 4x^2 + 4x^2 - 16x + 16x - 64$$
$$\qquad\qquad\qquad \text{Multiplying monomials}$$
$$= x^3 - 64 \qquad \text{Collecting like terms}$$

20. $y^3 + 27$

21.
$$(x + y)(x^2 - xy + y^2)$$
$$= (x + y)x^2 + (x + y)(-xy) + (x + y)(y^2)$$
$$= x(x^2) + y(x^2) + x(-xy) + y(-xy) + x(y^2) + y(y^2)$$
$$= x^3 + x^2y - x^2y - xy^2 + xy^2 + y^3$$
$$= x^3 + y^3$$

22. $a^3 - b^3$

23.

$$
\begin{array}{r}
a^2 + a - 1 \\
a^2 + 4a - 5 \\
\hline
-5a^2 - 5a + 5 \quad \text{Multiplying by } -5 \\
4a^3 + 4a^2 - 4a \qquad\quad \text{Multiplying by } 4a \\
a^4 + a^3 - a^2 \qquad\qquad\quad \text{Multiplying by } a^2 \\
\hline
a^4 + 5a^3 - 2a^2 - 9a + 5 \quad \text{Adding}
\end{array}
$$

24. $x^4 - x^3 + x^2 - 3x + 2$

25.

$$
\begin{array}{r}
3a^2b - 2ab + 3b^2 \\
ab - 2b + a \\
\hline
3a^3b - 2a^2b + 3ab^2 \qquad (1) \\
-6b^3 \qquad\qquad +4ab^2 - 6a^2b^2 \qquad (2) \\
3ab^3 \qquad\qquad\qquad -2a^2b^2 + 3a^3b^2 \quad (3) \\
\hline
3ab^3 - 6b^3 + 3a^3b - 2a^2b + 7ab^2 - 8a^2b^2 + 3a^3b^2 \quad (4)
\end{array}
$$

(1) Multiplying by a

(2) Multiplying by $-2b$

(3) Multiplying by ab

(4) Adding

26. $-4x^3y + 3xy^3 - x^2y^2 - 2y^4 + 2x^4$

27.
$$\left(x - \frac{1}{2}\right)\left(x - \frac{1}{4}\right)$$
$$= x^2 - \frac{1}{4}x - \frac{1}{2}x + \frac{1}{8} \qquad \text{FOIL}$$
$$= x^2 - \frac{1}{4}x - \frac{2}{4}x + \frac{1}{8}$$
$$= x^2 - \frac{3}{4}x + \frac{1}{8}$$

28. $b^2 - \frac{2}{3}b + \frac{1}{9}$

29.
$$(1.2x - 3y)(2.5x + 5y)$$
$$= 3x^2 + 6xy - 7.5xy - 15y^2 \qquad \text{FOIL}$$
$$= 3x^2 - 1.5xy - 15y^2$$

30. $12a^2 + 399.928ab - 2.4b^2$

31.
$$P(x) \cdot Q(x) = (3x^2 - 5)(4x^2 - 7x + 1)$$
$$= (3x^2 - 5)(4x^2) + (3x^2 - 5)(-7x) + (3x^2 - 5)(1)$$
$$= 12x^4 - 20x^2 - 21x^3 + 35x + 3x^2 - 5$$
$$= 12x^4 - 21x^3 - 17x^2 + 35x - 5$$

32. $x^5 + 6x^2 - 5x + 5$

33.
$$(a + 4)(a + 5)$$
$$= a^2 + 5a + 4a + 20 \qquad \text{FOIL}$$
$$= a^2 + 9a + 20$$

34. $x^2 + 5x + 6$

35.
$$(y - 8)(y + 3)$$
$$= y^2 + 3y - 8y - 24$$
$$= y^2 - 5y - 24$$

36. $y^2 + 4y - 5$

37.
$$(x + 5)^2$$
$$= x^2 + 2 \cdot x \cdot 5 + 5^2 \quad (A + B)^2 = A^2 + 2AB + B^2$$
$$= x^2 + 10x + 25$$

38. $y^2 - 14y + 49$

39.
$$(x - 2y)^2$$
$$= x^2 - 2(x)(2y) + (2y)^2$$
$$\qquad\qquad (A - B)^2 = A^2 - 2AB + B^2$$
$$= x^2 - 4xy + 4y^2$$

40. $4s^2 + 12st + 9t^2$

41.
$$(2x + 9)(x + 2)$$
$$= 2x^2 + 4x + 9x + 18 \qquad \text{FOIL}$$
$$= 2x^2 + 13x + 18$$

42. $6b^2 - 11b - 10$

43.
$$(10a - 0.12b)^2$$
$$= (10a)^2 - 2(10a)(0.12b) + (0.12b)^2$$
$$\qquad\qquad (A - B)^2 = A^2 - 2AB + B^2$$
$$= 100a^2 - 2.4ab + 0.0144b^2$$

44. $100p^4 + 46p^2q + 5.29q^2$

45. $(2x - 3y)(2x + y)$

$= 4x^2 + 2xy - 6xy - 3y^2$ FOIL

$= 4x^2 - 4xy - 3y^2$

46. $4a^2 - 8ab + 3b^2$

47. $(2x^3 - 3y^2)^2$

$= (2x^3)^2 - 2(2x^3)(3y^2) + (3y^2)^2$

$(A - B)^2 = A^2 - 2AB + B^2$

$= 4x^6 - 12x^3y^2 + 9y^4$

48. $9s^4 + 24s^2t^3 + 16t^6$

49. $(a^2b^2 + 1)^2$

$= (a^2b^2)^2 + 2(a^2b^2) \cdot 1 + 1^2$

$(A + B)^2 = A^2 + 2AB + B^2$

$= a^4b^4 + 2a^2b^2 + 1$

50. $x^4y^2 - 2x^3y^3 + x^2y^4$

51. $P(x) \cdot P(x)$

$= (4x - 1)(4x - 1)$

$= (4x)^2 - 2(4x)(1) + 1^2$

$(A - B)^2 = A^2 - 2AB + B^2$

$= 16x^2 - 8x + 1$

52. $9x^4 + 6x^2 + 1$

53. $[F(x)]^2$

$= \left(2x - \dfrac{1}{3}\right)^2$

$= (2x)^2 - 2 \cdot 2x \cdot \dfrac{1}{3} + \left(\dfrac{1}{3}\right)^2$

$(A - B)^2 = A^2 - 2AB + B^2$

$= 4x^2 - \dfrac{4}{3}x + \dfrac{1}{9}$

54. $25x^2 - 5x + \dfrac{1}{4}$

55. $(c + 2)(c - 2)$

$= c^2 - 2^2$ $(A + B)(A - B) = A^2 - B^2$

$= c^2 - 4$

56. $x^2 - 9$

57. $(4x + 1)(4x - 1)$

$= (4x)^2 - 1^2$ $(A + B)(A - B) = A^2 - B^2$

$= 16x^2 - 1$

58. $9 - 4x^2$

59. $(3m - 2n)(3m + 2n)$

$= (3m)^2 - (2n)^2$ $(A + B)(A - B) = A^2 - B^2$

$= 9m^2 - 4n^2$

60. $9x^2 - 25y^2$

61. $(x^3 + yz)(x^3 - yz)$

$= (x^3)^2 - (yz)^2$ $(A + B)(A - B) = A^2 - B^2$

$= x^6 - y^2z^2$

62. $16a^6 - 25a^2b^2$

63. $(-mn + m^2)(mn + m^2)$

$= (m^2 - mn)(m^2 + mn)$

$= (m^2)^2 - (mn)^2$ $(A + B)(A - B) = A^2 - B^2$

$= m^4 - m^2n^2$

64. $-9b^2 + a^4$, or $a^4 - 9b^2$

65. $(x + 1)(x - 1)(x^2 + 1)$

$= (x^2 - 1^2)(x^2 + 1)$

$= (x^2 - 1)(x^2 + 1)$

$= (x^2)^2 - 1^2$

$= x^4 - 1$

66. $y^4 - 16$

67. $(a - b)(a + b)(a^2 - b^2)$

$= (a^2 - b^2)(a^2 - b^2)$

$= (a^2 - b^2)^2$

$= (a^2)^2 - 2(a^2)(b^2) + (b^2)^2$

$= a^4 - 2a^2b^2 + b^4$

68. $16x^4 - 8x^2y^2 + y^4$

69. $(a + b + 1)(a + b - 1)$

$= [(a + b) + 1][(a + b) - 1]$

$= (a + b)^2 - 1^2$

$= a^2 + 2ab + b^2 - 1$

70. $m^2 + 2mn + n^2 - 4$

71. $(2x + 3y + 4)(2x + 3y - 4)$

$= [(2x + 3y) + 4][(2x + 3y) - 4]$

$= (2x + 3y)^2 - 4^2$

$= 4x^2 + 12xy + 9y^2 - 16$

72. $9a^2 - 12ab + 4b^2 - c^2$

73. $A = P(1 + i)^2$

$A = P(1 + 2i + i^2)$

$A = P + 2Pi + Pi^2$

74. $A = P + Pi + \dfrac{Pi^2}{4}$

75. a) Replace x with $t - 1$.

$$f(t - 1) = (t - 1)^2 + 5$$
$$= t^2 - 2t + 1 + 5$$
$$= t^2 - 2t + 6$$

b)
$$f(a + h) - f(a)$$
$$= [(a + h)^2 + 5] - (a^2 + 5)$$
$$= a^2 + 2ah + h^2 + 5 - a^2 - 5$$
$$= 2ah + h^2$$

c)
$$f(a) - f(a - h)$$
$$= (a^2 + 5) - [(a - h)^2 + 5]$$
$$= a^2 + 5 - (a^2 - 2ah + h^2 + 5)$$
$$= a^2 + 5 - a^2 + 2ah - h^2 - 5$$
$$= 2ah - h^2$$

76. a) $p^2 + 2p + 8$

b) $2ah + h^2$

c) $2ah - h^2$

77. *Writing Exercise*

78. *Writing Exercise*

79. $ab + ac = d$

$$a(b + c) = d$$
$$a = \frac{d}{b + c}$$

80. $y = \dfrac{w}{x + z}$

81. $mn + m = p$

$$m(n + 1) = p$$
$$m = \frac{p}{n + 1}$$

82. $s = \dfrac{t}{r + 1}$

83. **Familiarize.** Let d, n, and q represent the number of rolls of dimes, nickels, and quarters, respectively, that Kacie has. The value of a roll of dimes is 50($0.10), or $5.00; the value of a roll of nickels is 40($0.05), or $2.00; and the value of a roll of quarters is 40($0.25), or $10. Then the value of the rolls of dimes, nickels, and quarters is $5d$, $2n$, and $10q$, respectively.

Translate.

Total number of rolls of coins is 13.

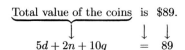

We have a system of equations.

$$d + n + q = 13,$$
$$5d + 2n + 10q = 89,$$
$$d = 3 + n$$

Carry out. Solving the system of equations, we get $(5, 2, 6)$.

Check. The total number of rolls of coins is $5 + 2 + 6$, or 13. The total value of the coins is $\$5 \cdot 5 + \$2 \cdot 2 + \$10 \cdot 6$, or $\$25 + \$4 + \$60$, or $\$89$. The number of rolls of dimes, 5, is three more than 2, the number of rolls of nickels. The answer checks.

State. Kacie has 5 rolls of dimes, 2 rolls of nickels, and 6 rolls of quarters.

84. 8 weekdays

85. *Writing Exercise*

86. *Writing Exercise*

87. $[(-x^a y^b)^4]^a = (x^{4a} y^{4b})^a = x^{4a^2} y^{4ab}$

88. z^{5n^5}

89. $(a^x b^{2y})\left(\dfrac{1}{2} a^{3x} b\right)^2 = (a^x b^{2y})\left(\dfrac{1}{4} a^{6x} b^2\right) = \dfrac{1}{4} a^{7x} b^{2y+2}$

90. $a^{xw+xz} b^{yw+yz}$

91.
$$y^3 z^n (y^{3n} z^3 - 4yz^{2n})$$
$$= y^3 z^n \cdot y^{3n} z^3 - y^3 z^n \cdot 4yz^{2n}$$
$$= y^{3n+3} z^{n+3} - 4y^4 z^{3n}$$

92. $-a^4 - 2a^3 b + 25a^2 + 2ab^3 - 25b^2 + b^4$

93.
$$(a - b + c - d)(a + b + c + d)$$
$$= [(a + c) - (b + d)][(a + c) + (b + d)]$$
$$= (a + c)^2 - (b + d)^2$$
$$= (a^2 + 2ac + c^2) - (b^2 + 2bd + d^2)$$
$$= a^2 + 2ac + c^2 - b^2 - 2bd - d^2$$

94. $\dfrac{4}{9} x^2 - \dfrac{1}{9} y^2 - \dfrac{2}{3} y - 1$

95.
$$(4x^2 + 2xy + y^2)(4x^2 - 2xy + y^2)$$
$$= [(4x^2 + y^2) + 2xy][(4x^2 + y^2) - 2xy]$$
$$= (4x^2 + y^2)^2 - (2xy)^2$$
$$= 16x^4 + 8x^2y^2 + y^4 - 4x^2y^2$$
$$= 16x^4 + 4x^2y^2 + y^4$$

96. $x^4 + x^2 + 25$

97.
$$(x^a + y^b)(x^a - y^b)(x^{2a} + y^{2b})$$
$$= (x^{2a} - y^{2b})(x^{2a} + y^{2b})$$
$$= x^{4a} - y^{4b}$$

98. $x^6 - 1$

99. $(x^{a-b})^{a+b} = x^{(a-b)(a+b)} = x^{a^2-b^2}$

100. $M^{x^2+2xy+y^2}$

101.
$$(x - a)(x - b)(x - c) \cdots (x - z)$$
$$= (x - a)(x - b) \cdots (x - x)(x - y)(x - z)$$
$$= (x - a)(x - b) \cdots 0 \cdot (x - y)(x - z)$$
$$= 0$$

102.

103. One method is as follows. For each equation, let y_1 represent the left-hand side and y_2 represent the right-hand side, and let $y_3 = y_2 - y_1$. Then use a graphing calculator to view the graph of y_3 and/or a table of values for y_3. If $y_3 = 0$, the equation is an identity. If $y_3 \neq 0$, the equation is not an identity.

a) Not an identity

b) Identity

c) Identity

d) Not an identity

e) Not an identity

Exercise Set 5.3

---

1. From the graph we see that $f(x) = 0$ when $x = -3$ or $x = 5$. These are the solutions.

2. $-5, -2$

3. From the graph we see that $f(x) = 0$ when $x = -2$ or $x = 0$. These are the zeros of the function.

4. $-1, 3$

5. From the graph we see that $f(x) = 3$ when $x = -3$ or $x = 1$. These are the solutions.

6. $-2, 2$

7. From the graph we see that $f(x) = 0$ when $x = -4$ or $x = 2$. These are the solutions.

8. 1

9. We can graph $y_1 = x^2$ and $y_2 = 5x$ and use the Intersect feature to find the first coordinates of the points of intersection, or we can begin by rewriting the equation so that one side is 0:
$$x^2 = 5x$$
$$x^2 - 5x = 0 \quad \text{Subtracting } 5x \text{ on both sides}$$

Then graph $y = x^2 - 5x$ and use the Zero feature to find the roots of the equation. In either case, we find that the solutions are 0 and 5.

10. 0, 10

11. We can graph $y_1 = 4x$ and $y_2 = x^2 + 3$ and use the Intersect feature to find the first coordinates of the points of intersection, or we can begin by rewriting the equation so that one side is 0:
$$4x = x^2 + 3$$
$$0 = x^2 + 3 - 4x \quad \text{Subtracting } 4x \text{ on both sides}$$

Then graph $y = x^2 + 3 - 4x$ and use the Zero feature to find the roots of the equation. In either case, we find that the solutions are 1 and 3.

12. $-1, 1$

13. We can graph $y_1 = x^2 + 150$ and $y_2 = 25x$ and use the Intersect feature to find the first coordinates of the points of intersection, or we can begin by rewriting the equation so that one side is 0:
$$x^2 + 150 = 25x$$
$$x^2 + 150 - 25x = 0 \quad \text{Subtracting } 25 \text{ on both sides}$$

Then graph $y = x^2 + 150 - 25x$ and use the Zero feature to find the roots of the equation. In either case, we find that the solutions are 10 and 15.

14. 0.5, 25

15. Graph $y = x^3 - 3x^2 - 2x$ and use the Zero feature to find the roots of the equation. The solutions are 0, 1, and 2.

16. $-2, -1, 1$

37. Graph $y = x^3 - 3x^2 - 198x + 1080$ and use the Zero feature to find the roots of the equation. The solutions are -15, 6, and 12.

38. -20, 1.5, 6

39. Graph $y = 21x^2 + 2x - 3$ and use the Zero feature to find the roots of the equation. The solutions are approximately -0.42857 and 0.33333.

40. -0.09091, 0.83333

41. Graph $y = x^2 - 4x + 45$ and use the Zero feature to find the zeros of the function. They are -5 and 9.

42. -5, 4

43. Graph $y = 2x^2 - 13x - 7$ and use the Zero feature to find the zeros of the function. They are -0.5 and 7.

44. -2.42013, -0.41320

45. Graph $y = x^3 - 2x^2 - 3x$ and use the Zero feature to find the zeros of the function. They are -1, 0, and 3.

46. -2, 0, 2

47. We see that $2x - 1 = 0$ when $x = 0.5$ and $3x + 1 = 0$ when $x = -0.\overline{3}$, so graph III corresponds to the given function.

48. II

49. We see that $4 - x = 0$ when $x = 4$ and $2x - 11 = 0$ when $x = 5.5$, so graph I corresponds to the given function.

50. IV

51. $x^2 + 6x + 9$ has no equals sign, so it is an expression.

52. Equation

53. $3x^2 = 3x$ has an equals sign, so it is an equation.

54. Expression

55. $2x^3 + x^2 = 0$ has an equals sign, so it is an equation.

56. Expression

57.
$$2t^2 + 8t$$
$$= 2t \cdot t + 2t \cdot 4$$
$$= 2t(t + 4)$$

58. $3y(y + 2)$

59.
$$y^3 + 9y^2$$
$$= y \cdot y^2 + 9 \cdot y^2$$
$$= y^2(y + 9)$$

40. $x^2(x + 8)$

41.
$$15x^2 - 5x^4$$
$$= 5x^2 \cdot 3 - 5x^2 \cdot x^2$$
$$= 5x^2(3 - x^2)$$

42. $4y^2(2 + y^2)$

43.
$$4x^2y - 12xy^2$$
$$= 4xy \cdot x - 4xy \cdot 3y$$
$$= 4xy(x - 3y)$$

44. $5x^2y^2(y + 3x)$

45.
$$3y^2 - 3y - 9$$
$$= 3 \cdot y^2 - 3 \cdot y - 3 \cdot 3$$
$$= 3(y^2 - y - 3)$$

46. $5(x^2 - x + 3)$

47.
$$6ab - 4ad + 12ac$$
$$= 2a \cdot 3b - 2a \cdot 2d + 2a \cdot 6c$$
$$= 2a(3b - 2d + 6c)$$

48. $2x(4y + 5z - 7w)$

49.
$$9x^3y^6z^2 - 12x^4y^4z^4 + 15x^2y^5z^3$$
$$= 3x^2y^4z^2 \cdot 3xy^2 - 3x^2y^4z^2 \cdot 4x^2z^2 + 3x^2y^4z^2 \cdot 5yz$$
$$= 3x^2y^4z^2(3xy^2 - 4x^2z^2 + 5yz)$$

50. $7a^3b^3c^3(2ac^2 + 3b^2c - 5ab)$

51. $-5x + 35 = -5(x - 7)$

52. $-5(x + 8)$

53. $-6y - 72 = -6(y + 12)$

54. $-8(t - 9)$

55. $-2x^2 + 4x - 12 = -2(x^2 - 2x + 6)$

56. $-2(x^2 - 6x - 20)$

57. $3y - 24 = -3(-y) - 3 \cdot 8 = -3(-y + 8)$, or $-3(8 - y)$

58. $-7(-x + 8y)$, or $-7(8y - x)$

59. $7s - 14t = -7(-s) - 7 \cdot 2t = -7(-s + 2t)$, or $-7(2t - s)$

60. $-5(-r + 2s)$, or $-5(2s - r)$

61. $-x^2 + 5x - 9 = -(x^2 - 5x + 9)$

62. $-(p^3 + 4p^2 - 11)$

63. $-a^4 + 2a^3 - 13a = -a(a^3 - 2a^2 + 13)$

64. $-(m^3 + m^2 - m + 2)$

65. $a(b-5) + c(b-5)$
$$= (b-5)(a+c)$$

66. $(t-3)(r-s)$

67. $(x+7)(x-1) + (x+7)(x-2)$
$$= (x+7)(x-1+x-2)$$
$$= (x+7)(2x-3)$$

68. $(a+5)(2a-1)$

69. $a^2(x-y) + 5(y-x)$
$$= a^2(x-y) + 5(-1)(x-y) \quad \text{Factoring out } -1$$
$$\qquad\qquad \text{to reverse the second subtraction}$$
$$= a^2(x-y) - 5(x-y) \qquad \text{Simplifying}$$
$$= (x-y)(a^2-5)$$

70. $(x-6)(5x^2-2)$

71. $ac + ad + bc + bd$
$$= a(c+d) + b(c+d)$$
$$= (c+d)(a+b)$$

72. $(y+z)(x+w)$

73. $b^3 - b^2 + 2b - 2$
$$= b^2(b-1) + 2(b-1)$$
$$= (b-1)(b^2+2)$$

74. $(y-1)(y^2+3)$

75. $a^3 - 3a^2 + 6 - 2a$
$$= a^2(a-3) + 2(3-a)$$
$$= a^2(a-3) + 2(-1)(a-3) \quad \text{Factoring out } -1$$
$$\qquad\qquad \text{to reverse the second subtraction}$$
$$= a^2(a-3) - 2(a-3)$$
$$= (a-3)(a^2-2)$$

76. $(t+6)(t^2-2)$

77. $72x^3 - 36x^2 + 24x$
$$= 12x \cdot 6x^2 - 12x \cdot 3x + 12x \cdot 2$$
$$= 12x(6x^2 - 3x + 2)$$

78. $3a^2(4a^2 - 7a - 3)$

79. $x^6 - x^5 - x^3 + x^4$
$$= x^3(x^3 - x^2 - 1 + x)$$
$$= x^3[x^2(x-1) + x - 1] \qquad (-1 + x = x - 1)$$
$$= x^3(x-1)(x^2+1)$$

80. $y(y-1)(y^2+1)$

81. $2y^4 + 6y^2 + 5y^2 + 15$
$$= 2y^2(y^2+3) + 5(y^2+3)$$
$$= (y^2+3)(2y^2+5)$$

82. $(2-x)(xy-3)$

83. a) $h(t) = -16t^2 + 72t$
$$h(t) = -8t(2t-9)$$

    b) Using $h(t) = -16t^2 + 72t$:
$$h(1) = -16 \cdot 1^2 + 72 \cdot 1 = -16 \cdot 1 + 72$$
$$= -16 + 72 = 56 \text{ ft}$$

    Using $h(t) = -8t(2t-9)$:
$$h(1) = -8(1)(2 \cdot 1 - 9) = -8(1)(-7) = 56 \text{ ft}$$

    The expressions have the same value for $t = 1$, so the factorization is probably correct.

84. a) $h(t) = -16t(t-6)$
    b) $h(1) = 80$ ft

85. $R(n) = n^2 - n$
$$R(n) = n(n-1)$$

86. $\pi r(2h + r)$

87. $P(x) = x^2 - 3x$
$$P(x) = x(x-3)$$

88. $P(t) = t(t-5)$

89. $R(x) = 280x - 0.4x^2$
$$R(x) = 0.4x(700 - x)$$

90. $C(x) = 0.6x(0.3 + x)$

91. $N(x) = \dfrac{1}{6}x^3 + \dfrac{1}{2}x^2 + \dfrac{1}{3}x$
$$N(x) = \frac{1}{6}(x^3 + 3x^2 + 2x) \quad \text{Factoring out } \frac{1}{6}$$

92. $f(n) = \dfrac{1}{2}(n^2 - n)$

93. $H(n) = \dfrac{1}{2}n^2 - \dfrac{1}{2}n$
$$H(n) = \frac{1}{2}n(n-1)$$

94. $P(n) = \dfrac{1}{2}(n^2 - 3n)$

95. $x(x + 1) = 0$

We use the principle of zero products.
$$x = 0 \;\; or \;\; x + 1 = 0$$
$$x = 0 \;\; or \;\;\;\;\;\;\;\; x = -1$$
The solutions are 0 and -1.

96. 0, 2

97. $x^2 - 3x = 0$
$$x(x - 3) = 0 \quad \text{Factoring}$$
$$x = 0 \;\; or \;\; x - 3 = 0 \quad \text{Using the}$$
$$\text{principle of zero products}$$
$$x = 0 \;\; or \;\;\;\;\;\;\;\; x = 3$$
The solutions are 0 and 3.

98. $-4, 0$

99. $-5x^2 = 15x$
$$0 = 5x^2 + 15x \quad \text{Adding } 5x^2 \text{ on both sides}$$
$$0 = 5x(x + 3) \quad \text{Factoring}$$
$$5x = 0 \;\; or \;\; x + 3 = 0 \quad \text{Using the}$$
$$\text{principle of zero products}$$
$$x = 0 \;\; or \;\;\;\;\;\;\; x = -3$$
The solutions are 0 and -3.

100. $0, \dfrac{1}{2}$

101.
$$12x^4 + 4x^3 = 0$$
$$4x^3(3x + 1) = 0 \quad \text{Factoring}$$
$$4x \cdot x \cdot x(3x + 1) = 0$$
$$4x = 0 \text{ or } x = 0 \text{ or } x = 0 \text{ or } 3x + 1 = 0$$
$$x = 0 \text{ or } x = 0 \text{ or } x = 0 \text{ or } \;\;\;\;\; 3x = -1$$
$$x = 0 \text{ or } x = 0 \text{ or } x = 0 \text{ or } \;\;\;\;\; x = -\dfrac{1}{3}$$
The solutions are 0 and $-\dfrac{1}{3}$.

102. $0, \dfrac{1}{3}$

103. *Writing Exercise*

104. *Writing Exercise*

105. $2(-3) + 4(-5) = -6 - 20 = -26$

106. -1

107. $4(-6) - 3(2) = -24 - 6 = -30$

108. -19

109. ***Familiarize***. Let $n = $ the first even number. Then $n + 2$ and $n + 4$ are the next two even numbers. Recall that the perimeter of a triangle is the sum of the lengths of the sides.

Translate.

$$\underbrace{\text{The perimeter}} \qquad \text{is} \quad 174.$$
$$n + (n + 2) + (n + 4) = 174$$

Carry out. We solve the equation.
$$n + (n + 2) + (n + 4) = 174$$
$$3n + 6 = 174$$
$$3n = 168$$
$$n = 56$$
When $n = 56$, then $n + 2 = 56 + 2$, or 58, and $n + 4 = 56 + 4$, or 60.

Check. The numbers 56, 58, and 60 are consecutive even integers. Also $56 + 58 + 60 = 174$. The answer checks.

State. The lengths of the sides of the triangle are 56, 58, and 60.

110. A: 75; B: 84; C: 63

111. *Writing Exercise*

112. *Writing Exercise*

113. We use the principle of zero products in reverse. Since the zeros of $f(x) = x^2 + 2x - 8$ are -4 and 2, we have
$$x = -4 \;\; or \;\;\;\;\;\; x = 2$$
$$x + 4 = 0 \;\; or \;\; x - 2 = 0,$$
so $x^2 + 2x - 8 = (x + 4)(x - 2)$.

114. $(x - 1)(x - 1)$

115. $x^5 y^4 + \underline{\quad\quad} = x^3 y(\underline{\quad\quad} + xy^5)$

The term that goes in the first blank is the product of $x^3 y$ and xy^5, or $x^4 y^6$.

The term that goes in the second blank is the expression that is multiplied with $x^3 y$ to obtain $x^5 y^4$, or $x^2 y^3$. Thus, we have
$$x^5 y^4 + x^4 y^6 = x^3 y(x^2 y^3 + xy^5).$$

116. $a^3 b^7 - a^2 b^3 c^2 = a^2 b^3(ab^4 - c^2)$

117.
$$rx^2 - rx + 5r + sx^2 - sx + 5s$$
$$= r(x^2 - x + 5) + s(x^2 - x + 5)$$
$$= (x^2 - x + 5)(r + s)$$

118. $(a^2 + 2a + 10)(3 + 7b)$

119.
$$a^4x^4 + a^4x^2 + 5a^4 + a^2x^4 + a^2x^2 + 5a^2 +$$
$$5x^4 + 5x^2 + 25$$
$$= a^4(x^4+x^2+5)+a^2(x^4+x^2+5)+5(x^4+x^2+5)$$
$$= (x^4+x^2+5)(a^4+a^2+5)$$

120. $x^{1/2}(1+5x)$

121. $x^{1/3} - 7x^{4/3} = x^{1/3} \cdot 1 - 7x \cdot x^{1/3} = x^{1/3}(1-7x)$

122. $x^{1/4}(x^{1/2}+x^{1/4}-1)$

123.
$$x^{1/3} - 5x^{1/2} + 3x^{3/4}$$
$$= x^{4/12} - 5x^{6/12} + 3x^{9/12}$$
$$= x^{4/12}(1 - 5x^{2/12} + 3x^{5/12})$$
$$= x^{1/3}(1 - 5x^{1/6} + 3x^{5/12})$$

124. $2x^a(x^{2a} + 4 + 2x^a)$

125.
$$3a^{n+1} + 6a^n - 15a^{n+2}$$
$$= 3a^n \cdot a + 3a^n \cdot 2 - 3a^n(5a^2)$$
$$= 3a^n(a + 2 - 5a^2)$$

126. $x^a(4x^b + 7x^{-b})$

127.
$$7y^{2a+b} - 5y^{a+b} + 3y^{a+2b}$$
$$= y^{a+b} \cdot 7y^a - y^{a+b}(5) + y^{a+b} \cdot 3y^b$$
$$= y^{a+b}(7y^a - 5 + 3y^b)$$

Exercise Set 5.4

1. $x^2 + 8x + 12$

We look for two numbers whose product is 12 and whose sum is 8. Since 12 and 8 are both positive, we need only consider positive factors.

Pair of Factors	Sum of Factors
1,12	13
2, 6	8

The numbers we need are 2 and 6. The factorization is $(x+2)(x+6)$.

2. $(x+1)(x+5)$

3. $t^2 + 8t + 15$

Since the constant term is positive and the coefficient of the middle term is also positive, we look for a factorization of 15 in which both factors are positive. Their sum must be 8.

Pair of Factors	Sum of Factors
1, 15	16
3, 5	8

The numbers we need are 3 and 5. The factorization is $(t+3)(t+5)$.

4. $(y+3)(y+9)$

5. $x^2 - 27 - 6x = x^2 - 6x - 27$

Since the constant term is negative, we look for a factorization of -27 in which one factor is positive and one factor is negative. Their sum must be -6, so the negative factor must have the larger absolute value. Thus we consider only pairs of factors in which the negative factor has the larger absolute value.

Pair of Factors	Sum of Factors
$-27,1$	-26
$-9,3$	-6

The numbers we need are -9 and 3. The factorization is $(x-9)(x+3)$.

6. $(t-5)(t+3)$

7.
$$2n^2 - 20n + 50$$
$$= 2(n^2 - 10n + 25) \quad \text{Removing the common}$$
$$\text{factor}$$

We now factor $n^2 - 10n + 25$. We look for two numbers whose product is 25 and whose sum is -10. Since the constant term is positive and the coefficient of the middle term is negative, we look for factorization of 25 in which both factors are negative.

Pair of Factors	Sum of Factors
$-1,-25$	-26
$-5, -5$	-10

The numbers we need are -5 and -5.
$$n^2 - 10n + 25 = (n-5)(n-5)$$

We must not forget to include the common factor 2.
$$2n^2 - 20n + 50 = 2(n-5)(n-5), \text{ or } 2(n-5)^2$$

8. $2(a-4)^2$

9.
$$a^3 - a^2 - 72a$$
$$= a(a^2 - a - 72) \quad \text{Removing the common factor}$$

We now factor $a^2 - a - 72$. Since the constant term is negative, we look for a factorization of -72 in which one factor is positive and one factor is negative. We consider only pairs of factors in which the negative factor has the larger absolute value, since the sum of the factors, -1, is negative.

Pair of Factors	Sum of Factors
$-72,1$	-71
$-36,2$	-34
$-18,4$	-14
$-9,8$	-1

The numbers we need are -9 and 8.
$$a^2 - a - 72 = (a-9)(a+8)$$

We must not forget to include the common factor a.
$$a^3 - a^2 - 72a = a(a - 9)(a + 8)$$

10. $x(x + 9)(x - 6)$

11. $14x + x^2 + 45 = x^2 + 14x + 45$

Since the constant term and the middle term are both positive, we look for a factorization of 45 in which both factors are positive. Their sum must be 14.

Pair of Factors	Sum of Factors
45,1	46
15,3	18
9,5	14

The numbers we need are 9 and 5. The factorization is $(x + 9)(x + 5)$.

12. $(y + 8)(y + 4)$

13. $3x + x^2 - 10 = x^2 + 3x - 10$

Since the constant term is negative, we look for a factorization of -10 in which one factor is positive and one factor is negative. We consider only pairs of factors in which the positive factor has the larger absolute value, since the sum of the factors, 3, is positive.

Pair of Factors	Sum of Factors
10,−1	9
5,−2	3

The numbers we need are 5 and -2. The factorization is $(x + 5)(x - 2)$.

14. $(x + 3)(x - 2)$

15.
$$3x^2 + 15x + 18$$
$$= 3(x^2 + 5x + 6) \quad \text{Removing the common factor}$$

We now factor $x^2 + 5x + 6$. We look for two numbers whose product is 6 and whose sum is 5. Since 6 and 5 are both positive, we need consider only positive factors.

Pair of Factors	Sum of Factors
1, 6	7
2, 3	5

The numbers we need are 2 and 3.
$$x^2 + 5x + 6 = (x + 2)(x + 3)$$

We must not forget to include the common factor 3.
$$3x^2 + 15x + 18 = 3(x + 2)(x + 3)$$

16. $5(y + 1)(y + 7)$

17. $56 + x - x^2 = -x^2 + x + 56 = -(x^2 - x - 56)$

We now factor $x^2 - x - 56$. Since the constant term is negative, we look for a factorization of -56 in which one factor is positive and one factor is negative. We consider only pairs of factors in which the negative factor has the

larger absolute value, since the sum of the factors, -1, is negative.

Pair of Factors	Sum of Factors
−56, 1	−55
−28, 2	−26
−14, 4	−10
−8, 7	−1

The numbers we need are -8 and 7. Thus, $x^2 - x - 56 = (x - 8)(x + 7)$. We must not forget to include the factor that was factored out earlier:
$$56 + x - x^2 = -(x - 8)(x + 7), \text{ or}$$
$$(-x + 8)(x + 7), \text{ or } (8 - x)(7 + x)$$

18. $(8 - y)(4 + y)$, or $-(y - 8)(y + 4)$, or $(-y + 8)(y + 4)$

19. $32y + 4y^2 - y^3$

There is a common factor, y. We also factor out -1 in order to make the leading coefficient positive.
$$32y + 4y^2 - y^3 = -y(-32 - 4y + y^2)$$
$$= -y(y^2 - 4y - 32)$$

Now we factor $y^2 - 4y - 32$. Since the constant term is negative, we look for a factorization of -32 in which one factor is positive and one factor is negative. We consider only pairs of factors in which the negative factor has the larger absolute value, since the sum of the factors, -4, is negative.

Pair of Factors	Sum of Factors
−32, 1	−31
−16, 2	−14
−8, 4	−4

The numbers we need are -8 and 4. Thus, $y^2 - 4y - 32 = (y - 8)(y + 4)$. We must not forget to include the common factor:
$$32y + 4y^2 - y^3 = -y(y - 8)(y + 4), \text{ or}$$
$$y(-y + 8)(y + 4), \text{ or } y(8 - y)(4 + y)$$

20. $x(8 - x)(7 + x)$, or $-x(x - 8)(x + 7)$, or $x(-x + 8)(x + 7)$

21.
$$x^4 + 11x^3 - 80x^2$$
$$= x^2(x^2 + 11x - 80) \quad \text{Removing the common factor}$$

We now factor $x^2 + 11x - 80$. We look for pairs of factors of -80, one positive and one negative, such that the positive factor has the larger absolute value and the sum of the factors is 11.

Pair of Factors	Sum of Factors
80, −1	79
40, −2	38
20, −4	16
16, −5	11
10, −8	2

The numbers we need are 16 and -5. Then $x^2 + 11x - 80 = (x+16)(x-5)$. We must not forget to include the common factor:

$$x^4 + 11x^3 - 80x^2 = x^2(x+16)(x-5)$$

22. $y^2(y+12)(y-7)$

23. $x^2 + 12x + 13$

There are no factors of 13 whose sum is 12. This trinomial is not factorable into binomials with integer coefficients. The polynomial is prime.

24. Prime

25. $p^2 - 5pq - 24q^2$

We look for numbers r and s such that $p^2 - 5pq - 24q^2 = (p + rq)(p + sq)$. Our thinking is much the same as if we were factoring $p^2 - 5p - 24$. We look for factors of -24 whose sum is -5, one positive and one negative, such that the negative factor has the larger absolute value.

Pair of Factors	Sum of Factors
$-24,\ 1$	-23
$-12,\ 2$	-10
$-8,\ 3$	-5

The numbers we need are -8 and 3. The factorization is $(p - 8q)(p + 3q)$.

26. $(x + 3y)(x + 9y)$

27. $y^2 + 8yz + 16z^2$

We look for numbers p and q such that $y^2 + 8yz + 16z^2 = (y + pz)(y + qz)$. Our thinking is much the same as if we factor $y^2 + 8y + 16$. Since the constant term is positive and the coefficient of the middle term is negative, we look for a factorization of 16 in which both factors are positive. Their sum must be 8.

Pair of Factors	Sum of Factors
$1,\ 16$	17
$2,\ 8$	10
$4,\ 4$	8

The numbers we need are 4 and 4. The factorization is $(y + 4z)(y + 4z)$, or $(y + 4z)^2$.

28. $(x - 7y)(x - 7y)$, or $(x - 7y)^2$

29. $p^4 + 80p^3 + 79p^2$

$\quad = p^2(p^2 + 80p + 79)$ Removing the common factor

We now factor $p^2 + 80p + 79$. We look for a pair of factors of 79 whose sum is 80. The only positive pair of factors is 1 and 79. These are the numbers we need. Then $p^2 + 80p + 79 = (p + 1)(p + 79)$. We must not forget to include the common factor:

$$p^4 + 80p^3 + 79p^2 = p^2(p + 1)(p + 79)$$

30. $x^2(x + 1)(x + 49)$

31. $x^2 + 8x + 12 = 0$

$\quad (x + 2)(x + 6) = 0$ From Exercise 1

$\quad x + 2 = 0 \quad or \quad x + 6 = 0$ Using the principle of zero products

$\quad\quad x = -2 \quad or \quad\quad x = -6$

The solutions are -2 and -6.

32. $-5, -1$

33. $2n^2 + 50 = 20n$

$\quad 2n^2 - 20n + 50 = 0$ Subtracting $20n$ from both sides

$\quad 2(n - 5)(n - 5) = 0$ From Exercise 7

$\quad n - 5 = 0 \quad or \quad n - 5 = 0$ Using the principle of zero products

$\quad\quad n = 5 \quad or \quad\quad n = 5$

The solution is 5.

34. $-4, 8$

35. $a^3 + a^2 = 72a$

$\quad a^3 + a^2 - 72a = 0$ Subtracting $72a$ on both sides

$\quad a(a + 9)(a - 8) = 0$ From Exercise 9

$\quad a = 0 \quad or \quad a + 9 = 0 \quad or \quad a - 8 = 0$

$\quad a = 0 \quad or \quad\quad a = -9 \quad or \quad\quad a = 8$

The solutions are 0, -9, and 8.

36. $-7, 0, 8$

37. The x-intercepts are $(-5, 0)$ and $(1, 0)$, so the solutions are -5 and 1.

Check: For -5:

$$\frac{x^2 + 4x - 5 = 0}{(-5)^2 + 4(-5) - 5\ ?\ 0}$$
$$\begin{array}{c|c} 25 - 20 - 5 & \\ 0 & 0 \quad \text{TRUE} \end{array}$$

For 1:

$$\frac{x^2 + 4x - 5 = 0}{1^2 + 4 \cdot 1 - 5\ ?\ 0}$$
$$\begin{array}{c|c} 1 + 4 - 5 & \\ 0 & 0 \quad \text{TRUE} \end{array}$$

Both numbers check, so they are the solutions.

38. $-2, 3$

39. The x-intercepts are $(-3, 0)$ and $(2, 0)$, so the solutions are -3 and 2.

Check: For -3:

$$\begin{array}{c} x^2 + x - 6 = 0 \\ \hline (-3)^2 + (-3) - 6 \ ? \ 0 \\ 9 - 3 - 6 \ \Big| \\ 0 \ \Big| \ 0 \quad \text{TRUE} \end{array}$$

For 2:

$$\begin{array}{c} x^2 + x - 6 = 0 \\ \hline 2^2 + 2 - 6 \ ? \ 0 \\ 4 + 2 - 6 \ \Big| \\ 0 \ \Big| \ 0 \quad \text{TRUE} \end{array}$$

Both numbers check, so they are the solutions.

40. $-5, -3$

41. The zeros of $f(x) = x^2 - 4x + 45$ are the solutions of the equation $x^2 - 4x + 45 = 0$. We factor and use the principle of zero products.

$$x^2 - 4x + 45 = 0$$
$$(x - 9)(x + 5) = 0$$
$$x - 9 = 0 \ \ or \ \ x + 5 = 0$$
$$x = 9 \ \ or \ \ \ \ \ x = -5$$

The zeros are 9 and -5.

42. $-5, 4$

43. The zeros of $r(x) = x^3 - 2x^2 - 3x$ are the solutions of the equation $x^3 - 2x^2 - 3x = 0$. We factor and use the principle of zero products.

$$x^3 - 2x^2 - 3x = 0$$
$$x(x^2 - 2x - 3) = 0$$
$$x(x + 1)(x - 3) = 0$$
$$x = 0 \ \ or \ \ x + 1 = 0 \ \ \ or \ \ x - 3 = 0$$
$$x = 0 \ \ or \ \ \ \ \ \ x = -1 \ \ or \ \ \ \ \ \ x = 3$$

The zeros are 0, -1, and 3.

44. $-5, -2$

45.
$$x^2 + 4x = 45$$
$$x^2 + 4x - 45 = 0$$
$$(x + 9)(x - 5) = 0$$
$$x + 9 = 0 \ \ \ or \ \ x - 5 = 0$$
$$x = -9 \ \ or \ \ \ \ \ x = 5$$

The solutions are -9 and 5.

46. $-4, 7$

47.
$$x^2 - 9x = 0$$
$$x(x - 9) = 0$$
$$x = 0 \ \ or \ \ x - 9 = 0$$
$$x = 0 \ \ or \ \ \ \ \ \ x = 9$$

The solutions are 0 and 9.

48. $-18, 0$

49.
$$a^3 - 3a^2 = 40a$$
$$a^3 - 3a^2 - 40a = 0$$
$$a(a^2 - 3a - 40) = 0$$
$$a(a - 8)(a + 5) = 0$$
$$a = 0 \ \ or \ \ a - 8 = 0 \ \ or \ \ a + 5 = 0$$
$$a = 0 \ \ or \ \ \ \ \ \ a = 8 \ \ or \ \ \ \ \ \ a = -5$$

The solutions are 0, 8, and -5.

50. $-7, 0, 9$

51.
$$(x - 3)(x + 2) = 14$$
$$x^2 - x - 6 = 14$$
$$x^2 - x - 20 = 0$$
$$(x - 5)(x + 4) = 0$$
$$x - 5 = 0 \ \ or \ \ x + 4 = 0$$
$$x = 5 \ \ or \ \ \ \ \ \ x = -4$$

The solutions are 5 and -4.

52. $-3, 1$

53.
$$35 - x^2 = 2x$$
$$35 - 2x - x^2 = 0$$
$$(7 + x)(5 - x) = 0$$
$$7 + x = 0 \ \ \ or \ \ 5 - x = 0$$
$$x = -7 \ \ or \ \ \ \ \ \ 5 = x$$

The solutions are -7 and 5.

54. $-5, 8$

55. From the graph we see that the zeros of $f(x) = x^2 + 10x - 264$ and -22 and 12. We also know that -22 is a zero of $g(x) = x + 22$ and 12 is a zero of $h(x) = x - 12$. Using the principle of zero products in reverse, we have
$$x^2 + 10x - 264 = (x + 22)(x - 12).$$

56. $(x + 28)(x - 12)$

57. Graph $y = x^2 + 40x + 384$ and find the zeros. They are -24 and -16. We know that -24 is a zero of $g(x) = x + 24$ and -16 is a zero of $h(x) = x + 16$. Using the principle of zero products in reverse, we have
$$x^2 + 40x + 384 = (x + 24)(x + 16).$$

58. $(x + 12)(x - 25)$

59. Graph $y = x^2 + 26x - 2432$ and find the zeros. They are -64 and 38. We know that -64 is a zero of $g(x) = x + 64$ and 38 is a zero of $h(x) = x - 38$. Using the principle of zero products in reverse, we have
$$x^2 + 26x - 2432 = (x + 64)(x - 38).$$

60. $(x - 18)(x - 28)$

61. We write a linear function for each zero:
$$-1 \text{ is a zero of } g(x) = x + 1;$$
$$2 \text{ is a zero of } h(x) = x - 2.$$
Then $f(x) = (x + 1)(x - 2)$, or $f(x) = x^2 - x - 2$.

62. $f(x) = x^2 - 7x + 10$

63. We write a linear function for each zero:
$$-7 \text{ is a zero of } g(x) = x + 7;$$
$$-10 \text{ is a zero of } h(x) = x + 10.$$
Then $f(x) = (x + 7)(x + 10)$, or $f(x) = x^2 + 17x + 70$.

64. $f(x) = x^2 - 5x - 24$

65. We write a linear function for each zero:
$$0 \text{ is a zero of } g(x) = x;$$
$$1 \text{ is a zero of } h(x) = x - 1;$$
$$2 \text{ is a zero of } k(x) = x - 2.$$
Then $f(x) = x(x - 1)(x - 2)$, or $f(x) = x^3 - 3x^2 + 2x$.

66. $f(x) = x^3 - 2x^2 - 15x$

67. *Writing Exercise*

68. *Writing Exercise*

69. $10x^3 - 35x^2 + 5x = 5x \cdot 2x^2 - 5x \cdot 7x + 5x \cdot 1 = 5x(2x^2 - 7x + 1)$

70. $4t(3t^2 - 10t - 2)$

71. $(5a^4)^3 = 5^3(a^4)^3 = 125a^{4\cdot3} = 125a^{12}$

72. $-32x^{10}$

73. $g(x) = -5x^2 - 7x$
$g(-3) = -5(-3)^2 - 7(-3) = -5 \cdot 9 + 21 = -45 + 21 = -24$

74. 880 ft; 960 ft; 1024 ft; 624 ft; 240 ft

75. *Writing Exercise*

76. *Writing Exercise*

77. The x-coordinates of the x-intercepts are -1 and 3. These are the solutions of $x^2 - 2x - 3 = 0$.

From the graph we see that the x-values for which $f(x) < 5$ are in the interval $(-2, 4)$. We could also express the solution set as $\{x| - 2 < x < 4\}$.

78. $\{-3, 1\}$; $[-4, 2]$, or $\{x| - 4 \le x \le 2\}$

79. Answers may vary. A polynomial function of lowest degree that meets the given criteria is of the form $f(x) = ax^3 + bx^2 + cx + d$. Substituting, we have
$$a \cdot 2^3 + b \cdot 2^2 + c \cdot 2 + d = 0,$$
$$a(-1)^3 + b(-1)^2 + c(-1) + d = 0,$$
$$a \cdot 3^3 + b \cdot 3^2 + c \cdot 3 + d = 0,$$
$$a \cdot 0^3 + b \cdot 0^2 + c \cdot 0 + d = 30, \text{ or}$$
$$8a + 4b + 2c + d = 0,$$
$$-a + b - c + d = 0,$$
$$27a + 9b + 3c + d = 0,$$
$$d = 30.$$
Solving the system of equations, we get $(5, -20, 5, 30)$, so the corresponding function is $f(x) = 5x^3 - 20x^2 + 5x + 30$.

80. $g(x) = 3x^3 - 9x^2 - 39x + 45$

81. Graph $y_1 = -x^2 + 13.80x$ and $y_2 = 47.61$ and use the Intersect feature to find the first coordinate of the point of intersection. The solution is 6.90.

82. $-3.33, 5.15$

83. Graph $y_1 = x^3 - 3.48x^2 + x$ and $y_2 = 3.48$ and use the Intersect feature to find the first coordinates of the points of intersection. The solution is 3.48.

84. No real-number solutions

85. $2a^4b^6 - 3a^2b^3 - 20ab^2 = ab^2(2a^3b^4 - 3ab - 20)$

The trinomial $2a^3b^4 - 3ab - 20$ cannot be factored as a product of two binomials. Thus, the complete factorization is $ab^2(2a^3b^4 - 3ab - 20)$.

86. $5(x^4y^3 + 4)(x^4y^3 + 3)$

87. $x^2 - \dfrac{4}{25} + \dfrac{3}{5}x = x^2 + \dfrac{3}{5}x - \dfrac{4}{25}$

We look for factors of $-\dfrac{4}{25}$ whose sum is $\dfrac{3}{5}$. The factors are $\dfrac{4}{5}$ and $-\dfrac{1}{5}$. The factorization is
$$\left(x + \frac{4}{5}\right)\left(x - \frac{1}{5}\right).$$

88. $\left(y + \dfrac{4}{7}\right)\left(y - \dfrac{2}{7}\right)$

89. $y^2 + 0.4y - 0.05$

We look for factors of -0.05 whose sum is 0.4. The factors are -0.1 and 0.5. The factorization is $(y - 0.1)(y + 0.5)$.

90. $(x^a + 8)(x^a - 3)$

91.
$$x^2 + ax + bx + ab$$
$$= x(x + a) + b(x + a)$$
$$= (x + a)(x + b)$$

92. $(bx + a)(dx + c)$

93. $a^2 p^{2a} + a^2 p^a - 2a^2 = a^2(p^{2a} + p^a - 2)$

Substitute u for p^a (and u^2 for p^{2a}). We factor $u^2 + u - 2$. Look for factors of -2 whose sum is 1. The factors are 2 and -1. We have $u^2 + u - 2 =$ $(u + 2)(u - 1)$. Replace u by p^a: $p^{2a} + p^a - 2 = (p^a + 2)(p^a - 1)$. We must include the common factor a^2 to get a factorization of the original trinomial:
$$a^2 p^{2a} + a^2 p^a - 2a^2 = a^2(p^a + 2)(p^a - 1)$$

94. $(x - 4)(x + 8)$

95. All such m are the sums of the factors of 75.

Pair of Factors	Sum of Factors
75, 1	76
$-75, -1$	-76
25, 3	28
$-25, -3$	-28
15, 5	20
$-15, -5$	-20

m can be 76, -76, 28, -28, 20, or -20.

96. 31, -31, 14, -14, 4, or -4

97. $20(-365) = -7300$ and $20 + (-365) = -345$ so the other factor is $(x - 365)$.

Exercise Set 5.5

1. $6x^2 - 5x - 25$

We will use the FOIL method.

1. There is no common factor (other than 1 or -1.)

2. Factor the first term, $6x^2$. The factors are $6x$, x and $3x$, $2x$. We have these possibilities:

$(6x+ \quad)(x+ \quad)$ or $(3x+ \quad)(2x+ \quad)$

3. Factor the last term, -25. The possibilities are $25(-1)$, $-25 \cdot 1$, and $-5 \cdot 5$.

4. We need factors for which the sum of the products (the "outer" and "inner" parts of FOIL) is the middle term, $-5x$. Try some possibilities and check by multiplying.

$(6x - 5)(x + 5) = 6x^2 + 25x - 25$

We try again.

$(3x + 5)(2x - 5) = 6x^2 - 5x - 25$

The factorization is $(3x + 5)(2x - 5)$.

2. $(3x + 2)(x - 6)$

3. $10y^3 - 12y - 7y^2 = 10y^3 - 7y^2 - 12y$

We will use the grouping method.

1. Look for a common factor. We factor out y:

$y(10y^2 - 7y - 12)$

2. Factor the trinomial $10y^2 - 7y - 12$. Multiply the leading coefficient, 10, and the constant, -12.

$10(-12) = -120$

3. Try to factor -120 so the sum of the factors is -7. We need only consider pairs of factors in which the negative factor has the larger absolute value, since their sum is negative.

Pair of Factors	Sum of Factors
$-120, 1$	-119
$-30, 4$	-26
$-15, 8$	-7

4. We split the middle term, $-12y$, using the results of step (3).

$-7y = -15y + 8y$

5. Factor by grouping:
$$10y^2 - 7y - 12 = 10y^2 - 15y + 8y - 12$$
$$= 5y(2y - 3) + 4(2y - 3)$$
$$= (2y - 3)(5y + 4)$$

We must include the common factor to get a factorization of the original trinomial:
$$10y^3 - 12y - 7y^2 = y(2y - 3)(5y + 4)$$

4. $x(3x - 5)(2x + 3)$

5. $24a^2 - 14a + 2$

We will use the FOIL method.

1. Factor out the common factor, 2:

$2(12a^2 - 7a + 1)$

2. Now we factor the trinomial $12a^2 - 7a + 1$. Factor the first term, $12a^2$. The factors are $12a$, a and $6a$, $2a$ and $4a$, $3a$. We have these possibilities: $(12a+ \quad)(a+ \quad)$, $(6a+ \quad)(2a+ \quad)$, $(4a+ \quad)(3a+ \quad)$.

3. Factor the last term, 1. The possibilities are $1 \cdot 1$ and $-1(-1)$.

4. Look for factors such that the sum of the products is the middle term, $-7a$. Trial and error leads us to the correct factorization:

$$12a^2 - 7a + 1 = (4a - 1)(3a - 1)$$

We must include the common factor to get a factorization of the original trinomial:

$$24a^2 - 14a + 2 = 2(4a - 1)(3a - 1)$$

6. $(3a - 4)(a - 2)$

7. $35y^2 + 34y + 8$

We will use the grouping method.

1. There is no common factor (other than 1 or -1).

2. Multiply the leading coefficient, 35, and the constant, 8: $35(8) = 280$

3. Try to factor 280 so the sum of the factors is 34. We need only consider pairs of positive factors since 280 and 34 are both positive.

Pair of Factors	Sum of Factors
280, 1	281
140, 2	142
70, 4	74
56, 5	61
40, 7	47
28, 10	38
20, 14	34

4. Split $34y$ using the results of step (3):
$$34y = 20y + 14y$$

5. Factor by grouping:
$$\begin{aligned} 35y^2 + 34y + 8 &= 35y^2 + 20y + 14y + 8 \\ &= 5y(7y + 4) + 2(7y + 4) \\ &= (7y + 4)(5y + 2) \end{aligned}$$

8. $(3a + 2)(3a + 4)$

9. $4t + 10t^2 - 6 = 10t^2 + 4t - 6$

We will use the FOIL method.

1. Factor out the common factor, 2:
$$2(5t^2 + 2t - 3)$$

2. Now we factor the trinomial $5t^2 + 2t - 3$. Factor the first term, $5t^2$. The factors are $5t$ and t. We have this possibility: $(5t+\ \ )(t+\ \ )$

3. Factor the last term, -3. The possibilities are $(1)(-3)$ and $(-1)3$ as well as $(-3)(1)$ and $3(-1)$.

4. Look for factors such that the sum of the products is the middle term, $2t$. Trial and error leads us to the correct factorization:

$$5t^2 + 2t - 3 = (5t - 3)(t + 1)$$

We must include the common factor to get a factorization of the original trinomial:

$$4t + 10t^2 - 6 = 2(5t - 3)(t + 1)$$

10. $2(5x + 3)(3x - 1)$

11. $8x^2 - 16 - 28x = 8x^2 - 28x - 16$

We will use the grouping method.

1. Factor out the common factor, 4:
$$4(2x^2 - 7x - 4)$$

2. Now we factor the trinomial $2x^2 - 7x - 4$. Multiply the leading coefficient, 2, and the constant, -4: $2(-4) = -8$

3. Factor -8 so the sum of the factors is -7. We need only consider pairs of factors in which the negative factor has the larger absolute value, since their sum is negative.

Pair of Factors	Sum of Factors
-4, 2	-2
-8, 1	-7

4. Split $-7x$ using the results of step (3):
$$-7x = -8x + x$$

5. Factor by grouping:
$$\begin{aligned} 2x^2 - 7x - 4 &= 2x^2 - 8x + x - 4 \\ &= 2x(x - 4) + (x - 4) \\ &= (x - 4)(2x + 1) \end{aligned}$$

We must include the common factor to get a factorization of the original trinomial:

$$8x^2 - 16 - 28x = 4(x - 4)(2x + 1)$$

12. $6(3x - 4)(x + 1)$

13. $14x^4 - 19x^3 - 3x^2$

We will use the grouping method.

1. Factor out the common factor, x^2:
$$x^2(14x^2 - 19x - 3)$$

2. Now we factor the trinomial $14x^2 - 19x - 3$. Multiply the leading coefficient, 14, and the constant, -3: $14(-3) = -42$

3. Factor -42 so the sum of the factors is -19. We need only consider pairs of factors in which the negative factor has the larger absolute value, since the sum is negative.

Pair of Factors	Sum of Factors
$-42,\ 1$	-41
$-21,\ 2$	-19
$-14,\ 3$	-11
$-7,\ 6$	-1

4. Split $-19x$ using the results of step (3):
$$-19x = -21x + 2x$$

5. Factor by grouping:
$$
\begin{aligned}
14x^2 - 19x - 3 &= 14x^2 - 21x + 2x - 3 \\
&= 7x(2x - 3) + 2x - 3 \\
&= (2x - 3)(7x + 1)
\end{aligned}
$$

We must include the common factor to get a factorization of the original trinomial:
$$14x^4 - 19x^3 - 3x^2 = x^2(2x - 3)(7x + 1)$$

14. $2x^2(5x - 2)(7x - 4)$

15. $12a^2 - 4a - 16$

We will use the FOIL method.

1. Factor out the common factor, 4:
$$4(3a^2 - a - 4)$$

2. We now factor the trinomial $3a^2 - a - 4$. Factor the first term, $3a^2$. The possibility is $(3a+\quad)(a+\quad)$.

3. Factor the last term, -4. The possibilities are $-4 \cdot 1$, $4(-1)$, and $-2 \cdot 2$.

4. We need factors for which the sum of the products is the middle term, $-a$. Trial and error leads us to the correct factorization:
$$3a^2 - a - 4 = (3a - 4)(a + 1)$$

We must include the common factor to get a factorization of the original trinomial:
$$12a^2 - 4a - 16 = 4(3a - 4)(a + 1)$$

16. $2(6a + 5)(a - 2)$

17. $9x^2 + 15x + 4$

We will use the grouping method.

1. There is no common factor (other than 1 or -1).

2. Multiply the leading coefficient and constant: $9(4) = 36$

3. Factor 36 so the sum of the factors is 15. We need only consider pairs of positive factors since 36 and 15 are both positive.

Pair of Factors	Sum of Factors
$36,\ 1$	37
$18,\ 2$	20
$12,\ 3$	15
$9,\ 4$	13
$6,\ 6$	12

4. Split $15x$ using the results of step (3):
$$15x = 12x + 3x$$

5. Factor by grouping:
$$
\begin{aligned}
9x^2 + 15x + 4 &= 9x^2 + 12x + 3x + 4 \\
&= 3x(3x + 4) + 3x + 4 \\
&= (3x + 4)(3x + 1)
\end{aligned}
$$

18. $(3y - 2)(2y + 1)$

19. $4x^2 + 15x + 9$

We will use the FOIL method.

1. There is no common factor (other than 1 or -1).

2. Factor the first term, $4x^2$. The possibilities are $(4x+\quad)(x+\quad)$ and $(2x+\quad)(2x+\quad)$.

3. Factor the last term, 9. We consider only positive factors since both the middle term and the last term are positive. The possibilities are $9 \cdot 1$ and $3 \cdot 3$.

4. We need factors for which the sum of products is the middle term, $15x$. Trial and error leads us to the correct factorization:
$$(4x + 3)(x + 3)$$

20. $(2y + 3)(y + 2)$

21. $-8t^2 - 8t + 30$

We will use the grouping method.

1. Factor out -2: $-2(4t^2 + 4t - 15)$

2. Now we factor the trinomial $4t^2 + 4t - 15$. Multiply the leading coefficient and the constant: $4(-15) = -60$

3. Factor -60 so the sum of the factors is 4. The desired factorization is $10(-6)$.

4. Split $4t$ using the results of step (3):
$$4t = 10t - 6t$$

5. Factor by grouping:
$$
\begin{aligned}
4t^2 + 4t - 15 &= 4t^2 + 10t - 6t - 15 \\
&= 2t(2t + 5) - 3(2t + 5) \\
&= (2t + 5)(2t - 3)
\end{aligned}
$$

We must include the common factor to get a factorization of the original trinomial:
$$-8t^2 - 8t + 30 = -2(2t + 5)(2t - 3)$$

22. $-3(4a-1)(3a-1)$

23. $8-6z-9z^2$

We will use the FOIL method.

1. There is no common factor (other than 1 or -1).

2. Factor the first term, 8. The possibilities are $(8+\quad)(1+\quad)$ and $(4+\quad)(2+\quad)$.

3. Factor the last term, $-9z^2$. The possibilities are $-9z\cdot z$, $-3z\cdot 3z$, and $9z(-z)$.

4. We need factors for which the sum of products is the middle term, $-6z$. Trial and error leads us to the correct factorization:
$$(4+3z)(2-3z)$$

24. $(3-a)(1+12a)$

25. $18xy^3+3xy^2-10xy$

We will use the FOIL method.

1. Factor out the common factor, xy.
$$xy(18y^2+3y-10)$$

2. We now factor the trinomial $18y^2+3y-10$. Factor the first term, $18y^2$. The possibilities are $(18y+\quad)(y+\quad)$, $(9y+\quad)(2y+\quad)$, and $(6y+\quad)(3y+\quad)$.

3. Factor the last term, -10. The possibilities are $-10\cdot 1$, $-5\cdot 2$, $10(-1)$ and $5(-2)$.

4. We need factors for which the sum of the products is the middle term, $3y$. Trial and error leads us to the correct factorization.
$$18y^2+3y-10=(6y+5)(3y-2)$$

We must include the common factor to get a factorization of the original trinomial:
$$18xy^3+3xy^2-10xy=xy(6y+5)(3y-2)$$

26. $xy^2(3x+1)(x-2)$

27. $24x^2-2-47x=24x^2-47x-2$

We will use the grouping method.

1. There is no common factor (other than 1 or -1).

2. Multiply the leading coefficient and the constant:
$24(-2)=-48$

3. Factor -48 so the sum of the factors is -47. The desired factorization is $-48\cdot 1$.

4. Split $-47x$ using the results of step (3):
$$-47x=-48x+x$$

5. Factor by grouping:
$$\begin{aligned}24x^2-47x-2&=24x^2-48x+x-2\\&=24x(x-2)+(x-2)\\&=(x-2)(24x+1)\end{aligned}$$

28. $(5z+1)(3z-10)$

29. $63x^3+111x^2+36x$

We will use the FOIL method.

1. Factor out the common factor, $3x$.
$$3x(21x^2+37x+12)$$

2. Now we will factor the trinomial $21x^2+37x+12$. Factor the first term, $21x^2$. The factors are $21x$, x and $7x$, $3x$. We have these possibilities: $(21x+\quad)(x+\quad)$ and $(7x+\quad)(3x+\quad)$.

3. Factor the last term, 12. The possibilities are $12\cdot 1$, $(-12)(-1)$, $6\cdot 2$, $(-6)(-2)$, $4\cdot 3$, and $(-4)(-3)$ as well as $1\cdot 12$, $(-1)(-12)$, $2\cdot 6$, $(-2)(-6)$, $3\cdot 4$, and $(-3)(-4)$.

4. Look for factors such that the sum of the products is the middle term, $37x$. Trial and error leads us to the correct factorization:
$$(7x+3)(3x+4)$$

We must include the common factor to get a factorization of the original trinomial:
$$63x^3+111x^2+36x=3x(7x+3)(3x+4)$$

30. $5t(5t+4)(2t+3)$

31. $48x^4+4x^3-30x^2$

We will use the grouping method.

1. We factor out the common factor, $2x^2$:
$$2x^2(24x^2+2x-15)$$

2. We now factor $24x^2+2x-15$. Multiply the leading coefficient and the constant:
$$24(-15)=-360$$

3. Factor -360 so the sum of the factors is 2. The desired factorization is $-18\cdot 20$.

4. Split $2x$ using the results of step (3):
$$2x=-18x+20x$$

5. Factor by grouping:
$$\begin{aligned}24x^2+2x-15&=24x^2-18x+20x-15\\&=6x(4x-3)+5(4x-3)\\&=(4x-3)(6x+5)\end{aligned}$$

We must not forget to include the common factor:
$$48x^4+4x^3-30x^2=2x^2(4x-3)(6x+5)$$

32. $4(5y^2+3)(2y^2-1)$

33. $12a^2 - 17ab + 6b^2$

We will use the FOIL method. (Our thinking is much the same as if we were factoring $12a^2 - 17a + 6$.)

1. There is no common factor (other than 1 or -1).

2. Factor the first term, $12a^2$. The factors are $12a$, a and $6a$, $2a$ and $4a$, $3a$. We have these possibilities: $(12a+ \quad)(a+ \quad)$ and $(6a+ \quad)(2a+ \quad)$ and $(4a+ \quad)(3a+ \quad)$.

3. Factor the last term, $6b^2$. The possibilities are $6b \cdot b$, $(-6b)(-b)$, $3b \cdot 2b$, and $(-3b)(-2b)$ as well as $b \cdot 6b$, $(-b)(-6b)$, $2b \cdot 3b$, and $(-2b)(-3b)$.

4. Look for factors such that the sum of the products is the middle term, $-17ab$. Trial and error leads us to the correct factorization:
$$(4a - 3b)(3a - 2b)$$

34. $(4p - 3q)(5p - 2q)$

35. $2x^2 + xy - 6y^2$

We will use the grouping method.

1. There is no common factor (other than 1 or -1).

2. Multiply the coefficients of the first and last terms: $2(-6) = -12$

3. Factor -12 so the sum of the factors is 1. The desired factorization is $4(-3)$.

4. Split xy using the results of step (3):
$$xy = 4xy - 3xy$$

5. Factor by grouping:
$$\begin{aligned}
2x^2 + xy - 6y^2 &= 2x^2 + 4xy - 3xy - 6y^2 \\
&= 2x(x + 2y) - 3y(x + 2y) \\
&= (x + 2y)(2x - 3y)
\end{aligned}$$

36. $(4m + 3n)(2m - 3n)$

37. $6x^2 - 29xy + 28y^2$

We will use the FOIL method.

1. There is no common factor (other than 1 or -1).

2. Factor the first term, $6x^2$. The factors are $6x$, x and $3x$, $2x$. We have these possibilities: $(6x+ \quad)(x+ \quad)$ and $(3x+ \quad)(2x+ \quad)$.

3. Factor the last term, $28y^2$. The possibilities are $28y \cdot y$, $(-28y)(-y)$, $14y \cdot 2y$, $(-14y)(-2y)$, $7y \cdot 4y$, and $(-7y)(-4y)$ as well as $y \cdot 28y$, $(-y)(-28y)$, $2y \cdot 14y$, $(-2y)(-14y)$, $4y \cdot 7y$, and $(-4y)(-7y)$.

4. Look for factors such that the sum of the products is the middle term, $-29xy$. Trial and error leads us to the correct factorization:
$$(3x - 4y)(2x - 7y)$$

38. $(2p + 3q)(5p - 4q)$

39. $9x^2 - 30xy + 25y^2$

We will use the grouping method.

1. There is no common factor (other than 1 or -1).

2. Multiply the coefficients of the first and last terms: $9(25) = 225$

3. Factor 225 so the sum of the factors is -30. The desired factorization is $-15(-15)$.

4. Split $-30xy$ using the results of step (3):
$$-30xy = -15xy - 15xy$$

5. Factor by grouping:
$$\begin{aligned}
9x^2 - 30xy + 25y^2 &= 9x^2 - 15xy - 15xy + 25y^2 \\
&= 3x(3x - 5y) - 5y(3x - 5y) \\
&= (3x - 5y)(3x - 5y), \text{ or} \\
&\quad (3x - 5y)^2
\end{aligned}$$

40. $(2p + 3q)(2p + 3q)$, or $(2p + 3q)^2$

41. $9x^2y^2 + 5xy - 4$

Let $u = xy$ and $u^2 = x^2y^2$. Factor $9u^2 + 5u - 4$. We will use the FOIL method.

1. There is no common factor (other than 1 or -1).

2. Factor the first term, $9u^2$. The factors are $9u$, u and $3u$, $3u$. We have these possibilities: $(9u+ \quad)(u+ \quad)$ and $(3u+ \quad)(3u+ \quad)$.

3. Factor the last term, -4. The possibilities are $-4 \cdot 1$, $-2 \cdot 2$, and $-1 \cdot 4$.

4. We need factors for which the sum of the products is the middle term, $5u$. Trial and error leads us to the factorization: $(9u - 4)(u + 1)$. Replace u by xy. We have $9x^2y^2 + 5xy - 4 = (9xy - 4)(xy + 1)$.

42. $(7ab + 6)(ab + 1)$

43.
$$\begin{aligned}
6x^2 - 5x - 25 &= 0 \\
(3x + 5)(2x - 5) &= 0 \quad \text{From Exercise 1}
\end{aligned}$$
$3x + 5 = 0 \quad or \quad 2x - 5 = 0 \quad$ Using the principle of zero products
$$\begin{aligned}
3x &= -5 \quad or \quad 2x = 5 \\
x &= -\frac{5}{3} \quad or \quad \quad x = \frac{5}{2}
\end{aligned}$$
The solutions are $-\dfrac{5}{3}$ and $\dfrac{5}{2}$.

44. $-\dfrac{2}{3}$, 6

45. $9z^2 + 6z = 8$

$$0 = 8 - 6z - 9z^2$$

$$0 = (4 + 3z)(2 - 3z) \quad \text{From Exercise 23}$$

$$4 + 3z = 0 \quad \text{or} \quad 2 - 3z = 0 \quad \text{Using the}$$
$$\text{principle of zero products}$$

$$3z = -4 \quad \text{or} \quad 2 = 3z$$

$$z = -\frac{4}{3} \quad \text{or} \quad \frac{2}{3} = z$$

The solutions are $-\dfrac{4}{3}$ and $\dfrac{2}{3}$.

46. $-\dfrac{1}{12}, 3$

47. $63x^3 + 111x^2 + 36x = 0$

$$3x(7x + 3)(3x + 4) = 0 \quad \text{From Exercise 29}$$

$$3x = 0 \quad \text{or} \quad 7x + 3 = 0 \quad \text{or} \quad 3x + 4 = 0$$

$$x = 0 \quad \text{or} \quad 7x = -3 \quad \text{or} \quad 3x = -4$$

$$x = 0 \quad \text{or} \quad x = -\frac{3}{7} \quad \text{or} \quad x = -\frac{4}{3}$$

The solutions are 0, $-\dfrac{3}{7}$, and $-\dfrac{4}{3}$.

48. $-\dfrac{3}{2}, -\dfrac{4}{5}, 0$

49. $3x^2 - 8x + 4 = 0$

$$(3x - 2)(x - 2) = 0 \quad \text{Factoring}$$

$$3x - 2 = 0 \quad \text{or} \quad x - 2 = 0$$

$$3x = 2 \quad \text{or} \quad x = 2$$

$$x = \frac{2}{3} \quad \text{or} \quad x = 2$$

The solutions are $\dfrac{2}{3}$ and 2.

50. $\dfrac{1}{3}, \dfrac{4}{3}$

51. $4t^3 + 11t^2 + 6t = 0$

$$t(4t^2 + 11t + 6) = 0$$

$$t(4t + 3)(t + 2) = 0$$

$$t = 0 \quad \text{or} \quad 4t + 3 = 0 \quad \text{or} \quad t + 2 = 0$$

$$t = 0 \quad \text{or} \quad 4t = -3 \quad \text{or} \quad t = -2$$

$$t = 0 \quad \text{or} \quad t = -\frac{3}{4} \quad \text{or} \quad t = -2$$

The solutions are 0, $-\dfrac{3}{4}$, and -2.

52. $-\dfrac{3}{4}, -\dfrac{1}{2}, 0$

53. $6x^2 = 13x + 5$

$$6x^2 - 13x - 5 = 0$$

$$(2x - 5)(3x + 1) = 0$$

$$2x - 5 = 0 \quad \text{or} \quad 3x + 1 = 0$$

$$2x = 5 \quad \text{or} \quad 3x = -1$$

$$x = \frac{5}{2} \quad \text{or} \quad x = -\frac{1}{3}$$

The solutions are $\dfrac{5}{2}$ and $-\dfrac{1}{3}$.

54. $-\dfrac{6}{5}, \dfrac{1}{8}$

55. $x(5 + 12x) = 28$

$$5x + 12x^2 = 28$$

$$5x + 12x^2 - 28 = 0$$

$$12x^2 + 5x - 28 = 0 \quad \text{Rearranging}$$

$$(4x + 7)(3x - 4) = 0$$

$$4x + 7 = 0 \quad \text{or} \quad 3x - 4 = 0$$

$$4x = -7 \quad \text{or} \quad 3x = 4$$

$$x = -\frac{7}{4} \quad \text{or} \quad x = \frac{4}{3}$$

The solutions are $-\dfrac{7}{4}$ and $\dfrac{4}{3}$.

56. $-\dfrac{5}{7}, \dfrac{2}{3}$

57. The zeros of $f(x) = 2x^2 - 13x - 7$ are the roots, or solutions, of the equation $2x^2 - 13x - 7 = 0$.

$$2x^2 - 13x - 7 = 0$$

$$(2x + 1)(x - 7) = 0$$

$$2x + 1 = 0 \quad \text{or} \quad x - 7 = 0$$

$$2x = -1 \quad \text{or} \quad x = 7$$

$$x = -\frac{1}{2} \quad \text{or} \quad x = 7$$

The zeros are $-\dfrac{1}{2}$ and 7.

58. $-\dfrac{3}{2}, -\dfrac{2}{3}$

59. We set $f(a)$ equal to 8.

$$a^2 + 12a + 40 = 8$$

$$a^2 + 12a + 32 = 0$$

$$(a + 8)(a + 4) = 0$$

$$a + 8 = 0 \quad \text{or} \quad a + 4 = 0$$

$$a = -8 \quad \text{or} \quad a = -4$$

The values of a for which $f(a) = 8$ are -8 and -4.

60. $-9, -5$

61. We set $g(a)$ equal to 12.

$$2a^2 + 5a = 12$$
$$2a^2 + 5a - 12 = 0$$
$$(2a - 3)(a + 4) = 0$$
$$2a - 3 = 0 \quad or \quad a + 4 = 0$$
$$2a = 3 \quad or \qquad a = -4$$
$$a = \frac{3}{2} \quad or \qquad a = -4$$

The values of a for which $g(a) = 12$ are $\frac{3}{2}$ and -4.

62. $\frac{1}{2}, 7$

63. We set $h(a)$ equal to -27.

$$12a + a^2 = -27$$
$$12a + a^2 + 27 = 0$$
$$a^2 + 12a + 27 = 0 \quad \text{Rearranging}$$
$$(a + 3)(a + 9) = 0$$
$$a + 3 = 0 \quad or \quad a + 9 = 0$$
$$a = -3 \quad or \qquad a = -9$$

The values of a for which $h(a) = -27$ are -3 and -9.

64. $-4, 8$

65. $f(x) = \dfrac{3}{x^2 - 4x - 5}$

$f(x)$ cannot be calculated for any x-value for which the denominator, $x^2 - 4x - 5$, is 0. To find the excluded values, we solve:

$$x^2 - 4x - 5 = 0$$
$$(x - 5)(x + 1) = 0$$
$$x - 5 = 0 \quad or \quad x + 1 = 0$$
$$x = 5 \quad or \qquad x = -1$$

The domain of f is $\{x | x$ is a real number and $x \neq 5$ and $x \neq -1\}$.

66. $\{x | x$ is a real number and $x \neq 6$ and $x \neq 1\}$

67. $f(x) = \dfrac{x - 5}{9x - 18x^2}$

$f(x)$ cannot be calculated for any x-value for which the denominator, $9x - 18x^2$, is 0. To find the excluded values, we solve:

$$9x - 18x^2 = 0$$
$$9x(1 - 2x) = 0$$
$$9x = 0 \quad or \quad 1 - 2x = 0$$
$$x = 0 \quad or \qquad -2x = -1$$
$$x = 0 \quad or \qquad x = \frac{1}{2}$$

The domain of f is $\left\{x | x$ is a real number and $x \neq 0$ and $x \neq \dfrac{1}{2}\right\}$.

68. $\left\{x | x$ is a real number and $x \neq 0$ and $x \neq \dfrac{1}{5}\right\}$

69. $f(x) = \dfrac{7}{5x^3 - 35x^2 + 50x}$

$f(x)$ cannot be calculated for any x-value for which the denominator, $5x^3 - 35x^2 + 50x$, is 0. To find the excluded values, we solve:

$$5x^3 - 35x^2 + 50x = 0$$
$$5x(x^2 - 7x + 10) = 0$$
$$5x(x - 2)(x - 5) = 0$$
$$5x = 0 \quad or \quad x - 2 = 0 \quad or \quad x - 5 = 0$$
$$x = 0 \quad or \qquad x = 2 \quad or \qquad x = 5$$

The domain of f is $\{x | x$ is a real number and $x \neq 0$ and $x \neq 2$ and $x \neq 5\}$.

70. $\{x | x$ is a real number and $x \neq 0$ and $x \neq 3$ and $x \neq -2\}$

71. *Writing Exercise.* Both are correct because $(a - b)(x - y)$ and $(b - a)(y - x)$ are equivalent expressions.

$$(a - b)(x - y)$$
$$= (-1)(a - b)(-1)(x - y) \qquad (-1)(-1) = 1$$
$$= (-a + b)(-x + y)$$
$$= (b - a)(y - x)$$

72. *Writing Exercise*

73. **Familiarize.** Let w represent the width and l represent the length of the rectangle. If the width is increased by 2 ft, the new width is $w + 2$. Also, recall the formulas for the perimeter and area of a rectangle:

$$P = 2l + 2w$$
$$A = lw$$

Translate.

Width is length less 7 feet.
$$w \quad = \quad l \quad - \quad 7$$

When the length is l and the width is $w + 2$, the perimeter is 66 ft, so we have

$$66 = 2l + 2(w + 2).$$

We have a system of equations:

$$w = l - 7,$$
$$66 = 2l + 2(w + 2)$$

Carry out. Solving the system we get $(19, 12)$.

Check. The width, 12 ft, is 7 ft less than the length, 19 ft. When the width is increased by 2 ft it becomes 14 ft, and the perimeter of the rectangle is $2 \cdot 19 + 2 \cdot 14 = 38 + 28 = 66$ ft. The numbers check.

State. The area of the original rectangle is 19 ft $\cdot$ 12 ft, or 228 ft^2.

74. -2

75. Write the equation in slope-intercept form, $y = mx + b$ where m is the slope and b is the y-intercept.

$$4x - 3y = 8$$
$$-3y = -4x + 8$$
$$y = \frac{4}{3}x - \frac{8}{3}$$

The slope is $\frac{4}{3}$, and the y-intercept is $\left(0, -\frac{8}{3}\right)$.

76. $\{27, -27\}$

77. $|5x - 6| \leq 39$

$$-39 \leq 5x - 6 \leq 39$$
$$-33 \leq 5x \leq 45$$
$$-\frac{33}{5} \leq x \leq 9$$

The solution set is $\left\{x \,\middle|\, -\frac{33}{5} \leq x \leq 9\right\}$, or $\left[-\frac{33}{5}, 9\right]$.

78. $\left\{x \,\middle|\, x < -\frac{33}{5} \text{ or } x > 9\right\}$, or $\left(-\infty, -\frac{33}{5}\right) \cup (9, \infty)$.

79. *Writing Exercise*

80. *Writing Exercise*

81. Graph $y = 4x^2 + 120x + 675$ and find the zeros. They are -7.5 and -22.5, or $-\frac{15}{2}$ and $-\frac{45}{2}$. We know that $-\frac{15}{2}$ is a zero of $g(x) = 2x + 15$ and $-\frac{45}{2}$ is a zero of $h(x) = 2x + 45$. We have $4x^2 + 120x + 675 = (2x + 15)(2x + 45)$.

82. $(2x + 63)(2x + 19)$

83. First factor out the largest common factor.

$3x^3 + 150x^2 - 3672x = 3x(x^2 + 50x - 1224)$.

Now graph $y = x^2 + 50x - 1224$ and find the zeros. They are -68, and 18. We know that -68 is a zero of $g(x) = x + 68$, and 18 is a zero of $h(x) = x - 18$.

We have $x^2 + 50x - 1224 = (x + 68)(x - 18)$, so $3x^3 + 150x^2 - 3672x = 3x(x + 68)(x - 18)$.

84. $5x^2(x + 20)(x - 16)$

85. $(8x + 11)(12x^2 - 5x - 2) = 0$

$(8x + 11)(3x - 2)(4x + 1) = 0$

$8x + 11 = 0 \quad$ or $\; 3x - 2 = 0 \;$ or $\; 4x + 1 = 0$

$8x = -11 \;$ or $\qquad 3x = 2 \;$ or $\qquad 4x = -1$

$x = -\dfrac{11}{8} \;$ or $\qquad x = \dfrac{2}{3} \;$ or $\qquad x = -\dfrac{1}{4}$

The solutions are $-\dfrac{11}{8}, \dfrac{2}{3}$, and $-\dfrac{1}{4}$.

86. $-2, 2$

87. $(x - 2)^3 = x^3 - 2$

$x^3 - 6x^2 + 12x - 8 = x^3 - 2$

$0 = 6x^2 - 12x + 6$

$0 = 6(x^2 - 2x + 1)$

$0 = 6(x - 1)(x - 1)$

$x - 1 = 0 \;$ or $\; x - 1 = 0$

$x = 1 \;$ or $\qquad x = 1$

The solution 1.

88. $(2x - 9)(3x - 22)$

89. $2a^4b^6 - 3a^2b^3 - 20$

Let $u = a^2b^3$ (and $u^2 = a^4b^6$). Factor $2u^2 - 3u - 20$. We will use the FOIL method.

1. There is no common factor (other than 1 or -1).

2. Factor the first term, $2u^2$. The factors are $2u$, u. The possibility is $(2u + \quad)(u + \quad)$.

3. Factor the last term, -20. The possibilities are $-20 \cdot 1$, $-10 \cdot 2$, $-5 \cdot 4$, $-4 \cdot 5$, $-2 \cdot 10$, and $-1 \cdot 20$.

4. We need factors for which the sum of the products is the middle term, $-3u$. Trial and error leads us to the factorization: $(2u + 5)(u - 4)$. Replace u by a^2b^3. We have $(2a^2b^3 + 5)(a^2b^3 - 4)$.

90. $5(x^4y^3 + 4)(x^4y^3 + 3)$

91. $4x^{2a} - 4x^a - 3$

Let $u = x^a$ (and $u^2 = x^{2a}$). Factor $4u^2 - 4u - 3$. We will use the grouping method. Multiply the leading coefficient and the constant: $4(-3) = -12$. Factor -12 so the sum of the factors is -4. The desired factorizations is $-6 \cdot 2$.

Split the middle term and factor by grouping.

$4u^2 - 4u - 3 = 4u^2 - 6u + 2u - 3$

$\qquad = 2u(2u - 3) + (2u - 3)$

$\qquad = (2u - 3)(2u + 1)$

Replace u by x^a. The factorization is $(2x^a - 3)(2x^a + 1)$.

92. $a(2r + s)(r + s)$

93. See the answer section in the text.

Exercise Set 5.6

1. $x^2 + 8x + 16 = (x + 4)^2$
  Find the square terms and write the quantities that were squared with a plus sign between them.

2. $(t + 3)^2$

3. $a^2 + 16a + 64 = (a + 8)^2$
  Find the square terms and write the quantities that were squared with a minus sign between them.

4. $(a - 7)^2$

5. $2a^2 + 8a + 8$
  $= 2(a^2 + 4a + 4)$ Factoring out the common factor
  $= 2(a + 2)^2$ Factoring the perfect-square trinomial

6. $4(a - 2)^2$

7. $y^2 + 36 - 12y$
  $= y^2 - 12y + 36$ Changing order
  $= (y - 6)^2$ Factoring the perfect-square trinomial

8. $(y + 6)^2$

9. $24a^2 + a^3 + 144a$
  $= a^3 + 24a^2 + 144a$ Changing order
  $= a(a^2 + 24a + 144)$ Factoring out the common factor
  $= a(a + 12)^2$ Factoring the perfect-square trinomial

10. $y(y - 9)^2$

11. $32x^2 + 48x + 18$
  $= 2(16x^2 + 24x + 9)$ Factoring out the common factor
  $= 2(4x + 3)^2$ Factoring the perfect-square trinomial

12. $2(x - 10)^2$

13. $64 + 25a^2 - 80a$
  $= 25a^2 - 80a + 64$ Changing order
  $= (5a - 8)^2$ Factoring the perfect-square trinomial

14. $(1 - 4d)^2$

15. $0.25x^2 + 0.30x + 0.09 = (0.5x + 0.3)^2$
  Find the square terms and write the quantities that were squared with a plus sign between them.

16. $(0.2x - 0.7)^2$

17. $p^2 - 2pq + q^2 = (p - q)^2$

18. $(m + n)^2$

19. $a^3 - 10a^2 + 25a$
  $= a(a^2 - 10a + 25)$
  $= a(a - 5)^2$

20. $y(y + 4)^2$

21. $25a^2 - 30ab + 9b^2 = (5a - 3b)^2$

22. $(7p - 6q)^2$

23. $4t^2 - 8tr + 4r^2$
  $= 4(t^2 - 2tr + r^2)$
  $= 4(t - r)^2$

24. $5(a - b)^2$

25. $x^2 - 16 = x^2 - 4^2 = (x + 4)(x - 4)$

26. $(y + 10)(y - 10)$

27. $p^2 - 49 = p^2 - 7^2 = (p + 7)(p - 7)$

28. $(m + 8)(m - 8)$

29. $a^2b^2 - 81 = (ab)^2 - 9^2 = (ab + 9)(ab - 9)$

30. $(pq + 5)(pq - 5)$

31. $6x^2 - 6y^2$
  $= 6(x^2 - y^2)$ Factoring out the common factor
  $= 6(x + y)(x - y)$ Factoring the difference of squares

32. $8(x + y)(x - y)$

33. $7xy^4 - 7xz^4$
  $= 7x(y^4 - z^4)$
  $= 7x[(y^2)^2 - (z^2)^2]$
  $= 7x(y^2 + z^2)(y^2 - z^2)$
  $= 7x(y^2 + z^2)(y + z)(y - z)$

34. $25a(b^2 + z^2)(b + z)(b - z)$

35. $4a^3 - 49a = a(4a^2 - 49)$
$$= a[(2a)^2 - 7^2]$$
$$= a(2a + 7)(2a - 7)$$

36. $x^2(3x + 5)(3x - 5)$

37. $3x^8 - 3y^8$
$$= 3(x^8 - y^8)$$
$$= 3[(x^4)^2 - (y^4)^2]$$
$$= 3(x^4 + y^4)(x^4 - y^4)$$
$$= 3(x^4 + y^4)[(x^2)^2 - (y^2)^2]$$
$$= 3(x^4 + y^4)(x^2 + y^2)(x^2 - y^2)$$
$$= 3(x^4 + y^4)(x^2 + y^2)(x + y)(x - y)$$

38. $a^2(3a + b)(3a - b)$

39. $9a^4 - 25a^2b^4 = a^2(9a^2 - 25b^4)$
$$= a^2[(3a)^2 - (5b^2)^2]$$
$$= a^2(3a + 5b^2)(3a - 5b^2)$$

40. $x^2(4x^2 + 11y^2)(4x^2 - 11y^2)$

41. $\dfrac{1}{25} - x^2 = \left(\dfrac{1}{5}\right)^2 - x^2$
$$= \left(\dfrac{1}{5} + x\right)\left(\dfrac{1}{5} - x\right)$$

42. $\left(\dfrac{1}{4} + y\right)\left(\dfrac{1}{4} - y\right)$

43. $(a + b)^2 - 9 = (a + b)^2 - 3^2$
$$= [(a + b) + 3][(a + b) - 3]$$
$$= (a + b + 3)(a + b - 3)$$

44. $(p + q + 5)(p + q - 5)$

45. $x^2 - 6x + 9 - y^2$
$$= (x^2 - 6x + 9) - y^2 \quad \text{Grouping as a difference of squares}$$
$$= (x - 3)^2 - y^2$$
$$= (x - 3 + y)(x - 3 - y)$$

46. $(a - 4 + b)(a - 4 - b)$

47. $m^2 - 2mn + n^2 - 25$
$$= (m^2 - 2mn + n^2) - 25 \quad \text{Grouping as a difference of squares}$$
$$= (m - n)^2 - 5^2$$
$$= (m - n + 5)(m - n - 5)$$

48. $(x + y + 3)(x + y - 3)$

49. $36 - (x + y)^2 = 6^2 - (x + y)^2$
$$= [6 + (x + y)][6 - (x + y)]$$
$$= (6 + x + y)(6 - x - y)$$

50. $(7 + a + b)(7 - a - b)$

51. $r^2 - 2r + 1 - 4s^2$
$$= (r^2 - 2r + 1) - 4s^2 \quad \text{Grouping as a difference of squares}$$
$$= (r - 1)^2 - (2s)^2$$
$$= (r - 1 + 2s)(r - 1 - 2s)$$

52. $(c + 2d + 3p)(c + 2d - 3p)$

53. $16 - a^2 - 2ab - b^2$
$$= 16 - (a^2 + 2ab + b^2) \quad \text{Grouping as a difference of squares}$$
$$= 4^2 - (a + b)^2$$
$$= [4 + (a + b)][4 - (a + b)]$$
$$= (4 + a + b)(4 - a - b)$$

54. $(3 + x + y)(3 - x - y)$

55. $m^3 - 7m^2 - 4m + 28$
$$= m^2(m - 7) - 4(m - 7) \quad \text{Factoring by grouping}$$
$$= (m - 7)(m^2 - 4)$$
$$= (m - 7)(m + 2)(m - 2) \quad \text{Factoring the difference of squares}$$

56. $(x + 8)(x + 1)(x - 1)$

57. $a^3 - ab^2 - 2a^2 + 2b^2$
$$= a(a^2 - b^2) - 2(a^2 - b^2) \quad \text{Factoring by grouping}$$
$$= (a^2 - b^2)(a - 2)$$
$$= (a + b)(a - b)(a - 2) \quad \text{Factoring the difference of squares}$$

58. $(p + 5)(p - 5)(q + 3)$

59. $x^2 + 8x + 16 = 0$
$$(x + 4)^2 = 0 \quad \text{From Exercise 1}$$
$$(x + 4)(x + 4) = 0$$
$$x + 4 = 0 \quad or \quad x + 4 = 0$$
$$x = -4 \quad or \quad x = -4$$
The solution is -4.

60. 7

61.
$$x^2 = 16$$
$$x^2 - 16 = 0$$
$$(x+4)(x+4) = 0 \quad \text{From Exercise 25}$$
$$x + 4 = 0 \quad \text{or} \quad x - 4 = 0$$
$$x = -4 \quad \text{or} \quad x = 4$$
The solutions are -4 and 4.

62. $-10, 10$

63.
$$a^2 + 1 = 2a$$
$$a^2 - 2a + 1 = 0$$
$$(a-1)(a-1) = 0$$
$$a - 1 = 0 \quad \text{or} \quad a - 1 = 0$$
$$a = 1 \quad \text{or} \quad a = 1$$
The solution is 1.

64. 4

65.
$$2x^2 - 24x + 72 = 0$$
$$2(x^2 - 12x + 36) = 0$$
$$2(x-6)(x-6) = 0$$
$$x - 6 = 0 \quad \text{or} \quad x - 6 = 0$$
$$x = 6 \quad \text{or} \quad x = 6$$
The solution is 6.

66. -8

67.
$$x^2 - 9 = 0$$
$$(x+3)(x-3) = 0$$
$$x + 3 = 0 \quad \text{or} \quad x - 3 = 0$$
$$x = -3 \quad \text{or} \quad x = 3$$
The solutions are -3 and 3.

68. $-8, 8$

69.
$$a^2 = \frac{1}{25}$$
$$a^2 - \frac{1}{25} = 0$$
$$\left(a + \frac{1}{5}\right)\left(a - \frac{1}{5}\right) = 0$$
$$a + \frac{1}{5} = 0 \quad \text{or} \quad a - \frac{1}{5} = 0$$
$$a = -\frac{1}{5} \quad \text{or} \quad a = \frac{1}{5}$$
The solutions are $-\frac{1}{5}$ and $\frac{1}{5}$.

70. $-\frac{1}{10}, \frac{1}{10}$

71.
$$8x^3 + 1 = 4x^2 + 2x$$
$$8x^3 - 4x^2 - 2x + 1 = 0$$
$$4x^2(2x - 1) - (2x - 1) = 0$$
$$(2x - 1)(4x^2 - 1) = 0$$
$$(2x - 1)(2x + 1)(2x - 1) = 0$$
$$2x - 1 = 0 \quad \text{or} \quad 2x + 1 = 0 \quad \text{or} \quad 2x - 1 = 0$$
$$2x = 1 \quad \text{or} \quad 2x = -1 \quad \text{or} \quad 2x = 1$$
$$x = \frac{1}{2} \quad \text{or} \quad x = -\frac{1}{2} \quad \text{or} \quad x = \frac{1}{2}$$
The solutions are $\frac{1}{2}$ and $-\frac{1}{2}$.

72. $-\frac{2}{3}, \frac{2}{3}$

73.
$$x^3 + 3 = 3x^2 + x$$
$$x^3 - 3x^2 - x + 3 = 0$$
$$x^2(x - 3) - (x - 3) = 0$$
$$(x - 3)(x^2 - 1) = 0$$
$$(x - 3)(x + 1)(x - 1) = 0$$
$$x - 3 = 0 \text{ or } x + 1 = 0 \quad \text{or} \quad x - 1 = 0$$
$$x = 3 \text{ or } \quad x = -1 \text{ or} \quad x = 1$$
The solutions are 3, -1, and 1.

74. $-4, -1, 4$

75. The polynomial $x^2 - 3x - 7$ is prime. We solve the equation by graphing $y = x^2 - 3x - 7$ and finding the zeros. They are approximately -1.541 and 4.541. These are the solutions.

76. $0.209, 4.791$

77. The polynomial $2x^2 + 8x + 1$ is prime. We solve the equation by graphing $y = 2x^2 + 8x + 1$ and finding the zeros. They are approximately -3.871 and -0.129. These are the solutions.

78. $-0.768, 0.434$

79. The polynomial $x^3 + 3x^2 + x - 1$ is prime. We solve the equation by graphing $y = x^3 + 3x^2 + x - 1$ and finding the zeros. They are approximately -2.414, -1, and approximately 0.414. These are the solutions.

80. 0.544

81. We set $f(a)$ equal to -36.
$$a^2 - 12a = -36$$
$$a^2 - 12a + 36 = 0$$
$$(a - 6)(a - 6) = 0$$
$$a - 6 = 0 \quad \text{or} \quad a - 6 = 0$$
$$a = 6 \quad \text{or} \quad a = 6$$
The value of a for which $f(a) = -36$ is 6.

82. $-12, 12$

83. To find the zeros of $f(x) = x^2 - 16$, we find the roots of the equation $x^2 - 16 = 0$.

$$x^2 - 16 = 0$$
$$(x+4)(x-4) = 0$$
$$x + 4 = 0 \quad or \quad x - 4 = 0$$
$$x = -4 \quad or \quad x = 4$$

The zeros are -4 and 4.

84. 4

85. To find the zeros of $f(x) = 2x^2 + 4x + 2$, we find the roots of the equation $2x^2 + 4x + 2 = 0$.

$$2x^2 + 4x + 2 = 0$$
$$2(x^2 + 2x + 1) = 0$$
$$2(x+1)(x+1) = 0$$
$$x + 1 = 0 \quad or \quad x + 1 = 0$$
$$x = -1 \quad or \quad x = -1$$

The zero is -1.

86. $-3, 3$

87. To find the zeros of $f(x) = x^3 - 2x^2 - x + 2$, we find the roots of the equation $x^3 - 2x^2 - x + 2 = 0$.

$$x^3 - 2x^2 - x + 2 = 0$$
$$x^2(x - 2) - (x - 2) = 0$$
$$(x - 2)(x^2 - 1) = 0$$
$$(x - 2)(x + 1)(x - 1) = 0$$
$$x - 2 = 0 \quad or \quad x + 1 = 0 \quad or \quad x - 1 = 0$$
$$x = 2 \quad or \quad x = -1 \quad or \quad x = 1$$

The zeros are 2, -1, and 1.

88. $-2, -1, 2$

89. *Writing Exercise*

90. *Writing Exercise*

91. $(2a^4b^5)^3 = 2^3(a^4)^3(b^5)^3 = 8a^{4\cdot3}b^{5\cdot3} = 8a^{12}b^{15}$

92. $125x^6y^{12}$

93.
$$(x+y)^3$$
$$= (x+y)(x+y)^2$$
$$= (x+y)(x^2 + 2xy + y^2)$$
$$= x(x^2 + 2xy + y^2) + y(x^2 + 2xy + y^2)$$
$$= x^3 + 2x^2y + xy^2 + x^2y + 2xy^2 + y^3$$
$$= x^3 + 3x^2y + 3xy^2 + y^3$$

94. $a^3 + 3a^2 + 3a + 1$

95.
$$x - y + z = 6, \quad (1)$$
$$2x + y - z = 0, \quad (2)$$
$$x + 2y + z = 3 \quad (3)$$

Add (1) and (2).
$$x - y + z = 6 \quad (1)$$
$$\underline{2x + y - z = 0} \quad (2)$$
$$3x \qquad\quad = 6 \quad \text{Adding}$$
$$x = 2$$

Add (2) and (3).
$$2x + \; y - z = 0 \quad (2)$$
$$\underline{x + 2y + z = 3} \quad (3)$$
$$3x + 3y \quad\; = 3 \quad (4)$$

Substitute 2 for x in (4).
$$3(2) + 3y = 3$$
$$6 + 3y = 3$$
$$3y = -3$$
$$y = -1$$

Substitute 2 for x and -1 for y in (1).
$$2 - (-1) + z = 6$$
$$3 + z = 6$$
$$z = 3$$

The solution is $(2, -1, 3)$.

96. $\left\{ x \,\middle|\, x \le -\dfrac{4}{7} \text{ or } x \ge 2 \right\}$, or $\left(-\infty, -\dfrac{4}{7} \right] \cup [2, \infty)$

97. $|5 - 7x| \le 9$
$$-9 \le 5 - 7x \le 9$$
$$-14 \le -7x \le 4$$
$$2 \ge x \ge -\frac{4}{7}$$

The solution set is $\left\{ x \,\middle|\, -\dfrac{4}{7} \le x \le 2 \right\}$, or $\left[-\dfrac{4}{7}, 2 \right]$.

98. $\left\{ x \,\middle|\, x < \dfrac{14}{19} \right\}$, or $\left(-\infty, \dfrac{14}{19} \right)$

99. *Writing Exercise*

100. *Writing Exercise*

101.
$$-\frac{8}{27}r^2 - \frac{10}{9}rs - \frac{1}{6}s^2 + \frac{2}{3}rs$$
$$= -\frac{8}{27}r^2 - \frac{4}{9}rs - \frac{1}{6}s^2$$
$$= -\frac{1}{54}(16r^2 + 24rs + 9s^2)$$
$$= -\frac{1}{54}(4r + 3s)^2$$

102. $\left(\dfrac{1}{6}x^4 + \dfrac{2}{3}\right)^2$

103. $0.09x^8 + 0.48x^4 + 0.64 = (0.3x^4 + 0.8)^2$, or

$\dfrac{1}{100}(3x^4 + 8)^2$

104. $(a + b + c - 3)(a + b - c + 3)$

105.
$$r^2 - 8r - 25 - s^2 - 10s + 16$$
$$= (r^2 - 8r + 16) - (s^2 + 10s + 25)$$
$$= (r - 4)^2 - (s + 5)^2$$
$$= [(r - 4) + (s + 5)][(r - 4) - (s + 5)]$$
$$= (r - 4 + s + 5)(r - 4 - s - 5)$$
$$= (r + s + 1)(r - s - 9)$$

106. $(x^a + y)(x^a - y)$

107. $x^{4a} - y^{2b} = (x^{2a})^2 - (y^b)^2 = (x^{2a} + y^b)(x^{2a} - y^b)$

108. $4(y^{2a} + 5)^2$

109.
$$25y^{2a} - (x^{2b} - 2x^b + 1)$$
$$= (5y^a)^2 - (x^b - 1)^2$$
$$= [5y^a + (x^b - 1)][5y^a - (x^b - 1)]$$
$$= (5y^a + x^b - 1)(5y^a - x^b + 1)$$

110. $8(a - 7)^2$

111.
$$3(x + 1)^2 + 12(x + 1) + 12$$
$$= 3[(x + 1)^2 + 4(x + 1) + 4]$$
$$= 3[(x + 1) + 2]^2$$
$$= 3(x + 3)^2$$

112. $5(c^{50} + 4d^{50})(c^{25} + 2d^{25})(c^{25} - 2d^{25})$

113.
$$9x^{2n} - 6x^n + 1 = (3x^n)^2 - 6x^n + 1$$
$$= (3x^n - 1)^2$$

114. $c(c^w + 1)^2$

115. If $P(x) = x^2$, then
$$P(a + h) - P(a)$$
$$= (a + h)^2 - a^2$$
$$= [(a + h) + a][(a + h) - a]$$
$$= (2a + h)h, \text{ or } h(2a + h)$$

116. $h(2a + h)(2a^2 + 2ah + h^2)$

117. a) $\pi R^2 h - \pi r^2 h = \pi h(R^2 - r^2)$
$$= \pi h(R + r)(R - r)$$

b) Note that 4 m = 400 cm.
$$\pi R^2 h - \pi r^2 h$$
$$= \pi(50)^2(400) - \pi(10)^2(400)$$
$$= 1,000,000\pi - 40,000\pi$$
$$= 960,000\pi \text{ cm}^3 \quad \text{(or } 0.96\pi \text{ m}^3\text{)}$$
$$\approx 3,014,400 \text{ cm}^3 \quad \text{Using 3.14 for } \pi$$

$$\pi h(R + r)(R - r)$$
$$= \pi(400)(50 + 10)(50 - 10)$$
$$= \pi(400)(60)(40)$$
$$= 960,000\pi \text{ cm}^3 \quad \text{(or } 0.96\pi \text{ m}^3\text{)}$$
$$\approx 3,014,400 \text{ cm}^3 \quad \text{Using 3.14 for } \pi$$

If we use the π key on a calculator, the result is approximately 3,015,929 cm^3.

118. Enter $y_1 = (x^2 - 3x + 2)^4$ and $y_2 = x^8 + 81x^4 + 16$ and look at a table of values. Observe that $y_1 \neq y_2$.

Exercise Set 5.7

1.
$$t^3 + 27 = t^3 + 3^3$$
$$= (t + 3)(t^2 - 3t + 9)$$
$$A^3 + B^3 = (A + B)(A^2 - AB + B^2)$$

2. $(x + 4)(x^2 - 4x + 16)$

3.
$$x^3 - 8 = x^3 - 2^3$$
$$= (x - 2)(x^2 + 2x + 4)$$
$$A^3 - B^3 = (A - B)(A^2 + AB + B^2)$$

4. $(z - 1)(z^2 + z + 1)$

5.
$$m^3 - 64 = m^3 - 4^3$$
$$= (m - 4)(m^2 + 4m + 16)$$
$$A^3 - B^3 = (A - B)(A^2 + AB + B^2)$$

6. $(x - 3)(x^2 + 3x + 9)$

7.
$$8a^3 + 1 = (2a)^3 + 1^3$$
$$= (2a + 1)(4a^2 - 2a + 1)$$
$$A^3 + B^3 = (A + B)(A^2 - AB + B^2)$$

8. $(3x + 1)(9x^2 - 3x + 1)$

9.
$$27 - 8t^3 = 3^3 - (2t)^3$$
$$= (3 - 2t)(9 + 6t + 4t^2)$$

10. $(4 - 5x)(16 + 20x + 25x^2)$

11.
$$8x^3 + 27 = (2x)^3 + 3^3$$
$$= (2x + 3)(4x^2 - 6x + 9)$$

12. $(3y + 4)(9y^2 - 12y + 16)$

13. $y^3 - z^3 = (y - z)(y^2 + yz + z^2)$

14. $(x - y)(x^2 + xy + y^2)$

15. $x^3 + \dfrac{1}{27} = x^3 + \left(\dfrac{1}{3}\right)^3$

$\qquad = \left(x + \dfrac{1}{3}\right)\left(x^2 - \dfrac{1}{3}x + \dfrac{1}{9}\right)$

16. $\left(a + \dfrac{1}{2}\right)\left(a^2 - \dfrac{1}{2}a + \dfrac{1}{4}\right)$

17. $2y^3 - 128 = 2(y^3 - 64)$

$\qquad = 2(y^3 - 4^3)$

$\qquad = 2(y - 4)(y^2 + 4y + 16)$

18. $8(t - 1)(t^2 + t + 1)$

19. $8a^3 + 1000 = 8(a^3 + 125)$

$\qquad = 8(a^3 + 5^3)$

$\qquad = 8(a + 5)(a^2 - 5a + 25)$

20. $2(3x + 1)(9x^2 - 3x + 1)$

21. $rs^3 + 64r = r(s^3 + 64)$

$\qquad = r(s^3 + 4^3)$

$\qquad = r(s + 4)(s^2 - 4s + 16)$

22. $a(b + 5)(b^2 - 5b + 25)$

23. $2y^3 - 54z^3 = 2(y^3 - 27z^3)$

$\qquad = 2[y^3 - (3z)^3]$

$\qquad = 2(y - 3z)(y^2 + 3yz + 9z^2)$

24. $5(x - 2z)(x^2 + 2xz + 4z^2)$

25. $y^3 + 0.125 = y^3 + (0.5)^3$

$\qquad = (y + 0.5)(y^2 - 0.5y + 0.25)$

26. $(x + 0.1)(x^2 - 0.1x + 0.01)$

27. $125c^6 - 8d^6 = (5c^2)^3 - (2d^2)^3$

$\qquad = (5c^2 - 2d^2)(25c^4 + 10c^2d^2 + 4d^4)$

28. $8(2x^2 - t^2)(4x^4 + 2x^2t^2 + t^4)$

29. $3z^5 - 3z^2 = 3z^2(z^3 - 1)$

$\qquad = 3z^2(z^3 - 1^3)$

$\qquad = 3z^2(z - 1)(z^2 + z + 1)$

30. $2y(y - 4)(y^2 + 4y + 16)$

31. $t^6 + 1 = (t^2)^3 + 1^3$

$\qquad = (t^2 + 1)(t^4 - t^2 + 1)$

32. $(z + 1)(z^2 - z + 1)(z - 1)(z^2 + z + 1)$

33. $\quad p^6 - q^6$

$= (p^3)^2 - (q^3)^2 \qquad$ Writing as a difference
of squares

$= (p^3 + q^3)(p^3 - q^3) \qquad$ Factoring a difference
of squares

$= (p + q)(p^2 - pq + q^2)(p - q)(p^2 + pq + q^2)$

$\qquad\qquad\qquad$ Factoring a sum and
a difference of cubes

34. $(t^2 + 4y^2)(t^4 - 4t^2y^2 + 16y^4)$

35. $\quad a^9 + b^{12}c^{15}$

$= (a^3)^3 + (b^4c^5)^3$

$= (a^3 + b^4c^5)(a^6 - a^3b^4c^5 + b^8c^{10})$

36. $(x^4 - yz^4)(x^8 + x^4yz^4 + y^2z^8)$

37. $\qquad\qquad x^3 + 1 = 0$

$(x + 1)(x^2 - x + 1) = 0$

$x + 1 = 0 \quad or \quad x^2 - x + 1 = 0$

$x = -1$

We cannot factor $x^2 - x + 1$. The only real-number solution
is -1.

38. 2

39. $\qquad\qquad\qquad 8x^3 = 27$

$8x^3 - 27 = 0$

$(2x - 3)(4x^2 + 6x + 9) = 0$

$2x - 3 = 0 \quad or \quad 4x^2 + 6x + 9 = 0$

$2x = 3$

$x = \dfrac{3}{2}$

We cannot factor $4x^2 + 6x + 9$. The only real-number
solution is $\dfrac{3}{2}$.

40. $-\dfrac{3}{4}$

41. $\qquad\qquad 2t^3 - 2000 = 0$

$2(t^3 - 1000) = 0$

$2(t - 10)(t^2 + 10t + 100) = 0$

$t - 10 = 0 \quad or \quad t^2 + 10t + 100 = 0$

$t = 10$

We cannot factor $t^2 + 10t + 100$. The only real-number
solution is 10.

42. $\dfrac{5}{2}$

43. *Writing Exercise*

44. *Writing Exercise*

45. $h(t) = -16t^2 + 80t + 224$

$h(0) = -16(0)^2 + 80(0) + 224 = 224$ ft

$h(1) = -16(1)^2 + 80(1) + 224 = 288$ ft

$h(3) = -16(3)^2 + 80(3) + 224 = 320$ ft

$h(4) = -16(4)^2 + 80(4) + 224 = 288$ ft

$h(6) = -16(6)^2 + 80(6) + 224 = 128$ ft

46. $f(x) = -\dfrac{3}{4}x - 5$

47. $y - y_1 = m(x - x_1)$

$y - (-5) = 3(x - 1)$

$y + 5 = 3x - 3$

$y = 3x - 8$

$f(x) = 3x - 8$ Using function notation

48. $f(x) = 2x - 8$

49. $3x - 5 = 0$

$3x = 5$ Adding 5 to both sides

$x = \dfrac{5}{3}$ Dividing both sides by 3

The solution is $\dfrac{5}{3}$.

50. $-\dfrac{7}{2}$

51. *Writing Exercise*

52. *Writing Exercise*

53. $x^{6a} - y^{3b} = (x^{2a})^3 - (y^b)^3$

$= (x^{2a} - y^b)(x^{4a} + x^{2a}y^b + y^{2b})$

54. $2(x^a + 2y^b)(x^{2a} - 2x^a y^b + 4y^{2b})$

55.

$(x + 5)^3 + (x - 5)^3$ Sum of cubes

$= [(x+5) + (x-5)][(x+5)^2 - (x+5)(x-5) + (x-5)^2]$

$= 2x[(x^2 + 10x + 25) - (x^2 - 25) + (x^2 - 10x + 25)]$

$= 2x(x^2 + 10x + 25 - x^2 + 25 + x^2 - 10x + 25)$

$= 2x(x^2 + 75)$

56. $\dfrac{1}{16}(x^a + 2y^{2a}z^{3b})(x^{2a} - 2x^a y^{2a}z^{3b} + 4y^{4a}z^{6b})$

57. $5x^3 y^6 - \dfrac{5}{8}$

$= 5\left(x^3 y^6 - \dfrac{1}{8}\right)$

$= 5\left(xy^2 - \dfrac{1}{2}\right)\left(x^2 y^4 + \dfrac{1}{2}xy^2 + \dfrac{1}{4}\right)$

58. $-y(3x^2 + 3xy + y^2)$

59. $x^{6a} - (x^{2a} + 1)^3$

$= [x^{2a} - (x^{2a}+1)][x^{4a} + x^{2a}(x^{2a}+1) + (x^{2a}+1)^2]$

$= (x^{2a} - x^{2a} - 1)(x^{4a} + x^{4a} + x^{2a} + x^{4a} + 2x^{2a} + 1)$

$= -(3x^{4a} + 3x^{2a} + 1)$

60. $-(3x^{4a} - 3x^{2a} + 1)$

61. $t^4 - 8t^3 - t + 8$

$= t^3(t - 8) - (t - 8)$

$= (t - 8)(t^3 - 1)$

$= (t - 8)(t - 1)(t^2 + t + 1)$

62. $h(3a^2 + 3ah + h^2)$

63. If $Q(x) = x^6$, then

$Q(a + h) - Q(a)$

$= (a + h)^6 - a^6$

$= [(a + h)^3 + a^3][(a + h)^3 - a^3]$

$= [(a + h) + a] \cdot [(a + h)^2 - (a + h)a + a^2] \cdot$

$\qquad [(a + h) - a] \cdot [(a + h)^2 + (a + h)a + a^2]$

$= (2a + h) \cdot (a^2 + 2ah + h^2 - a^2 - ah + a^2) \cdot (h) \cdot$

$\qquad (a^2 + 2ah + h^2 + a^2 + ah + a^2)$

$= h(2a + h)(a^2 + ah + h^2)(3a^2 + 3ah + h^2)$

64.

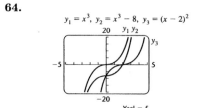

$y_1 = x^3, \ y_2 = x^3 - 8, \ y_3 = (x - 2)^2$

Yscl = 5

Exercise Set 5.8

1. Familiarize. Let x represent the number.

Translate.

Square of number plus number is 132.

$$\underbrace{x^2}_{} \quad \underbrace{+}_{} \quad \underbrace{x}_{} \quad = \quad 132$$

Carry out. We solve the equation:

$$x^2 + x = 132$$
$$x^2 + x - 132 = 0$$
$$(x + 12)(x - 11) = 0$$
$$x + 12 = 0 \quad or \quad x - 11 = 0$$
$$x = -12 \quad or \quad x = 11$$

Check. The square of -12, which is 144, plus -12 is 132. The square of 11, which is 121, plus 11 is 132. Both numbers check.

State. The number is -12 or 11.

2. $-13, 12$

3. *Familiarize*. We let w represent the width and $w + 5$ represent the length. We make a drawing and label it.

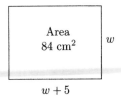

Recall that the formula for the area of a rectangle is $A = $ length $\times$ width.

Translate.

Area is 84 cm^2.
$$w(w + 5) = 84$$

Carry out. We solve the equation:
$$w(w + 5) = 84$$
$$w^2 + 5w = 84$$
$$w^2 + 5w - 84 = 0$$
$$(w + 12)(w - 7) = 0$$
$$w + 12 = 0 \quad or \quad w - 7 = 0$$
$$w = -12 \quad or \quad w = 7$$

Check. The number -12 is not a solution, because width cannot be negative. If the width is 7 cm and the length is 5 cm more, or 12 cm, then the area is $12 \cdot 7$, or 84 cm^2. This is a solution.

State. The length is 12 cm, and the width is 7 cm.

4. Length: 12 cm; width: 8 cm

5. *Familiarize*. We make a drawing and label it. We let x represent the length of a side of the original square, in meters.

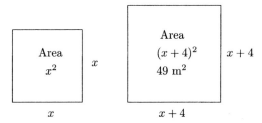

Translate.

Area of new square is 49 m^2.
$$(x + 4)^2 = 49$$

Carry out. We solve the equation:
$$(x + 4)^2 = 49$$
$$x^2 + 8x + 16 = 49$$
$$x^2 + 8x - 33 = 0$$
$$(x - 3)(x + 11) = 0$$
$$x - 3 = 0 \quad or \quad x + 11 = 0$$
$$x = 3 \quad or \quad x = -11$$

Check. We check only 3 since the length of a side cannot be negative. If we increase the length by 4, the new length is $3 + 4$, or 7 m. Then the new area is $7 \cdot 7$, or 49 m^2. We have a solution.

State. The length of a side of the original square is 3 m.

6. 6 cm

7. *Familiarize*. We make a drawing and label it with both known and unknown information. We let x represent the width of the frame.

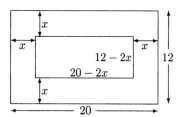

The length and width of the picture that shows are represented by $20 - 2x$ and $12 - 2x$. The area of the picture that shows is 84 cm^2.

Translate. Using the formula for the area of a rectangle, $A = l \cdot w$, we have
$$84 = (20 - 2x)(12 - 2x).$$

Carry out. We solve the equation:
$$84 = 240 - 64x + 4x^2$$
$$84 = 4(60 - 16x + x^2)$$
$$21 = 60 - 16x + x^2 \qquad \text{Dividing by 4}$$
$$0 = x^2 - 16x + 39$$
$$0 = (x - 3)(x - 13)$$

$x - 3 = 0 \quad or \quad x - 13 = 0$

$x = 3 \quad or \qquad x = 13$

Check. We see that 13 is not a solution because when $x = 13$, $20 - 2x = -6$ and $12 - 2x = -14$, and the length and width of the frame cannot be negative. We check 3. When $x = 3$, $20 - 2x = 14$ and $12 - 2x = 6$ and $14 \cdot 6 = 84$. The area is 84. The value checks.

State. The width of the frame is 3 cm.

8. 2 cm

9. Familiarize. We let x represent the width of the walkway. We make a drawing and label it with both the known and unknown information.

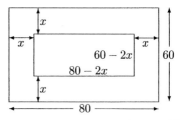

The area of the new lawn is $(80 - 2x)(60 - 2x)$.

Translate.

$$\underbrace{\text{Area of new lawn}}_{(80 - 2x)(60 - 2x)} \quad \text{is} \quad \underbrace{2400 \text{ ft}^2}_{2400}.$$

$(80 - 2x)(60 - 2x) = 2400$

Carry out. We solve the equation:

$(80 - 2x)(60 - 2x) = 2400$

$4800 - 280x + 4x^2 = 2400$

$4x^2 - 280x + 2400 = 0$

$x^2 - 70x + 600 = 0 \qquad \text{Dividing by 4}$

$(x - 10)(x - 60) = 0$

$x - 10 = 0 \quad or \quad x - 60 = 0$

$x = 10 \quad or \qquad x = 60$

Check. If the sidewalk is 10 ft wide, the length of the new lawn will be $80 - 2 \cdot 10$, or 60 ft, and its width will be $60 - 2 \cdot 10$, or 40 ft. Then the area of the new lawn will be $60 \cdot 40$, or 2400 ft^2. This answer checks.

If the sidewalk is 60 ft wide, the length of the new lawn will be $80 - 2 \cdot 60$, or -40 ft. Since the length cannot be negative, 60 is not a solution.

State. The sidewalk is 10 ft wide.

10. 5 ft

11. Familiarize. Let x represent the first integer, $x + 2$ the second, and $x + 4$ the third.

Translate.

$$\underbrace{\text{Square of}}_{(x + 4)^2} \text{ is } \underbrace{76}_{= 76} \underbrace{\text{more}}_{+} \underbrace{\text{square of}}_{(x + 2)^2}$$
the third   than   the second.

Carry out. We solve the equation:

$(x + 4)^2 = 76 + (x + 2)^2$

$x^2 + 8x + 16 = 76 + x^2 + 4x + 4$

$x^2 + 8x + 16 = x^2 + 4x + 80$

$4x = 64$

$x = 16$

Check. We check the integers 16, 18, and 20. The square of 20, or 400, is 76 more than 324, the square of 18. The answer checks.

State. The integers are 16, 18, and 20.

12. $-10, -8, -6$ or 6, 8, 10

13. Familiarize. Let x represent the base of the triangle and $x + 2$ represent the height. Recall that the formula for the area of the triangle with base b and height h is $\frac{1}{2}bh$.

Translate.

$$\underbrace{\text{The area}}_{\frac{1}{2}x(x + 2)} \quad \text{is} \quad \underbrace{12 \text{ ft}^2}_{12}.$$

Carry out. We solve the equation:

$\frac{1}{2}x(x + 2) = 12$

$x(x + 2) = 24 \qquad \text{Multiplying by 2}$

$x^2 + 2x = 24$

$x^2 + 2x - 24 = 0$

$(x + 6)(x - 4) = 0$

$x + 6 = 0 \quad or \quad x - 4 = 0$

$x = -6 \quad or \qquad x = 4$

Check. We check only 4 since the length of the base cannot be negative. If the base is 4 ft, then the height is $4 + 2$, or 6 ft, and the area is $\frac{1}{2} \cdot 4 \cdot 6$, or 12 ft^2. The answer checks.

State. The height is 6 ft, and the base is 4 ft.

14. Distance d: 12 ft; tower height: 16 ft

15. Familiarize. Let b represent the base of the sail. Then $b + 9$ represents the height. Recall that the formula for the area of a triangle is $A = \frac{1}{2} \times$ base $\times$ height.

Translate.

The area is 56 m^2.

$\frac{1}{2}b(b + 9) = 56$

Carry out. We solve the equation:

$$\frac{1}{2}b(b+9) = 56$$

$$b(b+9) = 112 \quad \text{Multiplying by 2}$$

$$b^2 + 9b = 112$$

$$b^2 + 9b - 112 = 0$$

$$(b+16)(b-7) = 0$$

$$b + 16 = 0 \quad or \quad b - 7 = 0$$

$$b = -16 \quad or \quad b = 7$$

Check. We check only 7, since the length of the base cannot be negative. If the base is 7 m, the height is $7+9$, or 16 m, and the area is $\frac{1}{2} \cdot 16 \cdot 7$, or 56 m^2. We have a solution.

State. The base is 7 m, and the height is 16 m.

16. Length: 200 ft; width: 150 ft

17. ***Familiarize***. We make a drawing. Let h = the height the ladder reaches on the wall. Then the length of the ladder is $h+1$.

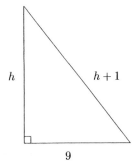

Translate. We use the Pythagorean theorem.

$$9^2 + h^2 = (h+1)^2$$

Carry out. We solve the equation:

$$81 + h^2 = h^2 + 2h + 1$$

$$80 = 2h$$

$$40 = h$$

Check. If $h = 40$, then $h+1 = 41$; $9^2 + 40^2 = 81 + 1600 = 1681 = 41^2$, so the answer checks.

State. The ladder is 41 ft long.

18. 24 ft

19. ***Familiarize***. Let w represent the width and $w + 25$ represent the length. Make a drawing.

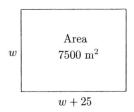

Recall that the formula for the area of a rectangle is $A = $ length $\times$ width.

Translate.

$$\underbrace{\text{Area}}_{w(w+25)} \quad \text{is} \quad \underbrace{7500 \text{ m}^2}_{7500}.$$

$$w(w+25) = 7500$$

Carry out. We solve the equation:

$$w(w+25) = 7500$$

$$w^2 + 25w = 7500$$

$$w^2 + 25w - 7500 = 0$$

$$(w+100)(w-75) = 0$$

$$w + 100 = 0 \quad or \quad w - 75 = 0$$

$$w = -100 \quad or \quad w = 75$$

Check. The number -100 is not a solution because width cannot be negative. If the width is 75 m and the length is 25 m more, or 100 m, then the area will be $100 \cdot 75$, or 7500 m^2. This is a solution.

State. The dimensions will be 100 m by 75 m.

20. Length: 12 m; width: 9 m

21. ***Familiarize***. The firm breaks even when the cost and the revenue are the same. We use the functions given in the text.

Translate.

$$\underbrace{\text{Cost}}_{x^2 - 2x + 10} \quad \text{equals} \quad \underbrace{\text{revenue.}}_{2x^2 + x}$$

Carry out. We solve the equation:

$$x^2 - 2x + 10 = 2x^2 + x$$

$$0 = x^2 + 3x - 10$$

$$0 = (x+5)(x-2)$$

$$x + 5 = 0 \quad or \quad x - 2 = 0$$

$$x = -5 \quad or \quad x = 2$$

Check. We check only 2 since the number of sets of cabinets cannot be negative. If 2 sets of cabinets are produced and sold, the cost is $2^2 - 2 \cdot 2 + 10 = 4 - 4 + 10 = \$10$ thousand and the revenue is $2 \cdot 2^2 + 2 = 8 + 2 = \$10$ thousand. The answer checks.

State. The company must produce and sell 2 sets of cabinets in order to break even.

22. 6 video cameras

23. *Familiarize*. We will use the formula in Example 5, $h(t) = -15t^2 + 75t + 10$. Note that t cannot be negative since it represents time after launch.

***Translate*.** We need to find the value of t for which $h(t) = 100$:

$$-15t^2 + 75t + 10 = 100$$

***Carry out*.** We solve the equation.

$$-15t^2 + 75t + 10 = 100$$
$$-15t^2 + 75t - 90 = 0$$
$$-15(t^2 - 5t + 6) = 0$$
$$-15(t - 2)(t - 3) = 0$$
$$t - 2 = 0 \ \ or \ \ t - 3 = 0$$
$$t = 2 \ \ or \ \ \ \ \ \ t = 3$$

***Check*.** We have:

$$h(2) = -15 \cdot 2^2 + 75 \cdot 2 + 10 = -60 + 150 + 10 = 100;$$
$$h(3) = -15 \cdot 3^2 + 75 \cdot 3 + 10 = -135 + 225 + 10 = 100.$$

Both numbers check. However, the problem states that the tee shirt is caught on the way up. Thus, we reject 3 since that would indicate that the shirt is caught on the way down, after peaking.

***State*.** The tee shirt was airborne for 2 sec before it was caught.

24. 5 sec

25. *Familiarize*. We will use the given formula, $h(t) = -16t^2 + 64t + 80$. Note that t cannot be negative since it represents time after launch.

***Translate*.** We need to find the value of t for which $h(t) = 0$. We have:

$$-16t^2 + 64t + 80 = 0$$

***Carry out*.** We solve the equation.

$$-16t^2 + 64t + 80 = 0$$
$$-16(t^2 - 4t - 5) = 0$$
$$-16(t - 5)(t + 1) = 0$$
$$t - 5 = 0 \ \ or \ \ t + 1 = 0$$
$$t = 5 \ \ or \ \ \ \ \ \ t = -1$$

***Check*.** Since t cannot be negative, we check only 5. $h(5) = -16 \cdot 5^2 + 64 \cdot 5 + 80 = -400 + 320 + 80 = 0$ The number 5 checks.

***State*.** The cardboard shell will reach the ground 5 sec after it is launched.

26. 7 sec

27. a) Enter the data in a graphing calculator, letting x represent the number of years after 1960. Then select item 7, quartic regression, from the STAT CALC menu to find the desired function. We have $E(x) = 0.0000256775x^4 - 0.0025507304x^3 + 0.0821803626x^2 - 0.9480449896x + 11.8202857$.

b) In 1990, $x = 30$.

$E(30) \approx 9.3$ cents per kilowatt-hour

In 2004, $x = 44$.

$E(44) \approx 8.2$ cents per kilowatt-hour

c) We solve $E(x) = 9$ graphically, looking for all the solutions between $x = 0$ and $x = 42$. Graph $y_1 = E(x)$ and $y_2 = 9$ and use the Intersect feature three times. We find that the first coordinates of the points of intersection are approximately 4, 17, and 32. Thus, the average retail cost of electricity was 9 cents per kilowatt-hour in 1964, 1977, and 1992.

28. a) $f(x) = 0.0000001957x^4 - 0.0000570879x^3 + 0.0042982745x^2 - 0.0159423087x + 1.45236808$

b) 1.986 million farms; 2.600 million farms

c) 1876, 1974

29. a) Enter the data in a graphing calculator, letting x represent the number of years after 1970. Then select item 5, quadratic regression, from the STAT CALC menu to find the desired function. We have $d(x) = -105.9592768x^2 + 3621.572598x + 42,307.79479$.

b) In 1999, $x = 29$.

$d(29) \approx 58,221$ degrees

In 2005, $x = 35$.

$d(35) \approx 39,263$ degrees

c) Graph $y_1 = d(x)$ and $y_2 = 60,000$ and use Intersect to find the first coordinates of the points of intersection of the graphs. They are approximately 6 and 28, so 60,000 engineering bachelor's degrees were earned in 1976 and in 1998.

30. a) $r(x) = -0.0085518407x^3 + 0.390736534x^2 - 4.819375298x + 69.4855836$

b) 61.3 births per 1000 women; 12.8 births per 1000 women

c) 1972, 1989, 1995

31. a) Enter the data in a graphing calculator, letting x represent the number of years after 1990. Then select item 7, quartic regression, from the STAT CALC menu to find the desired function. We have $b(x) = -1.143939394x^2 + 9.583333333x + 25.79545455$.

b) In 1992, $x = 2$.

$b(2) \approx 40\%$

In 2000, $x = 10$.

$b(10) \approx 7\%$

c) Graph $y_1 = b(x)$ and $y_2 = 30$ and use Intersect to find the first coordinates of the points of intersection of the graphs. They are approximately 0 and 8. Only 8 represents a year after 1991, so 30% of bottles were recycled in 1998.

32. a) $n(x) = 0.0113125x^4 - 0.9444166667x^3 + 24.93375x^2 - 222.4083333x + 8174$

b) 7860 newspapers; 8985 newspapers

c) 1961

33. *Writing Exercise*

34. *Writing Exercise*

35. $\dfrac{5 - 10 \cdot 3}{-4 + 11 \cdot 4} = \dfrac{5 - 30}{-4 + 44} = \dfrac{-25}{40} = -\dfrac{5}{8}$

36. 2

37. Familiarize. Let r represent the speed of the faster car and d represent its distance. Then $r - 15$ and $651 - d$ represent the speed and distance of the slower car, respectively. We organize the information in a table.

	Speed	Time	Distance
Faster car	r	7	d
Slower car	$r - 15$	7	$651 - d$

Translate. We use the formula $rt = d$. Each row of the table gives us an equation.

$$7r = d$$
$$7(r - 15) = 651 - d$$

Carry out. We use the substitution method substituting $7r$ for d in the second equation and solving for r.

$$7(r - 15) = 651 - 7r \quad \text{Substituting}$$
$$7r - 105 = 651 - 7r$$
$$14r = 756$$
$$r = 54$$

Check. If $r = 54$, then the speed of the faster car is 54 mph and the speed of the slower car is $54 - 15$, or 39 mph. The distance the faster car travels is $54 \cdot 7$, or 378 miles. The distance the slower car travels is $39 \cdot 7$, or 273 miles. The total of the two distances is $378 + 273$, or 651 miles. The result checks.

State. The speed of the faster car is 54 mph. The speed of the slower car is 39 mph.

38. Conventional: 36; surround-sound: 42

39. $2x - 14 + 9x > -8x + 16 + 10x$

$\quad\quad 11x - 14 > 2x + 16 \quad$ Collecting like terms

$\quad\quad\quad 9x - 14 > 16 \quad\quad$ Adding $-2x$

$\quad\quad\quad\quad 9x > 30 \quad\quad$ Adding 14

$\quad\quad\quad\quad x > \dfrac{10}{3} \quad\quad$ Multiplying by $\dfrac{1}{9}$

The solution set is $\left\{ x \middle| x > \dfrac{10}{3} \right\}$, or $\left(\dfrac{10}{3}, \infty \right)$.

40. $(2, -2, -4)$

41. *Writing Exercise*

42. *Writing Exercise*

43. Familiarize. Using the labels on the drawing in the text, we let x represent the width of the piece of tin and $2x$ represent the length. Then the width and length of the base of the box are represented by $x - 4$ and $2x - 4$, respectively. Recall that the formula for the volume of a rectangular solid with length l, width w, and height h is $l \cdot w \cdot h$.

Translate.

The volume is 480 cm^3.

$$(2x - 4)(x - 4)(2) = 480$$

Carry out. We solve the equation:

$$(2x - 4)(x - 4)(2) = 480$$
$$(2x - 4)(x - 4) = 240 \quad \text{Dividing by 2}$$
$$2x^2 - 12x + 16 = 240$$
$$2x^2 - 12x - 224 = 0$$
$$x^2 - 6x - 112 = 0 \quad \text{Dividing by 2}$$
$$(x + 8)(x - 14) = 0$$
$$x + 8 = 0 \quad or \quad x - 14 = 0$$
$$x = -8 \quad or \quad x = 14$$

Check. We check only 14 since the width cannot be negative. If the width of the piece of tin is 14 cm, then its length is $2 \cdot 14$, or 28 cm, and the dimensions of the base of the box are $14 - 4$, or 10 cm by $28 - 4$, or 24 cm. The volume of the box is $24 \cdot 10 \cdot 2$, or 480 cm^3. The answer checks.

State. The dimensions of the piece of tin are 14 cm by 28 cm.

44. 50-year-old drivers

45. Familiarize. Let $r =$ the speed of the tugboat and $r - 7 =$ the speed of the freighter. After 4 hr they have traveled $4r$ km and $4(r - 7)$ km, respectively.

Translate. We use the Pythagorean theorem.

$$(4r)^2 + [4(r - 7)]^2 = 68^2$$

Carry out. We solve the equation.

$$(4r)^2 + [4(r-7)]^2 = 68^2$$
$$16r^2 + 16(r^2 - 14r + 49) = 4624$$
$$16r^2 + 16r^2 - 224r + 784 = 4624$$
$$32r^2 - 224r - 3840 = 0$$
$$32(r^2 - 7r - 120) = 0$$
$$32(r-15)(r+8) = 0$$
$$r - 15 = 0 \quad or \quad r + 8 = 0$$
$$r = 15 \quad or \qquad r = -8$$

Check. Since the speed cannot be negative, we check only 15. If $r = 15$, then $r - 7 = 15 - 7$, or 8. At a rate of 15 km/h, in 4 hr the tugboat travels $4 \cdot 15$ or 60 km. At a rate of 8 km/h, in 4 hr the freighter travels $4 \cdot 8$, or 32 km. Since $60^2 + 32^2 = 4624 = 68^2$, the answer checks.

State. The speed of the tugboat is 15 km/h, and the speed of the freighter is 8 km/h.

46. About 5.7 sec

Chapter 6

Rational Expressions, Equations, and Functions

1. $H(t) = \dfrac{t^2 + 3t}{2t + 3}$

$H(5) = \dfrac{5^2 + 3 \cdot 5}{2 \cdot 5 + 3} = \dfrac{25 + 15}{10 + 3} = \dfrac{40}{13}$ hr, or $3\dfrac{1}{13}$ hr

2. $\dfrac{70}{17}$ hr, or $4\dfrac{2}{17}$ hr

3. $v(t) = \dfrac{4t^2 - 5t + 2}{t + 3}$

$v(0) = \dfrac{4 \cdot 0^2 - 5 \cdot 0 + 2}{0 + 3} = \dfrac{0 - 0 + 2}{0 + 3} = \dfrac{2}{3}$

$v(-2) = \dfrac{4(-2)^2 - 5(-2) + 2}{-2 + 3} = \dfrac{16 + 10 + 2}{-2 + 3} = 28$

$v(7) = \dfrac{4 \cdot 7^2 - 5 \cdot 7 + 2}{7 + 3} = \dfrac{196 - 35 + 2}{7 + 3} = \dfrac{163}{10}$

4. $-2; -\dfrac{11}{7}; 15$

5. $g(x) = \dfrac{2x^3 - 9}{x^2 - 4x + 4}$

$g(0) = \dfrac{2 \cdot 0^3 - 9}{0^2 - 4 \cdot 0 + 4} = \dfrac{0 - 9}{0 - 0 + 4} = -\dfrac{9}{4}$

$g(2) = \dfrac{2 \cdot 2^3 - 9}{2^2 - 4 \cdot 2 + 4} = \dfrac{16 - 9}{4 - 8 + 4} = \dfrac{7}{0}$

Since division by zero is not defined, $g(2)$ does not exist.

$g(-1) = \dfrac{2(-1)^3 - 9}{(-1)^2 - 4(-1) + 4} = \dfrac{-2 - 9}{1 + 4 + 4} = -\dfrac{11}{9}$

6. $0; \dfrac{2}{5};$ does not exist

7. $\dfrac{4x}{4x} \cdot \dfrac{x - 3}{x + 2} = \dfrac{4x(x - 3)}{4x(x + 2)}$

8. $\dfrac{(3 - a^2)(-1)}{(a - 7)(-1)}$

9. $\dfrac{t - 2}{t + 3} \cdot \dfrac{-1}{-1} = \dfrac{(t - 2)(-1)}{(t + 3)(-1)}$

10. $\dfrac{(x - 4)(x - 5)}{(x + 5)(x - 5)}$

11. $\dfrac{15x}{5x^2}$

$= \dfrac{5x \cdot 3}{5x \cdot x}$ Factoring; the greatest common factor is $5x$.

$= \dfrac{5x}{5x} \cdot \dfrac{3}{x}$ Factoring the rational expression

$= 1 \cdot \dfrac{3}{x}$ $\dfrac{5x}{5x} = 1$

$= \dfrac{3}{x}$ Removing a factor equal to 1

12. $\dfrac{a^2}{3}$

13. $\dfrac{18t^3}{27t^7}$

$= \dfrac{9t^3 \cdot 2}{9t^3 \cdot 3t^4}$ Factoring the numerator and the denominator

$= \dfrac{9t^3}{9t^3} \cdot \dfrac{2}{3t^4}$ Factoring the rational expression

$= \dfrac{2}{3t^4}$ Removing a factor equal to 1

14. $\dfrac{2}{y^4}$

15. $\dfrac{2a - 10}{2} = \dfrac{2(a - 5)}{2 \cdot 1} = \dfrac{2}{2} \cdot \dfrac{a - 5}{1} = a - 5$

16. $a + 4$

17. $\dfrac{15}{25a - 30} = \dfrac{5 \cdot 3}{5(5a - 6)} = \dfrac{5}{5} \cdot \dfrac{3}{5a - 6} = \dfrac{3}{5a - 6}$

18. $\dfrac{7}{2x - 3}$

19. $\dfrac{3x - 12}{3x + 15} = \dfrac{3(x - 4)}{3(x + 5)} = \dfrac{3}{3} \cdot \dfrac{x - 4}{x + 5} = \dfrac{x - 4}{x + 5}$

20. $\dfrac{y - 5}{y + 3}$

21. $\dfrac{5x + 20}{x^2 + 4x} = \dfrac{5(x + 4)}{x(x + 4)} = \dfrac{5}{x} \cdot \dfrac{x + 4}{x + 4} = \dfrac{5}{x}$

22. $\dfrac{3}{x}$

23. $\dfrac{3a-1}{2-6a}$

$= \dfrac{3a-1}{2(1-3a)}$

$= \dfrac{-1(1-3a)}{2(1-3a)}$ Factoring out -1 in the numerator reverses the subtraction.

$= \dfrac{-1}{2} \cdot \dfrac{1-3a}{1-3a}$

$= -\dfrac{1}{2}$

24. $-\dfrac{1}{2}$

25. $\dfrac{8t-16}{t^2-4} = \dfrac{8(t-2)}{(t+2)(t-2)} = \dfrac{8}{t+2} \cdot \dfrac{t-2}{t-2} = \dfrac{8}{t+2}$

26. $\dfrac{t-3}{5}$

27. $\dfrac{2t-1}{1-4t^2}$

$= \dfrac{2t-1}{(1+2t)(1-2t)}$

$= \dfrac{-1(1-2t)}{(1+2t)(1-2t)}$ Factoring out -1 in the numerator reverses the subtraction

$= \dfrac{-1}{1+2t} \cdot \dfrac{1-2t}{1-2t}$

$= -\dfrac{1}{1+2t}$

28. $-\dfrac{1}{2+3a}$

29. $\dfrac{12-6x}{5x-10} = \dfrac{-6(-2+x)}{5(x-2)} = \dfrac{-6(x-2)}{5(x-2)} =$

$\dfrac{-6}{5} \cdot \dfrac{x-2}{x-2} = -\dfrac{6}{5}$

30. $-\dfrac{7}{3}$

31. $\dfrac{a^2-25}{a^2+10a+25} = \dfrac{(a+5)(a-5)}{(a+5)(a+5)} =$

$\dfrac{a+5}{a+5} \cdot \dfrac{a-5}{a+5} = \dfrac{a-5}{a+5}$

32. $\dfrac{a+4}{a-4}$

33. $\dfrac{x^2+9x+8}{x^2-3x-4} = \dfrac{(x+1)(x+8)}{(x+1)(x-4)} = \dfrac{x+1}{x+1} \cdot \dfrac{x+8}{x-4} =$

$\dfrac{x+8}{x-4}$

34. $\dfrac{t-9}{t+4}$

35. $\dfrac{16-t^2}{t^2-8t+16} = \dfrac{16-t^2}{16-8t+t^2} = \dfrac{(4+t)(4-t)}{(4-t)(4-t)} =$

$\dfrac{4+t}{4-t} \cdot \dfrac{4-t}{4-t} = \dfrac{4+t}{4-t}$

36. $\dfrac{5-p}{5+p}$

37. $\dfrac{x^3-1}{x^2-1} = \dfrac{(x-1)(x^2+x+1)}{(x+1)(x-1)} = \dfrac{x-1}{x-1} \cdot \dfrac{x^2+x+1}{x+1} =$

$\dfrac{x^2+x+1}{x+1}$

38. $\dfrac{a^2-2a+4}{a-2}$

39. $\dfrac{3y^3+24}{y^2-2y+4} = \dfrac{3(y^3+8)}{y^2-2y+4} = \dfrac{3(y+2)(y^2-2y+4)}{y^2-2y+4} =$

$\dfrac{y^2-2y+4}{y^2-2y+4} \cdot \dfrac{3(y+2)}{1} = 3(y+2)$

40. $\dfrac{x-3}{5}$

41. First we simplify the rational expression describing the function.

$\dfrac{3x-12}{3x+15} = \dfrac{3(x-4)}{3(x+5)} = \dfrac{3}{3} \cdot \dfrac{x-4}{x+5} = \dfrac{x-4}{x+5}$

$x+5 = 0$ when $x = -5$. Thus, the vertical asymptote is $x = -5$.

42. $x = -3$

43. First we simplify the rational expression describing the function.

$\dfrac{12-6x}{5x-10} = \dfrac{-6(-2+x)}{5(x-2)} = \dfrac{-6(x-2)}{5(x-2)} =$

$\dfrac{-6}{5} \cdot \dfrac{x-2}{x-2} = -\dfrac{6}{5}$

The denominator of the simplified expression is not equal to 0 for any value of x, so there are no vertical asymptotes.

44. No vertical asymptotes

45. First we simplify the rational expression describing the function.

$\dfrac{x^3+3x^2}{x^2+6x+9} = \dfrac{x^2(x+3)}{(x+3)(x+3)} = \dfrac{x^2}{x+3} \cdot \dfrac{x+3}{x+3} =$

$\dfrac{x^2}{x+3}$

$x+3 = 0$ when $x = -3$. Thus, the vertical asymptote is $x = -3$.

46. $x = \dfrac{1}{2}$

47. First we simplify the rational expression describing the function.

$$\frac{x^2 - x - 6}{x^2 - 6x + 8} = \frac{(x-3)(x+2)}{(x-4)(x-2)}$$

We cannot remove a factor equal to 1. Observe that $x-4 = 0$ when $x = 4$ and $x-2 = 0$ when $x = 2$. Thus, the vertical asymptotes are $x = 4$ and $x = 2$.

48. $x = 1$

49. The vertical asymptote of $h(x) = \dfrac{1}{x}$ is $x = 0$. Observe that $h(x) > 0$ for $x > 0$ and $h(x) < 0$ for $x < 0$. Thus, graph (b) corresponds to this function.

50. (e)

51. The vertical asymptote of $f(x) = \dfrac{x}{x-3}$ is $x = 3$. Thus, graph (f) corresponds to this function.

52. (d)

53. $\dfrac{4x-2}{x^2 - 2x + 1} = \dfrac{2(2x-1)}{(x-1)(x-1)}$

The vertical asymptote of $r(x)$ is $x = 1$. Thus, graph (a) corresponds to this function.

54. (c)

55.

$$\frac{5a^3}{3b} \cdot \frac{7b^3}{10a^7}$$

$$= \frac{5a^3 \cdot 7b^3}{3b \cdot 10a^7} \qquad \text{Multiplying the numerators and also the denominators}$$

$$= \frac{5 \cdot a^3 \cdot 7 \cdot b \cdot b^2}{3 \cdot b \cdot 2 \cdot 5 \cdot a^3 \cdot a^4} \qquad \text{Factoring the numerator and the denominator}$$

$$= \frac{\cancel{5} \cdot \cancel{a^3} \cdot 7 \cdot \cancel{b} \cdot b^2}{3 \cdot \cancel{b} \cdot 2 \cdot \cancel{5} \cdot \cancel{a^3} \cdot a^4} \qquad \text{Removing a factor equal to 1}$$

$$= \frac{7b^2}{6a^4}$$

56. $\dfrac{5}{3ab^3}$

57.

$$\frac{8x-16}{5x} \cdot \frac{x^3}{5x-10} = \frac{(8x-16)(x^3)}{5x(5x-10)}$$

$$= \frac{8(x-2)(x)(x^2)}{5 \cdot x \cdot 5(x-2)}$$

$$= \frac{8\cancel{(x-2)}\cancel{(x)}(x^2)}{5 \cdot \cancel{x} \cdot 5\cancel{(x-2)}}$$

$$= \frac{8x^2}{25}$$

58. $\dfrac{3t^2}{4}$

59.

$$\frac{y^2 - 16}{4y + 12} \cdot \frac{y+3}{y-4} = \frac{(y^2-16)(y+3)}{(4y+12)(y-4)}$$

$$= \frac{(y+4)(y-4)(y+3)}{4(y+3)(y-4)}$$

$$= \frac{(y+4)\cancel{(y-4)}\cancel{(y+3)}}{4\cancel{(y+3)}\cancel{(y-4)}}$$

$$= \frac{y+4}{4}$$

60. $\dfrac{m+n}{4}$

61.

$$\frac{x^2 - 16}{x^2} \cdot \frac{x^2 - 4x}{x^2 - x - 12} = \frac{(x^2-16)(x^2-4x)}{x^2(x^2-x-12)}$$

$$= \frac{(x+4)(x-4)(x)(x-4)}{x \cdot x(x-4)(x+3)}$$

$$= \frac{(x+4)\cancel{(x-4)}\cancel{(x)}(x-4)}{\cancel{x} \cdot x\cancel{(x-4)}(x+3)}$$

$$= \frac{(x+4)(x-4)}{x(x+3)}$$

62. $\dfrac{y(y+5)}{y-3}$

63.

$$\frac{7a-14}{4-a^2} \cdot \frac{5a^2 + 6a + 1}{35a + 7}$$

$$= \frac{(7a-14)(5a^2+6a+1)}{(4-a^2)(35a+7)}$$

$$= \frac{7(a-2)(5a+1)(a+1)}{(2+a)(2-a)(7)(5a+1)}$$

$$= \frac{7(-1)(2-a)(5a+1)(a+1)}{(2+a)(2-a)(7)(5a+1)}$$

$$= \frac{\cancel{7}(-1)\cancel{(2-a)}\cancel{(5a+1)}(a+1)}{(2+a)\cancel{(2-a)}\cancel{(7)}\cancel{(5a+1)}}$$

$$= \frac{-1(a+1)}{2+a}$$

$$= \frac{-a-1}{2+a}, \text{ or } -\frac{a+1}{2+a}$$

64. $-\dfrac{3(a+1)}{a+6}$

65.

$$\frac{t^3 - 4t}{t - t^4} \cdot \frac{t^4 - t}{4t - t^3}$$

$$= \frac{t^3 - 4t}{t - t^4} \cdot \frac{-1(t - t^4)}{-1(t^3 - 4t)}$$

$$= \frac{(t^3 - 4t)(-1)(t - t^4)}{(t - t^4)(-1)(t^3 - 4t)}$$

$$= 1$$

66. $\dfrac{x^2(x+3)(x-3)}{-4}$

67. $\dfrac{x^2-2x-35}{2x^3-3x^2}\cdot\dfrac{4x^3-9x}{7x-49}$

$= \dfrac{(x^2-2x-35)(4x^3-9x)}{(2x^3-3x^2)(7x-49)}$

$= \dfrac{(x-7)(x+5)(x)(2x+3)(2x-3)}{x^2(2x-3)(7)(x-7)}$

$= \dfrac{(x-7)(x+5)(x)(2x+3)(2x-3)}{x\cdot x(2x-3)(7)(x-7)}$

$= \dfrac{(x+5)(2x+3)}{7x}$

68. $\dfrac{1-y}{y+4}$

69. $\dfrac{c^3+8}{c^5-4c^3}\cdot\dfrac{c^6-4c^5+4c^4}{c^2-2c+4}$

$= \dfrac{(c^3+8)(c^6-4c^5+4c^4)}{(c^5-4c^3)(c^2-2c+4)}$

$= \dfrac{(c+2)(c^2-2c+4)(c^4)(c-2)(c-2)}{c^3(c+2)(c-2)(c^2-2c+4)}$

$= \dfrac{c^3(c+2)(c^2-2c+4)(c-2)}{c^3(c+2)(c^2-2c+4)(c-2)}\cdot\dfrac{c(c-2)}{1}$

$= c(c-2)$

70. $\dfrac{x(x-3)^2}{x+3}$

71. $\dfrac{a^3-b^3}{3a^2+9ab+6b^2}\cdot\dfrac{a^2+2ab+b^2}{a^2-b^2}$

$= \dfrac{(a^3-b^3)(a^2+2ab+b^2)}{(3a^2+9ab+6b^2)(a^2-b^2)}$

$= \dfrac{(a-b)(a^2+ab+b^2)(a+b)(a+b)}{3(a+b)(a+2b)(a+b)(a-b)}$

$= \dfrac{(a-b)(a^2+ab+b^2)(a+b)(a+b)}{3(a+b)(a+2b)(a+b)(a-b)}$

$= \dfrac{a^2+ab+b^2}{3(a+2b)}$

72. $\dfrac{x^2-xy+y^2}{3(x+3y)}$

73. $\dfrac{4x^2-9y^2}{8x^3-27y^3}\cdot\dfrac{4x^2+6xy+9y^2}{4x^2+12xy+9y^2}$

$= \dfrac{(4x^2-9y^2)(4x^2+6xy+9y^2)}{(8x^3-27y^3)(4x^2+12xy+9y^2)}$

$= \dfrac{(2x+3y)(2x-3y)(4x^2+6xy+9y^2)\cdot1}{(2x-3y)(4x^2+6xy+9y^2)(2x+3y)(2x+3y)}$

$= \dfrac{(2x+3y)(2x-3y)(4x^2+6xy+9y^2)}{(2x+3y)(2x-3y)(4x^2+6xy+9y^2)}\cdot\dfrac{1}{2x+3y}$

$= \dfrac{1}{2x+3y}$

74. $\dfrac{(x-y)(2x+3y)}{2(x+y)(9x^2+6xy+4y^2)}$

75. $\dfrac{9x^5}{8y^2}\div\dfrac{3x}{16y^9}$

$= \dfrac{9x^5}{8y^2}\cdot\dfrac{16y^9}{3x}\qquad$ Multiplying by the reciprocal of the divisor

$= \dfrac{9x^5(16y^9)}{8y^2(3x)}$

$= \dfrac{3\cdot3\cdot x\cdot x^4\cdot2\cdot8\cdot y^2\cdot y^7}{8\cdot y^2\cdot3\cdot x}$

$= \dfrac{3\cdot3\cdot x\cdot x^4\cdot2\cdot8\cdot y^2\cdot y^7}{8\cdot y^2\cdot3\cdot x\cdot1}$

$= 6x^4y^7$

76. $\dfrac{4a^4}{b^4}$

77. $\dfrac{5x+10}{x^8}\div\dfrac{x+2}{x^3}=\dfrac{5x+10}{x^8}\cdot\dfrac{x^3}{x+2}$

$= \dfrac{(5x+10)(x^3)}{x^8(x+2)}$

$= \dfrac{5(x+2)(x^3)}{x^3\cdot x^5(x+2)}$

$= \dfrac{5(x+2)(x^3)}{x^3\cdot x^5(x+2)}$

$= \dfrac{5}{x^5}$

78. $\dfrac{3}{y^5}$

79. $\dfrac{x^2-4}{x^3}\div\dfrac{x^5-2x^4}{x+4}=\dfrac{x^2-4}{x^3}\cdot\dfrac{x+4}{x^5-2x^4}$

$= \dfrac{(x^2-4)(x+4)}{x^3(x^5-2x^4)}$

$= \dfrac{(x+2)(x-2)(x+4)}{x^3(x^4)(x-2)}$

$= \dfrac{(x+2)(x-2)(x+4)}{x^3(x^4)(x-2)}$

$= \dfrac{(x+2)(x+4)}{x^7}$

80. $\dfrac{(y-3)(y+2)}{y^6}$

81. $\dfrac{25x^2-4}{x^2-9} \div \dfrac{2-5x}{x+3} = \dfrac{25x^2-4}{x^2-9} \cdot \dfrac{x+3}{2-5x}$

$= \dfrac{(25x^2-4)(x+3)}{(x^2-9)(2-5x)}$

$= \dfrac{(5x+2)(5x-2)(x+3)}{(x+3)(x-3)(-1)(5x-2)}$

$= \dfrac{(5x+2)\cancel{(5x-2)}\cancel{(x+3)}}{\cancel{(x+3)}(x-3)(-1)\cancel{(5x-2)}}$

$= \dfrac{5x+2}{-x+3}, \text{ or } -\dfrac{5x+2}{x-3}$

82. $\dfrac{-2a-1}{a+2}$

83. $\dfrac{5y-5x}{15y^3} \div \dfrac{x^2-y^2}{3x+3y} = \dfrac{5y-5x}{15y^3} \cdot \dfrac{3x+3y}{x^2-y^2}$

$= \dfrac{(5y-5x)(3x+3y)}{15y^3(x^2-y^2)}$

$= \dfrac{5(y-x)(3)(x+y)}{5\cdot 3\cdot y^3(x+y)(x-y)}$

$= \dfrac{5(-1)(x-y)(3)(x+y)}{5\cdot 3\cdot y^3(x+y)(x-y)}$

$= \dfrac{\cancel{5}(-1)\cancel{(x-y)}\cancel{(3)}\cancel{(x+y)}}{\cancel{5}\cdot \cancel{3}\cdot y^3\cancel{(x+y)}\cancel{(x-y)}}$

$= \dfrac{-1}{y^3}, \text{ or } -\dfrac{1}{y^3}$

84. $-x^2$

85. $\dfrac{x^2-16}{x^2-10x+25} \div \dfrac{3x-12}{x^2-3x-10}$

$= \dfrac{x^2-16}{x^2-10x+25} \cdot \dfrac{x^2-3x-10}{3x-12}$

$= \dfrac{(x^2-16)(x^2-3x-10)}{(x^2-10x+25)(3x-12)}$

$= \dfrac{(x+4)(x-4)(x-5)(x+2)}{(x-5)(x-5)(3)(x-4)}$

$= \dfrac{(x+4)\cancel{(x-4)}\cancel{(x-5)}(x+2)}{\cancel{(x-5)}(x-5)(3)\cancel{(x-4)}}$

$= \dfrac{(x+4)(x+2)}{3(x-5)}$

86. $\dfrac{(y+6)(y+3)}{3(y-4)}$

87. $\dfrac{y^3+3y}{y^2-9} \div \dfrac{y^2+5y-14}{y^2+4y-21}$

$= \dfrac{y^3+3y}{y^2-9} \cdot \dfrac{y^2+4y-21}{y^2+5y-14}$

$= \dfrac{(y^3+3y)(y^2+4y-21)}{(y^2-9)(y^2+5y-14)}$

$= \dfrac{y(y^2+3)(y+7)(y-3)}{(y+3)(y-3)(y+7)(y-2)}$

$= \dfrac{y(y^2+3)\cancel{(y+7)}\cancel{(y-3)}}{(y+3)\cancel{(y-3)}\cancel{(y+7)}(y-2)}$

$= \dfrac{y(y^2+3)}{(y+3)(y-2)}$

88. $\dfrac{a(a^2+4)}{(a+4)(a+3)}$

89. $\dfrac{x^3-64}{x^3+64} \div \dfrac{x^2-16}{x^2-4x+16}$

$= \dfrac{x^3-64}{x^3+64} \cdot \dfrac{x^2-4x+16}{x^2-16}$

$= \dfrac{(x^3-64)(x^2-4x+16)}{(x^3+64)(x^2-16)}$

$= \dfrac{(x-4)(x^2+4x+16)(x^2-4x+16)}{(x+4)(x^2-4x+16)(x+4)(x-4)}$

$= \dfrac{(x-4)(x^2-4x+16)}{(x-4)(x^2-4x+16)} \cdot \dfrac{x^2+4x+16}{(x+4)(x+4)}$

$= \dfrac{x^2+4x+16}{(x+4)(x+4)}, \text{ or } \dfrac{x^2+4x+16}{(x+4)^2}$

90. $\dfrac{4y^2+6y+9}{(4y-1)(2y+3)}$

91. $\dfrac{8a^3+b^3}{2a^2+3ab+b^2} \div \dfrac{8a^2-4ab+2b^2}{4a^2+4ab+b^2}$

$= \dfrac{8a^3+b^3}{2a^2+3ab+b^2} \cdot \dfrac{4a^2+4ab+b^2}{8a^2-4ab+2b^2}$

$= \dfrac{(8a^3+b^3)(4a^2+4ab+b^2)}{(2a^2+3ab+b^2)(8a^2-4ab+2b^2)}$

$= \dfrac{(2a+b)(4a^2-2ab+b^2)(2a+b)(2a+b)}{(2a+b)(a+b)(2)(4a^2-2ab+b^2)}$

$= \dfrac{(2a+b)(4a^2-2ab+b^2)}{(2a+b)(4a^2-2ab+b^2)} \cdot \dfrac{(2a+b)(2a+b)}{(a+b)(2)}$

$= \dfrac{(2a+b)(2a+b)}{2(a+b)}, \text{ or } \dfrac{(2a+b)^2}{2(a+b)}$

92. $\dfrac{2(2x-y)}{x}$

93. *Writing Exercise*

94. *Writing Exercise*

95. $\dfrac{3}{10} - \dfrac{8}{15} = \dfrac{3}{10} \cdot \dfrac{3}{3} - \dfrac{8}{15} \cdot \dfrac{2}{2}$

$= \dfrac{9}{30} - \dfrac{16}{30}$

$= \dfrac{9 - 16}{30} = \dfrac{-7}{30}$

$= -\dfrac{7}{30}$

96. $-\dfrac{13}{40}$

97. $\dfrac{2}{3} \cdot \dfrac{5}{7} - \dfrac{5}{7} \cdot \dfrac{1}{6} = \dfrac{10}{21} - \dfrac{5}{42}$

$= \dfrac{10}{21} \cdot \dfrac{2}{2} - \dfrac{5}{42}$

$= \dfrac{20}{42} - \dfrac{5}{42}$

$= \dfrac{15}{42}$

$= \dfrac{3 \cdot 5}{3 \cdot 14} = \dfrac{\cancel{3} \cdot 5}{\cancel{3} \cdot 14}$

$= \dfrac{5}{14}$

98. $\dfrac{1}{35}$

99. $(8x^3 - 5x^2 + 6x + 2) - (4x^3 + 2x^2 - 3x + 7)$

$= 8x^3 - 5x^2 + 6x + 2 - 4x^3 - 2x^2 + 3x - 7$

$= 4x^3 - 7x^2 + 9x - 5$

100. $-2t^4 + 11t^3 - t^2 + 10t - 3$

101. *Writing Exercise*

102. *Writing Exercise*

103. $g(x) = \dfrac{2x + 3}{4x - 1}$

a) $g(x + h) = \dfrac{2(x + h) + 3}{4(x + h) - 1} = \dfrac{2x + 2h + 3}{4x + 4h - 1}$

b) $g(2x - 2) \cdot g(x) = \dfrac{2(2x - 2) + 3}{4(2x - 2) - 1} \cdot \dfrac{2x + 3}{4x - 1}$

$= \dfrac{4x - 1}{8x - 9} \cdot \dfrac{2x + 3}{4x - 1}$

$= \dfrac{2x + 3}{8x - 9}$

c) $g\left(\dfrac{1}{2}x + 1\right) \cdot g(x) = \dfrac{2\left(\dfrac{1}{2}x + 1\right) + 3}{4\left(\dfrac{1}{2}x + 1\right) - 1} \cdot \dfrac{2x + 3}{4x - 1}$

$= \dfrac{x + 5}{2x + 3} \cdot \dfrac{2x + 3}{4x - 1}$

$= \dfrac{x + 5}{4x - 1}$

104.

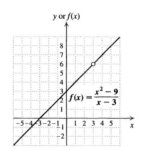

105. $\left[\dfrac{r^2 - 4s^2}{r + 2s} \div (r + 2s)\right] \cdot \dfrac{2s}{r - 2s}$

$= \left[\dfrac{r^2 - 4s^2}{r + 2s} \cdot \dfrac{1}{r + 2s}\right] \cdot \dfrac{2s}{r - 2s}$

$= \dfrac{(r^2 - 4s^2)(2s)}{(r + 2s)(r + 2s)(r - 2s)}$

$= \dfrac{(r + 2s)(r - 2s)(2s)}{(r + 2s)(r + 2s)(r - 2s)}$

$= \dfrac{(r + 2s)(r - 2s)(2s)}{(r + 2s)(r + 2s)(r - 2s)}$

$= \dfrac{2s}{r + 2s}$

106. $\dfrac{(d - 1)(d - 5)}{5d(d + 5)}$

107. $\left[\dfrac{6t^2 - 26t + 30}{8t^2 - 15t + 21} \cdot \dfrac{5t^2 - 9t - 15}{6t^2 - 14t - 20}\right] \div \dfrac{5t^2 - 9t - 15}{6t^2 - 14t - 20}$

$= \dfrac{(6t^2 - 26t + 30)(5t^2 - 9t - 15)}{(8t^2 - 15t + 21)(6t^2 - 14t - 20)} \div \dfrac{5t^2 - 9t - 15}{6t^2 - 14t - 20}$

$= \dfrac{(6t^2 - 26t + 30)(5t^2 - 9t - 15)}{(8t^2 - 15t + 21)(6t^2 - 14t - 20)} \cdot \dfrac{6t^2 - 14t - 20}{5t^2 - 9t - 15}$

$= \dfrac{(6t^2 - 26t + 30)(5t^2 - 9t - 15)(6t^2 - 14t - 20)}{(8t^2 - 15t + 21)(6t^2 - 14t - 20)(5t^2 - 9t - 15)}$

$= \dfrac{6t^2 - 26t + 30}{8t^2 - 15t + 21} \cdot \dfrac{(5t^2 - 9t - 15)(6t^2 - 14t - 20)}{(6t^2 - 14t - 20)(5t^2 - 9t - 15)}$

$= \dfrac{6t^2 - 26t + 30}{8t^2 - 15t + 21} \cdot 1$

$= \dfrac{6t^2 - 26t + 30}{8t^2 - 15t + 21},$ or $\dfrac{2(3t^2 - 13t + 15)}{8t^2 - 15t + 21}$

108. $\dfrac{x - 3}{(x + 3)(x + 1)}$

109.

$$\frac{m^2 - t^2}{m^2 + t^2 + m + t + 2mt}$$

$$= \frac{m^2 - t^2}{(m^2 + 2mt + t^2) + (m + t)}$$

$$= \frac{(m + t)(m - t)}{(m + t)^2 + (m + t)}$$

$$= \frac{(m + t)(m - t)}{(m + t)[(m + t) + 1]}$$

$$= \frac{(\cancel{m + t})(m - t)}{(\cancel{m + t})(m + t + 1)}$$

$$= \frac{m - t}{m + t + 1}$$

110. $\dfrac{a^2 + 2}{a^2 - 3}$

111.

$$\frac{x^3 + x^2 - y^3 - y^2}{x^2 - 2xy + y^2}$$

$$= \frac{(x^3 - y^3) + (x^2 - y^2)}{x^2 - 2xy + y^2}$$

$$= \frac{(x - y)(x^2 + xy + y^2) + (x + y)(x - y)}{(x - y)^2}$$

$$= \frac{(x - y)[(x^2 + xy + y^2) + (x + y)]}{(x - y)(x - y)}$$

$$= \frac{(\cancel{x - y})(x^2 + xy + y^2 + x + y)}{(\cancel{x - y})(x - y)}$$

$$= \frac{x^2 + xy + y^2 + x + y}{x - y}$$

112. $\dfrac{(u^2 - uv + v^2)^2}{u - v}$

113.

$$\frac{x^5 - x^3 + x^2 - 1 - (x^3 - 1)(x + 1)^2}{(x^2 - 1)^2}$$

$$= \frac{x^5 - x^3 + (x^2 - 1) - [(x^3 - 1)(x + 1)^2]}{(x^2 - 1)^2}$$

$$= \frac{x^3(x^2 - 1) + (x^2 - 1) - [(x - 1)(x^2 + x + 1)(x + 1)(x + 1)]}{(x^2 - 1)^2}$$

$$= \frac{x^3(x^2 - 1) + (x^2 - 1) - [(x^2 - 1)(x + 1)(x^2 + x + 1)]}{(x^2 - 1)^2}$$

$$= \frac{(x^2 - 1)[x^3 + 1 - (x + 1)(x^2 + x + 1)]}{(x^2 - 1)^2}$$

$$= \frac{(x^2 - 1)[x^3 + 1 - (x^3 + x^2 + x + x^2 + x + 1)]}{(x^2 - 1)(x^2 - 1)}$$

$$= \frac{(\cancel{x^2 - 1})(-2x^2 - 2x)}{(\cancel{x^2 - 1})(x^2 - 1)}$$

$$= \frac{-2x^2 - 2x}{x^2 - 1}$$

$$= \frac{-2x(x + 1)}{(x + 1)(x - 1)}$$

$$= \frac{-2x(\cancel{x + 1})}{(\cancel{x + 1})(x - 1)}$$

$$= \frac{-2x}{x - 1}, \text{ or } -\frac{2x}{x - 1}$$

114. a) $\dfrac{16(x + 1)}{(x - 1)^2(x^2 + x + 1)}$

b) $\dfrac{x^2 + x + 1}{(x + 1)^3}$

c) $\dfrac{(x + 1)^3}{x^2 + x + 1}$

115. From the graph we see that the domain consists of all real numbers except -2 and 1, so the domain is $(-\infty, -2) \cup (-2, 1) \cup (1, \infty)$. We also see that the range consists of all real numbers except 2 and 3, so the range is $(-\infty, 2) \cup (2, 3) \cup (3, \infty)$.

116. Domain: $(-\infty, -1) \cup (-1, 0) \cup (0, 1) \cup (1, \infty)$;

range: $(-\infty, -3) \cup (-3, -1) \cup (-1, 0) \cup (0, \infty)$

117. From the graph we see that the domain consists of all real numbers except -1 and 1, so the domain is $(-\infty, -1) \cup (-1, 1) \cup (1, \infty)$. We also see that the range consists of all real numbers less than or equal to -1 or greater than 0. Thus, the range is $(-\infty, -1] \cup (0, \infty)$.

Exercise Set 6.2

1. $\dfrac{5}{3a} + \dfrac{7}{3a}$

$= \dfrac{12}{3a}$ Adding the numerators. The denominator is unchanged.

$= \dfrac{3 \cdot 4}{3 \cdot a}$

$= \dfrac{\cancel{3} \cdot 4}{\cancel{3} \cdot a}$

$= \dfrac{4}{a}$

2. $\dfrac{4}{y}$

3. $\dfrac{1}{4a^2b} - \dfrac{5}{4a^2b} = \dfrac{-4}{4a^2b} = \dfrac{-1 \cdot 4}{4a^2b} = \dfrac{-1 \cdot \cancel{4}}{\cancel{4}a^2b} = -\dfrac{1}{a^2b}$

4. $\dfrac{1}{3m^2n^2}$

5. $\dfrac{a - 5b}{a + b} + \dfrac{a + 7b}{a + b} = \dfrac{2a + 2b}{a + b}$

$= \dfrac{2(a + b)}{a + b}$

$= \dfrac{2(\cancel{a+b})}{1(\cancel{a+b})}$

$= 2$

6. 2

7. $\dfrac{4y + 2}{y - 2} - \dfrac{y - 3}{y - 2} = \dfrac{4y + 2 - (y - 3)}{y - 2}$

$= \dfrac{4y + 2 - y + 3}{y - 2}$

$= \dfrac{3y + 5}{y - 2}$

8. $\dfrac{2t + 4}{t - 4}$

9. $\dfrac{3x - 4}{x^2 - 5x + 4} + \dfrac{3 - 2x}{x^2 - 5x + 4} = \dfrac{3x - 4 + 3 - 2x}{x^2 - 5x + 4}$

$= \dfrac{x - 1}{(x - 4)(x - 1)}$

$= \dfrac{1 \cdot (\cancel{x-1})}{(x - 4)(\cancel{x-1})}$

$= \dfrac{1}{x - 4}$

10. $\dfrac{1}{x - 7}$

11. $\dfrac{3a - 2}{a^2 - 25} - \dfrac{4a - 7}{a^2 - 25} = \dfrac{3a - 2 - (4a - 7)}{a^2 - 25}$

$= \dfrac{3a - 2 - 4a + 7}{a^2 - 25}$

$= \dfrac{-a + 5}{a^2 - 25}$

$= \dfrac{-1(a - 5)}{(a + 5)(a - 5)}$

$= \dfrac{-1(\cancel{a-5})}{(a + 5)(\cancel{a-5})}$

$= \dfrac{-1}{a + 5}, \text{ or } -\dfrac{1}{a + 5}$

12. $-\dfrac{1}{a + 3}$

13. $\dfrac{a^2}{a - b} + \dfrac{b^2}{b - a} = \dfrac{a^2}{a - b} + \dfrac{-1}{-1} \cdot \dfrac{b^2}{b - a}$

$= \dfrac{a^2}{a - b} + \dfrac{-b^2}{a - b}$

$= \dfrac{a^2 - b^2}{a - b} = \dfrac{(a + b)(a - b)}{a - b}$

$= \dfrac{(a + b)(\cancel{a-b})}{1 \cdot (\cancel{a-b})} = a + b$

14. $-(s + r)$

15. $\dfrac{7}{x} - \dfrac{8}{-x} = \dfrac{7}{x} + (-1) \cdot \dfrac{8}{-x} = \dfrac{7}{x} + \dfrac{1}{-1} \cdot \dfrac{8}{-x} =$

$\dfrac{7}{x} + \dfrac{8}{x} = \dfrac{15}{x}$

16. $\dfrac{7}{a}$

17. $\dfrac{x - 7}{x^2 - 16} - \dfrac{x - 1}{16 - x^2} = \dfrac{x - 7}{x^2 - 16} + (-1) \cdot \dfrac{x - 1}{16 - x^2}$

$= \dfrac{x - 7}{x^2 - 16} + \dfrac{1}{-1} \cdot \dfrac{x - 1}{16 - x^2}$

$= \dfrac{x - 7}{x^2 - 16} + \dfrac{x - 1}{x^2 - 16}$

$= \dfrac{2x - 8}{x^2 - 16}$

$= \dfrac{2(x - 4)}{(x + 4)(x - 4)}$

$= \dfrac{2(\cancel{x-4})}{(x + 4)(\cancel{x-4})}$

$= \dfrac{2}{x + 4}$

18. $-\dfrac{1}{y + 5}$

19. $\dfrac{t^2+3}{t^4-16}+\dfrac{7}{16-t^4}=\dfrac{t^2+3}{t^4-16}+\dfrac{-1}{-1}\cdot\dfrac{7}{16-t^4}$

$\qquad=\dfrac{t^2+3}{t^4-16}+\dfrac{-7}{t^4-16}$

$\qquad=\dfrac{t^2-4}{t^4-16}$

$\qquad=\dfrac{(t+2)(t-2)}{(t^2+4)(t+2)(t-2)}$

$\qquad=\dfrac{1\cdot\cancel{(t+2)}\cancel{(t-2)}}{(t^2+4)\cancel{(t+2)}\cancel{(t-2)}}$

$\qquad=\dfrac{1}{t^2+4}$

20. $\dfrac{1}{y^2+9}$

21. $\dfrac{m-3n}{m^3-n^3}-\dfrac{2n}{n^3-m^3}=\dfrac{m-3n}{m^3-n^3}+\dfrac{1}{-1}\cdot\dfrac{2n}{n^3-m^3}$

$\qquad=\dfrac{m-3n}{m^3-n^3}+\dfrac{2n}{m^3-n^3}$

$\qquad=\dfrac{m-n}{m^3-n^3}$

$\qquad=\dfrac{m-n}{(m-n)(m^2+mn+n^2)}$

$\qquad=\dfrac{1\cdot\cancel{(m-n)}}{\cancel{(m-n)}(m^2+mn+n^2)}$

$\qquad=\dfrac{1}{m^2+mn+n^2}$

22. $\dfrac{1}{r^2+rs+s^2}$

23. $\dfrac{a+2}{a-4}+\dfrac{a-2}{a+3}$

[LCD is $(a-4)(a+3)$.]

$\qquad=\dfrac{a+2}{a-4}\cdot\dfrac{a+3}{a+3}+\dfrac{a-2}{a+3}\cdot\dfrac{a-4}{a-4}$

$\qquad=\dfrac{(a^2+5a+6)+(a^2-6a+8)}{(a-4)(a+3)}$

$\qquad=\dfrac{2a^2-a+14}{(a-4)(a+3)}$

24. $\dfrac{2a^2+22}{(a-5)(a+4)}$

25. $4+\dfrac{x-3}{x+1}=\dfrac{2}{1}+\dfrac{x-3}{x+1}$

[LCD is $x+1$.]

$\qquad=\dfrac{4}{1}\cdot\dfrac{x+1}{x+1}+\dfrac{x-3}{x+1}$

$\qquad=\dfrac{(4x+4)+(x-3)}{x+1}$

$\qquad=\dfrac{5x+1}{x+1}$

26. $\dfrac{4y-13}{y-5}$

27. $\dfrac{4xy}{x^2-y^2}+\dfrac{x-y}{x+y}$

$\qquad=\dfrac{4xy}{(x+y)(x-y)}+\dfrac{x-y}{x+y}$

LCD is $(x+y)(x-y)$.]

$\qquad=\dfrac{4xy}{(x+y)(x-y)}+\dfrac{x-y}{x+y}\cdot\dfrac{x-y}{x-y}$

$\qquad=\dfrac{4xy+x^2-2xy+y^2}{(x+y)(x-y)}$

$\qquad=\dfrac{x^2+2xy+y^2}{(x+y)(x-y)}=\dfrac{(x+y)(x+y)}{(x+y)(x-y)}$

$\qquad=\dfrac{\cancel{(x+y)}(x+y)}{\cancel{(x+y)}(x-y)}=\dfrac{x+y}{x-y}$

28. $\dfrac{a^2+7ab+b^2}{(a+b)(a-b)}$

29. $\dfrac{8}{2x^2-7x+5}+\dfrac{3x+2}{2x^2-x-10}$

$\qquad=\dfrac{8}{(2x-5)(x-1)}+\dfrac{3x+2}{(2x-5)(x+2)}$

[LCD is $(2x-5)(x-1)(x+2)$.]

$\qquad=\dfrac{8}{(2x-5)(x-1)}\cdot\dfrac{x+2}{x+2}+\dfrac{3x+2}{(2x-5)(x+2)}\cdot\dfrac{x-1}{x-1}$

$\qquad=\dfrac{8x+16+3x^2-x-2}{(2x-5)(x-1)(x+2)}$

$\qquad=\dfrac{3x^2+7x+14}{(2x-5)(x-1)(x+2)}$

30. $\dfrac{3y-4}{(y-1)(y-2)}$

31. $\dfrac{4}{x+1}+\dfrac{x+2}{x^2-1}+\dfrac{3}{x-1}$

$\qquad=\dfrac{4}{x+1}+\dfrac{x+2}{(x+1)(x-1)}+\dfrac{3}{x-1}$

[LCD is $(x+1)(x-1)$.]

$\qquad=\dfrac{4}{x+1}\cdot\dfrac{x-1}{x-1}+\dfrac{x+2}{(x+1)(x-1)}+\dfrac{3}{x-1}\cdot\dfrac{x+1}{x+1}$

$\qquad=\dfrac{4x-4+x+2+3x+3}{(x+1)(x-1)}$

$\qquad=\dfrac{8x+1}{(x+1)(x-1)}$

32. $\dfrac{4y+17}{(y+2)(y-2)}$

33. $\dfrac{x+6}{5x+10} - \dfrac{x-2}{4x+8}$

$= \dfrac{x+6}{5(x+2)} - \dfrac{x-2}{4(x+2)}$

[LCD is $5 \cdot 4(x+2)$.]

$= \dfrac{x+6}{5(x+2)} \cdot \dfrac{4}{4} - \dfrac{x-2}{4(x+2)} \cdot \dfrac{5}{5}$

$= \dfrac{4(x+6) - 5(x-2)}{5 \cdot 4(x+2)}$

$= \dfrac{4x + 24 - 5x + 10}{5 \cdot 4(x+2)}$

$= \dfrac{-x+34}{5 \cdot 4(x+2)}, \text{ or } \dfrac{-x+34}{20(x+2)}$

34. $\dfrac{-2a+14}{15(a+5)}$

35. $\dfrac{5ab}{a^2 - b^2} - \dfrac{a-b}{a+b}$

$= \dfrac{5ab}{(a+b)(a-b)} - \dfrac{a-b}{a+b}$

[LCD is $(a+b)(a-b)$.]

$= \dfrac{5ab}{(a+b)(a-b)} - \dfrac{a-b}{a+b} \cdot \dfrac{a-b}{a-b}$

$= \dfrac{5ab - (a^2 - 2ab + b^2)}{(a+b)(a-b)}$

$= \dfrac{5ab - a^2 + 2ab - b^2}{(a+b)(a-b)}$

$= \dfrac{-a^2 + 7ab - b^2}{(a+b)(a-b)}$

36. $\dfrac{-x^2 + 4xy - y^2}{(x+y)(x-y)}$

37. $\dfrac{x}{x^2 + 9x + 20} - \dfrac{4}{x^2 + 7x + 12}$

$= \dfrac{x}{(x+5)(x+4)} - \dfrac{4}{(x+3)(x+4)}$

[LCD is $(x+5)(x+4)(x+3)$.]

$= \dfrac{x}{(x+5)(x+4)} \cdot \dfrac{x+3}{x+3} - \dfrac{4}{(x+3)(x+4)} \cdot \dfrac{x+5}{x+5}$

$= \dfrac{x^2 + 3x - (4x + 20)}{(x+5)(x+4)(x+3)}$

$= \dfrac{x^2 + 3x - 4x - 20}{(x+5)(x+4)(x+3)}$

$= \dfrac{x^2 - x - 20}{(x+5)(x+4)(x+3)}$

$= \dfrac{(x-5)(x+4)}{(x+5)(x+4)(x+3)}$

$= \dfrac{(x-5)\cancel{(x+4)}}{(x+5)\cancel{(x+4)}(x+3)}$

$= \dfrac{x-5}{(x+5)(x+3)}$

38. $\dfrac{x-6}{(x+6)(x+4)}$

39. $\dfrac{3y}{y^2 - 7y + 10} - \dfrac{2y}{y^2 - 8y + 15}$

$= \dfrac{3y}{(y-5)(y-2)} - \dfrac{2y}{(y-5)(y-3)}$

[LCD is $(y-5)(y-2)(y-3)$.]

$= \dfrac{3y}{(y-5)(y-2)} \cdot \dfrac{y-3}{y-3} - \dfrac{2y}{(y-5)(y-3)} \cdot \dfrac{y-2}{y-2}$

$= \dfrac{3y^2 - 9y - (2y^2 - 4y)}{(y-5)(y-2)(y-3)}$

$= \dfrac{3y^2 - 9y - 2y^2 + 4y}{(y-5)(y-2)(y-3)}$

$= \dfrac{y^2 - 5y}{(y-5)(y-2)(y-3)} = \dfrac{y(y-5)}{(y-5)(y-2)(y-3)}$

$= \dfrac{y\cancel{(y-5)}}{\cancel{(y-5)}(y-2)(y-3)} = \dfrac{y}{(y-2)(y-3)}$

40. $\dfrac{2x^2 + 21x}{(x-4)(x-2)(x+3)}$

41. $\dfrac{2x+1}{x-y} + \dfrac{5x^2 - 5xy}{x^2 - 2xy + y^2}$

$= \dfrac{2x+1}{x-y} + \dfrac{5x(x-y)}{(x-y)(x-y)}$

$= \dfrac{2x+1}{x-y} + \dfrac{5x\cancel{(x-y)}}{\cancel{(x-y)}(x-y)}$

$= \dfrac{2x+1}{x-y} + \dfrac{5x}{x-y}$

$= \dfrac{7x+1}{x-y}$

42. $\dfrac{2}{a-b}$

43. $\dfrac{3y+2}{y^2 + 5y - 24} + \dfrac{7}{y^2 + 4y - 32}$

$= \dfrac{3y+2}{(y+8)(y-3)} + \dfrac{7}{(y+8)(y-4)}$

[LCD is $(y+8)(y-3)(y-4)$.]

$= \dfrac{3y+2}{(y+8)(y-3)} \cdot \dfrac{y-4}{y-4} + \dfrac{7}{(y+8)(y-4)} \cdot \dfrac{y-3}{y-3}$

$= \dfrac{3y^2 - 10y - 8 + 7y - 21}{(y+8)(y-3)(y-4)}$

$= \dfrac{3y^2 - 3y - 29}{(y+8)(y-3)(y-4)}$

44. $\dfrac{5x^2 - 11x - 6}{(x-5)(x-2)(x-3)}$

45. $\dfrac{a-3}{a^2-16} - \dfrac{3a-2}{a^2+2a-24}$

$= \dfrac{a-3}{(a+4)(a-4)} - \dfrac{3a-2}{(a+6)(a-4)}$

[LCD is $(a+4)(a-4)(a+6)$.]

$= \dfrac{a-3}{(a+4)(a-4)} \cdot \dfrac{a+6}{a+6} - \dfrac{3a-2}{(a+6)(a-4)} \cdot \dfrac{a+4}{a+4}$

$= \dfrac{(a-3)(a+6) - (3a-2)(a+4)}{(a+4)(a-4)(a+6)}$

$= \dfrac{a^2+3a-18 - (3a^2+10a-8)}{(a+4)(a-4)(a+6)}$

$= \dfrac{a^2+3a-18 - 3a^2-10a+8}{(a+4)(a-4)(a+6)}$

$= \dfrac{-2a^2-7a-10}{(a+4)(a-4)(a+6)}$

46. $\dfrac{-2t^2+13t-7}{(t+3)(t-3)(t-1)}$

47. $\dfrac{2}{a^2-5a+4} + \dfrac{-2}{a^2-4}$

$= \dfrac{2}{(a-4)(a-1)} + \dfrac{-2}{(a+2)(a-2)}$

[LCD is $(a-4)(a-1)(a+2)(a-2)$.]

$= \dfrac{2}{(a-4)(a-1)} \cdot \dfrac{(a+2)(a-2)}{(a+2)(a-2)} +$

$\dfrac{-2}{(a+2)(a-2)} \cdot \dfrac{(a-4)(a-1)}{(a-4)(a-1)}$

$= \dfrac{2(a^2-4) - 2(a^2-5a+4)}{(a-4)(a-1)(a+2)(a-2)}$

$= \dfrac{2a^2-8 - 2a^2+10a-8}{(a-4)(a-1)(a+2)(a-2)}$

$= \dfrac{10a-16}{(a-4)(a-1)(a+2)(a-2)}$

48. $\dfrac{21a-45}{(a-6)(a-1)(a+3)(a-3)}$

49. $5 + \dfrac{t}{t+2} - \dfrac{8}{t^2-4} = \dfrac{5}{1} + \dfrac{t}{t+2} - \dfrac{8}{(t+2)(t-2)}$

[LCD is $(t+2)(t-2)$.]

$= \dfrac{5}{1} \cdot \dfrac{(t+2)(t-2)}{(t+2)(t-2)} + \dfrac{t}{t+2} \cdot \dfrac{t-2}{t-2} - \dfrac{8}{(t+2)(t-2)}$

$= \dfrac{5t^2-20 + t^2-2t-8}{(t+2)(t-2)}$

$= \dfrac{6t^2-2t-28}{(t+2)(t-2)}$

$= \dfrac{2(3t-7)(t+2)}{(t+2)(t-2)}$

$= \dfrac{2(3t-7)\cancel{(t+2)}}{\cancel{(t+2)}(t-2)}$

$= \dfrac{2(3t-7)}{t-2}$

50. $\dfrac{3(t+4)}{t+3}$

51. $\dfrac{2y-6}{y^2-9} - \dfrac{y}{y-1} + \dfrac{y^2+2}{y^2+2y-3}$

$= \dfrac{2\cancel{(y-3)}}{(y+3)\cancel{(y-3)}} - \dfrac{y}{y-1} + \dfrac{y^2+2}{(y+3)(y-1)}$

$= \dfrac{2}{y+3} - \dfrac{y}{y-1} + \dfrac{y^2+2}{(y+3)(y-1)}$

[LCD is $(y+3)(y-1)$.]

$= \dfrac{2}{y+3} \cdot \dfrac{y-1}{y-1} - \dfrac{y}{y-1} \cdot \dfrac{y+3}{y+3} + \dfrac{y^2+2}{(y+3)(y-1)}$

$= \dfrac{2(y-1) - y(y+3) + y^2+2}{(y+3)(y-1)}$

$= \dfrac{2y-2 - y^2-3y + y^2+2}{(y+3)(y-1)}$

$= \dfrac{-y}{(y+3)(y-1)}$, or $-\dfrac{y}{(y+3)(y-1)}$

52. 0

53. $\dfrac{5y}{1-4y^2} - \dfrac{2y}{2y+1} + \dfrac{5y}{4y^2-1}$

Observe that $\dfrac{5y}{1-4y^2}$ and $\dfrac{5y}{4y^2-1}$ are opposites, so their sum is 0. Then the result is the remaining expression, $-\dfrac{2y}{2y+1}$.

54. $\dfrac{-3x^2-3x-4}{(x+1)(x-1)}$

55.
$$\frac{2}{x^2-5x+6}-\frac{4}{x^2-2x-3}+\frac{2}{x^2+4x+3}$$
$$=\frac{2}{(x-3)(x-2)}-\frac{4}{(x-3)(x+1)}+\frac{2}{(x+3)(x+1)}$$
[LCD is $(x-3)(x-2)(x+1)(x+3)$.]
$$=\frac{2}{(x-3)(x-2)}\cdot\frac{(x+1)(x+3)}{(x+1)(x+3)}-$$
$$\frac{4}{(x-3)(x+1)}\cdot\frac{(x-2)(x+3)}{(x-2)(x+3)}+$$
$$\frac{2}{(x+3)(x+1)}\cdot\frac{(x-3)(x-2)}{(x-3)(x-2)}$$
$$=\frac{2(x+1)(x+3)-4(x-2)(x+3)+2(x-3)(x-2)}{(x-3)(x-2)(x+1)(x+3)}$$
$$=\frac{2x^2+8x+6-4x^2-4x+24+2x^2-10x+12}{(x-3)(x-2)(x+1)(x+3)}$$
$$=\frac{-6x+42}{(x-3)(x-2)(x+1)(x+3)}$$

56. $\dfrac{-2}{t^2+3t+2}$

57. *Writing Exercise*

58. *Writing Exercise*

59.
$$\frac{15x^{-7}y^{12}z^4}{35x^{-2}y^6z^{-3}}=\frac{15}{35}x^{-7-(-2)}y^{12-6}z^{4-(-3)}$$
$$=\frac{3}{7}x^{-5}y^6z^7=\frac{3}{7}\cdot\frac{1}{x^5}\cdot y^6z^7$$
$$=\frac{3y^6z^7}{7x^5}$$

60. $\dfrac{7b^{11}c^7}{9a^2}$

61.
$$\frac{34s^9t^{-40}r^{30}}{10s^{-3}t^{20}r^{-10}}=\frac{34}{10}s^{9-(-3)}t^{-40-20}r^{30-(-10)}$$
$$=\frac{17}{5}s^{12}t^{-60}r^{40}$$
$$=\frac{17}{5}s^{12}\cdot\frac{1}{t^{60}}\cdot r^{40}$$
$$=\frac{17s^{12}r^{40}}{5t^{60}}$$

62. $y=\dfrac{5}{4}x+3$

63. *Familiarize*. We let x, y, and z represent the number of rolls of dimes, nickels, and quarters, respectively.

Coins	Number of rolls	Value per roll	Total Value
Dimes	x	$50\times0.10,$ or 5.00	$5x$
Nickels	y	$40\times0.05,$ or 2.00	$2y$
Quarters	z	$40\times0.25,$ or 10.00	$10z$
Total	12		$70.00

Translate. The number of rolls of nickels is three more than the number of rolls of dimes. This gives us one equation.
$$y=x+3$$
From the table we get two more equations.
$$x+y+z=12$$
$$5x+2y+10z=70$$

Carry out. Solving the system we get the ordered triple $(2,5,5)$.

Check.

2 rolls of dimes $=2\times\$5$, or \$10

5 rolls of nickels $=5\times\$2$, or \$10

5 rolls of quarters $=5\times\$10$, or \$50

The total value is $\$10+\$10+\$50=\$70$.

The total number of rolls of coins is $2+5+5$, or 12, and the number of rolls of nickels, 5, is three more than the number of rolls of dimes, 2. The numbers check.

State. Robert has 2 rolls of dimes, 5 rolls of nickels, and 5 rolls of quarters.

64. 4 30-min tapes; 8 60-min tapes

65. *Writing Exercise*

66. *Writing Exercise*

67. We find the least common multiple of 14(2 weeks = 14 days), 20, and 30.
$$14=2\cdot7$$
$$20=2\cdot2\cdot5$$
$$30=2\cdot3\cdot5$$
$$\text{LCM}=2\cdot2\cdot3\cdot5\cdot7=420$$
It will be 420 days until Corinna can refill all three prescriptions on the same day.

68. Every 420 years

69. The smallest number of parts possible is the least common multiple of 6 and 4.

$$6 = 2 \cdot 3$$
$$4 = 2 \cdot 2$$
$$\text{LCM} = 2 \cdot 3 \cdot 2, \text{ or } 12$$

A measure should be divided into 12 parts.

70. 2060

71. $x^8 - x^4 = x^4(x^2 + 1)(x + 1)(x - 1)$

$x^5 - x^2 = x^2(x - 1)(x^2 + x + 1)$

$x^5 - x^3 = x^3(x + 1)(x - 1)$

$x^5 + x^2 = x^2(x + 1)(x^2 - x + 1)$

The LCM is

$x^4(x^2 + 1)(x + 1)(x - 1)(x^2 + x + 1)(x^2 - x + 1).$

72. $2ab(a^2 + ab + b^2)(a + b)^2(a^2 - ab + b^2)(a - b)(-2b - 3a)$

73. The LCM is $8a^4b^7$.

One expression is $2a^3b^7$.

Then the other expression must contain 8, a^4, and one of the following:

no factor of b, b, b^2, b^3, b^4, b^5, b^6, or b^7.

Thus, all the possibilities for the other expression are $8a^4$, $8a^4b$, $8a^4b^2$, $8a^4b^3$, $8a^4b^4$, $8a^4b^5$, $8a^4b^6$, $8a^4b^7$.

74. Domain: $\{x | x \text{ is a real number } and \ x \neq -2 \ and \ x \neq 1\}$;

range: $\{y | y \text{ is a real number } and \ y \neq 2 \ and \ y \neq 3\}$

75. $(f + g)(x) = \dfrac{x^3}{x^2 - 4} + \dfrac{x^2}{x^2 + 3x - 10}$

$= \dfrac{x^3}{(x + 2)(x - 2)} + \dfrac{x^2}{(x + 5)(x - 2)}$

$= \dfrac{x^3(x + 5) + x^2(x + 2)}{(x + 2)(x - 2)(x + 5)}$

$= \dfrac{x^4 + 5x^3 + x^3 + 2x^2}{(x + 2)(x - 2)(x + 5)}$

$= \dfrac{x^4 + 6x^3 + 2x^2}{(x + 2)(x - 2)(x + 5)}$

76. $\dfrac{x^4 + 4x^3 - 2x^2}{(x + 2)(x - 2)(x + 5)}$

77. $(f \cdot g)(x) = \dfrac{x^3}{x^2 - 4} \cdot \dfrac{x^2}{x^2 + 3x - 10}$

$= \dfrac{x^5}{(x^2 - 4)(x^2 + 3x - 10)}$

78. $\dfrac{x(x + 5)}{x + 2}$

(Note that $x \neq 0$, $x \neq -5$, and $x \neq 2$ are additional restrictions, since $g(0) = 0$, -5 is not in the domain of g, and 2 is not in the domain of either f or g.)

79. The denominator of $f + g$ is 0 when $x = -2$, $x = 2$, or $x = -5$. Thus the domain of $f + g$ is $\{x | x \text{ is a real number } and \ x \neq -2 \ and \ x \neq 2 \ and \ x \neq -5\}$, or $(-\infty, -5) \cup (-5, -2) \cup (-2, 2) \cup (2, \infty)$.

80. $\{x | x \text{ is a real number } and \ x \neq -2 \ and \ x \neq 0 \ and \ x \neq -5 \ and \ x \neq 2\}$

81. $\quad 5(x - 3)^{-1} + 4(x + 3)^{-1} - 2(x + 3)^{-2}$

$= \dfrac{5}{x - 3} + \dfrac{4}{x + 3} - \dfrac{2}{(x + 3)^2}$

[LCD is $(x - 3)(x + 3)^2$.]

$= \dfrac{5(x + 3)^2 + 4(x - 3)(x + 3) - 2(x - 3)}{(x - 3)(x + 3)^2}$

$= \dfrac{5x^2 + 30x + 45 + 4x^2 - 36 - 2x + 6}{(x - 3)(x + 3)^2}$

$= \dfrac{9x^2 + 28x + 15}{(x - 3)(x + 3)^2}$

82. $\dfrac{5y + 23}{5 - 2y}$

83. $\quad \dfrac{x + 4}{6x^2 - 20x}\left(\dfrac{x}{x^2 - x - 20} + \dfrac{2}{x + 4}\right)$

$= \dfrac{x + 4}{2x(3x - 10)}\left(\dfrac{x}{(x - 5)(x + 4)} + \dfrac{2}{x + 4}\right)$

$= \dfrac{x + 4}{2x(3x - 10)}\left(\dfrac{x + 2(x - 5)}{(x - 5)(x + 4)}\right)$

$= \dfrac{x + 4}{2x(3x - 10)}\left(\dfrac{x + 2x - 10}{(x - 5)(x + 4)}\right)$

$= \dfrac{(x + 4)(3x - 10)}{2x(3x - 10)(x - 5)(x + 4)}$

$= \dfrac{(x + 4)(3x - 10)(1)}{2x(3x - 10)(x - 5)(x + 4)}$

$= \dfrac{1}{2x(x - 5)}$

84. $\dfrac{3}{x + 8}$

85.
$$\frac{8t^5}{2t^2-10t+12} \div \left(\frac{2t}{t^2-8t+15} - \frac{3t}{t^2-7t+10}\right)$$

$$= \frac{8t^5}{2t^2-10t+12} \div \left(\frac{2t}{(t-5)(t-3)} - \frac{3t}{(t-5)(t-2)}\right)$$

$$= \frac{8t^5}{2t^2-10t+12} \div \left(\frac{2t(t-2) - 3t(t-3)}{(t-5)(t-3)(t-2)}\right)$$

$$= \frac{8t^5}{2t^2-10t+12} \div \left(\frac{2t^2 - 4t - 3t^2 + 9t}{(t-5)(t-3)(t-2)}\right)$$

$$= \frac{8t^5}{2t^2-10t+12} \div \frac{-t^2 + 5t}{(t-5)(t-3)(t-2)}$$

$$= \frac{8t^5}{2(t-3)(t-2)} \cdot \frac{(t-5)(t-3)(t-2)}{-t(t-5)}$$

$$= \frac{2 \cdot 4 \cdot t \cdot t^4 \cancel{(t-5)}\cancel{(t-3)}\cancel{(t-2)}}{\cancel{2}\cancel{(t-3)}\cancel{(t-2)}(-1)\cancel{(t)}\cancel{(t-5)}}$$

$$= -4t^4$$

86. $\dfrac{3t^2(t+3)}{-2t^2 + 13t - 7}$

87.

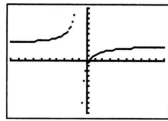

From the graph (shown in the standard window) we see that the domain of the function consists of all real numbers except -1, so the domain of f is $\{x|x$ is a real number and $x \neq -1\}$, or $(-\infty, -1) \cup (-1, \infty)$. We also see that the range consists of all real numbers except 3, so the range of f is $\{y|y$ is a real number and $y \neq 3\}$, or $(-\infty, 3) \cup (3, \infty)$.

88. Domain: $(-\infty, -1) \cup (-1, \infty)$; range: $(5, \infty)$

89.

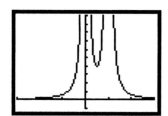

From the graph (shown in the window $[-3, 3, -2, 20]$, Yscl $= 2$), we see that the domain consists of all real numbers except 0 and 1, so the domain of r is $\{x|x$ is a real number an $x \neq 0$ and $x \neq 1\}$, or $(-\infty, 0) \cup (0, 1) \cup (1, \infty)$. We also see that the range consists of all real numbers greater than 0, so the range of r is $\{y|y > 0\}$, or $(0, \infty)$.

Exercise Set 6.3

1. $\dfrac{7 + \dfrac{1}{a}}{\dfrac{1}{a} - 3} = \dfrac{7 + \dfrac{1}{a}}{\dfrac{1}{a} - 3} \cdot \dfrac{a}{a}$ Multiplying by 1, using the LCD

$$= \frac{\left(7 + \dfrac{1}{a}\right)a}{\left(\dfrac{1}{a} - 3\right)a}$$ Multiplying the numerators and the denominator

$$= \frac{7 \cdot a + \dfrac{1}{a} \cdot a}{\dfrac{1}{a} \cdot a - 3 \cdot a}$$

$$= \frac{7a + \dfrac{\cancel{a}}{\cancel{a}} \cdot 1}{\dfrac{\cancel{a}}{\cancel{a}} \cdot 1 - 3a}$$ Removing factors equal to 1

$$= \frac{7a + 1}{1 - 3a}$$ Simplifying

2. $\dfrac{1 + 2y}{1 - 3y}$

3. $\dfrac{x - x^{-1}}{x + x^{-1}} = \dfrac{x - \dfrac{1}{x}}{x + \dfrac{1}{x}}$ Rewriting with positive exponents

$$= \frac{x - \dfrac{1}{x}}{x + \dfrac{1}{x}} \cdot \frac{x}{x}$$ Multiplying by 1, using the LCD

$$= \frac{x \cdot x - \dfrac{1}{x} \cdot x}{x \cdot x + \dfrac{1}{x} \cdot x}$$

$$= \frac{x^2 - 1}{x^2 + 1}$$

(Although the numerator can be factored, doing so does not lead to further simplification.)

4. $\dfrac{y^2 + 1}{y^2 - 1}$

5. $\dfrac{\dfrac{6}{x}+\dfrac{7}{y}}{\dfrac{7}{x}-\dfrac{6}{y}}=\dfrac{\dfrac{6}{x}+\dfrac{7}{y}}{\dfrac{7}{x}-\dfrac{6}{y}}\cdot\dfrac{xy}{xy}$ Multiplying by 1, using the LCD

$=\dfrac{\dfrac{6}{x}\cdot xy+\dfrac{7}{y}\cdot xy}{\dfrac{7}{x}\cdot xy-\dfrac{6}{y}\cdot xy}$

$=\dfrac{6y+7x}{7y-6x}$

6. $\dfrac{5y+2z}{4y-z}$

7. $\dfrac{\dfrac{x^2-y^2}{xy}}{\dfrac{x-y}{y}}=\dfrac{x^2-y^2}{xy}\cdot\dfrac{y}{x-y}$ Multiplying by the reciprocal of the divisor

$=\dfrac{(x+y)(x-y)\cdot y}{xy(x-y)}$

$=\dfrac{(x+y)(x\!-\!y)\cdot y}{xy(x\!-\!y)}$

$=\dfrac{x+y}{x}$

8. $\dfrac{a+b}{a}$

9. $\dfrac{\dfrac{3x}{y}-x}{2y-\dfrac{y}{x}}=\dfrac{\dfrac{3x}{y}-x}{2y-\dfrac{y}{x}}\cdot\dfrac{xy}{xy}$ Multiplying by 1, using the LCD

$=\dfrac{\dfrac{3x}{y}\cdot xy-x\cdot xy}{2y\cdot xy-\dfrac{y}{x}\cdot xy}$

$=\dfrac{3x^2-x^2y}{2xy^2-y^2}$

(Although both the numerator and the denominator can be factored, doing so does not lead to further simplification.)

10. $\dfrac{3}{3x+2}$

11. $\dfrac{a^{-1}+b^{-1}}{\dfrac{a^2-b^2}{ab}}=\dfrac{\dfrac{1}{a}+\dfrac{1}{b}}{\dfrac{a^2-b^2}{ab}}$

$=\dfrac{\dfrac{1}{a}+\dfrac{1}{b}}{\dfrac{a^2-b^2}{ab}}\cdot\dfrac{ab}{ab}$ Multiplying by 1, using the LCD

$=\dfrac{\dfrac{1}{a}\cdot ab+\dfrac{1}{b}\cdot ab}{\dfrac{a^2-b^2}{ab}\cdot ab}$

$=\dfrac{b+a}{a^2-b^2}=\dfrac{b+a}{(a+b)(a-b)}$

$=\dfrac{(a\!+\!b)\cdot(1)}{(a\!+\!b)(a-b)}$ $(b+a=a+b)$

$=\dfrac{1}{a-b}$

12. $\dfrac{1}{x-y}$

13. $\dfrac{8+\dfrac{8}{d}}{1+\dfrac{1}{d}}=\dfrac{8\left(1+\dfrac{1}{d}\right)}{1+\dfrac{1}{d}}$

$=\dfrac{8}{1}\cdot\dfrac{1+\dfrac{1}{d}}{1+\dfrac{1}{d}}$

$=8$

14. 1

15. $\dfrac{\dfrac{1}{x+h}-\dfrac{1}{x}}{h}=\dfrac{\dfrac{1}{x+h}\cdot\dfrac{x}{x}-\dfrac{1}{x}\cdot\dfrac{x+h}{x+h}}{h}$ Adding in the numerator

$=\dfrac{\dfrac{x-x-h}{x(x+h)}}{h}=\dfrac{\dfrac{-h}{x(x+h)}}{h}$

$=\dfrac{-h}{x(x+h)}\cdot\dfrac{1}{h}$ Multiplying by the reciprocal of the divisor

$=\dfrac{-1\cdot h\cdot 1}{x(x+h)(h)}$ $(-h=-1\cdot h)$

$=-\dfrac{1}{x(x+h)}$

16. $\dfrac{1}{a(a-h)}$

17.
$$\dfrac{\dfrac{x^2 - x - 12}{x^2 - 2x - 15}}{\dfrac{x^2 + 8x + 12}{x^2 - 5x - 14}}$$

$$= \dfrac{x^2 - x - 12}{x^2 - 2x - 15} \cdot \dfrac{x^2 - 5x - 14}{x^2 + 8x + 12} \quad \text{Multiplying by}$$
the reciprocal of the divisor

$$= \dfrac{(x-4)(x+3)}{(x-5)(x+3)} \cdot \dfrac{(x-7)(x+2)}{(x+6)(x+2)}$$

$$= \dfrac{(x-4)(x+3)(x-7)(x+2)}{(x-5)(x+3)(x+6)(x+2)}$$

$$= \dfrac{(x-4)(x{+}3)(x-7)(x{+}2)}{(x-5)(x{+}3)(x+6)(x{+}2)}$$

$$= \dfrac{(x-4)(x-7)}{(x-5)(x+6)}$$

18. $\dfrac{(a-2)(a-7)}{(a+1)(a-6)}$

19.
$$\dfrac{\dfrac{1}{x-2} + \dfrac{3}{x-1}}{\dfrac{2}{x-1} + \dfrac{5}{x-2}}$$

$$= \dfrac{\dfrac{1}{x-2} + \dfrac{3}{x-1}}{\dfrac{2}{x-1} + \dfrac{5}{x-2}} \cdot \dfrac{(x-2)(x-1)}{(x-2)(x-1)}$$

Multiplying by 1, using the LCD

$$= \dfrac{\dfrac{1}{x-2} \cdot (x-2)(x-1) + \dfrac{3}{x-1} \cdot (x-2)(x-1)}{\dfrac{2}{x-1} \cdot (x-2)(x-1) + \dfrac{5}{x-2} \cdot (x-2)(x-1)}$$

$$= \dfrac{x - 1 + 3(x-2)}{2(x-2) + 5(x-1)}$$

$$= \dfrac{x - 1 + 3x - 6}{2x - 4 + 5x - 5}$$

$$= \dfrac{4x - 7}{7x - 9}$$

20. $\dfrac{3y - 1}{7y - 5}$

21.
$$\dfrac{a(a+3)^{-1} - 2(a-1)^{-1}}{a(a+3)^{-1} - (a-1)^{-1}}$$

$$= \dfrac{\dfrac{a}{a+3} - \dfrac{2}{a-1}}{\dfrac{a}{a+3} - \dfrac{1}{a-1}}$$

$$= \dfrac{\dfrac{a}{a+3} - \dfrac{2}{a-1}}{\dfrac{a}{a+3} - \dfrac{1}{a-1}} \cdot \dfrac{(a+3)(a-1)}{(a+3)(a-1)}$$

Multiplying by 1, using the LCD

$$= \dfrac{\dfrac{a}{a+3} \cdot (a+3)(a-1) - \dfrac{2}{a-1} \cdot (a+3)(a-1)}{\dfrac{a}{a+3} \cdot (a+3)(a-1) - \dfrac{1}{a-1} \cdot (a+3)(a-1)}$$

$$= \dfrac{a(a-1) - 2(a+3)}{a(a-1) - (a+3)}$$

$$= \dfrac{a^2 - a - 2a - 6}{a^2 - a - a - 3} = \dfrac{a^2 - 3a - 6}{a^2 - 2a - 3}$$

(Although the denominator can be factored, doing so does not lead to further simplification.)

22. $\dfrac{a^2 - 6a - 6}{a^2 - 4a - 2}$

23.
$$\dfrac{\dfrac{x}{x^2 + 3x - 4} - \dfrac{1}{x^2 + 3x - 4}}{\dfrac{x}{x^2 + 6x + 8} + \dfrac{3}{x^2 + 6x + 8}}$$

$$= \dfrac{\dfrac{x-1}{x^2 + 3x - 4}}{\dfrac{x+3}{x^2 + 6x + 8}} \quad \begin{array}{l}\text{Adding in the numerator} \\ \text{and the denominator}\end{array}$$

$$= \dfrac{x-1}{x^2 + 3x - 4} \cdot \dfrac{x^2 + 6x + 8}{x+3}$$

$$= \dfrac{(x-1)(x+4)(x+2)}{(x+4)(x-1)(x+3)}$$

$$= \dfrac{(x{-}1)(x{+}4)(x+2)}{(x{+}4)(x{-}1)(x+3)} = \dfrac{x+2}{x+3}$$

24. $\dfrac{x-4}{x-2}$

25.
$$\dfrac{\dfrac{2}{a^2-1}+\dfrac{1}{a+1}}{\dfrac{3}{a^2-1}+\dfrac{2}{a-1}}$$

$$=\dfrac{\dfrac{2}{(a+1)(a-1)}+\dfrac{1}{a+1}}{\dfrac{3}{(a+1)(a-1)}+\dfrac{2}{a-1}}$$

$$=\dfrac{\dfrac{2}{(a+1)(a-1)}+\dfrac{1}{a+1}}{\dfrac{3}{(a+1)(a-1)}+\dfrac{2}{a-1}}\cdot\dfrac{(a+1)(a-1)}{(a+1)(a-1)}$$

Multiplying by 1, using the LCD

$$=\dfrac{\dfrac{2}{(a+1)(a-1)}\cdot(a+1)(a-1)+\dfrac{1}{a+1}\cdot(a+1)(a-1)}{\dfrac{3}{(a+1)(a-1)}\cdot(a+1)(a-1)+\dfrac{2}{a-1}\cdot(a+1)(a-1)}$$

$$=\dfrac{2+a-1}{3+2(a+1)}=\dfrac{a+1}{3+2a+2}=\dfrac{a+1}{2a+5}$$

26. $\dfrac{2a-3}{a+1}$

27.
$$\dfrac{\dfrac{5}{x^2-4}-\dfrac{3}{x-2}}{\dfrac{4}{x^2-4}-\dfrac{2}{x+2}}$$

$$=\dfrac{\dfrac{5}{(x+2)(x-2)}-\dfrac{3}{x-2}}{\dfrac{4}{(x+2)(x-2)}-\dfrac{2}{x+2}}$$

$$=\dfrac{\dfrac{5}{(x+2)(x-2)}-\dfrac{3}{x-2}}{\dfrac{4}{(x+2)(x-2)}-\dfrac{2}{x+2}}\cdot\dfrac{(x+2)(x-2)}{(x+2)(x-2)}$$

Multiplying by 1, using the LCD

$$=\dfrac{\dfrac{5}{(x+2)(x-2)}\cdot(x+2)(x-2)-\dfrac{3}{x-2}\cdot(x+2)(x-2)}{\dfrac{4}{(x+2)(x-2)}\cdot(x+2)(x-2)-\dfrac{2}{x+2}\cdot(x+2)(x-2)}$$

$$=\dfrac{5-3(x+2)}{4-2(x-2)}=\dfrac{5-3x-6}{4-2x+4}=\dfrac{-1-3x}{8-2x},\text{ or}$$

$$\dfrac{3x+1}{2x-8}$$

28. $\dfrac{7-3x}{3-2x}$, or $\dfrac{3x-7}{2x-3}$

29.
$$\dfrac{\dfrac{y}{y^2-4}+\dfrac{5}{4-y^2}}{\dfrac{y^2}{y^2-4}+\dfrac{25}{4-y^2}}$$

$$=\dfrac{\dfrac{y}{y^2-4}+\dfrac{-1}{-1}\cdot\dfrac{5}{4-y^2}}{\dfrac{y^2}{y^2-4}+\dfrac{-1}{-1}\cdot\dfrac{25}{4-y^2}}$$

$$=\dfrac{\dfrac{y}{y^2-4}-\dfrac{5}{y^2-4}}{\dfrac{y^2}{y^2-4}-\dfrac{25}{y^2-4}}$$

$$=\dfrac{\dfrac{y-5}{y^2-4}}{\dfrac{y^2-25}{y^2-4}}\qquad\text{Adding in the numerator and the denominator}$$

$$=\dfrac{y-5}{y^2-4}\cdot\dfrac{y^2-4}{y^2-25}\qquad\text{Multiplying by the reciprocal of the divisor}$$

$$=\dfrac{(y-5)(y^2-4)}{(y^2-4)(y+5)(y-5)}$$

$$=\dfrac{(y-5)(y^2-4)(1)}{(y^2-4)(y+5)(y-5)}$$

$$=\dfrac{1}{y+5}$$

30. $\dfrac{1}{y+3}$

31.
$$\dfrac{\dfrac{y^2}{y^2-9}-\dfrac{y}{y+3}}{\dfrac{y}{y^2-9}-\dfrac{1}{y-3}}$$

$$=\dfrac{\dfrac{y^2}{(y+3)(y-3)}-\dfrac{y}{y+3}}{\dfrac{y}{(y+3)(y-3)}-\dfrac{1}{y-3}}$$

$$=\dfrac{\dfrac{y^2}{(y+3)(y-3)}-\dfrac{y}{y+3}}{\dfrac{y}{(y+3)(y-3)}-\dfrac{1}{y-3}}\cdot\dfrac{(y+3)(y-3)}{(y+3)(y-3)}$$

Multiplying by 1, using the LCD

$$=\dfrac{\dfrac{y^2}{(y+3)(y-3)}\cdot(y+3)(y-3)-\dfrac{y}{y+3}\cdot(y+3)(y-3)}{\dfrac{y}{(y+3)(y-3)}\cdot(y+3)(y-3)-\dfrac{1}{y-3}\cdot(y+3)(y-3)}$$

$$=\dfrac{y^2-y(y-3)}{y-(y+3)}=\dfrac{y^2-y^2+3y}{y-y-3}=\dfrac{3y}{-3}$$

$$=\dfrac{3y}{-1\cdot3}=-y$$

32. $-y$

33.

$$\dfrac{\dfrac{a}{a+3}+\dfrac{4}{5a}}{\dfrac{a}{2a+6}+\dfrac{3}{a}}$$

$$=\dfrac{\dfrac{a}{a+3}+\dfrac{4}{5a}}{\dfrac{a}{2(a+3)}+\dfrac{3}{a}}$$

$$=\dfrac{\dfrac{a}{a+3}+\dfrac{4}{5a}}{\dfrac{a}{2(a+3)}+\dfrac{3}{a}}\cdot\dfrac{10a(a+3)}{10a(a+3)}$$

        Multiplying by 1, using the LCD

$$=\dfrac{\dfrac{a}{a+3}\cdot 10a(a+3)+\dfrac{4}{5a}\cdot 10a(a+3)}{\dfrac{a}{2(a+3)}\cdot 10a(a+3)+\dfrac{3}{a}\cdot 10a(a+3)}$$

$$=\dfrac{10a^2+8(a+3)}{5a^2+30(a+3)}=\dfrac{10a^2+8a+24}{5a^2+30a+90}$$

$$=\dfrac{2(5a^2+4a+12)}{5(a^2+6a+18)}$$

34. $\dfrac{6a^2+30a+60}{3a^2+2a+4}$

35.

$$\dfrac{c+\dfrac{8}{c^2}}{1+\dfrac{2}{c}}$$

$$=\dfrac{c+\dfrac{8}{c^2}}{1+\dfrac{2}{c}}\cdot\dfrac{c^2}{c^2}\qquad\begin{array}{l}\text{Multiplying by 1,}\\\text{using the LCD}\end{array}$$

$$=\dfrac{c\cdot c^2+\dfrac{8}{c^2}\cdot c^2}{1\cdot c^2+\dfrac{2}{c}\cdot c^2}$$

$$=\dfrac{c^3+8}{c^2+2c}$$

$$=\dfrac{(c+2)(c^2-2c+4)}{c(c+2)}$$

$$=\dfrac{(c\!\!\!\!\diagup+2)(c^2-2c+4)}{c(c\!\!\!\!\diagup+2)}$$

$$=\dfrac{c^2-2c+4}{c}$$

36. $\dfrac{x^2y^2}{y^2-yx+x^2}$

37.

$$\dfrac{x^2+xy+y^2}{\dfrac{x^2}{y}-\dfrac{y^2}{x}}$$

$$=\dfrac{x^2+xy+y^2}{\dfrac{x^2}{y}-\dfrac{y^2}{x}}\cdot\dfrac{xy}{xy}$$

$$=\dfrac{xy(x^2+xy+y^2)}{\dfrac{x^2}{y}\cdot xy-\dfrac{y^2}{x}\cdot xy}$$

$$=\dfrac{xy(x^2+xy+y^2)}{x^3-y^3}$$

$$=\dfrac{xy(x^2+xy+y^2)}{(x-y)(x^2+xy+y^2)}$$

$$=\dfrac{xy}{x-y}\cdot\dfrac{x^2+xy+y^2}{x^2+xy+y^2}$$

$$=\dfrac{xy}{x-y}$$

38. $\dfrac{a+b}{ab}$

39.

$$\dfrac{\dfrac{1}{x^2-3x+2}+\dfrac{1}{x^2-4}}{\dfrac{1}{x^2+4x+4}+\dfrac{1}{x^2-4}}$$

$$=\dfrac{\dfrac{1}{(x-1)(x-2)}+\dfrac{1}{(x+2)(x-2)}}{\dfrac{1}{(x+2)(x+2)}+\dfrac{1}{(x+2)(x-2)}}$$

$$=\dfrac{\dfrac{1}{(x-1)(x-2)}+\dfrac{1}{(x+2)(x-2)}}{\dfrac{1}{(x+2)(x+2)}+\dfrac{1}{(x+2)(x-2)}}\cdot$$

$$\dfrac{(x-1)(x-2)(x+2)(x+2)}{(x-1)(x-2)(x+2)(x+2)}$$

      Multiplying by 1, using the LCD

$$=\dfrac{(x+2)(x+2)+(x-1)(x+2)}{(x-1)(x-2)+(x-1)(x+2)}$$

$$=\dfrac{x^2+4x+4+x^2+x-2}{x^2-3x+2+x^2+x-2}$$

$$=\dfrac{2x^2+5x+2}{2x^2-2x}$$

(Although both the numerator and the denominator can be factored, doing so will not lead to further simplification.)

40. $\dfrac{2x^2-5x-3}{2x^2+2x-4}$

41.

$$\dfrac{\dfrac{3}{a^2-4a+3}+\dfrac{3}{a^2-5a+6}}{\dfrac{3}{a^2-3a+2}+\dfrac{3}{a^2+3a-10}}$$

$$=\dfrac{\dfrac{3}{(a-1)(a-3)}+\dfrac{3}{(a-2)(a-3)}}{\dfrac{3}{(a-1)(a-2)}+\dfrac{3}{(a+5)(a-2)}}$$

$$=\dfrac{\dfrac{3}{(a-1)(a-3)}+\dfrac{3}{(a-2)(a-3)}}{\dfrac{3}{(a-1)(a-2)}+\dfrac{3}{(a+5)(a-2)}}\cdot$$

$$\dfrac{(a-1)(a-3)(a-2)(a+5)}{(a-1)(a-3)(a-2)(a+5)}$$

Multiplying by 1, using the LCD

$$=\dfrac{3(a-2)(a+5)+3(a-1)(a+5)}{3(a-3)(a+5)+3(a-1)(a-3)}$$

$$=\dfrac{3[(a-2)(a+5)+(a-1)(a+5)]}{3[(a-3)(a+5)+(a-1)(a-3)]}$$

$$=\dfrac{\cancel{3}[(a-2)(a+5)+(a-1)(a+5)]}{\cancel{3}[(a-3)(a+5)+(a-1)(a-3)]}$$

$$=\dfrac{a^2+3a-10+a^2+4a-5}{a^2+2a-15+a^2-4a+3}$$

$$=\dfrac{2a^2+7a-15}{2a^2-2a-12}$$

(Although both the numerator and the denominator can be factored, doing so will not lead to further simplification.)

42. $\dfrac{-a^2-21a-8}{a^2+3a-34}$

43. $\dfrac{\dfrac{y}{y^2-4}-\dfrac{2y}{y^2+y-6}}{\dfrac{2y}{y^2+y-6}-\dfrac{y}{y^2-4}}$

Observe that $\dfrac{y}{y^2-4}-\dfrac{2y}{y^2+y-6}=$

$-\left(\dfrac{2y}{y^2+y-6}-\dfrac{y}{y^2-4}\right)$. Then, the numerator and denominator are opposites and thus their quotient is -1.

44. $\dfrac{-(y-3)(2y-7)}{2(y-5)(y+4)}$

45. *Writing Exercise*

46. *Writing Exercise*

47.
$$2(3x-1)+5(4x-3)=3(2x+1)$$
$$6x-2+20x-15=6x+3$$
$$26x-17=6x+3$$
$$20x-17=3$$
$$20x=20$$
$$x=1$$

The solution is 1.

48. $\dfrac{11}{4}$

49.
$$\dfrac{t}{s+y}=r$$
$$(s+y)\cdot\dfrac{t}{s+y}=r(s+y)$$
$$t=rs+ry$$
$$t-rs=ry$$
$$\dfrac{t-rs}{r}=y$$

50. $\{-2,5\}$

51. **Familiarize.** We let l and w represent the length and width of the second frame, in centimeters. Then $l-3$ and $w-4$ represent the length and width of the first frame, in centimeters. The perimeter of the second frame is $2l+2w$; the perimeter of the first frame is $2(l-3)+2(w-4)$, or $2l+2w-14$.

Translate.

Perimeter of second frame	$=$	2	$\cdot$	perimeter of first frame	$-$	1
$2l+2w$	$=$	2	$\cdot$	$(2l+2w-14)$	$-$	1

Carry out. We first solve for $2l+2w$.
$$2l+2w=2(2l+2w-14)-1$$
$$2l+2w=4l+4w-28-1$$
$$2l+2w=4l+4w-29$$
$$29=2l+2w$$

If $2l+2w=29$, then $2l+2w-14=29-14$, or 15.

Check. The perimeter of the second frame is 1 cm less than twice the perimeter of the first frame:
$$29=2\cdot15-1$$

The values check.

State. The perimeter of the first frame is 15 cm; the perimeter of the second frame is 29 cm.

52. $34

53. *Writing Exercise*

54. *Writing Exercise*

55.
$$\frac{5x^{-2} + 10x^{-1}y^{-1} + 5y^{-2}}{3x^{-2} - 3y^{-2}}$$

$$= \frac{\dfrac{5}{x^2} + \dfrac{10}{xy} + \dfrac{5}{y^2}}{\dfrac{3}{x^2} - \dfrac{3}{y^2}}$$

$$= \frac{\dfrac{5}{x^2} + \dfrac{10}{xy} + \dfrac{5}{y^2}}{\dfrac{3}{x^2} - \dfrac{3}{y^2}} \cdot \frac{x^2 y^2}{x^2 y^2}$$

$$= \frac{5y^2 + 10xy + 5x^2}{3y^2 - 3x^2}$$

$$= \frac{5(y^2 + 2xy + x^2)}{3(y^2 - x^2)}$$

$$= \frac{5(y + x)(y + x)}{3(y + x)(y - x)}$$

$$= \frac{5(\cancel{y + x})(y + x)}{3(\cancel{y + x})(y - x)}$$

$$= \frac{5(y + x)}{3(y - x)}$$

56. $\dfrac{b - a}{ab}$

57. Substitute $\dfrac{c}{4}$ for both v_1 and v_2.

$$\frac{\dfrac{c}{4} + \dfrac{c}{4}}{1 + \dfrac{\dfrac{c}{4} \cdot \dfrac{c}{4}}{c^2}}$$

$$= \frac{\dfrac{2c}{4}}{1 + \dfrac{\dfrac{c^2}{16}}{c^2}}$$

$$= \frac{\dfrac{c}{2}}{1 + \dfrac{c^2}{16} \cdot \dfrac{1}{c^2}}$$

$$= \frac{\dfrac{c}{2}}{1 + \dfrac{1}{16}}$$

$$= \frac{\dfrac{c}{2}}{\dfrac{17}{16}}$$

$$= \frac{c}{2} \cdot \frac{16}{17}$$

$$= \frac{8c}{17}$$

The observed speed is $\dfrac{8c}{17}$, or $\dfrac{8}{17}$ the speed of light.

58. $\dfrac{-2(2x + h)}{x^2(x + h)^2}$

59. $f(x) = \dfrac{3}{x}$, $f(x + h) = \dfrac{3}{x + h}$

$$\frac{f(x + h) - f(x)}{h} = \frac{\dfrac{3}{x + h} - \dfrac{3}{x}}{h}$$

$$= \frac{\dfrac{3x - 3(x + h)}{x(x + h)}}{h}$$

$$= \frac{3x - 3(x + h)}{x(x + h)} \cdot \frac{1}{h}$$

$$= \frac{3x - 3x - 3h}{xh(x + h)}$$

$$= \frac{-3h}{xh(x + h)}$$

$$= \frac{-3\cancel{h}}{x\cancel{h}(x + h)}$$

$$= \frac{-3}{x(x + h)}$$

60. $\dfrac{1}{(1 - x - h)(1 - x)}$

61. $f(x) = \dfrac{2x}{1 + x}$, $f(x + h) = \dfrac{2(x + h)}{1 + x + h}$

$$\frac{f(x + h) - f(x)}{h}$$

$$= \frac{\dfrac{2(x + h)}{1 + x + h} - \dfrac{2x}{1 + x}}{h}$$

$$= \frac{\dfrac{2(x + h)(1 + x) - 2x(1 + x + h)}{(1 + x + h)(1 + x)}}{h}$$

$$= \frac{2(x + h)(1 + x) - 2x(1 + x + h)}{(1 + x + h)(1 + x)} \cdot \frac{1}{h}$$

$$= \frac{2x + 2x^2 + 2h + 2hx - 2x - 2x^2 - 2xh}{(1 + x + h)(1 + x)h}$$

$$= \frac{2 \cdot \cancel{h}}{(1 + x + h)(1 + x)\cancel{h}}$$

$$= \frac{2}{(1 + x + h)(1 + x)}$$

62. $\{x | x$ is a real number $and\ x \neq 0\ and\ x \neq -2\ and\ x \neq 2\}$

63. Division by zero occurs in $\dfrac{1}{x^2 - 1}$ when $x = 1$ or $x = -1$.

Division by zero occurs in $\dfrac{1}{x^2 - 16}$ when $x = 4$ or $x = -4$.
To avoid division in the complex fraction we solve:

$$\frac{1}{9} - \frac{1}{x^2 - 16} = 0$$

$$x^2 - 16 - 9 = 0 \quad \text{Multiplying by } 9(x^2 - 16)$$

$$x^2 - 25 = 0$$

$$(x + 5)(x - 5) = 0$$

$$x + 5 = 0 \quad or \quad x - 5 = 0$$

$$x = -5 \quad or \quad x = 5.$$

The domain of $G = \{x | x$ is a real number *and* $x \neq$ 1 *and* $x \neq -1$ *and* $x \neq 4$ *and* $x \neq -4$ *and* $x \neq 5$ *and* $x \neq -5\}$.

64. $\dfrac{x^2}{x^4 + x^3 + x^2 + x + 1}$

65. $\quad f(x) = \dfrac{2}{2 + x}$

$$f(a) = \frac{2}{2 + a}$$

$$f(f(a)) = \cfrac{2}{2 + \cfrac{2}{2 + a}}$$

$$= \cfrac{2}{2 + \cfrac{2}{2 + a}} \cdot \frac{2 + a}{2 + a}$$

$$= \frac{2(2 + a)}{2(2 + a) + \cfrac{2}{2 + a} \cdot 2 + a}$$

$$= \frac{4 + 2a}{4 + 2a + 2}$$

$$= \frac{4 + 2a}{6 + 2a} = \frac{2(2 + a)}{2(3 + a)}$$

$$= \frac{\not{2}(2 + a)}{\not{2}(3 + a)} = \frac{2 + a}{3 + a}$$

66. a

67.
$$\left[\cfrac{\cfrac{x + 3}{x - 3} + 1}{\cfrac{x + 3}{x - 3} - 1} \right]^4$$

$$= \left[\cfrac{\cfrac{x + 3}{x - 3} + 1}{\cfrac{x + 3}{x - 3} - 1} \cdot \frac{x - 3}{x - 3} \right]^4$$

$$= \left[\frac{x + 3 + x - 3}{x + 3 - x + 3} \right]^4$$

$$= \left(\frac{2x}{6} \right)^4 = \left(\frac{x}{3} \right)^4 = \frac{x^4}{81}$$

Division by zero occurs in both the numerator and the denominator of the original fraction when $x = 3$. To avoid division by zero in the complex fraction we solve:

$$\frac{x + 3}{x - 3} - 1 = 0$$

$$\frac{x + 3}{x - 3} = 1$$

$$x + 3 = x - 3$$

$$3 = -3$$

The equation has no solution, so the denominator of the complex fraction cannot be zero. Thus, the domain of $f = \{x | x$ is a real number *and* $x \neq 3\}$, or $(-\infty, 3) \cup (3, \infty)$.

68. $168.61

Exercise Set 6.4

1. $\qquad \dfrac{4}{5} + \dfrac{1}{3} = \dfrac{x}{9}, \quad$ LCD is 45

$$45\left(\frac{4}{5} + \frac{1}{3} \right) = 45 \cdot \frac{x}{9}$$

$$45 \cdot \frac{4}{5} + 45 \cdot \frac{1}{3} = 45 \cdot \frac{x}{9}$$

$$36 + 15 = 5x$$

$$51 = 5x$$

$$\frac{51}{5} = x$$

Check: $\qquad \dfrac{4}{5} + \dfrac{1}{3} = \dfrac{x}{9}$

$$\frac{4}{5} + \frac{1}{3} \; ? \; \frac{51/5}{9}$$

$$\frac{12}{15} + \frac{5}{15} \; \bigg| \; \frac{51}{5} \cdot \frac{1}{9}$$

$$\frac{17}{15} \; \bigg| \; \frac{17}{15} \qquad \text{TRUE}$$

The solution is $\dfrac{51}{5}$.

2. $\dfrac{51}{2}$

3. $\qquad \dfrac{x}{3} - \dfrac{x}{4} = 12, \quad$ LCD is 12

$$12\left(\frac{x}{3} - \frac{x}{4} \right) = 12 \cdot 12$$

$$12 \cdot \frac{x}{3} - 12 \cdot \frac{x}{4} = 12 \cdot 12$$

$$4x - 3x = 144$$

$$x = 144$$

Check:　$\dfrac{x}{3} - \dfrac{x}{4} = 12$

$$\dfrac{144}{3} - \dfrac{144}{4} \;\overset{?}{|}\; 12$$

$$48 - 36 \;\Big|$$

$$12 \;\Big|\; 12 \quad \text{TRUE}$$

The solution is 144.

4. $-\dfrac{225}{2}$

5. $\dfrac{1}{3} - \dfrac{1}{x} = \dfrac{5}{6}$

Because $\dfrac{1}{x}$ is undefined when x is 0, we note at the outset that $x \neq 0$. Then we multiply both sides by the LCD, $x \cdot 6$, or $6x$.

$$6x\left(\dfrac{1}{3} - \dfrac{1}{x}\right) = 6x \cdot \dfrac{5}{6}$$

$$6x \cdot \dfrac{1}{3} - 6x \cdot \dfrac{1}{x} = 6x \cdot \dfrac{5}{6}$$

$$2x - 6 = 5x$$

$$-6 = 3x$$

$$-2 = x$$

Check:　$\dfrac{1}{3} - \dfrac{1}{x} = \dfrac{5}{6}$

$$\dfrac{1}{3} - \dfrac{1}{-2} \;\overset{?}{|}\; \dfrac{5}{6}$$

$$\dfrac{1}{3} + \dfrac{1}{2} \;\Big|$$

$$\dfrac{2}{6} + \dfrac{3}{6} \;\Big|$$

$$\dfrac{5}{6} \;\Big|\; \dfrac{5}{6} \quad \text{TRUE}$$

The solution is -2.

6. $\dfrac{40}{9}$

7. $\dfrac{1}{2} - \dfrac{2}{7} = \dfrac{3}{2x}$

Because $\dfrac{3}{2x}$ is undefined when x is 0, we note at the outset that $x \neq 0$. Then we multiply both sides by the LCD, $2 \cdot 7 \cdot x$, or $14x$.

$$14x\left(\dfrac{1}{2} - \dfrac{2}{7}\right) = 14x \cdot \dfrac{3}{2x}$$

$$14x \cdot \dfrac{1}{2} - 14x \cdot \dfrac{2}{7} = 14x \cdot \dfrac{3}{2x}$$

$$7x - 4x = 21$$

$$3x = 21$$

$$x = 7$$

Check:　$\dfrac{1}{2} - \dfrac{2}{7} = \dfrac{3}{2x}$

$$\dfrac{1}{2} - \dfrac{2}{7} \;\overset{?}{|}\; \dfrac{3}{2 \cdot 7}$$

$$\dfrac{7}{14} - \dfrac{4}{14} \;\Big|\; \dfrac{3}{14}$$

$$\dfrac{3}{14} \;\Big|\; \dfrac{3}{14} \quad \text{TRUE}$$

The solution is 7.

8. 5

9. $\dfrac{12}{15} - \dfrac{1}{3x} = \dfrac{4}{5}$

Because $\dfrac{1}{3x}$ is undefined when x is 0, we note at the outset that $x \neq 0$. Then we multiply both sides by the LCD, $3 \cdot 5 \cdot x$, or $15x$.

$$15x\left(\dfrac{12}{15} - \dfrac{1}{3x}\right) = 15x \cdot \dfrac{4}{5}$$

$$15x \cdot \dfrac{12}{15} - 15x \cdot \dfrac{1}{3x} = 15x \cdot \dfrac{4}{5}$$

$$12x - 5 = 12x$$

$$-5 = 0$$

We get a false equation. The given equation has no solution.

10. No solution

11. $\dfrac{4}{3y} - \dfrac{3}{y} = \dfrac{10}{3}$

To assure that neither denominator on the left side is 0, we note at the outset that $y \neq 0$. Then we multiply on both sides by the LCD, $3 \cdot y$, or $3y$.

$$3y\left(\dfrac{4}{3y} - \dfrac{3}{y}\right) = 3y \cdot \dfrac{10}{3}$$

$$3y \cdot \dfrac{4}{3y} - 3y \cdot \dfrac{3}{y} = 3y \cdot \dfrac{10}{3}$$

$$4 - 9 = 10y$$

$$-5 = 10y$$

$$-\dfrac{1}{2} = y$$

This value checks. The solution is $-\dfrac{1}{2}$.

12. $-4, -1$

13. $\dfrac{x-2}{x-4} = \dfrac{2}{x-4}$

To assure that neither denominator is 0, we note at the outset that $x \neq 4$. Then we multiply both sides by the LCD, $x - 4$.

$$(x-4) \cdot \frac{x-2}{x-4} = (x-4) \cdot \frac{2}{x-4}$$
$$x - 2 = 2$$
$$x = 4$$

Recall that, because of the restriction above, 4 cannot be a solution. A check confirms this.

Check: $\dfrac{x-2}{x-4} = \dfrac{2}{x-4}$

$$\begin{array}{c|c} \dfrac{4-2}{4-4} \ ? & \dfrac{2}{4-4} \\[2mm] \dfrac{2}{0} & \dfrac{2}{0} \qquad \text{UNDEFINED} \end{array}$$

The equation has no solution.

14. No solution

15. $\dfrac{5}{4t} = \dfrac{7}{5t-2}$

To assure that neither denominator is 0, we note at the outset that $t \neq 0$ and $t \neq \dfrac{2}{5}$. Then we multiply both sides by the LCD, $4t(5t-2)$.

$$4t(5t-2) \cdot \frac{5}{4t} = 4t(5t-2) \cdot \frac{7}{5t-2}$$
$$5(5t-2) = 4t \cdot 7$$
$$25t - 10 = 28t$$
$$-10 = 3t$$
$$-\frac{10}{3} = t$$

This value checks. The solution is $-\dfrac{10}{3}$.

16. 11

17. $\dfrac{x^2+4}{x-1} = \dfrac{5}{x-1}$

To assure that neither denominator is 0, we note at the outset that $x \neq 1$. Then we multiply both sides by the LCD, $x - 1$.

$$(x-1) \cdot \frac{x^2+4}{x-1} = (x-1) \cdot \frac{5}{x-1}$$
$$x^2 + 4 = 5$$
$$x^2 - 1 = 0$$
$$(x+1)(x-1) = 0$$
$$x + 1 = 0 \quad \text{or} \quad x - 1 = 0$$
$$x = -1 \quad \text{or} \qquad x = 1$$

Recall that, because of the restriction above, 1 cannot be a solution. The number -1 checks and is the solution.

We might also observe that since the denominators are the same, the numerators must be the same. Solving $x^2 + 4 = 5$, we get $x = -1$ or $x = 1$ as shown above. Again, because of the restriction $x \neq 1$, only -1 is a solution of the equation.

18. 2

19. $\dfrac{6}{a+1} = \dfrac{a}{a-1}$

To assure that neither denominator is 0, we note at the outset that $a \neq -1$ and $a \neq 1$. Then we multiply both sides by the LCD, $(a+1)(a-1)$.

$$(a+1)(a-1) \cdot \frac{6}{a+1} = (a+1)(a-1) \cdot \frac{a}{a-1}$$
$$6(a-1) = a(a+1)$$
$$6a - 6 = a^2 + a$$
$$0 = a^2 - 5a + 6$$
$$0 = (a-2)(a-3)$$
$$a - 2 = 0 \quad \text{or} \quad a - 3 = 0$$
$$a = 2 \quad \text{or} \qquad a = 3$$

Both values check. The solutions are 2 and 3.

20. 2, 3

21. $\dfrac{60}{t-5} - \dfrac{18}{t} = \dfrac{40}{t}$

To assure that none of the denominators is 0, we note at the outset that $t \neq 5$ and $t \neq 0$. Then we multiply on both sides by the LCD, $t(t-5)$.

$$t(t-5)\left(\frac{60}{t-5} - \frac{18}{t}\right) = t(t-5) \cdot \frac{40}{t}$$
$$60t - 18(t-5) = 40(t-5)$$
$$60t - 18t + 90 = 40t - 200$$
$$2t = -290$$
$$t = -145$$

This value checks. The solution is -145.

22. -23

23. $\dfrac{3}{x} + \dfrac{x}{x+2} = \dfrac{4}{x^2+2x}$

$$\frac{3}{x} + \frac{x}{x+2} = \frac{4}{x(x+2)}$$

To assure that none of the denominators is 0, we note at the outset that $x \neq 0$ and $x \neq -2$. Then we multiply both sides by the LCD, $x(x+2)$.

$$x(x+2)\left(\frac{3}{x} + \frac{x}{x+2}\right) = x(x+2) \cdot \frac{4}{x(x+2)}$$
$$3(x+2) + x \cdot x = 4$$
$$3x + 6 + x^2 = 4$$
$$x^2 + 3x + 2 = 0$$
$$(x+1)(x+2) = 0$$
$$x + 1 = 0 \quad \text{or} \quad x + 2 = 0$$
$$x = -1 \quad \text{or} \qquad x = -2$$

Recall that, because of the restrictions above, -2 cannot be a solution. The number -1 checks. The solution is -1.

24. -4

25. We find all values of a for which $2a - \dfrac{15}{a} = 7$. First note that $a \neq 0$. Then multiply on both sides by the LCD, a.

$$a\left(2a - \dfrac{15}{a}\right) = a \cdot 7$$

$$a \cdot 2a - a \cdot \dfrac{15}{a} = 7a$$

$$2a^2 - 15 = 7a$$

$$2a^2 - 7a - 15 = 0$$

$$(2a + 3)(a - 5) = 0$$

$$a = -\dfrac{3}{2} \ or \ a = 5$$

Both values check. The solutions are $-\dfrac{3}{2}$ and 5.

26. $-\dfrac{3}{2}, 2$

27. We find all values of a for which $\dfrac{a-5}{a+1} = \dfrac{3}{5}$. First note that $a \neq -1$. Then multiply on both sides by the LCD, $5(a+1)$.

$$5(a+1) \cdot \dfrac{a-5}{a+1} = 5(a+1) \cdot \dfrac{3}{5}$$

$$5(a-5) = 3(a+1)$$

$$5a - 25 = 3a + 3$$

$$2a = 28$$

$$a = 14$$

This value checks. The solution is 14.

28. $\dfrac{17}{4}$

29. We find all values of a for which $\dfrac{12}{a} - \dfrac{12}{2a} = 8$. First note that $a \neq 0$. Then multiply on both sides by the LCD, $2a$.

$$2a\left(\dfrac{12}{a} - \dfrac{12}{2a}\right) = 2a \cdot 8$$

$$2a \cdot \dfrac{12}{a} - 2a \cdot \dfrac{12}{2a} = 16a$$

$$24 - 12 = 16a$$

$$12 = 16a$$

$$\dfrac{3}{4} = a$$

This value checks. The solution is $\dfrac{3}{4}$.

30. $\dfrac{3}{5}$

31.
$$\dfrac{5}{x+2} - \dfrac{3}{x-2} = \dfrac{2x}{4-x^2}$$

$$\dfrac{5}{x+2} - \dfrac{3}{x-2} = \dfrac{2x}{(2+x)(2-x)}$$

$$\dfrac{5}{x+2} + \dfrac{3}{2-x} = \dfrac{2x}{(2+x)(2-x)} \qquad \left(-\dfrac{3}{x-2} = \dfrac{3}{2-x}\right)$$

First note that $x \neq -2$ and $x \neq 2$. Then multiply on both sides by the LCD, $(2+x)(2-x)$.

$$(2+x)(2-x)\left(\dfrac{5}{x+2} + \dfrac{3}{2-x}\right) =$$
$$(2+x)(2-x) \cdot \dfrac{2x}{(2+x)(2-x)}$$

$$5(2-x) + 3(2+x) = 2x$$

$$10 - 5x + 6 + 3x = 2x$$

$$16 - 2x = 2x$$

$$16 = 4x$$

$$4 = x$$

This value checks. The solution is 4.

32. -3

33.
$$\dfrac{2}{a+4} + \dfrac{2a-1}{a^2+2a-8} = \dfrac{1}{a-2}$$

$$\dfrac{2}{a+4} + \dfrac{2a-1}{(a+4)(a-2)} = \dfrac{1}{a-2}$$

First note that $a \neq -4$ and $a \neq 2$. Then multiply on both sides by the LCD, $(a+4)(a-2)$.

$$(a+4)(a-2)\left(\dfrac{2}{a+4} + \dfrac{2a-1}{(a+4)(a-2)}\right) =$$
$$(a+4)(a-2) \cdot \dfrac{1}{a-2}$$

$$2(a-2) + 2a - 1 = a + 4$$

$$2a - 4 + 2a - 1 = a + 4$$

$$4a - 5 = a + 4$$

$$3a = 9$$

$$a = 3$$

This value checks. The solution is 3.

34. $-6, 5$

35.
$$\dfrac{2}{x+3} - \dfrac{3x+5}{x^2+4x+3} = \dfrac{5}{x+1}$$

$$\dfrac{2}{x+3} - \dfrac{3x+5}{(x+3)(x+1)} = \dfrac{5}{x+1}$$

Note that $x \neq -3$ and $x \neq -1$. Then multiply on both sides by the LCD, $(x+3)(x+1)$.

$$(x+3)(x+1)\left(\frac{2}{x+3} - \frac{3x+5}{(x+3)(x+1)}\right) =$$

$$(x+3)(x+1) \cdot \frac{5}{x+1}$$

$$2(x+1) - (3x+5) = 5(x+3)$$

$$2x + 2 - 3x - 5 = 5x + 15$$

$$-x - 3 = 5x + 15$$

$$-18 = 6x$$

$$-3 = x$$

Recall that, because of the restriction above, -3 is not a solution. Thus, the equation has no solution.

36. No solution

37.
$$\frac{x-1}{x^2 - 2x - 3} + \frac{x+2}{x^2 - 9} = \frac{2x+5}{x^2 + 4x + 3}$$

$$\frac{x-1}{(x-3)(x+1)} + \frac{x+2}{(x+3)(x-3)} = \frac{2x+5}{(x+3)(x+1)}$$

Note that $x \neq 3$ and $x \neq -1$ and $x \neq -3$. Then multiply on both sides by the LCD, $(x-3)(x+1)(x+3)$.

$$(x-3)(x+1)(x+3)\left(\frac{x-1}{(x-3)(x+1)} + \frac{x+2}{(x+3)(x-3)}\right) =$$

$$(x-3)(x+1)(x+3) \cdot \frac{2x+5}{(x+3)(x+1)}$$

$$(x+3)(x-1) + (x+1)(x+2) = (x-3)(2x+5)$$

$$x^2 + 2x - 3 + x^2 + 3x + 2 = 2x^2 - x - 15$$

$$2x^2 + 5x - 1 = 2x^2 - x - 15$$

$$5x - 1 = -x - 15$$

$$6x = -14$$

$$x = -\frac{7}{3}$$

This value checks. The solution is $-\frac{7}{3}$.

38. $\dfrac{5}{14}$

39.
$$\frac{3}{x^2 - x - 12} + \frac{1}{x^2 + x - 6} = \frac{4}{x^2 + 3x - 10}$$

$$\frac{3}{(x-4)(x+3)} + \frac{1}{(x+3)(x-2)} = \frac{4}{(x+5)(x-2)}$$

Note that $x \neq 4$ and $x \neq -3$ and $x \neq 2$ and $x \neq -5$. Then multiply on both sides by the LCD, $(x-4)(x+3)(x-2)(x+5)$.

$$(x-4)(x+3)(x-2)(x+5)\left(\frac{3}{(x-4)(x+3)} + \frac{1}{(x+3)(x-2)}\right) =$$

$$(x-4)(x+3)(x-2)(x+5) \cdot \frac{4}{(x+5)(x-2)}$$

$$3(x-2)(x+5) + (x-4)(x+5) = 4(x-4)(x+3)$$

$$3(x^2 + 3x - 10) + x^2 + x - 20 = 4(x^2 - x - 12)$$

$$3x^2 + 9x - 30 + x^2 + x - 20 = 4x^2 - 4x - 48$$

$$4x^2 + 10x - 50 = 4x^2 - 4x - 48$$

$$10x - 50 = -4x - 48$$

$$14x = 2$$

$$x = \frac{1}{7}$$

This value checks. The solution is $\dfrac{1}{7}$.

40. $\dfrac{3}{5}$

41. *Writing Exercise*

42. *Writing Exercise*

43. *Familiarize.* Let x, y, and z represent the number of multiple-choice, true-false and fill-in questions, respectively.

Translate. The total number of questions is 70.

$$x + y + z = 70$$

The number of true-false is twice the number of fill-ins.

$$y = 2z$$

The number of multiple-choice is 5 less than the number of true-false.

$$x = y - 5$$

Carry out. Solving the system of three equations we get $(25, 30, 15)$.

Check. The sum of 25, 30, and 15 is 70. The number of true-false, 30, is twice the number of fill-ins, 15. The number of multiple-choice, 25, is 5 less than the number of true-false, 30.

State. On the test there are 25 multiple-choice, 30 true-false, and 15 fill-in questions.

44. 132 adult plates; 118 children's plates

45. *Familiarize.* Let l represent the length and w represent the width, in meters. Recall that perimeter $P = 2l + 2w$, and area $A = lw$.

Translate.

$$\underbrace{\text{The perimeter}}_{2l + 2w} \ \underbrace{\text{is}}_{=} \ \underbrace{628 \text{ m.}}_{628}$$

$$\underbrace{\text{The length}}_{l} \ \underbrace{\text{is}}_{=} \ \underbrace{6 \text{ m}}_{6} \ \underbrace{\text{more than}}_{+} \ \underbrace{\text{the width.}}_{w}$$

We have a system of equations:

$$2l + 2w = 628, \quad (1)$$
$$l = 6 + w \qquad (2)$$

Carry out. Use the substitution method to solve the system of equations and then compute the area. Substitute $6 + w$ for l in (1).

$$2(6 + w) + 2w = 628$$
$$12 + 2w + 2w = 628$$
$$12 + 4w = 628$$
$$4w = 616$$
$$w = 154$$

Substitute 154 for w in (2).

$$l = 6 + 154 = 160$$
$$A = lw = 160(154) = 24,640$$

Check. $2l + 2w = 2 \cdot 160 + 2 \cdot 154 = 320 + 308 = 628$. 160 is 6 more than 154. The numbers check.

State. The area is 24,640 m^2.

46. 16 and 18

47. a) $\quad 2x - 3y = 4, \quad (1)$
$\qquad\quad 4x - 6y = 7 \quad (2)$

Multiply Equation (1) by -2 and add.

$$\begin{array}{r} -4x + 6y = -8 \\ \underline{4x - 6y = 7} \\ 0 = -1 \end{array}$$

We get a false equation, so the system has no solution. It is inconsistent.

b) $\quad x + 3y = 2, \quad (1)$
$\quad\; 2x - 3y = 1 \quad (2)$

First we add the equations.

$$\begin{array}{r} x + 3y = 2 \\ \underline{2x - 3y = 1} \\ 3x = 3 \\ x = 1 \end{array}$$

Now substitute 1 for x in one of the equations and solve for y.

$$x + 3y = 2 \quad (1)$$
$$1 + 3y = 2$$
$$3y = 1$$
$$y = \frac{1}{3}$$

The solution is $\left(1, \dfrac{1}{3}\right)$. The system is consistent.

48. $\{x | x < -1 \; or \; x > 5\}$, or $(-\infty, -1) \cup (5, \infty)$

49. *Writing Exercise*

50. *Writing Exercise*

51.
$$f(a) = g(a)$$

$$\frac{a - \dfrac{2}{3}}{a + \dfrac{1}{2}} = \frac{a + \dfrac{2}{3}}{a - \dfrac{3}{2}}$$

$$\frac{a - \dfrac{2}{3}}{a + \dfrac{1}{2}} \cdot \frac{6}{6} = \frac{a + \dfrac{2}{3}}{a - \dfrac{3}{2}} \cdot \frac{6}{6}$$

$$\frac{6a - \dfrac{2}{3} \cdot 6}{6a + \dfrac{1}{2} \cdot 6} = \frac{6a + \dfrac{2}{3} \cdot 6}{6a - \dfrac{3}{2} \cdot 6}$$

$$\frac{6a - 4}{6a + 3} = \frac{6a + 4}{6a - 9}$$

$$\frac{6a - 4}{3(2a + 1)} = \frac{6a + 4}{3(2a - 3)}$$

To assure that neither denominator is 0, we note at the outset that $a \neq -\dfrac{1}{2}$ and $a \neq \dfrac{3}{2}$. Then we multiply both sides by the LCD, $3(2a + 1)(2a - 3)$.

$$3(2a + 1)(2a - 3) \cdot \frac{6a - 4}{3(2a + 1)} =$$
$$\qquad\qquad 3(2a + 1)(2a - 3) \cdot \frac{6a + 4}{3(2a - 3)}$$
$$(2a - 3)(6a - 4) = (2a + 1)(6a + 4)$$
$$12a^2 - 26a + 12 = 12a^2 + 14a + 4$$
$$-26a + 12 = 14a + 4$$
$$-40a + 12 = 4$$
$$-40a = -8$$
$$a = \frac{1}{5}$$

This number checks. For $a = \dfrac{1}{5}$, $f(a) = g(a)$.

52. $-8, 8$

53. $\dfrac{a + 3}{a + 2} - \dfrac{a + 4}{a + 3} = \dfrac{a + 5}{a + 4} - \dfrac{a + 6}{a + 5}$

Note that $a \neq -2$ and $a \neq -3$ and $a \neq -4$ and $a \neq -5$.

$$(a + 2)(a + 3)(a + 4)(a + 5)\left(\frac{a + 3}{a + 2} - \frac{a + 4}{a + 3}\right) =$$
$$(a + 2)(a + 3)(a + 4)(a + 5)\left(\frac{a + 5}{a + 4} - \frac{a + 6}{a + 5}\right)$$
$$(a+3)(a+4)(a+5)(a+3) - (a+2)(a+4)(a+5)(a+4) =$$
$$\quad (a+2)(a+3)(a+5)(a+5) - (a+2)(a+3)(a+4)(a+6)$$
$$a^4 + 15a^3 + 83a^2 + 201a + 180 -$$

$$(a^4+15a^3+82a^2+192a+160) =$$
$$a^4+15a^3+81a^2+185a+150-$$
$$(a^4+15a^3+80a^2+180a+144)$$
$$a^2 + 9a + 20 = a^2 + 5a + 6$$
$$4a = -14$$
$$a = -\frac{7}{2}$$

This value checks. When $a = -\dfrac{7}{2}$, $f(a) = g(a)$.

54. $\{x | x$ is a real number $and\ x \neq -1\ and\ x \neq 1\}$

55. Set $f(a)$ equal to $g(a)$ and solve for a.
$$\frac{0.793}{a} + 18.15 = \frac{6.034}{a} - 43.17$$
Note that $a \neq 0$. Then multiply on both sides by the LCD, a.
$$a\left(\frac{0.793}{a} + 18.15\right) = a\left(\frac{6.034}{a} - 43.17\right)$$
$$0.793 + 18.15a = 6.034 - 43.17a$$
$$61.32a = 5.241$$
$$a \approx 0.0854697$$

This value checks. When $a \approx 0.0854697$, $f(a) = g(a)$.

56. -2.955341202

57. $\dfrac{x^2 + 6x - 16}{x - 2} = x + 8, x \neq 2$
$$\frac{(x+8)(x-2)}{x-2} = x + 8$$
$$\frac{(x+8)(\cancel{x-2})}{\cancel{x-2}} = x + 8$$
$$x + 8 = x + 8$$
$$8 = 8$$

Since $8 = 8$ is true for all values of x, the original equation is true for any possible replacements of the variable. It is an identity.

58. Yes

Exercise Set 6.5

1. *Familiarize*. Let $x =$ the number.

Translate.

The reciprocal of 3	plus	the reciprocal of 6	is	the reciprocal of the number.
↓	↓	↓	↓	↓
$\frac{1}{3}$	$+$	$\frac{1}{6}$	$=$	$\frac{1}{x}$

Carry out. We solve the equation.
$$\frac{1}{3} + \frac{1}{6} = \frac{1}{x}, \text{ LCD is } 6x$$
$$6x\left(\frac{1}{3} + \frac{1}{6}\right) = 6x \cdot \frac{1}{x}$$
$$2x + x = 6$$
$$3x = 6$$
$$x = 2$$

Check. $\dfrac{1}{3} + \dfrac{1}{6} = \dfrac{2}{6} + \dfrac{1}{6} = \dfrac{3}{6} = \dfrac{1}{2}$. This is the reciprocal of 2, so the result checks.

State. The number is 2.

2. $\dfrac{35}{12}$

3. *Familiarize*. We let $x =$ the number.

Translate.

A number	plus	6	times	its reciprocal	is	-5.
↓	↓	↓	↓	↓	↓	↓
x	$+$	6	$\cdot$	$\frac{1}{x}$	$=$	-5

Carry out. We solve the equation.
$$x + \frac{6}{x} = -5, \text{ LCD is } x$$
$$x\left(x + \frac{6}{x}\right) = x(-5)$$
$$x^2 + 6 = -5x$$
$$x^2 + 5x + 6 = 0$$
$$(x + 3)(x + 2) = 0$$
$$x = -3 \text{ or } x = -2$$

Check. The possible solutions are -3 and -2. We check -3 in the conditions of the problem.

Number:	-3
6 times the reciprocal of the number:	$6\left(-\dfrac{1}{3}\right) = -2$
Sum of the number and 6 times its reciprocal:	$-3 + (-2) = -5$

The number -3 checks.

Now we check -2:

Number:	-2
6 times the reciprocal of the number:	$6\left(-\dfrac{1}{2}\right) = -3$
Sum of the number and 6 times its reciprocal:	$-2 + (-3) = -5$

The number -2 also checks.

State. The number is -3 or -2.

4. $-3, -7$

5. Familiarize. We let $x =$ the first integer. Then $x + 1 =$ the second, and their product $= x(x+1)$.

Translate.

$$\underbrace{\text{Reciprocal of the product}}_{\downarrow} \; \text{is} \; \overset{\downarrow}{\frac{1}{42}}.$$

$$\frac{1}{x(x+1)} = \frac{1}{42}$$

Carry out. We solve the equation.

$$\frac{1}{x(x+1)} = \frac{1}{42}, \text{ LCD is } 42x(x+1)$$

$$42x(x+1) \cdot \frac{1}{x(x+1)} = 42x(x+1) \cdot \frac{1}{42}$$

$$42 = x(x+1)$$

$$42 = x^2 + x$$

$$0 = x^2 + x - 42$$

$$0 = (x+7)(x-6)$$

$$x = -7 \text{ or } x = 6$$

Check. When $x = -7$, then $x+1 = -6$ and $-7(-6) = 42$. The reciprocal of this product is $\frac{1}{42}$.

When $x = 6$, then $x + 1 = 7$ and $6 \cdot 7 = 42$. The reciprocal of this product is also $\frac{1}{42}$. Both possible solutions check.

State. The integers are -7 and -6 or 6 and 7.

6. -9 and -8 or 8 and 9

7. Familiarize. The job takes Cedric 8 hours working alone and Carolyn 6 hours working alone. Then in 1 hour, Cedric does $\frac{1}{8}$ of the job and Carolyn does $\frac{1}{6}$ of the job. Working together, they can do $\frac{1}{8} + \frac{1}{6}$ of the job in 1 hour. Let t represent the number of hours required for Cedric and Carolyn, working together, to do the job.

Translate. We want to find t such that

$$t\left(\frac{1}{8}\right) + t\left(\frac{1}{6}\right) = 1, \text{ or } \frac{t}{8} + \frac{t}{6} = 1,$$

where 1 represents one entire job.

Carry out. We solve the equation.

$$\frac{t}{8} + \frac{t}{6} = 1, \text{ LCD is } 24$$

$$24\left(\frac{t}{8} + \frac{t}{6}\right) = 24 \cdot 1$$

$$3t + 4t = 24$$

$$7t = 24$$

$$t = \frac{24}{7}$$

Check. In $\frac{24}{7}$ hours, Cedric will do $\frac{1}{8} \cdot \frac{24}{7}$, or $\frac{3}{7}$ of the job and Carolyn will do $\frac{1}{6} \cdot \frac{24}{7}$, or $\frac{4}{7}$ of the job. Together, they do $\frac{3}{7} + \frac{4}{7}$, or 1 entire job. The answer checks.

State. It will take $\frac{24}{7}$ hr, or $3\frac{3}{7}$ hr, for Cedric and Carolyn, together, to refinish the floor.

8. $3\frac{3}{14}$ hr

9. Familiarize. The tank can be filled in 18 hours by only the town office well and in 22 hours with only the high school well. Then in 1 hour, the office well fills $\frac{1}{18}$ of the tank, and the high school well fills $\frac{1}{22}$ of the tank. Using both the wells, $\frac{1}{18} + \frac{1}{22}$ of the tank can be filled in 1 hour. Suppose that it takes t hours to fill the tank using both the town office well and the high school well.

Translate. We want to find t such that

$$t\left(\frac{1}{18}\right) + t\left(\frac{1}{22}\right) = 1, \text{ or } \frac{t}{18} + \frac{t}{22} = 1,$$

where 1 represents one entire job.

Carry out. We solve the equation. We multiply both sides by the LCD, 198.

$$198\left(\frac{t}{18} + \frac{t}{22}\right) = 198 \cdot 1$$

$$11t + 9t = 198$$

$$20t = 198$$

$$t = \frac{99}{10}$$

Check. The possible solution is $\frac{99}{10}$ hours. If the town office well is used $\frac{99}{10}$ hours, it fills $\frac{1}{18} \cdot \frac{99}{10}$, or $\frac{11}{20}$ of the tank. If the high school well is used $\frac{99}{10}$ hours, it fills $\frac{1}{22} \cdot \frac{99}{10}$, or $\frac{9}{20}$ of the tank. Using both, $\frac{11}{20} + \frac{9}{20}$ of the tank, or all of it, will be filled in $\frac{99}{10}$ hours.

State. Using both the town office well and the high school well, it will take $\frac{99}{10}$, or $9\frac{9}{10}$ hours, to fill the tank.

10. $8\frac{4}{7}$ hr

11. Familiarize. In 1 minute the HQ17 does $\frac{1}{10}$ of the job and the HQ174 does $\frac{1}{6}$ of the job. Working together, they can do $\frac{1}{10} + \frac{1}{6}$ of the job in 1 minute. Suppose it takes them t minutes, working together, to do the job.

Translate. We find t such that
$$t\left(\frac{1}{10}\right) + t\left(\frac{1}{6}\right) = 1, \text{ or } \frac{t}{10} + \frac{t}{6} = 1.$$

Carry out. We solve the equation. We multiply both sides by the LCD, 30.
$$30\left(\frac{t}{10} + \frac{t}{6}\right) = 30 \cdot 1$$
$$3t + 5t = 30$$
$$8t = 30$$
$$t = \frac{15}{4}$$

Check. In $\frac{15}{4}$ min the HQ17 will do $\frac{15}{4} \cdot \frac{1}{10}$, or $\frac{3}{8}$ of the job and the HQ174 will do $\frac{15}{4} \cdot \frac{1}{6}$, or $\frac{5}{8}$ of the job. Together they will do $\frac{3}{8} + \frac{5}{8}$, or 1 entire job. The answer checks.

State. It would take the two machines $\frac{15}{4}$ min, or $3\frac{3}{4}$ min, to clean the air working together.

12. 2.475 hr

13. Familiarize. Let t represent the number of hours it would take Skyler to do the job working alone. Then $t - 6$ represents the time it would take Jake to do the job alone.

In 1 hr Skyler does $\frac{1}{t}$ of the job and Jake does $\frac{1}{t-6}$ of the job.

Translate. Working together, they can do the entire job in 4 hr, so we want to find t such that
$$4\left(\frac{1}{t}\right) + 4\left(\frac{1}{t-6}\right) = 1, \text{ or } \frac{4}{t} + \frac{4}{t-6} = 1.$$

Carry out. We solve the equation.
$$\frac{4}{t} + \frac{4}{t-6} = 1, \text{ LCD is } t(t-6)$$
$$t(t-6)\left(\frac{4}{t} + \frac{4}{t-6}\right) = t(t-6) \cdot 1$$
$$4(t-6) + 4t = t^2 - 6t$$
$$4t - 24 + 4t = t^2 - 6t$$
$$8t - 24 = t^2 - 6t$$
$$0 = t^2 - 14t + 24$$
$$0 = (t-2)(t-12)$$
$$t - 2 = 0 \ \text{ or } \ t - 12 = 0$$
$$t = 2 \ \text{ or } \ \ \ \ \ \ t = 12$$

Check. When $t = 2$, then $t - 6 = 2 - 6 = -4$, so 2 cannot be a solution of the original problem. If Skyler does the job in 12 hr, then Jake does the job in $12 - 6$, or 6 hr. In 4 hr, Skyler does $4 \cdot \frac{1}{12}$, or $\frac{1}{3}$ of the job and Jake does

$4 \cdot \frac{1}{6}$, or $\frac{2}{3}$ of the job. Together they do $\frac{1}{3} + \frac{2}{3}$, or 1 entire job in 4 hr. The number 12 checks.

State. It would take Skyler 12 hours and it would take Jake 6 hours to do the job, working alone.

14. $3\frac{9}{52}$ hr

15. Familiarize. Let t represent the time, in minutes, that it takes the EV25 to clean the air, working alone. Then $2t$ represents the time, in minutes, it takes the HQ17 to clean the same volume of air, working alone. In 1 minute, the EV25 does $\frac{1}{t}$ of the job and the HQ17 does $\frac{1}{2t}$ of the job.

Translate. Working together, they can do the entire job in 10 min, so we want to find t such that
$$10\left(\frac{1}{t}\right) + 10\left(\frac{1}{2t}\right) = 1, \text{ or } \frac{10}{t} + \frac{5}{t} = 1.$$

Carry out. We solve the equation.
$$\frac{10}{t} + \frac{5}{t} = 1, \text{ LCD is } t$$
$$t\left(\frac{10}{t} + \frac{5}{t}\right) = t \cdot 1$$
$$10 + 5 = t$$
$$15 = t$$

Check. If the EV25 does the job in 15 min, then in 10 min it does $10 \cdot \frac{1}{15}$, or $\frac{2}{3}$ of the job. If it takes HQ17 $2 \cdot 15$, or 30 min, to do the job, then in 10 min it does $10\left(\frac{1}{30}\right)$, or $\frac{1}{3}$ of the job. Working together they do $\frac{2}{3} + \frac{1}{3}$, or 1 entire job in 10 min. The answer checks.

State. Working alone, it takes the EV25 15 min and the HQ17 30 min to do the job.

16. Office Jet G85: $22\frac{1}{2}$ hr; Laser Jet II: 45 hr

17. Familiarize. Let t represent the number of hours it takes Kate to paint the floor. Then $t + 3$ represents the time it takes Sara to paint the floor. In 1 hour, Kate does $\frac{1}{t}$ of the job and Sara does $\frac{1}{t+3}$.

Translate. Working together, it takes them 2 hr to do the job, so we want to find t such that
$$2\left(\frac{1}{t}\right) + 2\left(\frac{1}{t+3}\right) = 1, \text{ or } \frac{2}{t} + \frac{2}{t+3} = 1.$$

Carry out. We solve the equation. We multiply by the LCD, $t(t+3)$.

$$t(t+3)\left(\frac{2}{t}+\frac{2}{t+3}\right) = t(t+3)(1)$$
$$2(t+3)+2t = t^2+3t$$
$$2t+6+2t = t^2+3t$$
$$4t+6 = t^2+3t$$
$$0 = t^2-t-6$$
$$0 = (t-3)(t+2)$$
$$t = 3 \ or \ t = -2$$

Check. We check only 3, since the time cannot be negative. If Kate does the job in 3 hr, then in 2 hr she does $2\left(\frac{1}{3}\right)$, or $\frac{2}{3}$ of the job. If Sara does the job in $3+3$ or 6 hr, then in 2 hr she does $2\left(\frac{1}{6}\right)$, or $\frac{1}{3}$ of the job. Together they do $\frac{2}{3}+\frac{1}{3}$, or 1 entire job in 2 hr. The result checks.

State. It would take Kate 3 hours to do the job and it would take Sara 6 hours to do the job working alone.

18. Claudia: 10 days; Jan: 40 days

19. Familiarize. Working alone, Rosita does $\frac{1}{2}$ of the job in 1 hr. Let t = the time it take Helga to wax the car, working alone. Then in 1 hr she does $\frac{1}{t}$ of the job. Represent 45 min as $\frac{3}{4}$ hr.

Translate. In $\frac{3}{4}$ hr they do 1 entire job, working together, so we have
$$\frac{3}{4}\left(\frac{1}{2}\right)+\frac{3}{4}\left(\frac{1}{t}\right) = 1, \text{ or } \frac{3}{8}+\frac{3}{4t} = 1.$$

Carry out. We solve the equation.
$$8t\left(\frac{3}{8}+\frac{3}{4t}\right) = 8t\cdot 1$$
$$3t+6 = 8t$$
$$6 = 5t$$
$$\frac{6}{5} = t$$

Check. In $\frac{3}{4}$ hr, Rosita will do $\frac{3}{4}\cdot\frac{1}{2}$, or $\frac{3}{8}$, of the job, and Helga will do $\frac{3}{4}\left(\frac{1}{6/5}\right)$, or $\frac{3}{4}\cdot\frac{5}{6}=\frac{5}{8}$, of the job. Together they do $\frac{3}{8}+\frac{5}{8}=1$ job. The answer checks.

State. It would take Helga $\frac{6}{5}$, or $1\frac{1}{5}$ hr, working alone.

20. Zsuzanna: $\frac{4}{3}$ hr; Stan: 4 hr

21. Familiarize. We will convert hours to minutes:

2 hr $= 2\cdot 60$ min $= 120$ min

2 hr 55 min $= 120$ min $+ 55$ min $= 175$ min

Let t = the number of minutes it takes Deb to do the job alone. Then $t+120$ = the number of minutes it takes John alone. In 1 hour (60 minutes) Deb does $\frac{1}{t}$ and John does $\frac{1}{t+120}$ of the job.

Translate. In 175 min John and Deb will complete one entire job, so we have
$$175\left(\frac{1}{t}\right)+175\left(\frac{1}{t+120}\right) = 1, \text{ or }$$
$$\frac{175}{t}+\frac{175}{t+120} = 1.$$

Carry out. We solve the equation. Multiply on both sides by the LCD, $t(t+120)$.
$$t(t+120)\left(\frac{175}{t}+\frac{175}{t+120}\right) = t(t+120)(1)$$
$$175(t+120)+175t = t^2+120t$$
$$175t+21,000+175t = t^2+120t$$
$$0 = t^2-230t-21,000$$
$$0 = (t-300)(t+70)$$
$$t = 300 \ or \ t = -70$$

Check. Since negative time has no meaning in this problem, -70 is not a solution of the original problem. If the job takes Deb 300 min and it takes John $300+120 = 420$ min, then in 175 min they would complete
$$175\left(\frac{1}{300}\right)+175\left(\frac{1}{420}\right) = \frac{7}{12}+\frac{5}{12} = 1 \text{ job.}$$
The result checks.

State. It would take Deb 300 min, or 5 hr, to do the job alone.

22. 8 hr

23. Familiarize. We first make a drawing. Let r = the kayak's speed in still water in mph. Then $r-3$ = the speed upstream and $r+3$ = the speed downstream.

Upstream 4 miles $r-3$ mph

10 miles $r+3$ mph Downstream

We organize the information in a table. The time is the same both upstream and downstream so we use t for each time.

	Distance	Speed	Time
Upstream	4	$r-3$	t
Downstream	10	$r+3$	t

Translate. Using the formula Time = Distance/Rate in each row of the table and the fact that the times are the same, we can write an equation.

$$\frac{4}{r-3} = \frac{10}{r+3}$$

Carry out. We solve the equation.

$$\frac{4}{r-3} = \frac{10}{r+3}, \text{ LCD is } (r-3)(r+3)$$

$$(r-3)(r+3) \cdot \frac{4}{r-3} = (r-3)(r+3) \cdot \frac{10}{r+3}$$

$$4(r+3) = 10(r-3)$$

$$4r + 12 = 10r - 30$$

$$42 = 6r$$

$$7 = r$$

Check. If $r = 7$ mph, then $r - 3$ is 4 mph and $r + 3$ is 10 mph. The time upstream is $\frac{4}{4}$, or 1 hour. The time downstream is $\frac{10}{10}$, or 1 hour. Since the times are the same, the answer checks.

State. The speed of the kayak in still water is 7 mph.

24. 12 mph

25. ***Familiarize***. We first make a drawing. Let $r =$ Camille's speed on a nonmoving sidewalk in ft/sec. Then her speed moving forward on the moving sidewalk is $r + 1.8$, and her speed in the opposite direction is $r - 1.8$.

Forward $r + 1.8$ 105 ft

                     Opposite
51 ft $r - 1.8$ direction

We organize the information in a table. The time is the same both forward and in the opposite direction so we use t for each time.

	Distance	Speed	Time
Forward	105	$r + 1.8$	t
Opposite direction	51	$r - 1.8$	t

Translate. Using the formula Time = Distance/Rate in each row of the table and the fact that the times are the same, we can write an equation.

$$\frac{105}{r+1.8} = \frac{51}{r-1.8}$$

Carry out. We solve the equation.

$$\frac{105}{r+1.8} = \frac{51}{r-1.8},$$

$$\text{LCD is } (r+1.8)(r-1.8)$$

$$(r+1.8)(r-1.8) \cdot \frac{105}{r+1.8} = (r+1.8)(r-1.8) \cdot \frac{51}{r-1.8}$$

$$105(r-1.8) = 51(r+1.8)$$

$$105r - 189 = 51r + 91.8$$

$$54r = 280.8$$

$$r = 5.2$$

Check. If Camille's speed on a nonmoving sidewalk is 5.2 ft/sec, then her speed moving forward on the moving sidewalk is $5.2 + 1.8$, or 7 ft/sec, and her speed moving in the opposite direction on the sidewalk is $5.2 - 1.8$, or 3.4 ft/sec. Moving 105 ft at 7 ft/sec takes $\frac{105}{7} = 15$ sec. Moving 51 ft at 3.4 ft/sec takes 15 sec. Since the times are the same, the answer checks.

State. Camille would be walking 5.2 ft/sec on a nonmoving sidewalk.

26. 4.3 ft/sec

27. ***Familiarize***. Let $r =$ the speed of the passenger train in mph. Then $r - 14 =$ the speed of the freight train in mph. We organize the information in a table. The time is the same for both trains so we use t for each time.

	Distance	Speed	Time
Passenger train	400	r	t
Freight train	330	$r - 14$	t

Translate. Using the formula Time = Distance/Rate in each row of the table and the fact that the times are the same, we can write an equation.

$$\frac{400}{r} = \frac{330}{r-14}$$

Carry out. We solve the equation.

$$\frac{400}{r} = \frac{330}{r-14}, \text{ LCD is } r(r-14)$$

$$r(r-14) \cdot \frac{400}{r} = r(r-14) \cdot \frac{330}{r-14}$$

$$400(r-14) = 330r$$

$$400r - 5600 = 330r$$

$$-5600 = -70r$$

$$80 = r$$

Check. If the passenger train's speed is 80 mph, then the freight train's speed is $80 - 14$, or 66 mph. Traveling 400 mi at 80 mph takes $\frac{400}{80} = 5$ hr. Traveling 330 mi at 66 mph takes $\frac{330}{66} = 5$ hr. Since the times are the same, the answer checks.

State. The speed of the passenger train is 80 mph; the speed of the freight train is 66 mph.

28. Rosanna: $3\frac{1}{3}$ mph; Simone: $5\frac{1}{3}$ mph

29. Note that 38 mi is 7 mi less than 45 mi and that the local bus travels 7 mph slower than the express. Then the express travels 45 mi in one hr, or 45 mph, and the local bus travels 38 mi in one hr, or 38 mph.

30. A: 46 mph; E: 58 mph

31. *Familiarize*. We let $r =$ the speed of the river, in km/h. Then $2 + r =$ the paddleboat's speed downstream, in km/h, and $2 - r =$ the speed upstream, in km/h. The times are the same. Let t represent the time. We organize the information in a table.

	Distance	Speed	Time
Downstream	4	$2 + r$	t
Upstream	1	$2 - r$	t

***Translate*.** Using the formula Time = Distance/Rate in each row of the table and the fact that the times are the same, we can write an equation.

$$\frac{4}{2+r} = \frac{1}{2-r}$$

***Carry out*.** We solve the equation.

$$\frac{4}{2+r} = \frac{1}{2-r},$$

LCD is $(2+r)(2-r)$

$$(2+r)(2-r) \cdot \frac{4}{2+r} = (2+r)(2-r) \cdot \frac{1}{2-r}$$
$$4(2-r) = 2+r$$
$$8 - 4r = 2 + r$$
$$6 = 5r$$
$$\frac{6}{5} = r$$

***Check*.** If $r = \frac{6}{5}$, then the speed downstream is $2 + \frac{6}{5}$, or $\frac{16}{5}$ km/h and the speed upstream is $2 - \frac{6}{5}$, or $\frac{4}{5}$ km/h. The time for the trip downstream is $\frac{4}{16/5}$, or $4 \cdot \frac{5}{16}$, or $\frac{5}{4}$ hr. The time for the trip upstream is $\frac{1}{4/5}$, or $1 \cdot \frac{5}{4}$, or $\frac{5}{4}$ hr. Since the times are the same, the answer checks.

***State*.** The speed of the river is $\frac{6}{5}$ km/h, or $1\frac{1}{5}$ km/h.

32. 9 km/h

33. *Familiarize*. Let $c =$ the speed of the current, in km/h. Then $7 + c =$ the speed downriver and $7 - c =$ the speed upriver. We organize the information in a table.

	Distance	Speed	Time
Downriver	45	$7 + c$	t_1
Upriver	45	$7 - c$	t_2

***Translate*.** Using the formula Time = Distance/Rate we see that $t_1 = \dfrac{45}{7+c}$ and $t_2 = \dfrac{45}{7-c}$. The total time upriver and back is 14 hr, so $t_1 + t_2 = 14$, or

$$\frac{45}{7+c} + \frac{45}{7-c} = 14.$$

***Carry out*.** We solve the equation. Multiply both sides by the LCD, $(7+c)(7-c)$.

$$(7+c)(7-c)\left(\frac{45}{7+c} + \frac{45}{7-c}\right) = (7+c)(7-c)14$$
$$45(7-c) + 45(7+c) = 14(49 - c^2)$$
$$315 - 45c + 315 + 45c = 686 - 14c^2$$
$$14c^2 - 56 = 0$$
$$14(c+2)(c-2) = 0$$
$$c + 2 = 0 \quad or \quad c - 2 = 0$$
$$c = -2 \quad or \quad \quad c = 2$$

***Check*.** Since speed cannot be negative in this problem, -2 cannot be a solution of the original problem. If the speed of the current is 2 km/h, the barge travels upriver at $7 - 2$, or 5 km/h. At this rate it takes $\frac{45}{5}$, or 9 hr, to travel 45 km. The barge travels downriver at $7 + 2$, or 9 km/h. At this rate it takes $\frac{45}{9}$, or 5 hr, to travel 45 km. The total travel time is $9 + 5$, or 14 hr. The answer checks.

***State*.** The speed of the current is 2 km/h.

34. Jaime: 23 km/h; Mara: 15 km/h

35. *Familiarize*. Let $w =$ the wind speed, in mph. Then the speed into the wind is $350 - w$, and the speed with the wind is $350 + w$. We organize the information in a table.

	Distance	Speed	Time
Into the wind	487.5	$350 - w$	t_1
With the wind	487.5	$350 + w$	t_2

***Translate*.** Using the formula Time = Distance/Rate we see that $t_1 = \dfrac{487.5}{350 - w}$ and $t_2 = \dfrac{487.5}{350 + w}$. The total time upstream and back is

2.8 hr, so $t_1 + t_2 = 2.8$, or

$$\frac{487.5}{350 - w} + \frac{487.5}{350 + w} = 2.8.$$

***Carry out*.** We solve the equation. Multiply on both sides by the LCD, $(350 - w)(350 + w)$.

$$(350 - w)(350 + w)\left(\frac{487.5}{350 - w} + \frac{487.5}{350 + w}\right) =$$
$$(350 - w)(350 + w)(2.8)$$
$$487.5(350 + w) + 487.5(350 - w) =$$
$$2.8(122{,}500 - w^2)$$
$$170{,}625 + 487.5w + 170{,}625 - 487.5w =$$
$$343{,}000 - 2.8w^2$$
$$341{,}250 =$$
$$343{,}000 - 2.8w^2$$
$$2.8w^2 - 1750 = 0$$
$$2.8(w^2 - 625) = 0$$
$$2.8(w + 25)(w - 25) = 0$$
$$w = -25 \ or \ w = 25$$

Check. We check only 25 since the wind speed cannot be negative. If the wind speed is 25 mph, then the plane's speed into the wind is $350 - 25$, or 325 mph, and the speed with the wind is $350 + 25$, or 375 mph. Flying 487.5 mi into the wind takes $\frac{478.5}{325}$, or 1.5 hr. Flying 487.5 mi with the wind takes $\frac{487.5}{375}$, or 1.3 hr. The total time is $1.5 + 1.3$, or 2.8 hr. The answer checks.

State. The wind speed is 25 mph.

36. 5 m per minute

37. Familiarize. Let $r =$ the speed at which the train actually traveled in mph, and let $t =$ the actual travel time in hours. We organize the information in a table.

	Distance	Speed	Time
Actual speed	120	r	t
Faster speed	120	$r + 10$	$t - 2$

Translate. From the first row of the table we have $120 = rt$, and from the second row we have $120 = (r + 10)(t - 2)$. Solving the first equation for t, we have $t = \frac{120}{r}$. Substituting for t in the second equation, we have
$$120 = (r + 10)\left(\frac{120}{r} - 2\right).$$

Carry out. We solve the equation.

$$120 = (r + 10)\left(\frac{120}{r} - 2\right)$$
$$120 = 120 - 2r + \frac{1200}{r} - 20$$
$$20 = -2r + \frac{1200}{r}$$
$$r \cdot 20 = r\left(-2r + \frac{1200}{r}\right)$$
$$20r = -2r^2 + 1200$$
$$2r^2 + 20r - 1200 = 0$$
$$2(r^2 + 10r - 600) = 0$$
$$2(r + 30)(r - 20) = 0$$
$$r = -30 \ or \ r = 20$$

Check. Since speed cannot be negative in this problem, -30 cannot be a solution of the original problem. If the speed is 20 mph, it takes $\frac{120}{20}$, or 6 hr, to travel 120 mi. If the speed is 10 mph faster, or 30 mph, it takes $\frac{120}{30}$, or 4 hr, to travel 120 mi. Since 4 hr is 2 hr less time than 6 hr, the answer checks.

State. The speed was 20 mph.

38. 12 mph

39. *Writing Exercise*

40. *Writing Exercise*

41. $\dfrac{35a^6b^8}{7a^2b^2} = \dfrac{35}{7}a^{6-2}b^{8-2} = 5a^4b^6$

42. $5x^6y^4$

43. $\dfrac{36s^{15}t^{10}}{9s^5t^2} = \dfrac{36}{9}s^{15-5}t^{10-2}$
$$= 4s^{10}t^8$$

44. $-2x^4 - 7x^2 + 11x$

45. $\quad 2(x^3 + 4x^2 - 5x + 7) - 5(2x^3 - 4x^2 + 3x - 1)$
$$= 2x^3 + 8x^2 - 10x + 14 - 10x^3 + 20x^2 - 15x + 5$$
$$= -8x^3 + 28x^2 - 25x + 19$$

46. $11x^4 + 7x^3 - 2x^2 - 4x - 10$

47. *Writing Exercise*

48. *Writing Exercise*

49. Familiarize. If the drainage gate is closed, $\frac{1}{9}$ of the bog is filled in 1 hr. If the bog is not being filled, $\frac{1}{11}$ of the bog is drained in 1 hr. If the bog is being filled with the drainage gate left open, $\frac{1}{9} - \frac{1}{11}$ of the bog is filled in 1 hr.

Let t = the time it takes to fill the bog with the drainage gate left open.

Translate. We want to find t such that

$$t\left(\frac{1}{9} - \frac{1}{11}\right) = 1, \text{ or } \frac{t}{9} - \frac{t}{11} = 1.$$

Carry out. We solve the equation. First we multiply by the LCD, 99.

$$99\left(\frac{t}{9} - \frac{t}{11}\right) = 99 \cdot 1$$
$$11t - 9t = 99$$
$$2t = 99$$
$$t = \frac{99}{2}$$

Check. In $\frac{99}{2}$ hr, we have $\frac{99}{2}\left(\frac{1}{9} - \frac{1}{11}\right) = \frac{11}{2} - \frac{9}{2} = \frac{2}{2} = $ 1 full bog.

State. It will take $\frac{99}{2}$, or $49\frac{1}{2}$ hr, to fill the bog.

50. 40 min

51. Monica's speed downstream is $12 + 4$, or 16 mph. Using Time = Distance/Rate, we find that the time it will take Monica to motor 3 mi downstream is 3/16 hr. We can convert this time to minutes:

$$\frac{3}{16} \text{ hr} = \frac{3}{16} \times 1 \text{ hr} = \frac{3}{16} \times 60 \text{ min} = 11.25 \text{ min}$$

52. 30 min

53. **Familiarize**. Let p = the number of people per hour moved by the 60 cm-wide escalator. Then $2p$ = the number of people per hour moved by the 100 cm-wide escalator. We convert 1575 people per 14 minutes to people per hour:

$$\frac{1575 \text{ people}}{14 \text{ min}} \cdot \frac{60 \text{ min}}{1 \text{ hr}} = 6750 \text{ people/hr}$$

Translate. We use the information that together the escalators move 6750 people per hour to write an equation.

$$p + 2p = 6750$$

Carry out. We solve the equation.

$$p + 2p = 6750$$
$$3p = 6750$$
$$p = 2250$$

Check. If the 60 cm-wide escalator moves 2250 people per hour, then the 100 cm-wide escalator moves $2 \cdot 2250$, or 4500 people per hour. Together, they move $2250 + 4500$, or 6750 people per hour. The answer checks.

State. The 60 cm-wide escalator moves 2250 people per hour.

54. 700 mi from the airport

55. **Familiarize**. Let d = the distance, in miles, the paddleboat can cruise upriver before it is time to turn around. The boat's speed upriver is $12 - 5$, or 7 mph, and its speed downriver is $12 + 5$, or 17 mph. We organize the information in a table.

	Distance	Speed	Time
Upriver	d	7	t_1
Downriver	d	17	t_2

Translate. Using the formula Time = Distance/Rate we see that $t_1 = \frac{d}{7}$ and $t_2 = \frac{d}{17}$. The time upriver and back is 3 hr, so $t_1 + t_2 = 3$, or

$$\frac{d}{7} + \frac{d}{17} = 3.$$

Carry out. We solve the equation.

$$7 \cdot 17\left(\frac{d}{7} + \frac{d}{17}\right) = 7 \cdot 17 \cdot 3$$
$$17d + 7d = 357$$
$$24d = 357$$
$$d = \frac{119}{8}$$

Check. Traveling $\frac{119}{8}$ mi upriver at a speed of 7 mph takes $\frac{119/8}{7} = \frac{17}{8}$ hr. Traveling $\frac{119}{8}$ mi downriver at a speed of 17 mph takes $\frac{119/8}{17} = \frac{7}{8}$ hr. The total time is $\frac{17}{8} + \frac{7}{8} = \frac{24}{8} = 3$ hr. The answer checks.

State. The pilot can go $\frac{119}{8}$, or $14\frac{7}{8}$ mi upriver before it is time to turn around.

56. $3\frac{3}{4}$ km/h

57. **Familiarize**. Let d = the distance, in miles, Melissa lives from work. Also let t = the travel time in hours, when Melissa arrives on time. Note that 1 min = $\frac{1}{60}$ hr and 5 min = $\frac{5}{60}$, or $\frac{1}{12}$ hr.

Translate. Melissa's travel time at 50 mph is $\frac{d}{50}$. This is $\frac{1}{60}$ hr more than t, so we write an equation using this information:

$$\frac{d}{50} = t + \frac{1}{60}$$

Her travel time at 60 mph, $\frac{d}{60}$, is $\frac{1}{12}$ hr less than t, so we write a second equation:

$$\frac{d}{60} = t - \frac{1}{12}$$

We have a system of equations:

$$\frac{d}{50} = t + \frac{1}{60},$$

$$\frac{d}{60} = t - \frac{1}{12}$$

Carry out. Solving the system of equations, we get $\left(30, \frac{7}{12}\right)$.

Check. Traveling 30 mi at 50 mph takes $\frac{30}{50}$, or $\frac{3}{5}$ hr. Since $\frac{7}{12} + \frac{1}{60} = \frac{36}{60} = \frac{3}{5}$, this time makes Melissa $\frac{1}{60}$ hr, or 1 min late. Traveling 30 mi at 60 mph takes $\frac{30}{60}$, or $\frac{1}{2}$ hr. Since $\frac{7}{12} - \frac{1}{12} = \frac{6}{12} = \frac{1}{2}$, this time makes Melissa $\frac{1}{12}$ hr, or 5 min early. The answer checks.

State. Melissa lives 30 mi from work.

58. $21\frac{9}{11}$ min after 4:00

59. Familiarize Express the position of the hands in terms of minute units on the face of the clock. At 10:30 the hour hand is at $\frac{10.5}{12}$ hr $\times \frac{60 \text{ min}}{1 \text{ hr}}$, or 52.5 minutes, and the minute hand is at 30 minutes. The rate of the minute hand is 12 times the rate of the hour hand. (When the minute hand moves 60 minutes, the hour hand moves 5 minutes.) Let $t =$ the number of minutes after 10:30 that the hands will first be perpendicular. After t minutes the minute hand has moved t units, and the hour hand has moved $\frac{t}{12}$ units. The position of the hour hand will be 15 units "ahead" of the position of the minute hand when they are first perpendicular.

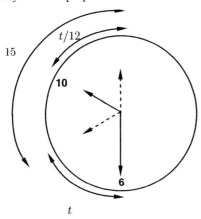

Translate.

Position of hour hand after t min	is	position of minute hand after t min	plus	15 min.
↓	↓	↓	↓	↓
$52.5 + \dfrac{t}{12}$	$=$	$30 + t$	$+$	15

Solve. We solve the equation.

$$52.5 + \frac{t}{12} = 30 + t + 15$$

$$52.5 + \frac{t}{12} = 45 + t, \text{ LCM is } 12$$

$$12\left(52.5 + \frac{t}{12}\right) = 12(45 + t)$$

$$630 + t = 540 + 12t$$

$$90 = 11t$$

$$\frac{90}{11} = t, \text{ or}$$

$$8\frac{2}{11} = t$$

Check. At $\frac{90}{11}$ min after 10:30, the position of the hour hand is at $52.5 + \frac{90/11}{12}$, or $53\frac{2}{11}$ min. The minute hand is at $30 + \frac{90}{11}$, or $38\frac{2}{11}$ min. The hour hand is 15 minutes ahead of the minute hand so the hands are perpendicular. The answer checks.

State. After 10:30 the hands of a clock will first be perpendicular in $8\frac{2}{11}$ min. The time is $10:38\frac{2}{11}$, or $21\frac{9}{11}$ min before 11:00.

60. 48 km/h

61. Familiarize. Let $r =$ the speed in mph Chip would have to travel for the last half of the trip in order to average a speed of 45 mph for the entire trip. We organize the information in a table.

	Distance	Speed	Time
First half	50	40	t_1
Last half	50	r	t_2

The total distance is $50 + 50$, or 100 mi. The total time is $t_1 + t_2$, or $\frac{50}{40} + \frac{50}{r}$, or $\frac{5}{4} + \frac{50}{r}$. The average speed is 45 mph.

Translate.

$$\text{Average speed} = \frac{\text{Total distance}}{\text{Total time}}$$

$$45 = \frac{100}{\dfrac{5}{4} + \dfrac{50}{r}}$$

Carry out. We solve the equation.

$$45 = \frac{100}{\dfrac{5}{4} + \dfrac{50}{r}}$$

$$45 = \frac{100}{\dfrac{5r + 200}{4r}}$$

$$45 = 100 \cdot \frac{4r}{5r + 200}$$

$$45 = \frac{400r}{5r + 200}$$

$$(5r + 200)(45) = (5r + 200) \cdot \frac{400r}{5r + 200}$$

$$225r + 9000 = 400r$$

$$9000 = 175r$$

$$\frac{360}{7} = r$$

Check. Traveling 50 mi at 40 mph takes $\dfrac{50}{40}$, or $\dfrac{5}{4}$ hr. Traveling 50 mi at $\dfrac{360}{7}$ mph takes $\dfrac{50}{360/7}$, or $\dfrac{35}{36}$ hr. Then the total time is $\dfrac{5}{4} + \dfrac{35}{36} = \dfrac{80}{36} = \dfrac{20}{9}$ hr.

The average speed when traveling 100 mi for $\dfrac{20}{9}$ hr is $\dfrac{100}{20/9} = 45$ mph. The answer checks.

State. Chip would have to travel at a speed of $\dfrac{360}{7}$, or $51\dfrac{3}{7}$ mph for the last half of the trip so that the average speed for the entire trip would be 45 mph.

Exercise Set 6.6

1.
$$\frac{34x^6 + 18x^5 - 28x^2}{6x^2}$$
$$= \frac{34x^6}{6x^2} + \frac{18x^5}{6x^2} - \frac{28x^2}{6x^2}$$
$$= \frac{17}{3}x^4 + 3x^3 - \frac{14}{3}$$

2. $6y^4 - 3y^2 + 8$

3.
$$\frac{21a^3 + 7a^2 - 3a - 14}{7a}$$
$$= \frac{21a^3}{7a} + \frac{7a^2}{7a} - \frac{3a}{7a} - \frac{14}{7a}$$
$$= 3a^2 + a - \frac{3}{7} - \frac{2}{a}$$

4. $-5x^2 + 4x - \dfrac{3}{5} + \dfrac{7}{5x}$

5.
$$\frac{14y^3 - 9y^2 - 8y}{2y^2}$$
$$= \frac{14y^3}{2y^2} - \frac{9y^2}{2y^2} - \frac{8y}{2y^2}$$
$$= 7y - \frac{9}{2} - \frac{4}{y}$$

6. $3a^3 + \dfrac{9a}{2} - \dfrac{4}{a}$

7.
$$\frac{15x^7 - 21x^4 - 3x^2}{-3x^2}$$
$$= \frac{15x^7}{-3x^2} + \frac{-21x^4}{-3x^2} + \frac{-3x^2}{-3x^2}$$
$$= -5x^5 + 7x^2 + 1$$

8. $-6y^5 + 3y^3 + 2y$

9.
$$(a^2b - a^3b^3 - a^5b^5) \div (a^2b)$$
$$= \frac{a^2b}{a^2b} - \frac{a^3b^3}{a^2b} - \frac{a^5b^5}{a^2b}$$
$$= 1 - ab^2 - a^3b^4$$

10. $x - xy - x^2$

11.
$$(6p^2q^2 - 9p^2q + 12pq^2) \div -3pq$$
$$= \frac{6p^2q^2}{-3pq} + \frac{-9p^2q}{-3pq} + \frac{12pq^2}{-3pq}$$
$$= -2pq + 3p - 4q$$

12. $4z - 2y^2z^3 + 3y^4z^2$

13.
$$(x^2 + 10x + 21) \div (x + 7)$$
$$= \frac{(x + 7)(x + 3)}{x + 7}$$
$$= \frac{(\cancel{x + 7})(x + 3)}{\cancel{x + 7}}$$
$$= x + 3$$

The answer is $x + 3$.

14. $y - 4$

15.
$$\begin{array}{r}
a - 12 \\
a + 4 \overline{\smash{\big)}\ a^2 - 8a - 16} \\
\underline{a^2 + 4a} \\
-12a - 16 \quad (a^2 - 8a) - (a^2 + 4a) = -12a \\
\underline{-12a - 48} \\
32 \quad (-12a - 16) - (-12a - 48) = 32
\end{array}$$

The answer is $a - 12$, R 32, or $a - 12 + \dfrac{32}{a + 4}$.

16. $y - 5 + \dfrac{-50}{y - 5}$

17.

$$
\begin{array}{r}
x - 4 \\
x - 5\overline{\smash{\big)}\ x^2 - 9x + 21} \\
\underline{x^2 - 5x} \\
-4x + 21 \\
\underline{-4x + 20} \\
1
\end{array}
$$

The answer is $x - 4$, R 1, or $x - 4 + \dfrac{1}{x - 5}$.

18. $x - 4 + \dfrac{-5}{x - 7}$

19. $(y^2 - 25) \div (y + 5) = \dfrac{y^2 - 25}{y + 5}$

$$
= \dfrac{(y + 5)(y - 5)}{y + 5}
$$

$$
= \dfrac{(\cancel{y + 5})(y - 5)}{\cancel{y + 5}}
$$

$$
= y - 5
$$

We could also find this quotient as follows.

$$
\begin{array}{r}
y - 5 \\
y + 5\overline{\smash{\big)}\ y^2 + 0y - 25} \\
\underline{y^2 + 5y} \\
-5y - 25 \\
\underline{-5y - 25} \\
0
\end{array}
$$
Writing in the missing term

The answer is $y - 5$.

20. $a + 9$

21.

$$
\begin{array}{r}
y^2 - 2y - 1 \\
y - 2\overline{\smash{\big)}\ y^3 - 4y^2 + 3y - 6} \\
\underline{y^3 - 2y^2} \\
-2y^2 + 3y \\
\underline{-2y^2 + 4y} \\
-y - 6 \\
\underline{-y + 2} \\
-8
\end{array}
$$

The answer is $y^2 - 2y - 1$, R -8, or

$$
y^2 - 2y - 1 + \dfrac{-8}{y - 2}.
$$

22. $x^2 - 2x - 2 + \dfrac{-13}{x - 3}$

23.

$$
\begin{array}{r}
2x^2 - x + 1 \\
x + 2\overline{\smash{\big)}\ 2x^3 + 3x^2 - x - 3} \\
\underline{2x^3 + 4x^2} \\
-x^2 - x \\
\underline{-x^2 - 2x} \\
x - 3 \\
\underline{x + 2} \\
-5
\end{array}
$$

The answer is $2x^2 - x + 1$, R -5, or

$$
2x^2 - x + 1 + \dfrac{-5}{x + 2}.
$$

24. $3x^2 + x - 1 + \dfrac{-4}{x - 2}$

25.

$$
\begin{array}{r}
a^2 + 4a + 15 \\
a - 4\overline{\smash{\big)}\ a^3 + 0a^2 - a + 10} \\
\underline{a^3 - 4a^2} \\
4a^2 - a \\
\underline{4a^2 - 16a} \\
15a + 10 \\
\underline{15a - 60} \\
70
\end{array}
$$

The answer is $a^2 + 4a + 15$, R 70, or

$$
a^2 + 4a + 15 + \dfrac{70}{a - 4}.
$$

26. $x^2 - 2x + 3$

27.

$$
\begin{array}{r}
2y^2 + 2y - 1 \\
5y - 2\overline{\smash{\big)}\ 10y^3 + 6y^2 - 9y + 10} \\
\underline{10y^3 - 4y^2} \\
10y^2 - 9y \\
\underline{10y^2 - 4y} \\
-5y + 10 \\
\underline{-5y + 2} \\
8
\end{array}
$$

The answer is $2y^2 + 2y - 1$, R 8, or

$$
2y^2 + 2y - 1 + \dfrac{8}{5y - 2}.
$$

28. $3x^2 - x + 4 + \dfrac{10}{2x - 3}$

29.

$$
\begin{array}{r}
2x^2 - x - 9 \\
x^2 + 2\overline{\smash{\big)}\ 2x^4 - x^3 - 5x^2 + x - 6} \\
\underline{2x^4 + 4x^2} \\
-x^3 - 9x^2 + x \\
\underline{-x^3 - 2x} \\
-9x^2 + 3x - 6 \\
\underline{-9x^2 - 18} \\
3x + 12
\end{array}
$$

The answer is $2x^2 - x - 9$, R $3x + 12$, or

$$
2x^2 - x - 9 + \dfrac{3x + 12}{x^2 + 2}.
$$

30. $3x^2 + 2x - 5 + \dfrac{2x - 5}{x^2 - 2}$

31.

$$
\begin{array}{r}
4x^2 - 6x + 9 \\
2x + 3\overline{\smash{\big)}\ 8x^3 + 27} \\
\underline{8x^3 + 12x^2} \\
-12x^2 + 0x \\
\underline{-12x^2 - 18x} \\
18x + 27 \\
\underline{18x + 27} \\
0
\end{array}
$$

Since $g(x)$ is 0 for $x = -\dfrac{3}{2}$, we have

$$
F(x) = 4x^2 - 6x + 9, \text{ provided } x \neq -\dfrac{3}{2}.
$$

32. $16x^2 + 8x + 4$, $x \neq \dfrac{1}{2}$

33. $F(x) = \dfrac{f(x)}{g(x)} = \dfrac{6x^2 - 11x - 10}{3x + 2}$

$$
\begin{array}{r}
2x - 5 \\
3x+2\,\overline{\smash{\big)}\,6x^2 - 11x - 10} \\
\underline{6x^2 + 4x } \\
-15x - 10 \\
\underline{-15x - 10} \\
0
\end{array}
$$

Since $g(x)$ is 0 for $x = -\dfrac{2}{3}$, we have

$F(x) = 2x - 5$, provided $x \neq -\dfrac{2}{3}$.

34. $4x + 3$, $x \neq \dfrac{7}{2}$

35.

$$
\begin{array}{r}
x^2 + 1 \\
x^2-25\,\overline{\smash{\big)}\,x^4 - 24x^2 - 25} \\
\underline{x^4 - 25x^2 } \\
x^2 - 25 \\
\underline{x^2 - 25} \\
0
\end{array}
$$

Since $g(x)$ is 0 for $x = -5$ or $x = 5$, we have
$F(x) = x^2 + 1$, provided $x \neq -5$ and $x \neq 5$.

36. $x^2 + 6$, $x \neq -3$, $x \neq 3$

37. We rewrite $f(x)$ in descending order.

$F(x) = \dfrac{f(x)}{g(x)} = \dfrac{2x^5 - 3x^4 - 2x^3 + 8x^2 - 5}{x^2 - 1}$

$$
\begin{array}{r}
2x^3 - 3x^2 + 5 \\
x^2-1\,\overline{\smash{\big)}\,2x^5 - 3x^4 - 2x^3 + 8x^2 - 5} \\
\underline{2x^5 - 2x^3 } \\
-3x^4 + 8x^2 \\
\underline{-3x^4 + 3x^2 } \\
5x^2 - 5 \\
\underline{5x^2 - 5} \\
0
\end{array}
$$

Since $g(x)$ is 0 for $x = -1$ or $x = 1$, we have
$F(x) = 2x^3 - 3x^2 + 5$, provided $x \neq -1$ and $x \neq 1$.

38. $3x^2 - x + 2$, $x \neq -2$, $x \neq 2$

39. *Writing Exercise*

40. *Writing Exercise*

41. $ab - cd = k$

$-cd = k - ab$

$c = \dfrac{k - ab}{-d}$, or $\dfrac{ab - k}{d}$

42. $\dfrac{xy - t}{w}$

43. **Familiarize.** Let x, $x + 1$, and $x + 2$ represent the three consecutive positive integers.

Translate. Rewording, we write an equation.

Product of first and second	is	product of second and third		less	26.

$$
\underbrace{x(x+1)}_{} \quad = \quad \underbrace{(x+1)(x+2)}_{} \quad - \quad 26
$$

Carry out. We solve the equation.

$$x^2 + x = x^2 + 3x + 2 - 26$$
$$x = 3x - 24$$
$$24 = 2x$$
$$12 = x$$

If the first integer is 12, the next two are 13 and 14.

Check. The product of 12 and 13 is 156. The product of 13 and 14 is 182, and $182 - 26 = 156$. The numbers check.

State. The three consecutive positive integers are 12, 13, and 14.

44. $-54a^3$

45. $|2x - 3| > 7$

$2x - 3 < -7$ *or* $2x - 3 > 7$

$2x < -4$ *or* $2x > 10$

$x < -2$ *or* $x > 5$

The solution set is $\{x | x < -2 \ or \ x > 5\}$, or $(-\infty, -2) \cup (5, \infty)$.

46. $\left|\left\{ x \middle| -\dfrac{7}{3} < x < 3 \right\}\right.$, or $\left(-\dfrac{7}{3}, 3 \right)$

47. *Writing Exercise*

48. *Writing Exercise*

49.

$$
\begin{array}{r}
a^2 + ab \\
a^2+3ab+2b^2\,\overline{\smash{\big)}\,a^4 + 4a^3b + 5a^2b^2 + 2ab^3} \\
\underline{a^4 + 3a^3b + 2a^2b^2 } \\
a^3b + 3a^2b^2 + 2ab^3 \\
\underline{a^3b + 3a^2b^2 + 2ab^3} \\
0
\end{array}
$$

The answer is $a^2 + ab$.

50. $x^2 + 2y$

51.

$$a + b \, \overline{\big)\, a^7 \qquad\qquad\qquad\qquad\qquad\qquad +b^7}$$

with quotient $a^6 - a^5b + a^4b^2 - a^3b^3 + a^2b^4 - ab^5 + b^6$

$$\begin{array}{l}
\underline{a^7 + a^6b} \\
-a^6b \\
\underline{-a^6b - a^5b^2} \\
a^5b^2 \\
\underline{a^5b^2 + a^4b^3} \\
-a^4b^3 \\
\underline{-a^4b^3 - a^3b^4} \\
a^3b^4 \\
\underline{a^3b^4 + a^2b^5} \\
-a^2b^5 \\
\underline{-a^2b^5 - ab^6} \\
ab^6 + b^7 \\
\underline{ab^6 + b^7} \\
0
\end{array}$$

The answer is $a^6 - a^5b + a^4b^2 - a^3b^3 + a^2b^4 - ab^5 + b^6$.

52. $\dfrac{14}{3}$

53.

$$x + 2 \, \overline{\big)\, x^2 - 3x + 2k} \qquad x - 5$$

$$\begin{array}{l}
\underline{x^2 + 2x} \\
-5x + 2k \\
\underline{-5x - 10} \\
2k + 10
\end{array}$$

The remainder is 7. Thus, we solve the following equation for k.

$$2k + 10 = 7$$
$$2k = -3$$
$$k = -\frac{3}{2}$$

54. a), b) $f(x) = 3 + \dfrac{1}{x + 2}$

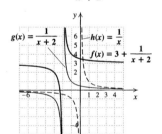

c) The graph of f looks like the graph of g, shifted up 3 units. The graph of g looks like the graph of h, shifted to the left 2 units.

55. *Writing Exercise*

Exercise Set 6.7

1. $(x^3 - 2x^2 + 2x - 7) \div (x + 1) =$
$(x^3 - 2x^2 + 2x - 7) \div [x - (-1)]$

$$\begin{array}{r|rrrr}
-1 & 1 & -2 & 2 & -7 \\
   &   & -1 & 3 & -5 \\
\hline
   & 1 & -3 & 5 \,|\, {-12}
\end{array}$$

The answer is $x^2 - 3x + 5$, R -12, or $x^2 - 3x + 5 + \dfrac{-12}{x + 1}$.

2. $x^2 - x + 1 + \dfrac{-6}{x - 1}$

3. $(a^2 + 8a + 11) \div (a + 3) =$
$(a^2 + 8a + 11) \div [a - (-3)]$

$$\begin{array}{r|rrr}
-3 & 1 & 8 & 11 \\
   &   & -3 & -15 \\
\hline
   & 1 & 5 \,|\, {-4}
\end{array}$$

The answer is $a + 5$, R -4, or $a + 5 + \dfrac{-4}{a + 3}$.

4. $a + 3 + \dfrac{-4}{a + 5}$

5. $(x^3 - 7x^2 - 13x + 3) \div (x - 2)$

$$\begin{array}{r|rrrr}
2 & 1 & -7 & -13 & 3 \\
  &   & 2 & -10 & -46 \\
\hline
  & 1 & -5 & -23 \,|\, {-43}
\end{array}$$

The answer is $x^2 - 5x - 23$, R -43, or
$x^2 - 5x - 23 + \dfrac{-43}{x - 2}$.

6. $x^2 - 9x + 5 + \dfrac{-7}{x + 2}$

7. $(3x^3 + 7x^2 - 4x + 3) \div (x + 3) =$
$(3x^3 + 7x^2 - 4x + 3) \div [x - (-3)]$

$$\begin{array}{r|rrrr}
-3 & 3 & 7 & -4 & 3 \\
   &   & -9 & 6 & -6 \\
\hline
   & 3 & -2 & 2 \,|\, {-3}
\end{array}$$

The answer is $3x^2 - 2x + 2$, R -3, or
$3x^2 - 2x + 2 + \dfrac{-3}{x + 3}$.

8. $3x^2 + 16x + 44 + \dfrac{135}{x - 3}$

9. $(y^3 - 3y + 10) \div (y - 2) =$

$(y^3 + 0y^2 - 3y + 10) \div (y - 2)$

$$\begin{array}{r|rrrr} 2 & 1 & 0 & -3 & 10 \\ & & 2 & 4 & 2 \\ \hline & 1 & 2 & 1 & |\;12 \end{array}$$

The answer is $y^2 + 2y + 1$, R 12, or

$y^2 + 2y + 1 + \dfrac{12}{y - 2}$.

10. $x^2 - 4x + 8 + \dfrac{-8}{x + 2}$

11. $(x^5 - 32) \div (x - 2) =$

$(x^5 + 0x^4 + 0x^3 + 0x^2 + 0x - 32) \div (x - 2)$

$$\begin{array}{r|rrrrrr} 2 & 1 & 0 & 0 & 0 & 0 & -32 \\ & & 2 & 4 & 8 & 16 & 32 \\ \hline & 1 & 2 & 4 & 8 & 16 & |\;0 \end{array}$$

The answer is $x^4 + 2x^3 + 4x^2 + 8x + 16$.

12. $y^4 + y^3 + y^2 + y + 1$

13. $\left(3x^3 + 1 - x + 7x^2\right) \div \left(x + \dfrac{1}{3}\right) =$

$(3x^3 + 7x^2 - x + 1) \div \left[x - \left(-\dfrac{1}{3}\right)\right]$

$$\begin{array}{r|rrrr} -\frac{1}{3} & 3 & 7 & -1 & 1 \\ & & -1 & -2 & 1 \\ \hline & 3 & 6 & -3 & |\;2 \end{array}$$

The answer is $3x^2 + 6x - 3$ R 2, or

$3x^2 + 6x - 3 + \dfrac{2}{x + \dfrac{1}{3}}$.

14. $8x^2 - 2x + 6 + \dfrac{2}{x - \dfrac{1}{2}}$

15. $$\begin{array}{r|rrrrr} -3 & 5 & 12 & 0 & 28 & 9 \\ & & -15 & 9 & -27 & -3 \\ \hline & 5 & -3 & 9 & 1 & |\;6 \end{array}$$

The remainder tells us that $f(-3) = 6$.

16. 0

17. $$\begin{array}{r|rrrrr} -1 & 6 & -1 & -7 & 1 & 2 \\ & & -6 & 7 & -1 & -1 \\ \hline & 6 & -7 & 0 & 0 & |\;1 \end{array}$$

The remainder tells us that $P(-1) = 1$.

18. 2

19. $$\begin{array}{r|rrrrr} 4 & 1 & -1 & -19 & 49 & -30 \\ & & 4 & 12 & -28 & 84 \\ \hline & 1 & 3 & -7 & 21 & |\;54 \end{array}$$

The remainder tells us that $f(4) = 54$.

20. 90

21. *Writing Exercise*

22. *Writing Exercise*

23. $\qquad 9 + cb = a - b$

$\qquad 9 + cb + b = a$

$\qquad cb + b = a - 9$

$\qquad b(c + 1) = a - 9$

$\qquad b = \dfrac{a - 9}{c + 1}$

24. $a = \dfrac{bd - 8}{c - b}$

25. $f(x) = \dfrac{5}{3x^2 - 75}$

$f(x)$ cannot be calculated for any x-value for which the denominator is 0. We solve an equation to find those values.

$$3x^2 - 75 = 0$$
$$3(x^2 - 25) = 0$$
$$3(x + 5)(x - 5) = 0$$
$$x + 5 = 0 \quad or \quad x - 5 = 0$$
$$x = -5 \quad or \qquad x = 5$$

The domain of f is $\{x | x$ is a real number *and* $x \neq -5$ *and* $x \neq 5\}$.

26. $\left\{x | x \text{ is a real number } and \; x \neq -\dfrac{9}{2} \; and \; x \neq 1\right\}$

27. Graph: $y - 2 = \dfrac{3}{4}(x + 1)$

This is the equation of the line with slope $\dfrac{3}{4}$ and passing through $(-1, 2)$. To graph this equation, start at $(-1, 2)$ and count off a slope of $\dfrac{3}{4}$ by going up 3 units and to the right 4 units (or down 3 units and to the left 4 units). Then draw the line.

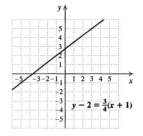

28.

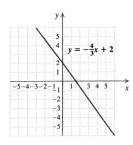

$y = -\frac{4}{3}x + 2$

29. *Writing Exercise*

30. *Writing Exercise*

31. a) The degree of the remainder must be less than the degree of the divisor. Thus, the degree of the remainder must be 0, so R must be a constant.

b) $P(x) = (x - r) \cdot Q(x) + R$
$P(r) = (r - r) \cdot Q(r) + R = 0 \cdot Q(r) + R = R$

32. $0; -3; -\frac{5}{2}; \frac{3}{2}$

33.
$$\begin{array}{r|rrrr} 4 & 6 & -13 & -79 & 140 \\ & & 24 & 44 & -140 \\ \hline & 6 & 11 & -35 & 0 \end{array}$$

The remainder tells us that $f(4) = 0$.

$f(x) = (x - 4)(6x^2 + 11x - 35) = (x - 4)(2x + 7)(3x - 5)$

Solve $f(x) = 0$:

$(x - 4)(2x + 7)(3x - 5) = 0$

$x - 4 = 0$ or $2x + 7 = 0$ or $3x - 5 = 0$

$x = 4$ or $2x = -7$ or $3x = 5$

$x = 4$ or $x = -\frac{7}{2}$ or $x = \frac{5}{3}$

The solutions are 4, $-\frac{7}{2}$, and $\frac{5}{3}$.

34. 0

35. $f(x) = 6x^3 - 13x^2 - 79x + 140$

$\quad = x(6x^2 - 13x - 79) + 140$

$\quad = x(x(6x - 13) - 79) + 140$

$f(4) = 4(4(6 \cdot 4 - 13) - 79) + 140$

$\quad = 4(4(24 - 13) - 79) + 140$

$\quad = 4(4 \cdot 11 - 79) + 140$

$\quad = 4(44 - 79) + 140$

$\quad = 4(-35) + 140$

$\quad = -140 + 140$

$\quad = 0$

Exercise Set 6.8

1. $\dfrac{W_1}{W_2} = \dfrac{d_1}{d_2}$

$\dfrac{d_2 W_1}{W_2} = d_1$ Multiplying by d_2

2. $W_1 = \dfrac{d_1 W_2}{d_2}$

3. $s = \dfrac{(v_1 + v_2)t}{2}$

$2s = (v_1 + v_2)t$ Multiplying by 2

$\dfrac{2s}{t} = v_1 + v_2$ Dividing by t

$\dfrac{2s}{t} - v_2 = v_1$

This result can also be expressed as $v_1 = \dfrac{2s - tv_2}{t}$.

4. $t = \dfrac{2s}{v_1 + v_2}$

5. $\dfrac{1}{f} = \dfrac{1}{d_i} + \dfrac{1}{d_o}$

$f d_i d_o \cdot \dfrac{1}{f} = f d_i d_o \left(\dfrac{1}{d_i} + \dfrac{1}{d_o} \right)$ Multiplying by the LCD

$d_i d_o = f d_i d_o \cdot \dfrac{1}{d_i} + f d_i d_o \cdot \dfrac{1}{d_o}$

$d_i d_o = f d_o + f d_i$

$d_i d_o = f(d_o + d_i)$ Factoring out f

$\dfrac{d_i d_o}{d_o + d_i} = f$ Multiplying by $\dfrac{1}{d_o + d_i}$

6. $R = \dfrac{r_1 r_2}{r_2 + r_1}$

7. $I = \dfrac{2V}{R + 2r}$

$I(R + 2r) = \dfrac{2V}{R + 2r} \cdot (R + 2r)$ Multiplying by the LCD

$I(R + 2r) = 2V$

$R + 2r = \dfrac{2V}{I}$

$R = \dfrac{2V}{I} - 2r$, or $\dfrac{2V - 2Ir}{I}$

8. $r = \dfrac{2V - IR}{2I}$

9.
$$R = \frac{gs}{g+s}$$

$$(g+s) \cdot R = (g+s) \cdot \frac{gs}{g+s} \qquad \text{Multiplying by the LCD}$$

$$Rg + Rs = gs$$

$$Rs = gs - Rg$$

$$Rs = g(s - R) \qquad \text{Factoring out } g$$

$$\frac{Rs}{s - R} = g \qquad \text{Multiplying by } \frac{1}{s - R}$$

10. $t = \dfrac{Kr}{r + K}$

11.
$$I = \frac{nE}{R + nr}$$

$$I(R + nr) = \frac{nE}{R + nr} \cdot (R + nr) \qquad \text{Multiplying by the LCD}$$

$$IR + Inr = nE$$

$$IR = nE - Inr$$

$$IR = n(E - Ir)$$

$$\frac{IR}{E - Ir} = n$$

12. $r = \dfrac{nE - IR}{In}$

13.
$$\frac{1}{p} + \frac{1}{q} = \frac{1}{f}$$

$$pqf\left(\frac{1}{p} + \frac{1}{q}\right) = pqf \cdot \frac{1}{f} \qquad \text{Multiplying by the LCD}$$

$$qf + pf = pq$$

$$pf = pq - qf$$

$$pf = q(p - f)$$

$$\frac{pf}{p - f} = q$$

14. $p = \dfrac{qf}{q - f}$

15.
$$S = \frac{H}{m(t_1 - t_2)}$$

$$(t_1 - t_2)S = \frac{H}{m} \qquad \text{Multiplying by } t_1 - t_2$$

$$t_1 - t_2 = \frac{H}{Sm} \qquad \text{Dividing by } S$$

$$t_1 = \frac{H}{Sm} + t_2, \text{ or } \frac{H + Smt_2}{Sm}$$

16. $H = m(t_1 - t_2)S$

17.
$$\frac{E}{e} = \frac{R + r}{r}$$

$$er \cdot \frac{E}{e} = er \cdot \frac{R + r}{r} \qquad \text{Multiplying by the LCD}$$

$$Er = e(R + r)$$

$$Er = eR + er$$

$$Er - er = eR$$

$$r(E - e) = eR$$

$$r = \frac{eR}{E - e}$$

18. $R = \dfrac{er}{E - e}$

19.
$$S = \frac{a}{1 - r}$$

$$(1 - r)S = a \qquad \text{Multiplying by the LCD, } 1 - r$$

$$1 - r = \frac{a}{S} \qquad \text{Dividing by } S$$

$$1 - \frac{a}{S} = r \qquad \text{Adding } r \text{ and } -\frac{a}{S}$$

This result can also be expressed as $r = \dfrac{S - a}{S}$.

20. $a = \dfrac{S - Sr}{1 - r^n}$

21.
$$c = \frac{f}{(a + b)c}$$

$$\frac{a + b}{c} \cdot c = \frac{a + b}{c} \cdot \frac{f}{(a + b)c}$$

$$a + b = \frac{f}{c^2}$$

22. $c + f = \dfrac{g}{d^2}$

23.
$$I_t = \frac{I_f}{1 - T}$$

$$(1 - T)I_t = I_f$$

$$1 - T = \frac{I_f}{I_t}$$

$$-T = \frac{I_f}{I_t} - 1$$

$$-1 \cdot (-T) = -1 \cdot \left(\frac{I_f}{I_t} - 1\right)$$

$$T = -\frac{I_f}{I_t} + 1, \text{ or } 1 - \frac{I_f}{I_t}, \text{ or } \frac{I_t - I_f}{I_t}$$

24. $r = \dfrac{A}{P} - 1, \text{ or } \dfrac{A - P}{P}$

25.
$$\frac{1}{R} = \frac{1}{r_1} + \frac{1}{r_2}$$

$$Rr_1r_2 \cdot \frac{1}{R} = Rr_1r_2\left(\frac{1}{r_1} + \frac{1}{r_2}\right)$$

$$r_1r_2 = Rr_1r_2 \cdot \frac{1}{r_1} + Rr_1r_2 \cdot \frac{1}{r_2}$$

$$r_1r_2 = Rr_2 + Rr_1$$

$$r_1r_2 - Rr_2 = Rr_1$$

$$r_2(r_1 - R) = Rr_1$$

$$r_2 = \frac{Rr_1}{r_1 - R}$$

26. $t = \dfrac{ab}{b+a}$

27.
$$a = \frac{v_2 - v_1}{t_2 - t_1}$$

$$(t_2 - t_1)a = v_2 - v_1$$

$$t_2 - t_1 = \frac{v_2 - v_1}{a}$$

$$-t_1 = \frac{v_2 - v_1}{a} - t_2$$

$$t_1 = t_2 - \frac{v_2 - v_1}{a}$$

28. $t_2 = \dfrac{d_2 - d_1}{v} + t_1$, or $\dfrac{d_2 - d_1 + t_1 v}{v}$

29.
$$A = \frac{2Tt + Qq}{2T + Q}$$

$$(2T + Q) \cdot A = (2T + Q) \cdot \frac{2Tt + Qq}{2T + Q}$$

$$2AT + AQ = 2Tt + Qq$$

$$AQ - Qq = 2Tt - 2AT \quad \text{Adding } -2AT$$
$$\text{and } -Qq$$

$$Q(A - q) = 2Tt - 2AT$$

$$Q = \frac{2Tt - 2AT}{A - q}$$

30. $D = \dfrac{dR}{L} + d$, or $\dfrac{dR + dL}{L}$

31. $y = kx$

$28 = k \cdot 4$ Substituting

$7 = k$

The variation constant is 7.
The equation of variation is $y = 7x$.

32. $k = \dfrac{5}{12}; y = \dfrac{5}{12}x$

33. $y = kx$

$3.4 = k \cdot 2$ Substituting

$1.7 = k$

The variation constant is 1.7.
The equation of variation is $y = 1.7x$.

34. $k = \dfrac{2}{5}; y = \dfrac{2}{5}x$

35. $y = kx$

$2 = k \cdot \dfrac{1}{3}$ Substituting

$6 = k$ Multiplying by 3

The variation constant is 6.
The equation of variation is $y = 6x$.

36. $k = 1.8; y = 1.8x$

37. *Familiarize.* Because of the phrase "d ... varies directly as ... m," we express the distance as a function of the mass. Thus we have $d(m) = km$. We know that $d(3) = 20$.

Translate. We find the variation constant and then find the equation of variation.

$$d(m) = km$$

$$d(3) = k \cdot 3 \quad \text{Replacing } m \text{ with } 3$$

$$20 = k \cdot 3 \quad \text{Replacing } d(3) \text{ with } 20$$

$$\frac{20}{3} = k \qquad \text{Variation constant}$$

The equation of variation is $d(m) = \dfrac{20}{3}m$.

Carry out. We compute $d(5)$.

$$d(m) = \frac{20}{3}m$$

$$d(5) = \frac{20}{3} \cdot 5 \quad \text{Replacing } m \text{ with } 5$$

$$d(5) = \frac{100}{3}, \text{ or } 33\frac{1}{3}$$

Check. Reexamine the calculations. Note that the answer seems reasonable since $\dfrac{3}{20}$ and $\dfrac{5}{100/3}$ are equal.

State. The spring is stretched $33\dfrac{1}{3}$ cm by a hanging object with mass 5 kg.

38. 6 amperes

39. *Familiarize.* Because N varies directly as the number of people P using the cans, we write N as a function of P: $N(P) = kP$. We know that $N(250) = 60,000$.

Translate.
$$N(P) = kP$$
$$N(250) = k \cdot 250 \qquad \text{Replacing } P \text{ with } 250$$
$$60,000 = k \cdot 250 \qquad \text{Replacing } N(250) \text{ with } 60,000$$
$$\frac{60,000}{250} = k$$
$$240 = k \qquad \text{Variation constant}$$
$$N(P) = 240P \qquad \text{Equation of variation}$$

Carry out. Find $N(1,008,000)$.
$$N(P) = 240P$$
$$N(1,008,000) = 240 \cdot 1,008,000$$
$$= 241,920,000$$

Check. Reexamine the calculation.

State. 241,920,000 aluminum cans are used each year in Dallas.

40. $4.29

41. Since we have direct variation and $48 = \dfrac{1}{2} \cdot 96$, then the result is $\dfrac{1}{2} \cdot 64$ kg, or 32 kg. We could also do this problem as follows.

Familiarize. Because W varies directly as the total mass, we write $W(m) = km$. We know that $W(96) = 64$.

Translate.
$$W(m) = km$$
$$W(96) = k \cdot 96 \qquad \text{Replacing } m \text{ with } 96$$
$$64 = k \cdot 96 \qquad \text{Replacing } W(96) \text{ with } 64$$
$$\frac{2}{3} = k \qquad \text{Variation constant}$$
$$W(m) = \frac{2}{3}m \qquad \text{Equation of variation}$$

Carry out. Find $W(48)$.
$$W(m) = \frac{2}{3}m$$
$$W(48) = \frac{2}{3} \cdot 48$$
$$= 32$$

Check. Reexamine the calculations.

State. There are 32 kg of water in a 64 kg person.

42. 40 lb

43. *Familiarize.* Because the f-stop varies directly as F, we write $f(F) = kF$. We know that $F(150) = 6.3$.

Translate.
$$f(F) = kF$$
$$f(150) = k \cdot 150 \qquad \text{Replacing } F \text{ with } 150$$
$$6.3 = k \cdot 150 \qquad \text{Replacing } f(150) \text{ with } 6.3$$
$$0.042 = k \qquad \text{Variation constant}$$
$$f(F) = 0.042F \qquad \text{Equation of variation}$$

Carry out. Find $f(80)$.
$$f(F) = 0.042F$$
$$f(80) = 0.042(80)$$
$$= 3.36$$

Check. Reexamine the calculations.

State. An 80 mm focal length has an f-stop of 3.36.

44. 7,700,000 tons

45.
$$y = \frac{k}{x}$$
$$3 = \frac{k}{20} \qquad \text{Substituting}$$
$$60 = k$$

The variation constant is 60.

The equation of variation is $y = \dfrac{60}{x}$.

46. $k = 64; y = \dfrac{64}{x}$

47.
$$y = \frac{k}{x}$$
$$28 = \frac{k}{4} \qquad \text{Substituting}$$
$$112 = k$$

The variation constant is 112.

The equation of variation is $y = \dfrac{112}{x}$.

48. $k = 45; y = \dfrac{45}{x}$

49.
$$y = \frac{k}{x}$$
$$27 = \frac{k}{\dfrac{1}{3}} \qquad \text{Substituting}$$
$$9 = k$$

The variation constant is 9.

The equation of variation is $y = \dfrac{9}{x}$.

50. $k = 9; y = \dfrac{9}{x}$

51. *Familiarize*. Because of the phrase "t varies inversely as $\dots u$," we write $t(u) = k/u$. We know that $t(4) = 70$.

Translate. We find the variation constant and then we find the equation of variation.

$$t(u) = \frac{k}{u}$$

$$t(4) = \frac{k}{4} \qquad \text{Replacing } u \text{ with } 4$$

$$70 = \frac{k}{4} \qquad \text{Replacing } t(4) \text{ with } 70$$

$$280 = k \qquad \text{Variation constant}$$

$$t(u) = \frac{280}{u} \qquad \text{Equation of variation}$$

Carry out. We find $t(14)$.

$$t(14) = \frac{280}{14} = 20$$

Check. Reexamine the calculations. Note that, as expected, as the UV rating increases, the time it takes to burn goes down.

State. It will take 20 min to burn when the UV rating is 14.

52. $\frac{2}{9}$ ampere

53. *Familiarize*. Because V varies inversely as P, we write $V(P) = k/P$. We know that $V(32) = 200$.

Translate.

$$V(P) = \frac{k}{P}$$

$$V(32) = \frac{k}{32} \qquad \text{Replacing } P \text{ with } 32$$

$$200 = \frac{k}{32} \qquad \text{Replacing } V(32) \text{ with } 200$$

$$6400 = k \qquad \text{Variation constant}$$

$$V(P) = \frac{6400}{P} \qquad \text{Equation of variation}$$

Carry out. Find $V(40)$.

$$V(40) = \frac{6400}{40}$$
$$= 160$$

Check. Reexamine the calculations.

State. The volume will be 160 cm^3.

54. 27 min

55. *Familiarize*. Because T varies inversely as P, we write $T(p) = k/p$. We know that $T(7) = 5$.

Translate. We find the variation constant and the equation of variation.

$$T(P) = \frac{k}{p}$$

$$T(7) = \frac{k}{7} \qquad \text{Replacing } P \text{ with } 7$$

$$5 = \frac{k}{7} \qquad \text{Replacing } T(P) \text{ with } 5$$

$$35 = k \qquad \text{Variation constant}$$

$$T(P) = \frac{35}{P} \qquad \text{Equation of variation}$$

Carry out. We find $T(10)$.

$$T(10) = \frac{35}{10}$$
$$= 3.5$$

Check. Reexamine the calculations.

State. It would take 3.5 hr for 10 volunteers to complete the job.

56. 450 meters

57. $y = kx^2$

$$6 = k \cdot 3^2 \qquad \text{Substituting}$$

$$6 = 9k$$

$$\frac{6}{9} = k$$

$$\frac{2}{3} = k \qquad \text{Variation constant}$$

The equation of variation is $y = \frac{2}{3}x^2$.

58. $y = 15x^2$

59. $y = \frac{k}{x^2}$

$$6 = \frac{k}{3^2} \qquad \text{Substituting}$$

$$6 = \frac{k}{9}$$

$$6 \cdot 9 = k$$

$$54 = k \qquad \text{Variation constant}$$

The equation of variation is $y = \frac{54}{x^2}$.

60. $y = \frac{0.0015}{x^2}$

61.
$$y = kxz^2$$
$$105 = k \cdot 14 \cdot 5^2 \quad \text{Substituting 105 for } y,$$
$$\text{14 for } x, \text{ and 5 for } z$$
$$105 = 350k$$
$$\frac{105}{350} = k$$
$$0.3 = k$$

The equation of variation is $y = 0.3xz^2$.

62. $y = \dfrac{xz}{w}$

63.
$$y = k \cdot \frac{wx^2}{z}$$
$$49 = k \cdot \frac{3 \cdot 7^2}{12} \quad \text{Substituting}$$
$$4 = k \qquad \text{Variation constant}$$

The equation of variation is $y = \dfrac{4wx^2}{z}$.

64. $y = \dfrac{6x}{wz^2}$

65. Familiarize. I varies inversely as d^2, so we write $I = k/d^2$. We know that $I = 90$ when $d = 5$.

Translate. Find k.
$$I = \frac{k}{d^2}$$
$$90 = \frac{k}{5^2}$$
$$2250 = k$$

$$I = \frac{2250}{d^2} \quad \text{Equation of variation}$$

Carry out. Substitute 7.5 for d and find for I.
$$I = \frac{2250}{(7.5)^2} = \frac{2250}{56.25} = 40$$

Check. Reexamine the calculations.

State. The intensity is 40 W/m^2 at a distance of 7.5 m from the bulb.

66. 72 ft

67. Familiarize. Because V varies directly as T and inversely as P, we write $V = kT/P$. We know that $V = 231$ when $T = 42$ and $P = 20$.

Translate. Find k and the equation of variation.
$$V = \frac{kT}{P}$$
$$231 = \frac{k \cdot 42}{20}$$
$$\frac{20}{42} \cdot 231 = k$$
$$110 = k$$

$$V = \frac{110T}{P} \quad \text{Equation of variation}$$

Carry out. Substitute 30 for T and 15 for P and find V.
$$V = \frac{110 \cdot 30}{15} = 220$$

Check. Reexamine the calculations.

State. The volume is 220 cm^3 when $T = 30°$ and $P = 15$ kg/cm^2.

68. 2.56 W/m^2

69. Familiarize. The drag W varies jointly as the surface area A and velocity v, so we write $W = kAv$. We know that $W = 222$ when $A = 37.8$ and $v = 40$.

Translate. Find k.
$$W = kAv$$
$$222 = k(37.8)(40)$$
$$\frac{222}{37.8(40)} = k$$
$$\frac{37}{252} = k$$

$$W = \frac{37}{252}Av \qquad \text{Equation of variation}$$

Carry out. Substitute 51 for A and 430 for W and solve for v.
$$430 = \frac{37}{252} \cdot 51 \cdot v$$
$$57.42 \text{ mph} \approx v$$

(If we had used the rounded value 0.1468 for k, the resulting speed would have been approximately 57.43 mph.)

Check. Reexamine the calculations.

State. The car must travel about 57.42 mph.

70. About 28.3 ft^2

71. a) We graph the data, letting x represent the UV index and y represent the safe exposure time, in minutes.

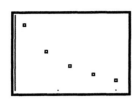

The points lie approximately on the graph of a function of the type $f(x) = \dfrac{k}{x}$, so it appears that the safe exposure time varies inversely as the UV index.

b) $y = \dfrac{k}{x}$

$50 = \dfrac{k}{6}$ Substituting 6 for x and 50 for y

$300 = k$ Variation constant

The equation of variation is $y = \dfrac{300}{x}$.

c) We substitute 3 for x in the equation of variation.

$y = \dfrac{300}{x}$

$y = \dfrac{300}{3}$

$y = 100$

When the UV index is 3, the safe exposure time for people with less sensitive skin is 100 min.

72. a) Inverse

b) $y = \dfrac{20}{x}$

c) 10 ft

73. a) We graph the data, letting x represent the number of persons ordering merchandise by mail, in millions, and y represent the number of persons ordering by phone, in millions.

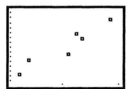

The points lie approximately on the graph of a function of the form $f(x) = kx$, so it appears that the number of people ordering by mail varies directly as the number of people ordering by phone.

b) $y = kx$

$12.775 = 11.813x$ Substituting 11.813 for x and 12.775 for y

$1.08 \approx x$ Variation constant

The equation of variation is $y = 1.08x$.

c) Substitute 8 for x in the equation of variation.

$y = 1.08x$

$y = 1.08(8)$

$y = 8.64$

If 8 million people order by mail, then 8.64 million people will order by phone.

74. a) Directly

b) $y = 1.66x$

c) 0.747 million automobiles

75. *Writing Exercise*

76. *Writing Exercise*

77. $f(x) = \dfrac{2x - 1}{x^2 + 1}$

$x^2 + 1 > 0$ for all real numbers x, so the domain of f is $\{x | x \text{ is a real number}\}$.

78. $\{x | x \text{ is a real number}\}$

79. Graph: $6x - y < 6$

First graph the line $6x - y = 6$. The intercepts are $(0, -6)$ and $(1, 0)$. We draw the line dashed since the inequality is $<$. Since the ordered pair $(0, 0)$ is a solution of the inequality ($6 \cdot 0 - 0 < 6$ is true), we shade the half-plane containing $(0, 0)$.

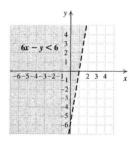

80. $8a^3 - 2a$

81. $t^3 + 8b^3 = t^3 + (2b)^3 = (t + 2b)(t^2 - 2tb + 4b^2)$

82. $-\dfrac{5}{3}, \dfrac{7}{2}$

83. *Writing Exercise*

84. *Writing Exercise*

85. Use the result of Example 2.

$$h = \frac{2R^2g}{V^2} - R$$

We have $V = 6.5$ mi/sec, $R = 3960$ mi, and $g = 32.2$ ft/sec^2. We must convert 32.2 ft/sec^2 to mi/sec^2 so all units of length are the same.

$$32.2 \frac{\cancel{ft}}{\sec^2} \cdot \frac{1 \text{ mi}}{5280 \cancel{ft}} \approx 0.0060984 \frac{\text{mi}}{\sec^2}$$

Now we substitute and compute.

$$h = \frac{2(3960)^2(0.0060984)}{(6.5)^2} - 3960$$

$$h \approx 567$$

The satellite is about 567 mi from the surface of the earth.

86. $M = \dfrac{2ab}{b+a}$

87. $c = \dfrac{a}{a+12} \cdot d$

$$c = \frac{2a}{2a+12} \cdot d \quad \text{Doubling } a$$

$$= \frac{\cancel{2}a}{\cancel{2}(a+6)} \cdot d$$

$$= \frac{a}{a+6} \cdot d \qquad \text{Simplifying}$$

The ratio of the larger dose to the smaller dose is

$$\frac{\dfrac{a}{a+6} \cdot d}{\dfrac{a}{a+12} \cdot d} = \frac{\dfrac{ad}{a+6}}{\dfrac{ad}{a+12}}$$

$$= \frac{ad}{a+6} \cdot \frac{a+12}{ad}$$

$$= \frac{\cancel{ad}\,(a+12)}{(a+6)\cancel{ad}}$$

$$= \frac{a+12}{a+6}.$$

The amount by which the dosage increases is

$$\frac{a}{a+6} \cdot d - \frac{a}{a+12} \cdot d$$

$$\frac{ad}{a+6} - \frac{ad}{a+12}$$

$$= \frac{ad}{a+6} \cdot \frac{a+12}{a+12} - \frac{ad}{a+12} \cdot \frac{a+6}{a+6}$$

$$= \frac{ad(a+12) - ad(a+6)}{(a+6)(a+12)}$$

$$= \frac{a^2d + 12ad - a^2d - 6ad}{(a+6)(a+12)}$$

$$= \frac{6ad}{(a+6)(a+12)}.$$

Then the percent by which the dosage increases is

$$\frac{\dfrac{6ad}{(a+6)(a+12)}}{\dfrac{a}{a+12} \cdot d} = \frac{\dfrac{6ad}{(a+6)(a+12)}}{\dfrac{ad}{a+12}}$$

$$= \frac{6ad}{(a+6)(a+12)} \cdot \frac{a+12}{ad}$$

$$= \frac{6 \cdot a\cancel{d} \cdot (a\cancel{+12})}{(a+6)(a\cancel{+12}) \cdot a\cancel{d}}$$

$$= \frac{6}{a+6}.$$

This is a decimal representation for the percent of increase. To give the result in percent notation we multiply by 100 and use a percent symbol. We have

$$\frac{6}{a+6} \cdot 100\%, \text{ or } \frac{600}{a+6}\%.$$

88. $x = pq \text{ or } x = 2pq$

89.

$$a = \frac{\dfrac{d_4 - d_3}{t_4 - t_3} - \dfrac{d_2 - d_1}{t_2 - t_1}}{t_4 - t_2}$$

$$a(t_4 - t_2) = \frac{d_4 - d_3}{t_4 - t_3} - \frac{d_2 - d_1}{t_2 - t_1} \quad \begin{array}{l}\text{Multiplying}\\ \text{by } t_4 - t_2\end{array}$$

$$a(t_4{-}t_2)(t_4{-}t_3)(t_2{-}t_1) = (d_4{-}d_3)(t_2{-}t_1){-}(d_2{-}d_1)(t_4{-}t_3)$$

$$\text{Multiplying by } (t_4 - t_3)(t_2 - t_1)$$

$$a(t_4 - t_2)(t_4 - t_3)(t_2 - t_1) - (d_4 - d_3)(t_2 - t_1) = $$
$$-(d_2 - d_1)(t_4 - t_3)$$

$$(t_2 - t_1)[a(t_4 - t_2)(t_4 - t_3) - (d_4 - d_3)] = $$
$$-(d_2 - d_1)(t_4 - t_3)$$

$$t_2 - t_1 = \frac{-(d_2 - d_1)(t_4 - t_3)}{a(t_4 - t_2)(t_4 - t_3) - (d_4 - d_3)}$$

$$t_2 + \frac{(d_2 - d_1)(t_4 - t_3)}{a(t_4 - t_2)(t_4 - t_3) + d_3 - d_4} = t_1$$

90. y is multiplied by 8.

91. $Q = \dfrac{kp^2}{q^3}$

Q varies directly as the square of p and inversely as the cube of q.

92. W varies jointly as m_1 and M_1 and inversely as the square of d.

93. **Familiarize.** We write $T = kml^2 f^2$. We know that $T = 100$ when $m = 5$, $l = 2$, and $f = 80$.

Translate. Find k.

$$T = kml^2 f^2$$

$$100 = k(5)(2)^2(80)^2$$

$$0.00078125 = k$$

$$T = 0.00078125ml^2 f^2$$

Carry out. Substitute 72 for T, 5 for m, and 80 for f and solve for l.

$$72 = 0.00078125(5)(l^2)(80)^2$$

$$2.88 = l^2$$

$$1.697 \approx l$$

Check. Recheck the calculations.

State. The string should be about 1.697 m long.

94. $7.20

95. ***Familiarize***. Because d varies inversely as s, we write $d(s) = k/s$. We know that $d(0.56) = 50$.

Translate.

$$d(s) = \frac{k}{s}$$

$$d(0.56) = \frac{k}{0.56} \quad \text{Replacing } s \text{ with } 0.56$$

$$50 = \frac{k}{0.56} \quad \text{Replacing } d(0.56) \text{ with } 50$$

$$28 = k$$

$$d(s) = \frac{28}{s} \quad \text{Equation of variation}$$

Carry out. Find $d(0.40)$.

$$d(0.40) = \frac{28}{0.40}$$

$$= 70$$

Check. Reexamine the calculations. Also observe that, as expected, when d decreases, then s increases.

State. The equation of variation is $d(s) = \dfrac{28}{s}$. The distance is 70 yd.

Chapter 7

Exponents and Radical Functions

Exercise Set 7.1

1. The square roots of 16 are 4 and -4, because $4^2 = 16$ and $(-4)^2 = 16$.

2. $7, -7$

3. The square roots of 144 are 12 and -12, because $12^2 = 144$ and $(-12)^2 = 144$.

4. $3, -3$

5. The square roots of 81 are 9 and -9, because $9^2 = 81$ and $(-9)^2 = 81$.

6. $20, -20$

7. The square roots of 900 are 30 and -30, because $30^2 = 900$ and $(-30)^2 = 900$.

8. $15, -15$

9. Using a calculator we find that the square roots of 7 are approximately 2.6458 and -2.6458.

10. $3.8730, -3.8730$

11. Using a calculator we find that the square roots of 23.7 are approximately 4.8683 and -4.8683.

12. $0.2236, -0.2236$

13. Using a calculator we find that the square roots of $\dfrac{3}{4}$ are approximately 0.8660 and -0.8660.

14. $0.5590, -0.5590$

15. $-\sqrt{\dfrac{49}{36}} = -\dfrac{7}{6}$ Since $\sqrt{\dfrac{49}{36}} = \dfrac{7}{6}$, $-\sqrt{\dfrac{49}{36}} = -\dfrac{7}{6}$.

16. $-\dfrac{19}{3}$

17. $\sqrt{441} = 21$ Remember, $\sqrt{}$ indicates the principle square root.

18. 14

19. $-\sqrt{\dfrac{16}{81}} = -\dfrac{4}{9}$ Since $\sqrt{\dfrac{16}{81}} = \dfrac{4}{9}$, $-\sqrt{\dfrac{16}{81}} = -\dfrac{4}{9}$.

20. $-\dfrac{3}{4}$

21. $\sqrt{0.09} = 0.3$

22. 0.6

23. $-\sqrt{0.0049} = -0.07$

24. 0.12

25. $5\sqrt{p^2 + 4}$

The radicand is the expression written under the radical sign, $p^2 + 4$.

Since the index is not written, we know it is 2.

26. $y^2 - 8$; 2

27. $x^2 y^2 \sqrt[3]{\dfrac{x}{y+4}}$

The radicand is the expression written under the radical sign, $\dfrac{x}{y+4}$.

The index is 3.

28. $\dfrac{a}{a^2 - b}$; 3

29. $f(t) = \sqrt{5t - 10}$

$f(6) = \sqrt{5 \cdot 6 - 10} = \sqrt{20}$

$f(2) = \sqrt{5 \cdot 2 - 10} = \sqrt{0} = 0$

$f(1) = \sqrt{5 \cdot 1 - 10} = \sqrt{-5}$

Since negative numbers do not have real-number square roots, $f(1)$ does not exist.

$f(-1) = \sqrt{5(-1) - 10} = \sqrt{-15}$

Since negative numbers do not have real-number square roots, $f(-1)$ does not exist.

30. $\sqrt{11}$; does not exist; $\sqrt{11}$; 12

31. $t(x) = -\sqrt{2x + 1}$

$t(4) = -\sqrt{2 \cdot 4 + 1} = -\sqrt{9} = -3$

$t(0) = -\sqrt{2 \cdot 0 + 1} = -\sqrt{1} = -1$

$t(-1) = -\sqrt{2(-1) + 1} = -\sqrt{-1}$;

$t(-1)$ does not exist.

$t\left(-\dfrac{1}{2}\right) = -\sqrt{2\left(-\dfrac{1}{2}\right) + 1} = -\sqrt{0} = 0$

32. $\sqrt{12}$; does not exist; $\sqrt{30}$; does not exist

33.
$$f(t) = \sqrt{t^2 + 1}$$
$$f(0) = \sqrt{0^2 + 1} = \sqrt{1} = 1$$
$$f(-1) = \sqrt{(-1)^2 + 1} = \sqrt{2}$$
$$f(-10) = \sqrt{(-10)^2 + 1} = \sqrt{101}$$

34. $-2; -5; -4$

35.
$$g(x) = \sqrt{x^3 + 9}$$
$$g(-2) = \sqrt{(-2)^3 + 9} = \sqrt{1} = 1$$
$$g(-3) = \sqrt{(-3)^3 + 9} = \sqrt{-18};$$
$$g(-3) \text{ does not exist.}$$
$$g(3) = \sqrt{3^3 + 9} = \sqrt{36} = 6$$

36. Does not exist; $\sqrt{17}$; $\sqrt{54}$

37. $\sqrt{36x^2} = \sqrt{(6x)^2} = |6x| = 6|x|$

Since x might be negative, absolute-value notation is necessary.

38. $5|t|$

39. $\sqrt{(-6b)^2} = |-6b| = |-6| \cdot |b| = 6|b|$

Since b might be negative, absolute-value notation is necessary.

40. $7|c|$

41. $\sqrt{(7-t)^2} = |7-t|$

Since $7 - t$ might be negative, absolute-value notation is necessary.

42. $|a + 1|$

43. $\sqrt{y^2 + 16y + 64} = \sqrt{(y+8)^2} = |y+8|$

Since $y + 8$ might be negative, absolute-value notation is necessary.

44. $|x - 2|$

45. $\sqrt{9x^2 - 30x + 25} = \sqrt{(3x-5)^2} = |3x-5|$

Since $3x - 5$ might be negative, absolute-value notation is necessary.

46. $|2x + 7|$

47. $-\sqrt[4]{625} = -5$ Since $5^4 = 625$

48. 4

49. $-\sqrt[5]{3^5} = -3$

50. -1

51. $\sqrt[5]{-\dfrac{1}{32}} = -\dfrac{1}{2}$ Since $\left(-\dfrac{1}{2}\right)^5 = -\dfrac{1}{32}$

52. $-\dfrac{2}{3}$

53. $\sqrt[8]{y^8} = |y|$

The index is even. Use absolute-value notation since y could have a negative value.

54. $|x|$

55. $\sqrt[4]{(7b)^4} = |7b| = 7|b|$

The index is even. Use absolute-value notation since b could have a negative value.

56. $5|a|$

57. $\sqrt[12]{(-10)^{12}} = |-10| = 10$

58. 6

59. $\sqrt[1976]{(2a + b)^{1976}} = |2a + b|$

The index is even. Use absolute-value notation since $2a+b$ could have a negative value.

60. $|a + b|$

61. $\sqrt{x^{10}} = |x^5|$ Note that $(x^5)^2 = x^{10}$; x^5 could have a negative value.

62. $|a^{11}|$

63. $\sqrt{a^{14}} = |a^7|$ Note that $(a^7)^2 = a^{14}$; a^7 could have a negative value.

64. x^8

65. $\sqrt{25t^2} = \sqrt{(5t)^2} = 5t$ Assuming t is nonnegative

66. $4x$

67. $\sqrt{(7c)^2} = 7c$ Assuming c is nonnegative

68. $6b$

69. $\sqrt{(5+b)^2} = 5 + b$ Assuming $5 + b$ is nonnegative

70. $a + 1$

71. $\sqrt{9x^2 + 36x + 36} = \sqrt{9(x^2 + 4x + 4)} = \sqrt{[3(x+2)]^2} = 3(x + 2)$, or $3x + 6$

72. $2(x + 1)$, or $2x + 2$

73. $\sqrt{25t^2 - 20t + 4} = \sqrt{(5t - 2)^2} = 5t - 2$

74. $3t - 2$

75. $-\sqrt[3]{64} = -4 \qquad (4^3 = 64)$

76. 3

77. $\sqrt[4]{81x^4} = \sqrt[4]{(3x)^4} = 3x$

78. $2x$

79. $-\sqrt[5]{-100,000} = -(-10) = 10 \quad [(-10)^5 = -100,000]$

80. -6

81. $-\sqrt[3]{-64x^3} = -(-4x) \qquad [(-4x)^3 = -64x^3]$
$$= 4x$$

82. $5y$

83. $\sqrt{a^{14}} = \sqrt{(a^7)^2} = a^7$

84. a^{11}

85. $\sqrt{(x+3)^{10}} = \sqrt{[(x+3)^5]^2} = (x+3)^5$

86. $(x-2)^4$

87. $\quad f(x) = \sqrt[3]{x+1}$
$$f(7) = \sqrt[3]{7+1} = \sqrt[3]{8} = 2$$
$$f(26) = \sqrt[3]{26+1} = \sqrt[3]{27} = 3$$
$$f(-9) = \sqrt[3]{-9+1} = \sqrt[3]{-8} = -2$$
$$f(-65) = \sqrt[3]{-65+1} = \sqrt[3]{-64} = -4$$

88. $1; 5; 3; -5$

89. $\quad g(t) = \sqrt[4]{t-3}$
$$g(19) = \sqrt[4]{19-3} = \sqrt[4]{16} = 2$$
$$g(-13) = \sqrt[4]{-13-3} = \sqrt[4]{-16};$$
$$\qquad\qquad g(-13) \text{ does not exist.}$$
$$g(1) = \sqrt[4]{1-3} = \sqrt[4]{-2};$$
$$\qquad\qquad g(1) \text{ does not exist.}$$
$$g(84) = \sqrt[4]{84-3} = \sqrt[4]{81} = 3$$

90. $1; 2;$ does not exist; 3

91. $f(x) = \sqrt{x-5}$

Since the index is even, the radicand, $x-5$, must be non-negative. We solve the inequality:
$$x - 5 \geq 0$$
$$x \geq 5$$
Domain of $f = \{x | x \geq 5\}$, or $[5, \infty)$

92. $\{x | x \geq -8\}$, or $[-8, \infty)$

93. $g(t) = \sqrt[4]{t+3}$

Since the index is even, the radicand, $t + 3$, must be non-negative. We solve the inequality:
$$t + 3 \geq 0$$
$$t \geq -3$$
Domain of $g = \{t | t \geq -3\}$, or $[-3, \infty)$

94. $\{x | x \geq 7\}$, or $[7, \infty)$

95. $g(x) = \sqrt[4]{5-x}$

Since the index is even, the radicand, $5 - x$, must be non-negative. We solve the inequality:
$$5 - x \geq 0$$
$$5 \geq x$$
Domain of $g = \{x | x \leq 5\}$, or $(-\infty, 5]$

96. $\{t | t$ is a real number$\}$, or $(-\infty, \infty)$

97. $f(t) = \sqrt[5]{2t+9}$

Since the index is odd, the radicand can be any real number.

Domain of $f = \{t | t$ is a real number$\}$, or $(-\infty, \infty)$

98. $\left\{ t | t \geq -\dfrac{5}{2} \right\}$, or $\left[-\dfrac{5}{2}, \infty \right)$

99. $h(z) = -\sqrt[6]{5z+3}$

Since the index is even, the radicand, $5z + 3$, must be nonnegative. We solve the inequality:
$$5z + 3 \geq 0$$
$$5z \geq -3$$
$$z \geq -\frac{3}{5}$$
Domain of $h = \left\{ z | z \geq -\dfrac{3}{5} \right\}$, or $\left[-\dfrac{3}{5}, \infty \right)$

100. $\left\{ x | x \geq \dfrac{5}{7} \right\}$, or $\left[\dfrac{5}{7}, \infty \right)$

101. $f(t) = 7 + \sqrt[8]{t^8}$

Since we can compute $7 + \sqrt[8]{t^8}$ for any real number t, the domain is the set of real numbers, or $\{x | x$ is a real number$\}$, or $(-\infty, \infty)$.

102. $\{x | x$ is a real number$\}$, or $(-\infty, \infty)$.

103. $f(x) = \sqrt{5-x}$

Find all values of x for which the radicand is nonnegative.
$$5 - x \geq 0$$
$$5 \geq x$$
The domain is $\{x | x \leq 5\}$, or $(-\infty, 5]$.

We graph the function in the standard window.

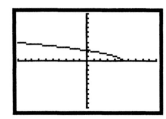

The range appears to be $\{y|y \geq 0\}$, or $[0, \infty)$.

04. Domain: $\left\{x\middle|x \geq -\frac{1}{2}\right\}$, or $\left[-\frac{1}{2}, \infty\right)$;

range: $\{y|y \geq 0\}$, or $[0, \infty)$

05. $f(x) = 1 - \sqrt{x+1}$

Find all values of x for which the radicand is nonnegative.

$$x + 1 \geq 0$$
$$x \geq -1$$

The domain is $\{x|x \geq -1\}$, or $[-1, \infty)$.

We graph the function in the window $[-10, 10, -5, 5]$.

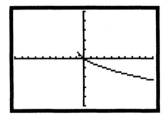

The range appears to be $\{y|y \leq 1\}$, or $(-\infty, 1]$.

06. Domain: $\left\{x\middle|x \geq \frac{5}{3}\right\}$, or $\left[\frac{5}{3}, \infty\right)$;

range: $\{y|y \geq 2\}$, or $[2, \infty)$

07. $g(x) = 3 + \sqrt{x^2 + 4}$

Since $x^2 + 4$ is positive for all values of x, the domain is $\{x|x \text{ is a real number}\}$, or $(-\infty, \infty)$.

We graph the function in the standard window.

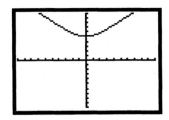

The range appears to be $\{y|y \geq 5\}$, or $[5, \infty)$.

08. Domain: $\{x|x \text{ is a real number}\}$, or $(-\infty, \infty)$;

range: $\{y|y \leq 4\}$, or $(-\infty, 4]$

109. For $f(x) = \sqrt{x-4}$, the domain is $[4, \infty)$ and all of the function values are nonnegative. Graph (c) corresponds to this function.

110. (a)

111. For $h(x) = \sqrt{x^2 + 4}$, the domain is $(-\infty, \infty)$. Graph (d) corresponds to this function.

112. (b)

113. A scatterplot of the data shows that it could be modeled with a radical function.

114. Yes

115. A scatterplot of the data shows that it could be modeled well with a radical function.

116. No

117. $h(x) = 10.681 + \sqrt{177.971284x - 1744.994255}$

In 1992, $x = 1992 - 1970 = 22$.

$$h(22) = 10.681 + \sqrt{177.971284(22) - 1744.994255}$$
$$\approx 57.3$$

In 1992, about 57.3 million households were served by cable television.

In 2001, $x = 2001 - 1970 = 31$.

$$h(31) = 10.681 + \sqrt{177.971284(31) - 1744.994255}$$
$$\approx 72.1$$

In 2001, about 72.1 million households were served by cable television.

118. 3.4 sec

119. *Writing Exercise*

120. *Writing Exercise*

121. $(a^3b^2c^5)^3 = a^{3\cdot3}b^{2\cdot3}c^{5\cdot3} = a^9b^6c^{15}$

122. $10a^{10}b^9$

123. $(2a^{-2}b^3c^{-4})^{-3} = 2^{-3}a^{-2(-3)}b^{3(-3)}c^{-4(-3)} = \frac{1}{2^3}a^6b^{-9}c^{12} = \frac{a^6c^{12}}{8b^9}$

124. $\frac{x^6y^2}{25z^4}$

125. $\frac{8x^{-2}y^5}{4x^{-6}z^{-2}} = \frac{8}{4}x^{-2-(-6)}y^5z^2 = 2x^4y^5z^2$

126. $\frac{5c^3}{a^4b^7}$

127. First find the slope of the function.
$$m = \frac{5419 - 3282}{16 - 0} = \frac{2137}{16} = 133.5625$$
The y-intercept is $(0, 3282)$, so the function is $n(t) = 133.5625t + 3282$, where t is the number of years after 1980.

128. $f(x) = 134.4737456x + 3212.856476$

129. *Writing Exercise*

130. *Writing Exercise*

131. *Writing Exercise*

132. *Writing Exercise*

133. $N = 2.5\sqrt{A}$
a) $N = 2.5\sqrt{25} = 2.5(5) = 12.5 \approx 13$
b) $N = 2.5\sqrt{36} = 2.5(6) = 15$
c) $N = 2.5\sqrt{49} = 2.5(7) = 17.5 \approx 18$
d) $N = 2.5\sqrt{64} = 2.5(8) = 20$

134. $\{x| -3 \le x < 2\}$, or $[-3, 2)$

135. $g(x) = \dfrac{\sqrt[4]{5 - x}}{\sqrt[6]{x + 4}}$

The radical expression in the numerator has an even index, so the radicand, $5 - x$, must be nonnegative. We solve the inequality:
$$5 - x \ge 0$$
$$5 \ge x$$
The radical expression in the denominator also has an even index, so the radicand, $x + 4$, must be nonnegative in order for $\sqrt[6]{x + 4}$ to exist. In addition, the denominator cannot be zero, so the radicand must be positive. We solve the inequality:
$$x + 4 > 0$$
$$x > -4$$
We have $x \le 5$ *and* $x > -4$ so
Domain of $g = \{x| -4 < x \le 5\}$, or $(-4, 5]$.

Exercise Set 7.2

1. $x^{1/4} = \sqrt[4]{x}$

2. $\sqrt[5]{y}$

3. $(16)^{1/2} = \sqrt{16} = 4$

4. 2

5. $81^{1/4} = \sqrt[4]{81} = 3$

6. 2

7. $9^{1/2} = \sqrt{9} = 3$

8. 5

9. $(xyz)^{1/3} = \sqrt[3]{xyz}$

10. $\sqrt[4]{ab}$

11. $(a^2b^2)^{1/5} = \sqrt[5]{a^2b^2}$

12. $\sqrt[4]{x^3y^3}$

13. $a^{2/3} = \sqrt[3]{a^2}$

14. $\sqrt{b^3}$

15. $16^{3/4} = \sqrt[4]{16^3} = (\sqrt[4]{16})^3 = 2^3 = 8$

16. 128

17. $49^{3/2} = \sqrt{49^3} = (\sqrt{49})^3 = 7^3 = 343$

18. 81

19. $9^{5/2} = \sqrt{9^5} = (\sqrt{9})^5 = 3^5 = 243$

20. 729

21. $(81x)^{3/4} = \sqrt[4]{(81x)^3} = \sqrt[4]{81^3x^3}$, or $\sqrt[4]{81^3} \cdot \sqrt[4]{x^3} = (\sqrt[4]{81})^3 \cdot \sqrt[4]{x^3} = 3^3\sqrt[4]{x^3} = 27\sqrt[4]{x^3}$

22. $25\sqrt[3]{a^2}$

23. $(25x^4)^{3/2} = \sqrt{(25x^4)^3} = \sqrt{25^3 \cdot x^{12}} = \sqrt{25^3} \cdot \sqrt{x^{12}} = (\sqrt{25})^3x^6 = 5^3x^6 = 125x^6$

24. $27y^9$

25. $\sqrt[3]{20} = 20^{1/3}$

26. $19^{1/3}$

27. $\sqrt{17} = 17^{1/2}$

28. $6^{1/2}$

29. $\sqrt{x^3} = x^{3/2}$

30. $a^{5/2}$

31. $\sqrt[5]{m^2} = m^{2/5}$

32. $n^{4/5}$

33. $\sqrt[4]{cd} = (cd)^{1/4}$ Parentheses are required.

34. $(xy)^{1/5}$

35. $\sqrt[5]{xy^2z} = (xy^2z)^{1/5}$

36. $(x^3 y^2 z^2)^{1/7}$

37. $(\sqrt{3mn})^3 = (3mn)^{3/2}$

38. $(7xy)^{4/3}$

39. $(\sqrt[7]{8x^2 y})^5 = (8x^2 y)^{5/7}$

40. $(2a^5 b)^{7/6}$

41. $\dfrac{2x}{\sqrt[3]{z^2}} = \dfrac{2x}{z^{2/3}}$

42. $\dfrac{3a}{c^{2/5}}$

43. $x^{-1/3} = \dfrac{1}{x^{1/3}}$

44. $\dfrac{1}{y^{1/4}}$

45. $(2rs)^{-3/4} = \dfrac{1}{(2rs)^{3/4}}$

46. $\dfrac{1}{(5xy)^{5/6}}$

47. $\left(\dfrac{1}{8}\right)^{-2/3} = \left(\dfrac{8}{1}\right)^{2/3} = (2^3)^{2/3} = 2^{\frac{3}{1}\cdot\frac{2}{3}} = 2^2 = 4$

48. 8

49. $\dfrac{1}{a^{-5/7}} = a^{5/7}$

50. $a^{3/5}$

51. $2a^{3/4} b^{-1/2} c^{2/3} = 2 \cdot a^{3/4} \cdot \dfrac{1}{b^{1/2}} \cdot c^{2/3} = \dfrac{2a^{3/4} c^{2/3}}{b^{1/2}}$

52. $\dfrac{5y^{4/5} z}{x^{2/3}}$

53. $2^{-1/3} x^4 y^{-2/7} = \dfrac{1}{2^{1/3}} \cdot x^4 \cdot \dfrac{1}{y^{2/7}} = \dfrac{x^4}{2^{1/3} y^{2/7}}$

54. $\dfrac{a^3}{3^{5/2} b^{7/3}}$

55. $\left(\dfrac{7x}{8yx}\right)^{-3/5} = \left(\dfrac{8yz}{7x}\right)^{3/5}$　　Finding the reciprocal
of the base and changing
the sign of the exponent

56. $\left(\dfrac{3c}{2ab}\right)^{5/6}$

57. $\dfrac{7x}{\sqrt[3]{z}} = \dfrac{7x}{z^{1/3}}$

58. $\dfrac{6a}{b^{1/4}}$

59. $\dfrac{5a}{3c^{-1/2}} = \dfrac{5a}{3} \cdot c^{1/2} = \dfrac{5ac^{1/2}}{3}$

60. $\dfrac{2x^{1/3} z}{5}$

61. $f(x) = \sqrt[4]{x+7} = (x+7)^{1/4}$

Enter $y = (x+7)\wedge(1/4)$, or $y = (x+7)\wedge 0.25$.

Since the index is even, the domain of the function is the set of all x for which the radicand is nonnegative, or $[-7, \infty)$. One good choice of a viewing window is $[-10, 25, -1, 5]$, Xscl $= 5$.

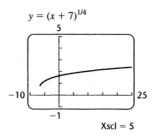

62.

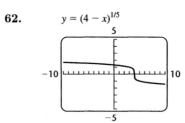

63. $r(x) = \sqrt[7]{3x-2} = (3x-2)^{1/7}$

Enter $y = (3x-2)\wedge(1/7)$. Since the index is odd the domain of the function is $(-\infty, \infty)$. One good choice of a viewing window is $[-10, 10, -5, 5]$.

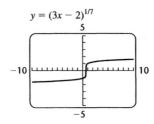

64.

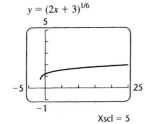

$y = (2x + 3)^{1/6}$

Xscl = 5

65. $f(x) = \sqrt[6]{x^3} = (x^3)^{1/6} = x^{3/6}$

Enter $y = x \wedge (3/6)$. The function is defined only for nonnegative value of x, so the domain is $[0, \infty)$. One good choice of a window is $[-5, 25, -1, 5]$, Xscl = 5.

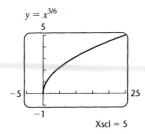

$y = x^{3/6}$

Xscl = 5

66.

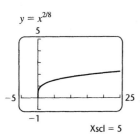

$y = x^{2/8}$

Xscl = 5

67. $\sqrt[5]{9} = 9^{1/5} = 9 \wedge (1/5) \approx 1.552$

68. 1.533

69. $\sqrt[4]{10} = 10^{1/4} = 10 \wedge (1/4) \approx 1.778$

70. -1.998

71. $\sqrt[3]{(-3)^5} = (-3)^{5/3} = (-3) \wedge (5/3) \approx -6.240$

72. 1.275

73. $5^{3/4} \cdot 5^{1/8} = 5^{3/4+1/8} = 5^{6/8+1/8} = 5^{7/8}$

We added exponents after finding a common denominator.

74. $11^{7/6}$

75. $\dfrac{3^{5/8}}{3^{-1/8}} = 3^{5/8-(-1/8)} = 3^{5/8+1/8} = 3^{6/8} = 3^{3/4}$

We subtracted exponents and simplified.

76. $8^{9/11}$

77. $\dfrac{4.1^{-1/6}}{4.1^{-2/3}} = 4.1^{-1/6-(-2/3)} = 4.1^{-1/6+2/3} =$
$4.1^{-1/6+4/6} = 4.1^{3/6} = 4.1^{1/2}$

We subtracted exponents after finding a common denominator. Then we simplified.

78. $\dfrac{1}{2.3^{1/10}}$

79. $(10^{3/5})^{2/5} = 10^{3/5 \cdot 2/5} = 10^{6/25}$

We multiplied exponents.

80. $5^{15/28}$

81. $a^{2/3} \cdot a^{5/4} = a^{2/3+5/4} = a^{8/12+15/12} = a^{23/12}$

We added exponents after finding a common denominator.

82. $x^{17/12}$

83. $(64^{3/4})^{4/3} = 64^{\frac{3}{4} \cdot \frac{4}{3}} = 64^1 = 64$

84. $\dfrac{1}{27}$

85. $(m^{2/3}n^{-1/4})^{1/2} = m^{2/3 \cdot 1/2}n^{-1/4 \cdot 1/2} = m^{1/3}n^{-1/8} =$
$m^{1/3} \cdot \dfrac{1}{n^{1/8}} = \dfrac{m^{1/3}}{n^{1/8}}$

86. $\dfrac{y^{1/10}}{x^{1/12}}$

87. $\sqrt[6]{a^2} = a^{2/6}$ Converting to exponential notation
       $= a^{1/3}$ Simplifying the exponent
       $= \sqrt[3]{a}$ Returning to radical notation

88. $\sqrt[3]{t^2}$

89. $\sqrt[3]{x^{15}} = x^{15/3}$ Converting to exponential notation
       $= x^5$ Simplifying

90. a^3

91. $\sqrt[6]{x^{18}} = x^{18/6}$ Converting to exponential notation
       $= x^3$ Simplifying

92. a^2

93. $(\sqrt[3]{ab})^{15} = (ab)^{15/3}$ Converting to exponential notation
       $= (ab)^5$ Simplifying the exponent
       $= a^5b^5$ Using the law of exponents

94. x^2y^2

95. $\sqrt[8]{(3x)^2} = (3x)^{2/8}$ Converting to exponential notation

          $= (3x)^{1/4}$ Simplifying the exponent

          $= \sqrt[4]{3x}$ Returning to radical notation

96. $\sqrt{7a}$

97. $(\sqrt[10]{3a})^5 = (3a)^{5/10}$ Converting to exponential notation

          $= (3a)^{1/2}$ Simplifying the exponent

          $= \sqrt{3a}$ Returning to radical notation

98. $\sqrt[4]{8x^3}$

99. $\sqrt[4]{\sqrt{x}} = \sqrt[4]{x^{1/2}}$ Converting to

          $= (x^{1/2})^{1/4}$ exponential notation

          $= x^{1/8}$ Using a law of exponents

          $= \sqrt[8]{x}$ Returning to radical notation

00. $\sqrt[18]{m}$

01. $\sqrt{(ab)^6} = (ab)^{6/2}$ Converting to exponential notation

          $= (ab)^3$ Using the laws

          $= a^3 b^3$ of exponents

02. $x^3 y^3$

03. $(\sqrt[3]{x^2 y^5})^{12} = (x^2 y^5)^{12/3}$ Converting to exponential notation

          $= (x^2 y^5)^4$ Simplifying the exponent

          $= x^8 y^{20}$ Using the laws of exponents

04. $a^6 b^{12}$

05. $\sqrt[3]{\sqrt[4]{xy}} = \sqrt[3]{(xy)^{1/4}}$ Converting to

          $= [(xy)^{1/4}]^{1/3}$ exponential notation

          $= (xy)^{1/12}$ Using a law of exponents

          $= \sqrt[12]{xy}$ Returning to radical notation

06. $\sqrt[10]{2a}$

07. *Writing Exercise*

08. *Writing Exercise*

09. $3x(x^3 - 2x^2) + 4x^2(2x^2 + 5x)$

     $= 3x^4 - 6x^3 + 8x^4 + 20x^3$

     $= 11x^4 + 14x^3$

110. $-3t^6 + 28t^5 - 20t^4$

111. $(3a - 4b)(5a + 3b)$

     $= 3a \cdot 5a + 3a \cdot 3b - 4b \cdot 5a - 4b \cdot 3b$

     $= 15a^2 + 9ab - 20ab - 12b^2$

     $= 15a^2 - 11ab - 12b^2$

112. $49x^2 - 14xy + y^2$

113. *Familiarize.* Let $p =$ the selling price of the home.

Translate.

    0.5% of the selling price   is   \$467.50

        $0.005p$ $=$ 467.50

Carry out. We solve the equation.

    $0.005p = 467.50$

       $p = 93,500$ Dividing by 0.005

Check. 0.5% of \$93,500 is 0.005(\$93,500), or \$467.50. The answer checks.

State. The selling price of the home was \$93,500.

114. 0, 1

115. *Writing Exercise*

116. *Writing Exercise*

117. $\sqrt[5]{x^2 y \sqrt{xy}} = \sqrt[5]{x^2 y (xy)^{1/2}} = \sqrt[5]{x^2 y x^{1/2} y^{1/2}} =$

   $\sqrt[5]{x^{5/2} y^{3/2}} = (x^{5/2} y^{3/2})^{1/5} = x^{5/10} y^{3/10} =$

   $(x^5 y^3)^{1/10} = \sqrt[10]{x^5 y^3}$

118. $\sqrt[6]{x^5}$

119. $\sqrt[4]{\sqrt[3]{8x^3 y^6}} = \sqrt[4]{(2^3 x^3 y^6)^{1/3}} = \sqrt[4]{2^{3/3} x^{3/3} y^{6/3}} =$

   $\sqrt[4]{2xy^2}$

120. $\sqrt[6]{p + q}$

121. $f(x) = 262 \cdot 2^{x/12}$

     $f(12) = 262 \cdot 2^{12/12}$

           $= 262 \cdot 2^1$

           $= 262 \cdot 2$

           $= 524$ cycles per second

122. 1760 cycles per second

123. $2^{7/12} \approx 1.498 \approx 1.5$ so the G that is 7 half steps above middle C has a frequency that is about 1.5 times that of middle C.

124. $2^{4/12} \approx 1.2599 \approx 1.25$ which is 25% greater than 1.

125. a) $L = \dfrac{(0.000169)60^{2.27}}{1} \approx 1.8$ m

b) $L = \dfrac{(0.000169)75^{2.27}}{0.9906} \approx 3.1$ m

c) $L = \dfrac{(0.000169)80^{2.27}}{2.4} \approx 1.5$ m

d) $L = \dfrac{(0.000169)100^{2.27}}{1.1} \approx 5.3$ m

126. About 7.937×10^{-13} to 1

127. $m = m_0(1 - v^2 c^{-2})^{-1/2}$

$$m = 8\left[1 - \left(\frac{9}{5} \times 10^8\right)^2 (3 \times 10^8)^{-2}\right]^{-1/2}$$

$$= 8\left[1 - \frac{\left(\frac{9}{5} \times 10^8\right)^2}{(3 \times 10^8)^2}\right]^{-1/2}$$

$$= 8\left[1 - \frac{\frac{81}{25} \times 10^{16}}{9 \times 10^6}\right]^{-1/2}$$

$$= 8\left[1 - \frac{81}{25} \cdot \frac{1}{9}\right]^{-1/2}$$

$$= 8\left[1 - \frac{9}{25}\right]^{-1/2}$$

$$= 8\left(\frac{16}{25}\right)^{-1/2}$$

$$= 8\left(\frac{25}{16}\right)^{1/2}$$

$$= 8 \cdot \frac{5}{4}$$

$$= 10$$

The particle's new mass is 10 mg.

128. $y_1 = x^{1/2}, \ y_2 = 3x^{2/5},$
$y_3 = x^{4/7}, \ y_4 = \frac{1}{5}x^{3/4}$

Exercise Set 7.3

1. $\sqrt{10}\sqrt{7} = \sqrt{10 \cdot 7} = \sqrt{70}$

2. $\sqrt{35}$

3. $\sqrt[3]{2}\sqrt[3]{5} = \sqrt[3]{2 \cdot 5} = \sqrt[3]{10}$

4. $\sqrt[3]{14}$

5. $\sqrt[4]{8}\sqrt[4]{9} = \sqrt[4]{8 \cdot 9} = \sqrt[4]{72}$

6. $\sqrt[4]{18}$

7. $\sqrt{5a}\sqrt{6b} = \sqrt{5a \cdot 6b} = \sqrt{30ab}$

8. $\sqrt{26xy}$

9. $\sqrt[5]{9t^2}\sqrt[5]{2t} = \sqrt[5]{9t^2 \cdot 2t} = \sqrt[5]{18t^3}$

10. $\sqrt[5]{80y^4}$

11. $\sqrt{x-a}\sqrt{x+a} = \sqrt{(x-a)(x+a)} = \sqrt{x^2 - a^2}$

12. $\sqrt{y^2 - b^2}$

13. $\sqrt[3]{0.5x}\sqrt[3]{0.2x} = \sqrt[3]{0.5x \cdot 0.2x} = \sqrt[3]{0.1x^2}$

14. $\sqrt[3]{0.21y^2}$

15. $\sqrt[4]{x-1}\sqrt[4]{x^2+x+1} = \sqrt[4]{(x-1)(x^2+x+1)} = \sqrt[4]{x^3 - 1}$

16. $\sqrt[5]{(x-2)^3}$

17. $\sqrt{\dfrac{x}{6}}\sqrt{\dfrac{7}{y}} = \sqrt{\dfrac{x}{6} \cdot \dfrac{7}{y}} = \sqrt{\dfrac{7x}{6y}}$

18. $\sqrt{\dfrac{7s}{11t}}$

19. $\sqrt[7]{\dfrac{x-3}{4}}\sqrt[7]{\dfrac{5}{x+2}} = \sqrt[7]{\dfrac{x-3}{4} \cdot \dfrac{5}{x+2}} = \sqrt[7]{\dfrac{5x-15}{4x+8}}$

20. $\sqrt[6]{\dfrac{3a}{b^2-4}}$

21. $\sqrt{50}$

$= \sqrt{25 \cdot 2}$ 25 is the largest perfect square factor of 50.

$= \sqrt{25} \cdot \sqrt{2}$

$= 5\sqrt{2}$

22. $3\sqrt{3}$

23. $\sqrt{28}$

$= \sqrt{4 \cdot 7}$ 4 is the largest perfect square factor of 28.

$= \sqrt{4} \cdot \sqrt{7}$

$= 2\sqrt{7}$

24. $3\sqrt{5}$

25. $\sqrt{8} = \sqrt{4 \cdot 2} = \sqrt{4} \cdot \sqrt{2} = 2\sqrt{2}$

26. $3\sqrt{2}$

27. $\sqrt{198} = \sqrt{9 \cdot 22} = \sqrt{9} \cdot \sqrt{22} = 3\sqrt{22}$

28. $5\sqrt{13}$

29. $\sqrt{36a^4 b}$

    $= \sqrt{36a^4 \cdot b}$ $36a^4$ is a perfect square.

    $= \sqrt{36a^4} \cdot \sqrt{b}$ Factoring into two radicals

    $= 6a^2\sqrt{b}$ Taking the square root of $36a^4$

30. $5y^4\sqrt{7}$

31. $\sqrt[3]{8x^3 y^2}$

    $= \sqrt[3]{8x^3 \cdot y^2}$ $8x^3$ is a perfect cube.

    $= \sqrt[3]{8x^3} \cdot \sqrt[3]{y^2}$ Factoring into two radicals

    $= 2x\sqrt[3]{y^2}$ Taking the cube root of $8x^3$

32. $3b^2\sqrt[3]{a}$

33. $\sqrt[3]{-16x^6}$

    $= \sqrt[3]{-8x^6 \cdot 2}$ $-8x^6$ is a perfect cube.

    $= \sqrt[3]{-8x^6} \cdot \sqrt[3]{2}$

    $= -2x^2\sqrt[3]{2}$ Taking the cube root of $-8x^6$

34. $-2a^2\sqrt[3]{4}$

35. $f(x) = \sqrt[3]{125x^5}$

        $= \sqrt[3]{125x^3 \cdot x^2}$

        $= \sqrt[3]{125x^3} \cdot \sqrt[3]{x^2}$

        $= 5x\sqrt[3]{x^2}$

36. $2x^2\sqrt[3]{2}$

37. $f(x) = \sqrt{49(x-3)^2}$ $49(x-3)^2$ is a perfect

                           square.

       $= |7(x-3)|$, or $7|x-3|$

38. $9|x-1|$

39. $f(x) = \sqrt{5x^2 - 10x + 5}$

        $= \sqrt{5(x^2 - 2x + 1)}$

        $= \sqrt{5(x-1)^2}$

        $= \sqrt{(x-1)^2} \cdot \sqrt{5}$

        $= |x-1|\sqrt{5}$

40. $|x+2|\sqrt{2}$

41. $\sqrt{a^3 b^4}$

    $= \sqrt{a^2 \cdot a \cdot b^4}$ Identifying the largest even

                     powers of a and b

    $= \sqrt{a^2}\sqrt{b^4}\sqrt{a}$ Factoring into several radicals

    $= ab^2\sqrt{a}$

42. $x^3 y^4\sqrt{y}$

43. $\sqrt[3]{x^5 y^6 z^{10}}$

    $= \sqrt[3]{x^3 \cdot x^2 \cdot y^6 \cdot z^9 \cdot z}$ Identifying the largest

                          perfect-cube powers of x, y, and z

    $= \sqrt[3]{x^3} \cdot \sqrt[3]{y^6} \cdot \sqrt[3]{z^9} \cdot \sqrt[3]{x^2 z}$ Factoring into

                                  several radicals

    $= xy^2 z^3\sqrt[3]{x^2 z}$

44. $a^2 b^2 c^4\sqrt[3]{bc}$

45. $\sqrt[5]{-32a^7 b^{11}} = \sqrt[5]{-32 \cdot a^5 \cdot a^2 \cdot b^{10} \cdot b} =$

    $\sqrt[5]{-32}\sqrt[5]{a^5}\sqrt[5]{b^{10}}\sqrt[5]{a^2 b} = -2ab^2\sqrt[5]{a^2 b}$

46. $2xy^2\sqrt[4]{xy^3}$

47. $\sqrt[5]{a^6 b^8 c^9} = \sqrt[5]{a^5 \cdot a \cdot b^5 \cdot b^3 \cdot c^5 \cdot c^4} =$

    $\sqrt[5]{a^5}\sqrt[5]{b^5}\sqrt[5]{c^5}\sqrt[5]{ab^3 c^4} =$

    $abc\sqrt[5]{ab^3 c^4}$

48. $x^2 yz^3\sqrt[5]{x^3 y^3 z^2}$

49. $\sqrt[4]{810x^9} = \sqrt[4]{81 \cdot 10 \cdot x^8 \cdot x} =$

    $\sqrt[4]{81} \cdot \sqrt[4]{x^8} \cdot \sqrt[4]{10x} = 3x^2\sqrt[4]{10x}$

50. $-2a^4\sqrt[3]{10a^2}$

51. $\sqrt{15}\sqrt{5} = \sqrt{15 \cdot 5} = \sqrt{75} = \sqrt{25 \cdot 3} = 5\sqrt{3}$

52. $3\sqrt{2}$

53. $\sqrt{10}\sqrt{14} = \sqrt{10 \cdot 14} = \sqrt{140} = \sqrt{4 \cdot 35} = 2\sqrt{35}$

54. $3\sqrt{35}$

55. $\sqrt[3]{2}\sqrt[3]{4} = \sqrt[3]{2 \cdot 4} = \sqrt[3]{8} = 2$

56. 3

57. $\sqrt{18a^3}\sqrt{18a^3} = \sqrt{(18a^3)^2} = 18a^3$

58. $75x^7$

59. $\sqrt[3]{5a^2}\sqrt[3]{2a} = \sqrt[3]{5a^2 \cdot 2a} = \sqrt[3]{10a^3} = \sqrt[3]{a^3 \cdot 10} = a\sqrt[3]{10}$

60. $x\sqrt[3]{21}$

61. $\sqrt{3x^5}\sqrt{15x^2} = \sqrt{45x^7} = \sqrt{9x^6 \cdot 5x} = 3x^3\sqrt{5x}$

62. $5a^5\sqrt{3}$

63. $\sqrt[3]{s^2 t^4}\sqrt[3]{s^4 t^6} = \sqrt[3]{s^6 t^{10}} = \sqrt[3]{s^6 t^9 \cdot t} = s^2 t^3\sqrt[3]{t}$

64. $xy^3\sqrt[3]{xy}$

65. $\sqrt[3]{(x+5)^2}\sqrt[3]{(x+5)^4} = \sqrt[3]{(x+5)^6} = (x+5)^2$

66. $(a - b)^4$

67. $\sqrt[4]{12a^3b^7}\,\sqrt[4]{4a^2b^5} = \sqrt[4]{48a^5b^{12}} = \sqrt[4]{16a^4b^{12} \cdot 3a} = 2ab^3\,\sqrt[4]{3a}$

68. $3x^2y^2\,\sqrt[4]{xy^3}$

69. $\sqrt[5]{x^3(y + z)^4}\,\sqrt[5]{x^3(y + z)^6} = \sqrt[5]{x^6(y + z)^{10}} = \sqrt[5]{x^5(y + z)^{10} \cdot x} = x(y + z)^2\,\sqrt[5]{x}$

70. $a^2(b - c)\,\sqrt[5]{(b - c)^3}$

71. *Writing Exercise*

72. *Writing Exercise*

73. $\dfrac{3x}{16y} + \dfrac{5y}{64x}$, LCD is $64xy$

$= \dfrac{3x}{16y} \cdot \dfrac{4x}{4x} + \dfrac{5y}{64x} \cdot \dfrac{y}{y}$

$= \dfrac{12x^2}{64xy} + \dfrac{5y^2}{64xy}$

$= \dfrac{12x^2 + 5y^2}{64xy}$

74. $\dfrac{2a + 6b^3}{a^4b^4}$

75. $\dfrac{4}{x^2 - 9} - \dfrac{7}{2x - 6}$

$= \dfrac{4}{(x + 3)(x - 3)} - \dfrac{7}{2(x - 3)}$, LCD is $2(x+3)(x-3)$

$= \dfrac{4}{(x + 3)(x - 3)} \cdot \dfrac{2}{2} - \dfrac{7}{2(x - 3)} \cdot \dfrac{x + 3}{x + 3}$

$= \dfrac{8}{2(x + 3)(x - 3)} - \dfrac{7(x + 3)}{2(x + 3)(x - 3)}$

$= \dfrac{8 - 7(x + 3)}{2(x + 3)(x - 3)}$

$= \dfrac{8 - 7x - 21}{2(x + 3)(x - 3)}$

$= \dfrac{-7x - 13}{2(x + 3)(x - 3)}$

76. $\dfrac{-3x + 1}{2(x + 5)(x - 5)}$

77. $\dfrac{9a^4b^7}{3a^2b^5} = \dfrac{9}{3}a^{4-2}b^{7-5} = 3a^2b^2$

78. $3ab^5$

79. *Writing Exercise*

80. *Writing Exercise*

81. $r(L) = 2\sqrt{5L}$

a) $r(L) = 2\sqrt{5 \cdot 20}$

$= 2\sqrt{100}$

$= 2 \cdot 10 = 20$ mph

b) $r(L) = 2\sqrt{5 \cdot 70}$

$= 2\sqrt{350}$

≈ 37.4 mph — Multiplying and rounding

c) $r(L) = 2\sqrt{5 \cdot 90}$

$= 2\sqrt{450}$

≈ 42.4 mph — Multiplying and rounding

82. a) 4.0 °F

b) $-10.3°$F

c) $-51.1°$F

d) $-78.5°$F

83. $(\sqrt{r^3t})^7 = \sqrt{(r^3t)^7} = \sqrt{r^{21}t^7} = \sqrt{r^{20} \cdot r \cdot t^6 \cdot t} = \sqrt{r^{20}}\sqrt{t^6}\sqrt{rt} = r^{10}t^3\sqrt{rt}$

84. $25x^5\,\sqrt[3]{25x}$

85. $(\sqrt[3]{a^2b^4})^5 = \sqrt[3]{(a^2b^4)^5} = \sqrt[3]{a^{10}b^{20}} = \sqrt[3]{a^9 \cdot a \cdot b^{18} \cdot b^2} = \sqrt[3]{a^9}\sqrt[3]{b^{18}}\sqrt[3]{ab^2} = a^3b^6\,\sqrt[3]{ab^2}$

86. $a^{10}b^{17}\sqrt{ab}$

87.

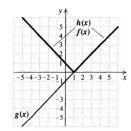

We see that $f(x) = h(x)$ and $f(x) \neq g(x)$.

88.

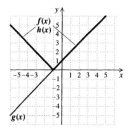

We see that $f(x) = h(x)$ and $f(x) \neq g(x)$.

89. $f(t) = \sqrt{t^2 - 3t - 4}$

We must have $t^2 - 3t - 4 \geq 0$, or $(t - 4)(t + 1) \geq 0$.

We graph $y = t^2 - 3t - 4$.

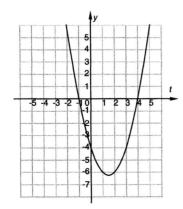

From the graph we see that $y \geq 0$ for $t \leq -1$ or $t \geq 4$, so the domain of f is $\{t | t \leq -1 \text{ or } t \geq 4\}$, or $(-\infty, -1] \cup [4, \infty)$.

90. $\{x | x \leq 2 \text{ or } x \geq 4\}$, or $(-\infty, 2] \cup [4, \infty)$

91. $\sqrt[3]{5x^{k+1}} \sqrt[3]{25x^k} = 5x^7$

$\sqrt[3]{5x^{k+1} \cdot 25x^k} = 5x^7$

$\sqrt[3]{125x^{2k+1}} = 5x^7$

$\sqrt[3]{125} \sqrt[3]{x^{2k+1}} = 5x^7$

$5\sqrt[3]{x^{2k+1}} = 5x^7$

$\sqrt[3]{x^{2k+1}} = x^7$

$(x^{2k+1})^{1/3} = x^7$

$x^{\frac{2k+1}{3}} = x^7$

Since the base is the same, the exponents must be equal. We have:

$\dfrac{2k + 1}{3} = 7$

$2k + 1 = 21$

$2k = 20$

$k = 10$

92. 6

93. *Writing Exercise*

Exercise Set 7.4

1. $\sqrt{\dfrac{25}{36}} = \dfrac{\sqrt{25}}{\sqrt{36}} = \dfrac{5}{6}$

2. $\dfrac{10}{9}$

3. $\sqrt[3]{\dfrac{64}{27}} = \dfrac{\sqrt[3]{64}}{\sqrt[3]{27}} = \dfrac{4}{3}$

4. $\dfrac{7}{10}$

5. $\sqrt{\dfrac{49}{y^2}} = \dfrac{\sqrt{49}}{\sqrt{y^2}} = \dfrac{7}{y}$

6. $\dfrac{11}{x}$

7. $\sqrt{\dfrac{25y^3}{x^4}} = \dfrac{\sqrt{25y^3}}{\sqrt{x^4}} = \dfrac{\sqrt{25y^2 \cdot y}}{\sqrt{x^4}} = \dfrac{\sqrt{25y^2} \sqrt{y}}{\sqrt{x^4}} = \dfrac{5y\sqrt{y}}{x^2}$

8. $\dfrac{6a^2 \sqrt{a}}{b^3}$

9. $\sqrt[3]{\dfrac{27a^4}{8b^3}} = \dfrac{\sqrt[3]{27a^4}}{\sqrt[3]{8b^3}} = \dfrac{\sqrt[3]{27a^3 \cdot a}}{\sqrt[3]{8b^3}} = \dfrac{\sqrt[3]{27a^3} \sqrt[3]{a}}{\sqrt[3]{8b^3}} = \dfrac{3a\sqrt[3]{a}}{2b}$

10. $\dfrac{2x^2 \sqrt[3]{x}}{3y^2}$

11. $\sqrt[4]{\dfrac{16a^4}{b^4c^8}} = \dfrac{\sqrt[4]{16a^4}}{\sqrt[4]{b^4c^8}} = \dfrac{2a}{bc^2}$

12. $\dfrac{3x}{y^2z}$

13. $\sqrt[4]{\dfrac{a^5b^8}{c^{10}}} = \dfrac{\sqrt[4]{a^5b^8}}{\sqrt[4]{c^{10}}} = \dfrac{\sqrt[4]{a^4b^8 \cdot a}}{\sqrt[4]{c^8 \cdot c^2}} = \dfrac{\sqrt[4]{a^4b^8} \sqrt[4]{a}}{\sqrt[4]{c^8} \sqrt[4]{c^2}} = \dfrac{ab^2 \sqrt[4]{a}}{c^2 \sqrt[4]{c^2}}$, or $\dfrac{ab^2}{c^2} \sqrt[4]{\dfrac{a}{c^2}}$

14. $\dfrac{x^2y^3}{z} \sqrt[4]{\dfrac{x}{z^2}}$

15. $\sqrt[5]{\dfrac{32x^6}{y^{11}}} = \dfrac{\sqrt[5]{32x^6}}{\sqrt[5]{y^{11}}} = \dfrac{\sqrt[5]{32x^5 \cdot x}}{\sqrt[5]{y^{10} \cdot y}} = \dfrac{\sqrt[5]{32x^5} \cdot \sqrt[5]{x}}{\sqrt[5]{y^{10}} \sqrt[5]{y}} = \dfrac{2x\sqrt[5]{x}}{y^2 \sqrt[5]{y}}$, or $\dfrac{2x}{y^2} \sqrt[5]{\dfrac{x}{y}}$

16. $\dfrac{3a}{b^2} \sqrt[5]{\dfrac{a^4}{b^3}}$

17. $\sqrt[6]{\dfrac{x^6y^8}{z^{15}}} = \dfrac{\sqrt[6]{x^6y^8}}{\sqrt[6]{z^{15}}} = \dfrac{\sqrt[6]{x^6y^6 \cdot y^2}}{\sqrt[6]{z^{12} \cdot z^3}} = \dfrac{\sqrt[6]{x^6y^6} \sqrt[6]{y^2}}{\sqrt[6]{z^{12}} \sqrt[6]{z^3}} = \dfrac{xy\sqrt[6]{y^2}}{z^2 \sqrt[6]{z^3}}$, or $\dfrac{xy}{z^2} \sqrt[6]{\dfrac{y^2}{z^3}}$

18. $\dfrac{ab^2}{c^2}\sqrt[6]{\dfrac{a^3}{c}}$

19. $\dfrac{\sqrt{35x}}{\sqrt{7x}} = \sqrt{\dfrac{35x}{7x}} = \sqrt{5}$

20. $\sqrt{7}$

21. $\dfrac{\sqrt[3]{270}}{\sqrt[3]{10}} = \sqrt[3]{\dfrac{270}{10}} = \sqrt[3]{27} = 3$

22. 2

23. $\dfrac{\sqrt{40xy^3}}{\sqrt{8x}} = \sqrt{\dfrac{40xy^3}{8x}} = \sqrt{5y^3} = \sqrt{y^2 \cdot 5y} =$ $\sqrt{y^2}\,\sqrt{5y} = y\sqrt{5y}$

24. $2b\sqrt{2b}$

25. $\dfrac{\sqrt[3]{96a^4b^2}}{\sqrt[3]{12a^2b}} = \sqrt[3]{\dfrac{96a^4b^2}{12a^2b}} = \sqrt[3]{8a^2b} = \sqrt[3]{8}\,\sqrt[3]{a^2b} =$ $2\sqrt[3]{a^2b}$

26. $3xy\sqrt[3]{y^2}$

27. $\dfrac{\sqrt{100ab}}{5\sqrt{2}} = \dfrac{1}{5}\dfrac{\sqrt{100ab}}{\sqrt{2}} = \dfrac{1}{5}\sqrt{\dfrac{100ab}{2}} = \dfrac{1}{5}\sqrt{50ab} =$ $\dfrac{1}{5}\sqrt{25 \cdot 2ab} = \dfrac{1}{5} \cdot 5\sqrt{2ab} = \sqrt{2ab}$

28. $\dfrac{5}{3}\sqrt{ab}$

29. $\dfrac{\sqrt[4]{48x^9y^{13}}}{\sqrt[4]{3xy^{-2}}} = \sqrt[4]{\dfrac{48x^9y^{13}}{3xy^{-2}}} = \sqrt[4]{16x^8y^{15}} =$ $\sqrt[4]{16x^8y^{12}}\,\sqrt[4]{y^3} = 2x^2y^3\sqrt[4]{y^3}$

30. $2a^2b^6$

31. $\dfrac{\sqrt[3]{x^3-y^3}}{\sqrt[3]{x-y}} = \sqrt[3]{\dfrac{x^3-y^3}{x-y}} =$ $\sqrt[3]{\dfrac{(x-y)(x^2+xy+y^2)}{x-y}} =$ $\sqrt[3]{\dfrac{(x\!\!\!\diagup\!\!-y)(x^2+xy+y^2)}{x\!\!\!\diagup\!\!-y}} = \sqrt[3]{x^2+xy+y^2}$

32. $\sqrt[3]{r^2-rs+s^2}$

33. $\sqrt{\dfrac{5}{7}} = \sqrt{\dfrac{5}{7} \cdot \dfrac{7}{7}} = \sqrt{\dfrac{35}{49}} = \dfrac{\sqrt{35}}{\sqrt{49}} = \dfrac{\sqrt{35}}{7}$

34. $\dfrac{\sqrt{66}}{6}$

35. $\dfrac{6\sqrt{5}}{5\sqrt{3}} = \dfrac{6\sqrt{5}}{5\sqrt{3}} \cdot \dfrac{\sqrt{3}}{\sqrt{3}} = \dfrac{6\sqrt{15}}{5 \cdot 3} = \dfrac{2\sqrt{15}}{5}$

36. $\dfrac{2\sqrt{10}}{3}$

37. $\sqrt[3]{\dfrac{16}{9}} = \sqrt[3]{\dfrac{16}{9} \cdot \dfrac{3}{3}} = \sqrt[3]{\dfrac{48}{27}} = \dfrac{\sqrt[3]{8 \cdot 6}}{\sqrt[3]{27}} = \dfrac{2\sqrt[3]{6}}{3}$

38. $\dfrac{\sqrt[3]{6}}{3}$

39. $\dfrac{\sqrt[3]{3a}}{\sqrt[3]{5c}} = \dfrac{\sqrt[3]{3a}}{\sqrt[3]{5c}} \cdot \dfrac{\sqrt[3]{5^2c^2}}{\sqrt[3]{5^2c^2}} = \dfrac{\sqrt[3]{75ac^2}}{\sqrt[3]{5^3c^3}} = \dfrac{\sqrt[3]{75ac^2}}{5c}$

40. $\dfrac{\sqrt[3]{63xy^2}}{3y}$

41. $\dfrac{\sqrt[3]{5y^4}}{\sqrt[3]{6x^4}} = \dfrac{\sqrt[3]{5y^4}}{\sqrt[3]{6x^4}} \cdot \dfrac{\sqrt[3]{36x^2}}{\sqrt[3]{36x^2}} = \dfrac{\sqrt[3]{y^3 \cdot 180x^2y}}{\sqrt[3]{216x^6}} =$ $\dfrac{y\sqrt[3]{180x^2y}}{6x^2}$

42. $\dfrac{a\sqrt[3]{147ab}}{7b}$

43. $\sqrt[3]{\dfrac{2}{x^2y}} = \sqrt[3]{\dfrac{2}{x^2y} \cdot \dfrac{xy^2}{xy^2}} = \sqrt[3]{\dfrac{2xy^2}{x^3y^3}} = \dfrac{\sqrt[3]{2xy^2}}{\sqrt[3]{x^3y^3}} =$ $\dfrac{\sqrt[3]{2xy^2}}{xy}$

44. $\dfrac{\sqrt[3]{5a^2b}}{ab}$

45. $\sqrt{\dfrac{7a}{18}} = \sqrt{\dfrac{7a}{18} \cdot \dfrac{2}{2}} = \sqrt{\dfrac{14a}{36}} = \dfrac{\sqrt{14a}}{\sqrt{36}} = \dfrac{\sqrt{14a}}{6}$

46. $\dfrac{\sqrt{30x}}{10}$

47. $\sqrt{\dfrac{9}{20x^2y}} = \sqrt{\dfrac{9}{20x^2y} \cdot \dfrac{5y}{5y}} = \sqrt{\dfrac{9 \cdot 5y}{100x^2y^2}} =$ $\dfrac{\sqrt{9 \cdot 5y}}{\sqrt{100x^2y^2}} = \dfrac{3\sqrt{5y}}{10xy}$

48. $\dfrac{\sqrt{14b}}{8ab}$

49. $\sqrt{\dfrac{10ab^2}{72a^3b}} = \sqrt{\dfrac{5b}{36a^2}} = \dfrac{\sqrt{5b}}{6a}$

50. $\dfrac{\sqrt{7x}}{5y^2}$

51. $\dfrac{\sqrt{5}}{\sqrt{7x}} = \dfrac{\sqrt{5}}{\sqrt{7x}} \cdot \dfrac{\sqrt{5}}{\sqrt{5}} = \dfrac{\sqrt{25}}{\sqrt{35x}} = \dfrac{5}{\sqrt{35x}}$

52. $\dfrac{10}{\sqrt{30x}}$

53. $\sqrt{\dfrac{14}{21}} = \sqrt{\dfrac{2}{3}} = \sqrt{\dfrac{2}{3} \cdot \dfrac{2}{2}} = \sqrt{\dfrac{4}{6}} = \dfrac{\sqrt{4}}{\sqrt{6}} = \dfrac{2}{\sqrt{6}}$

54. $\dfrac{2}{\sqrt{5}}$

55. $\dfrac{4\sqrt{13}}{3\sqrt{7}} = \dfrac{4\sqrt{13}}{3\sqrt{7}} \cdot \dfrac{\sqrt{13}}{\sqrt{13}} = \dfrac{4\sqrt{169}}{3\sqrt{91}} = \dfrac{4 \cdot 13}{3\sqrt{91}} = \dfrac{52}{3\sqrt{91}}$

56. $\dfrac{105}{2\sqrt{105}}$

57. $\dfrac{\sqrt[3]{7}}{\sqrt[3]{2}} = \dfrac{\sqrt[3]{7}}{\sqrt[3]{2}} \cdot \dfrac{\sqrt[3]{7^2}}{\sqrt[3]{7^2}} = \dfrac{\sqrt[3]{7^3}}{\sqrt[3]{98}} = \dfrac{7}{\sqrt[3]{98}}$

58. $\dfrac{5}{\sqrt[3]{100}}$

59. $\sqrt{\dfrac{7x}{3y}} = \sqrt{\dfrac{7x}{3y} \cdot \dfrac{7x}{7x}} = \dfrac{\sqrt{(7x)^2}}{\sqrt{21xy}} = \dfrac{7x}{\sqrt{21xy}}$

60. $\dfrac{6a}{\sqrt{30ab}}$

61. $\sqrt[3]{\dfrac{2a^5}{5b}} = \sqrt[3]{\dfrac{2a^5}{5b} \cdot \dfrac{4a}{4a}} = \sqrt[3]{\dfrac{8a^6}{20ab}} = \dfrac{2a^2}{\sqrt[3]{20ab}}$

62. $\dfrac{2a^2}{\sqrt[3]{28a^2b}}$

63. $\sqrt{\dfrac{x^3y}{2}} = \sqrt{\dfrac{x^3y}{2} \cdot \dfrac{xy}{xy}} = \sqrt{\dfrac{x^4y^2}{2xy}} = \dfrac{\sqrt{x^4y^2}}{\sqrt{2xy}} = \dfrac{x^2y}{\sqrt{2xy}}$

64. $\dfrac{ab^3}{\sqrt{3ab}}$

65. *Writing Exercise*

66. *Writing Exercise*

67. $\dfrac{3}{x-5} \cdot \dfrac{x-1}{x+5} = \dfrac{3(x-1)}{(x-5)(x+5)}$

68. $\dfrac{7(x-2)}{(x+4)(x-4)}$

69. $\dfrac{a^2 - 8a + 7}{a^2 - 49} = \dfrac{(a-1)(a-7)}{(a+7)(a-7)}$

$= \dfrac{(a-1)(a\!\!-\!\!7)}{(a+7)(a\!\!-\!\!7)}$

$= \dfrac{a-1}{a+7}$

70. $\dfrac{t+11}{t+2}$

71. $(5a^3b^4)^3 = 5^3(a^3)^3(b^4)^3 = 125a^{3\cdot3}b^{4\cdot3} = 125a^9b^{12}$

72. $225x^{10}y^6$

73. *Writing Exercise*

74. *Writing Exercise*

75. a) $T = 2\pi\sqrt{\dfrac{65}{980}} \approx 1.62$ sec

  b) $T = 2\pi\sqrt{\dfrac{98}{980}} \approx 1.99$ sec

  c) $T = 2\pi\sqrt{\dfrac{120}{980}} \approx 2.20$ sec

76. a^3bxy^2

77. $\dfrac{(\sqrt[3]{81mn^2})^2}{(\sqrt[3]{mn})^2} = \dfrac{\sqrt[3]{(81mn^2)^2}}{\sqrt[3]{(mn)^2}}$

$= \dfrac{\sqrt[3]{6561m^2n^4}}{\sqrt[3]{m^2n^2}}$

$= \sqrt[3]{\dfrac{6561m^2n^4}{m^2n^2}}$

$= \sqrt[3]{6561n^2}$

$= \sqrt[3]{729 \cdot 9n^2}$

$= \sqrt[3]{729}\,\sqrt[3]{9n^2}$

$= 9\sqrt[3]{9n^2}$

78. $2yz\sqrt{2z}$

79. $\sqrt{a^2-3} - \dfrac{a^2}{\sqrt{a^2-3}}$

$= \sqrt{a^2-3} - \dfrac{a^2}{\sqrt{a^2-3}} \cdot \dfrac{\sqrt{a^2-3}}{\sqrt{a^2-3}}$

$= \sqrt{a^2-3} - \dfrac{a^2\sqrt{a^2-3}}{a^2-3}$

$= \sqrt{a^2-3} \cdot \dfrac{a^2-3}{a^2-3} - \dfrac{a^2\sqrt{a^2-3}}{a^2-3}$

$= \dfrac{a^2\sqrt{a^2-3} - 3\sqrt{a^2-3} - a^2\sqrt{a^2-3}}{a^2-3}$

$= \dfrac{-3\sqrt{a^2-3}}{a^2-3}$, or $\dfrac{-3}{\sqrt{a^2-3}}$

80. $\dfrac{(5x+4y-3)\sqrt{xy}}{xy}$

81. Step 1: $\sqrt[n]{x} = x^{1/n}$, by definition;

Step 2: $\left(\dfrac{x}{y}\right)^n = \dfrac{x^n}{y^n}$, raising a quotient to a power;

Step 3: $x^{1/n} = \sqrt[n]{x}$, by definition

82. A number c is the nth root of a/b if $c^n = a/b$. Let $c = \sqrt[n]{a}/\sqrt[n]{b}$.

$$c^n = \left(\dfrac{\sqrt[n]{a}}{\sqrt[n]{b}}\right)^n = \left(\dfrac{a^{1/n}}{b^{1/n}}\right)^n = \dfrac{(a^{1/n})^n}{(b^{1/n})^n} = \dfrac{a}{b}$$

83. $f(x) = \sqrt{18x^3}$, $g(x) = \sqrt{2x}$

$$(f/g)(x) = \dfrac{f(x)}{g(x)} = \dfrac{\sqrt{18x^3}}{\sqrt{2x}} = \sqrt{\dfrac{18x^3}{2x}} = \sqrt{9x^2} = 3x$$

$\sqrt{2x}$ is defined for $2x \geq 0$, or $x \geq 0$. To avoid division by 0, we must exclude 0 from the domain. Thus, the domain of $f/g = \{x | x$ is a real number $and\ x > 0\}$, or $(0, \infty)$.

84. $(f/g)(t) = \dfrac{1}{5t}$;

$\{t | t$ is a real number $and\ t > 0\}$, or $(0, \infty)$

85. $f(x) = \sqrt{x^2 - 9}$, $g(x) = \sqrt{x - 3}$

$$(f/g)(x) = \dfrac{f(x)}{g(x)} = \dfrac{\sqrt{x^2 - 9}}{\sqrt{x - 3}} = \sqrt{\dfrac{x^2 - 9}{x - 3}} =$$

$$\sqrt{\dfrac{(x + 3)(x - 3)}{x - 3}} = \sqrt{x + 3}$$

$\sqrt{x - 3}$ is defined for $x - 3 \geq 0$, or $x \geq 3$. To avoid division by 0 we must exclude 3 from the domain. Thus, the domain of $f/g = \{x | x$ is a real number $and\ x > 3\}$, or $(3, \infty)$.

Exercise Set 7.5

1. $3\sqrt{7} + 2\sqrt{7} = (3 + 2)\sqrt{7} = 5\sqrt{7}$

2. $17\sqrt{5}$

3. $9\sqrt[3]{5} - 6\sqrt[3]{5} = (9 - 6)\sqrt[3]{5} = 3\sqrt[3]{5}$

4. $8\sqrt[5]{2}$

5. $4\sqrt[3]{y} + 9\sqrt[3]{y} = (4 + 9)\sqrt[3]{y} = 13\sqrt[3]{y}$

6. $6\sqrt[4]{t}$

7. $8\sqrt{2} - 6\sqrt{2} + 5\sqrt{2} = (8 - 6 + 5)\sqrt{2} = 7\sqrt{2}$

8. $7\sqrt{6}$

9. $9\sqrt[3]{7} - \sqrt{3} + 4\sqrt[3]{7} + 2\sqrt{3} =$
$(9 + 4)\sqrt[3]{7} + (-1 + 2)\sqrt{3} = 13\sqrt[3]{7} + \sqrt{3}$

10. $6\sqrt{7} + \sqrt[4]{11}$

11. $\quad 8\sqrt{27} - 3\sqrt{3}$
$= 8\sqrt{9 \cdot 3} - 3\sqrt{3}$ Factoring the
$= 8\sqrt{9} \cdot \sqrt{3} - 3\sqrt{3}$ first radical
$= 8 \cdot 3\sqrt{3} - 3\sqrt{3}$ Taking the square root of 9
$= 24\sqrt{3} - 3\sqrt{3}$
$= 21\sqrt{3}$ Combining like radicals

12. $41\sqrt{2}$

13. $\quad 3\sqrt{45} + 7\sqrt{20}$
$= 3\sqrt{9 \cdot 5} + 7\sqrt{4 \cdot 5}$ Factoring the
$= 3\sqrt{9} \cdot \sqrt{5} + 7\sqrt{4} \cdot \sqrt{5}$ radicals
$= 3 \cdot 3\sqrt{5} + 7 \cdot 2\sqrt{5}$ Taking the square roots
$= 9\sqrt{5} + 14\sqrt{5}$
$= 23\sqrt{5}$ Combining like radicals

14. $58\sqrt{3}$

15. $3\sqrt[3]{16} + \sqrt[3]{54} = 3\sqrt[3]{8 \cdot 2} + \sqrt[3]{27 \cdot 2} =$
$3\sqrt[3]{8} \cdot \sqrt[3]{2} + \sqrt[3]{27} \cdot \sqrt[3]{2} = 3 \cdot 2\sqrt[3]{2} + 3\sqrt[3]{2} =$
$6\sqrt[3]{2} + 3\sqrt[3]{2} = 9\sqrt[3]{2}$

16. -7

17. $\sqrt{5a} + 2\sqrt{45a^3} = \sqrt{5a} + 2\sqrt{9a^2 \cdot 5a} =$
$\sqrt{5a} + 2\sqrt{9a^2} \cdot \sqrt{5a} = \sqrt{5a} + 2 \cdot 3a\sqrt{5a} =$
$\sqrt{5a} + 6a\sqrt{5a} = (1 + 6a)\sqrt{5a}$

18. $(4x - 2)\sqrt{3x}$

19. $\sqrt[3]{6x^4} + \sqrt[3]{48x} = \sqrt[3]{x^3 \cdot 6x} + \sqrt[3]{8 \cdot 6x} =$
$\sqrt[3]{x^3} \cdot \sqrt[3]{6x} + \sqrt[3]{8} \cdot \sqrt[3]{6x} = x\sqrt[3]{6x} + 2\sqrt[3]{6x} =$
$(x + 2)\sqrt[3]{6x}$

20. $(3 - x)\sqrt[3]{2x}$

21. $\sqrt{4a - 4} + \sqrt{a - 1} = \sqrt{4(a - 4)} + \sqrt{a - 1} =$
$\sqrt{4}\sqrt{a - 1} + \sqrt{a - 1} = 2\sqrt{a - 1} + \sqrt{a - 1} = 3\sqrt{a - 1}$

22. $4\sqrt{y + 3}$

23. $\sqrt{x^3 - x^2} + \sqrt{9x - 9} = \sqrt{x^2(x - 1)} + \sqrt{9(x - 1)} =$
$\sqrt{x^2} \cdot \sqrt{x - 1} + \sqrt{9} \cdot \sqrt{x - 1} =$
$x\sqrt{x - 1} + 3\sqrt{x - 1} = (x + 3)\sqrt{x - 1}$

24. $(2 - x)\sqrt{x - 1}$

25. $\sqrt{7}(3 - \sqrt{7}) = \sqrt{7} \cdot 3 - \sqrt{7} \cdot \sqrt{7} = 3\sqrt{7} - 7$

26. $4\sqrt{3} + 3$

27. $4\sqrt{2}(\sqrt{3} - \sqrt{5}) = 4\sqrt{2} \cdot \sqrt{3} - 4\sqrt{2} \cdot \sqrt{5} = 4\sqrt{6} - 4\sqrt{10}$

28. $15 - 3\sqrt{10}$

29. $\sqrt{3}(2\sqrt{5} - 3\sqrt{4}) = \sqrt{3}(2\sqrt{5} - 3 \cdot 2) =$
$\sqrt{3} \cdot 2\sqrt{5} - \sqrt{3} \cdot 6 = 2\sqrt{15} - 6\sqrt{3}$

30. $6\sqrt{5} - 4$

31. $\sqrt[3]{2}(\sqrt[3]{4} - 2\sqrt[3]{32}) = \sqrt[3]{2} \cdot \sqrt[3]{4} - \sqrt[3]{2} \cdot 2\sqrt[3]{32} =$
$\sqrt[3]{8} - 2\sqrt[3]{64} = 2 - 2 \cdot 4 = 2 - 8 = -6$

32. $3 - 4\sqrt[3]{63}$

33. $\sqrt[3]{a}(\sqrt[3]{a^2} + \sqrt[3]{24a^2}) = \sqrt[3]{a} \cdot \sqrt[3]{a^2} + \sqrt[3]{a}\sqrt[3]{24a^2} =$
$\sqrt[3]{a^3} + \sqrt[3]{24a^3} = \sqrt[3]{a^3} + \sqrt[3]{8a^3 \cdot 3} =$
$a + 2a\sqrt[3]{3}$

34. $-2x\sqrt[3]{3}$

35. $(5 + \sqrt{6})(5 - \sqrt{6}) = 5^2 - (\sqrt{6})^2 = 25 - 6 = 19$

36. -1

37. $(3 - 2\sqrt{7})(3 + 2\sqrt{7}) = 3^2 - (2\sqrt{7})^2 = 9 - 4 \cdot 7 =$
$9 - 28 = -19$

38. -2

39. $(3 + \sqrt{5})^2 = 3^2 + 2 \cdot 3 \cdot \sqrt{5} + (\sqrt{5})^2 = 9 + 6\sqrt{5} + 5 =$
$14 + 6\sqrt{5}$

40. $52 + 14\sqrt{3}$

41. $(2\sqrt{7} - 4\sqrt{2})(3\sqrt{7} + 6\sqrt{2}) =$
$2\sqrt{7} \cdot 3\sqrt{7} + 2\sqrt{7} \cdot 6\sqrt{2} - 4\sqrt{2} \cdot 3\sqrt{7} - 4\sqrt{2} \cdot 6\sqrt{2} =$
$6 \cdot 7 + 12\sqrt{14} - 12\sqrt{14} - 24 \cdot 2 =$
$42 + 12\sqrt{14} - 12\sqrt{14} - 48 = -6$

42. $24 - 7\sqrt{15}$

43. $(2\sqrt[3]{3} - \sqrt[3]{2})(\sqrt[3]{3} + 2\sqrt[3]{2}) =$
$2\sqrt[3]{3} \cdot \sqrt[3]{3} + 2\sqrt[3]{3} \cdot 2\sqrt[3]{2} - \sqrt[3]{2} \cdot \sqrt[3]{3} - \sqrt[3]{2} \cdot 2\sqrt[3]{2} =$
$2\sqrt[3]{9} + 4\sqrt[3]{6} - \sqrt[3]{6} - 2\sqrt[3]{4} = 2\sqrt[3]{9} + 3\sqrt[3]{6} - 2\sqrt[3]{4}$

44. $6\sqrt[4]{63} - 9\sqrt[4]{42} + 2\sqrt[4]{54} - 3\sqrt[4]{36}$

45. $(\sqrt{3x} + \sqrt{y})^2$
$= (\sqrt{3x})^2 + 2 \cdot \sqrt{3x} \cdot \sqrt{y} + (\sqrt{y})^2$ Squaring a
                                                      binomial
$= 3x + 2\sqrt{3xy} + y$

46. $t - 2\sqrt{2rt} + 2r$

47. $\dfrac{2}{3 + \sqrt{5}} = \dfrac{2}{3 + \sqrt{5}} \cdot \dfrac{3 - \sqrt{5}}{3 - \sqrt{5}} =$

$\dfrac{2(3 - \sqrt{5})}{(3 + \sqrt{5})(3 - \sqrt{5})} = \dfrac{6 - 2\sqrt{5}}{3^2 - (\sqrt{5})^2} =$

$\dfrac{6 - 2\sqrt{5}}{9 - 5} = \dfrac{6 - 2\sqrt{5}}{4} = \dfrac{\cancel{2}(3 - \sqrt{5})}{\cancel{2} \cdot 2} =$

$\dfrac{3 - \sqrt{5}}{2}$

48. $\dfrac{4 + \sqrt{7}}{3}$

49. $\dfrac{2 + \sqrt{5}}{6 - \sqrt{3}} = \dfrac{2 + \sqrt{5}}{6 - \sqrt{3}} \cdot \dfrac{6 + \sqrt{3}}{6 + \sqrt{3}} =$

$\dfrac{(2 + \sqrt{5})(6 + \sqrt{3})}{(6 - \sqrt{3})(6 + \sqrt{3})} = \dfrac{12 + 2\sqrt{3} + 6\sqrt{5} + \sqrt{15}}{36 - 3} =$

$\dfrac{12 + 2\sqrt{3} + 6\sqrt{5} + \sqrt{15}}{33}$

50. $\dfrac{3 - \sqrt{5} + 3\sqrt{2} - \sqrt{10}}{4}$

51. $\dfrac{\sqrt{a}}{\sqrt{a} + \sqrt{b}} = \dfrac{\sqrt{a}}{\sqrt{a} + \sqrt{b}} \cdot \dfrac{\sqrt{a} - \sqrt{b}}{\sqrt{a} - \sqrt{b}} =$

$\dfrac{\sqrt{a}(\sqrt{a} - \sqrt{b})}{(\sqrt{a} + \sqrt{b})(\sqrt{a} - \sqrt{b})} = \dfrac{a - \sqrt{ab}}{a - b}$

52. $\dfrac{\sqrt{xz} + z}{x - z}$

53. $\dfrac{\sqrt{7} - \sqrt{3}}{\sqrt{3} - \sqrt{7}} = \dfrac{-1(\sqrt{3} - \sqrt{7})}{\sqrt{3} - \sqrt{7}} = -1 \cdot \dfrac{\sqrt{3} - \sqrt{7}}{\sqrt{3} - \sqrt{7}} =$
$-1 \cdot 1 = -1$

54. $\dfrac{\sqrt{35} - \sqrt{14} + 5 - \sqrt{10}}{3}$

55. $\dfrac{3\sqrt{2} - \sqrt{7}}{4\sqrt{2} + \sqrt{5}} = \dfrac{3\sqrt{2} - \sqrt{7}}{4\sqrt{2} + \sqrt{5}} \cdot \dfrac{4\sqrt{2} - \sqrt{5}}{4\sqrt{2} - \sqrt{5}} =$

$\dfrac{(3\sqrt{2} - \sqrt{7})(4\sqrt{2} - \sqrt{5})}{(4\sqrt{2} + \sqrt{5})(4\sqrt{2} - \sqrt{5})} =$

$\dfrac{12 \cdot 2 - 3\sqrt{10} - 4\sqrt{14} + \sqrt{35}}{16 \cdot 2 - 5} =$

$\dfrac{24 - 3\sqrt{10} - 4\sqrt{14} + \sqrt{35}}{32 - 5} =$

$\dfrac{24 - 3\sqrt{10} - 4\sqrt{14} + \sqrt{35}}{27}$

56. $\dfrac{-30 - 25\sqrt{6} + 2\sqrt{33} + 5\sqrt{22}}{38}$

57. $\dfrac{5\sqrt{3}-3\sqrt{2}}{3\sqrt{2}-2\sqrt{3}} = \dfrac{5\sqrt{3}-3\sqrt{2}}{3\sqrt{2}-2\sqrt{3}} \cdot \dfrac{3\sqrt{2}+2\sqrt{3}}{3\sqrt{2}+2\sqrt{3}} =$

$\dfrac{15\sqrt{6}+10\cdot 3 - 9\cdot 2 - 6\sqrt{6}}{9\cdot 2 - 4\cdot 3} =$

$\dfrac{15\sqrt{6}+30-18-6\sqrt{6}}{18-12} = \dfrac{9\sqrt{6}+12}{6} =$

$\dfrac{3(3\sqrt{6}+4)}{3\cdot 2} = \dfrac{3\sqrt{6}+4}{2}$

58. $\dfrac{4\sqrt{6}+9}{3}$

59. $\dfrac{\sqrt{7}+2}{5} = \dfrac{\sqrt{7}+2}{5} \cdot \dfrac{\sqrt{7}-2}{\sqrt{7}-2} =$

$\dfrac{(\sqrt{7}+2)(\sqrt{7}-2)}{5(\sqrt{7}-2)} = \dfrac{(\sqrt{7})^2-2^2}{5\sqrt{7}-10} =$

$\dfrac{7-4}{5\sqrt{7}-10} = \dfrac{3}{5\sqrt{7}-10}$

60. $\dfrac{1}{2\sqrt{3}-2}$

61. $\dfrac{\sqrt{6}-2}{\sqrt{3}+7} = \dfrac{\sqrt{6}-2}{\sqrt{3}+7} \cdot \dfrac{\sqrt{6}+2}{\sqrt{6}+2} =$

$\dfrac{(\sqrt{6}-2)(\sqrt{6}+2)}{(\sqrt{3}+7)(\sqrt{6}+2)} = \dfrac{6-4}{\sqrt{18}+2\sqrt{3}+7\sqrt{6}+14} =$

$\dfrac{2}{3\sqrt{2}+2\sqrt{3}+7\sqrt{6}+14}$

62. $\dfrac{6}{-2\sqrt{5}+4\sqrt{2}+3\sqrt{10}-12}$

63. $\dfrac{\sqrt{x}-\sqrt{y}}{\sqrt{x}+\sqrt{y}} = \dfrac{\sqrt{x}-\sqrt{y}}{\sqrt{x}+\sqrt{y}} \cdot \dfrac{\sqrt{x}+\sqrt{y}}{\sqrt{x}+\sqrt{y}} =$

$\dfrac{(\sqrt{x}-\sqrt{y})(\sqrt{x}+\sqrt{y})}{(\sqrt{x}+\sqrt{y})(\sqrt{x}+\sqrt{y})} = \dfrac{x-y}{x+2\sqrt{xy}+y}$

64. $\dfrac{a-b}{a-2\sqrt{ab}+b}$

65. $\sqrt{a}\,\sqrt[4]{a^3}$

$= a^{1/2}\cdot a^{3/4}$ Converting to exponential notation

$= a^{5/4}$ Adding exponents

$= a^{1+1/4}$ Writing 5/4 as a mixed number

$= a\cdot a^{1/4}$ Factoring

$= a\sqrt[4]{a}$ Returning to radical notation

66. $x\sqrt{x}$

67. $\sqrt[5]{b^2}\,\sqrt{b^3}$

$= b^{2/5}\cdot b^{3/2}$ Converting to exponential notation

$= b^{19/10}$ Adding exponents

$= b^{1+9/10}$ Writing 19/10 as a mixed number

$= b\cdot b^{9/10}$ Factoring

$= b\,\sqrt[10]{b^9}$ Returning to radical notation

68. $a\,\sqrt[12]{a^5}$

69. $\sqrt{xy^3}\,\sqrt[3]{x^2y} = (xy^3)^{1/2}(x^2y)^{1/3}$

$= (xy^3)^{3/6}(x^2y)^{2/6}$

$= [(xy^3)^3(x^2y)^2]^{1/6}$

$= \sqrt[6]{x^3y^9 \cdot x^4y^2}$

$= \sqrt[6]{x^7y^{11}}$

$= \sqrt[6]{x^6y^6 \cdot xy^5}$

$= xy\,\sqrt[6]{xy^5}$

70. $a\,\sqrt[10]{ab^7}$

71. $\sqrt[4]{9ab^3}\,\sqrt{3a^4b} = (9ab^3)^{1/4}(3a^4b)^{1/2}$

$= (9ab^3)^{1/4}(3a^4b)^{2/4}$

$= [(9ab^3)(3a^4b)^2]^{1/4}$

$= \sqrt[4]{9ab^3 \cdot 9a^8b^2}$

$= \sqrt[4]{81a^9b^5}$

$= \sqrt[4]{81a^8b^4 \cdot ab}$

$= 3a^2b\,\sqrt[4]{ab}$

72. $2xy^2\,\sqrt[6]{2x^5y}$

73. $\sqrt[3]{xy^2z}\,\sqrt{x^3yz^2} = (xy^2z)^{1/3}(x^3yz^2)^{1/2}$

$= (xy^2z)^{2/6}(x^3yz^2)^{3/6}$

$= [(xy^2z)^2(x^3yz^2)^3]^{1/6}$

$= \sqrt[6]{x^2y^4z^2 \cdot x^9y^3z^6}$

$= \sqrt[6]{x^{11}y^7z^8}$

$= \sqrt[6]{x^6y^6z^6 \cdot x^5yz^2}$

$= xyz\,\sqrt[6]{x^5yz^2}$

74. $a^2b^2c^2\,\sqrt[6]{a^2bc^2}$

75. $\dfrac{\sqrt[3]{x^2}}{\sqrt[5]{x}}$

$= \dfrac{x^{2/3}}{x^{1/5}}$ Converting to exponential notation

$= x^{2/3-1/5}$ Subtracting exponents

$= x^{7/15}$ Converting back

$= \sqrt[15]{x^7}$ to radical notation

76. $\sqrt[12]{a^5}$

77.
$$\frac{\sqrt[5]{a^4 b}}{\sqrt[3]{ab}}$$

$= \dfrac{(a^4 b)^{1/5}}{(ab)^{1/3}}$ Converting to exponential notation

$= \dfrac{a^{4/5} b^{1/5}}{a^{1/3} b^{1/3}}$ Using the product and power rules

$= a^{4/5 - 1/3} b^{1/5 - 1/3}$ Subtracting exponents

$= a^{7/15} b^{-2/15}$

$= (a^7 b^{-2})^{1/15}$ Converting back

$= \sqrt[15]{a^7 b^{-2}}$, or   to radical notation

$\sqrt[15]{\dfrac{a^7}{b^2}}$

78. $\sqrt[12]{x^2 y^5}$

79.
$$\frac{\sqrt[5]{x^3 y^4}}{\sqrt{xy}}$$

$= \dfrac{(x^3 y^4)^{1/5}}{(xy)^{1/2}}$ Converting to exponential notation

$= \dfrac{x^{3/5} y^{4/5}}{x^{1/2} y^{1/2}}$

$= x^{3/5 - 1/2} y^{4/5 - 1/2}$ Subtracting exponents

$= x^{1/10} y^{3/10}$

$= (xy^3)^{1/10}$ Converting back to

$= \sqrt[10]{xy^3}$ radical notation

80. $\sqrt[10]{ab^9}$

81.
$$\frac{\sqrt[3]{(2+5x)^2}}{\sqrt[4]{2+5x}}$$

$= \dfrac{(2+5x)^{2/3}}{(2+5x)^{1/4}}$ Converting to exponential notation

$= (2+5x)^{2/3 - 1/4}$ Subtracting exponents

$= (2+5x)^{5/12}$ Converting back to

$= \sqrt[12]{(2+5x)^5}$ radical notation

82. $\sqrt[20]{(3x-1)^3}$

83.
$$\frac{\sqrt[4]{(5+3x)^3}}{\sqrt[3]{(5+3x)^2}}$$

$= \dfrac{(5+3x)^{3/4}}{(5+3x)^{2/3}}$ Converting to exponential notation

$= (5+3x)^{3/4 - 2/3}$ Subtracting exponents

$= (5+3x)^{1/12}$ Converting back

$= \sqrt[12]{5+3x}$ to radical notation

84. $\sqrt[15]{(2x+1)^4}$

85.
$\sqrt[3]{x^2 y}\left(\sqrt{xy} - \sqrt[5]{xy^3}\right)$

$= (x^2 y)^{1/3}\left[(xy)^{1/2} - (xy^3)^{1/5}\right]$

$= x^{2/3} y^{1/3}\left(x^{1/2} y^{1/2} - x^{1/5} y^{3/5}\right)$

$= x^{2/3} y^{1/3} x^{1/2} y^{1/2} - x^{2/3} y^{1/3} x^{1/5} y^{3/5}$

$= x^{2/3 + 1/2} y^{1/3 + 1/2} - x^{2/3 + 1/5} y^{1/3 + 3/5}$

$= x^{7/6} y^{5/6} - x^{13/15} y^{14/15}$

$= x^{1\frac{1}{6}} y^{\frac{5}{6}} - x^{13/15} y^{14/15}$

                Writing a mixed numeral

$= x \cdot x^{1/6} y^{5/6} - x^{13/15} y^{14/15}$

$= x(xy^5)^{1/6} - (x^{13} y^{14})^{1/15}$

$= x\sqrt[6]{xy^5} - \sqrt[15]{x^{13} y^{14}}$

86. $a\sqrt[12]{a^2 b^7} - \sqrt[20]{a^{18} b^{13}}$

87.
$(m + \sqrt[3]{n^2})(2m + \sqrt[4]{n})$

$= (m + n^{2/3})(2m + n^{1/4})$ Converting to exponential notation

$= 2m^2 + mn^{1/4} + 2mn^{2/3} + n^{2/3} n^{1/4}$ Using FOIL

$= 2m^2 + mn^{1/4} + 2mn^{2/3} + n^{2/3 + 1/4}$ Adding exponents

$= 2m^2 + mn^{1/4} + 2mn^{2/3} + n^{11/12}$

$= 2m^2 + m\sqrt[4]{n} + 2m\sqrt[3]{n^2} + \sqrt[12]{n^{11}}$ Converting back to radical notation

88. $3r^2 - r\sqrt[5]{s} - 3r\sqrt[4]{s^3} + \sqrt[20]{s^{19}}$

89. $f(x) = \sqrt[4]{x},\ g(x) = \sqrt[4]{2x} - \sqrt[4]{x^{11}}$

$(f \cdot g)(x) = \sqrt[4]{x}\left(\sqrt[4]{2x} - \sqrt[4]{x^{11}}\right)$

$= \sqrt[4]{2x^2} - \sqrt[4]{x^{12}}$

$= \sqrt[4]{2x^2} - x^3$

90. $x^2 + \sqrt[4]{3x^3}$

91. $f(x) = x + \sqrt{7},\ g(x) = x - \sqrt{7}$

$(f \cdot g)(x) = (x + \sqrt{7})(x - \sqrt{7})$

$= x^2 - (\sqrt{7})^2$

$= x^2 - 7$

92. $x^2 + x\sqrt{6} - x\sqrt{2} - 2\sqrt{3}$

93. $f(x) = x^2$
$$f(5 - \sqrt{2}) = (5 - \sqrt{2})^2 = 25 - 10\sqrt{2} + (\sqrt{2})^2 =$$
$$25 - 10\sqrt{2} + 2 = 27 - 10\sqrt{2}$$

94. $52 + 14\sqrt{3}$

95. $f(x) = x^2$
$$f(\sqrt{3} + \sqrt{5}) = (\sqrt{3} + \sqrt{5})^2 =$$
$$(\sqrt{3})^2 + 2 \cdot \sqrt{3} \cdot \sqrt{5} + (\sqrt{5})^2 =$$
$$3 + 2\sqrt{15} + 5 = 8 + 2\sqrt{15}$$

96. $9 - 6\sqrt{2}$

97. *Writing Exercise*

98. *Writing Exercise*

99.
$$\frac{12x}{x-4} - \frac{3x^2}{x+4} = \frac{384}{x^2 - 16}$$
$$\frac{12x}{x-4} - \frac{3x^2}{x+4} = \frac{384}{(x+4)(x-4)},$$
$$\text{LCM is } (x+4)(x-4).$$

Note that $x \neq -4$ and $x \neq 4$.
$$(x+4)(x-4)\left[\frac{12x}{x-4} - \frac{3x^2}{x+4}\right] =$$
$$(x+4)(x-4) \cdot \frac{384}{(x+4)(x-4)}$$
$$12x(x+4) - 3x^2(x-4) = 384$$
$$12x^2 + 48x - 3x^3 + 12x^2 = 384$$
$$-3x^3 + 24x^2 + 48x - 384 = 0$$
$$-3(x^3 - 8x^2 - 16x + 128) = 0$$
$$-3[x^2(x-8) - 16(x-8)] = 0$$
$$-3(x-8)(x^2 - 16) = 0$$
$$-3(x-8)(x+4)(x-4) = 0$$
$$x - 8 = 0 \text{ or } x + 4 = 0 \quad \text{or } x - 4 = 0$$
$$x = 8 \text{ or } \quad x = -4 \text{ or } \quad x = 4$$

Check: For 8:
$$\frac{12x}{x-4} - \frac{3x^2}{x+4} = \frac{384}{x^2 - 16}$$

$\dfrac{12 \cdot 8}{8-4} - \dfrac{3 \cdot 8^2}{8+4}$	$\dfrac{384}{8^2 - 16}$
$\dfrac{96}{4} - \dfrac{192}{12}$	$\dfrac{384}{48}$
$24 - 16$	8
8	TRUE

8 is a solution.

For -4:
$$\frac{12x}{x-4} - \frac{3x^2}{x+4} = \frac{384}{x^2 - 16}$$

$\dfrac{12(-4)}{-4-4} - \dfrac{3(-4)^2}{-4+4}$	$\dfrac{384}{(-4)^2 - 16}$
$\dfrac{-48}{-8} - \dfrac{48}{0}$	$\dfrac{384}{16 - 16}$ UNDEFINED

-4 is not a solution.

For 4:
$$\frac{12x}{x-4} - \frac{3x^2}{x+4} = \frac{384}{x^2 - 16}$$

$\dfrac{12 \cdot 4}{4-4} - \dfrac{3 \cdot 4^2}{4+4}$	$\dfrac{384}{4^2 - 16}$
$\dfrac{48}{0} - \dfrac{48}{8}$	$\dfrac{384}{16 - 16}$ UNDEFINED

4 is not a solution.

The checks confirm that -4 and 4 are not solutions. The solution is 8.

100. $\dfrac{15}{2}$

101. **Familiarize.** Let x and y represent the width and length of the rectangle, respectively.

Translate. We write two equations.

$$\underbrace{\text{The width}}_{\downarrow} \; \underbrace{\text{is}}_{\downarrow} \; \underbrace{\text{one-fourth}}_{\downarrow} \; \underbrace{\text{the length.}}_{\downarrow}$$
$$x \quad = \quad \tfrac{1}{4} \cdot \quad y$$

$$\underbrace{\text{The area}}_{\downarrow} \; \underbrace{\text{is}}_{\downarrow} \; \underbrace{\text{twice}}_{\downarrow} \; \underbrace{\text{the perimeter.}}_{\downarrow}$$
$$xy \quad = \quad 2 \cdot \quad (2x + 2y)$$

Carry out. Solving the system of equations we get $(5, 20)$.

Check. The width, 5, is one-fourth the length, 20. The area is $5 \cdot 20$, or 100. The perimeter is $2 \cdot 5 + 2 \cdot 20$, or 50. Since $100 = 2 \cdot 50$, the area is twice the perimeter. The values check.

State. The width is 5 units, and the length is 20 units.

102. $-5, 4$

103.
$$5x^2 - 6x + 1 = 0$$
$$(5x - 1)(x - 1) = 0$$
$$5x - 1 = 0 \text{ or } x - 1 = 0$$
$$5x = 1 \text{ or } \quad x = 1$$
$$x = \frac{1}{5} \text{ or } \quad x = 1$$

The solutions are $\dfrac{1}{5}$ and 1.

104. $\dfrac{1}{7}, 1$

105. *Writing Exercise*

106. *Writing Exercise*

107. To add radical expressions, the <u>indices</u> and the <u>radicands</u> must be the same.

108. indices

109. To add rational expressions, the <u>denominators</u> must be the same.

110. bases

111.
$$
\begin{aligned}
f(x) &= \sqrt{20x^2 + 4x^3} - 3x\sqrt{45 + 9x} + \sqrt{5x^2 + x^3} \\
&= \sqrt{4x^2(5 + x)} - 3x\sqrt{9(5 + x)} + \sqrt{x^2(5 + x)} \\
&= \sqrt{4x^2}\sqrt{5 + x} - 3x\sqrt{9}\sqrt{5 + x} + \sqrt{x^2}\sqrt{5 + x} \\
&= 2x\sqrt{5 + x} - 3x \cdot 3\sqrt{5 + x} + x\sqrt{5 + x} \\
&= 2x\sqrt{5 + x} - 9x\sqrt{5 + x} + x\sqrt{5 + x} \\
&= -6x\sqrt{5 + x}
\end{aligned}
$$

112. $f(x) = 2x\sqrt{x - 1}$

113.
$$
\begin{aligned}
f(x) &= \sqrt[4]{x^5 - x^4} + 3\sqrt[4]{x^9 - x^8} \\
&= \sqrt[4]{x^4(x - 1)} + 3\sqrt[4]{x^8(x - 1)} \\
&= \sqrt[4]{x^4} \cdot \sqrt[4]{x - 1} + 3\sqrt[4]{x^8}\sqrt[4]{x - 1} \\
&= x\sqrt[4]{x - 1} + 3x^2\sqrt[4]{x - 1} \\
&= (x + 3x^2)\sqrt[4]{x - 1}
\end{aligned}
$$

114. $f(x) = 2x(1 - x)\sqrt[4]{1 + x}$

115.
$$
\begin{aligned}
&\frac{1}{2}\sqrt{36a^5bc^4} - \frac{1}{2}\sqrt[3]{64a^4bc^6} + \frac{1}{6}\sqrt{144a^3bc^6} = \\
&\frac{1}{2}\sqrt{36a^4c^4 \cdot ab} - \frac{1}{2}\sqrt[3]{64a^3c^6 \cdot ab} + \frac{1}{6}\sqrt{144a^2c^6 \cdot ab} = \\
&\frac{1}{2}(6a^2c^2)\sqrt{ab} - \frac{1}{2}(4ac^2)\sqrt[3]{ab} + \frac{1}{6}(12ac^3)\sqrt{ab} = \\
&3a^2c^2\sqrt{ab} - 2ac^2\sqrt[3]{ab} + 2ac^3\sqrt{ab} \\
&(3a^2c^2 + 2ac^3)\sqrt{ab} - 2ac^2\sqrt[3]{ab}, \text{ or} \\
&ac^2[(3a + 2c)\sqrt{ab} - 2\sqrt[3]{ab}]
\end{aligned}
$$

116. $(7x^2 - 2y^2)\sqrt{x + y}$

117.
$$
\begin{aligned}
&\sqrt{27a^5(b + 1)}\,\sqrt[3]{81a(b + 1)^4} \\
&= [27a^5(b + 1)]^{1/2}[81a(b + 1)^4]^{1/3} \\
&= [27a^5(b + 1)]^{3/6}[81a(b + 1)^4]^{2/6} \\
&= \{[3^3a^5(b + 1)]^3[3^4a(b + 1)^4]^2\}^{1/6} \\
&= \sqrt[6]{3^9a^{15}(b + 1)^3 \cdot 3^8a^2(b + 1)^8} \\
&= \sqrt[6]{3^{17}a^{17}(b + 1)^{11}} \\
&= \sqrt[6]{3^{12}a^{12}(b + 1)^6 \cdot 3^5a^5(b + 1)^5} \\
&= 3^2a^2(b + 1)\sqrt[6]{3^5a^5(b + 1)^5}, \text{ or} \\
&\quad 9a^2(b + 1)\sqrt[6]{243a^5(b + 1)^5}
\end{aligned}
$$

118. $4x(y + z)^3\sqrt[6]{2x(y + z)}$

119.
$$
\begin{aligned}
&\frac{\dfrac{1}{\sqrt{w}} - \sqrt{w}}{\dfrac{\sqrt{w} + 1}{\sqrt{w}}} = \frac{\dfrac{1}{\sqrt{w}} - \sqrt{w}}{\dfrac{\sqrt{w} + 1}{\sqrt{w}}} \cdot \frac{\sqrt{w}}{\sqrt{w}} = \frac{1 - w}{\sqrt{w} + 1} = \\
&\frac{1 - w}{\sqrt{w} + 1} \cdot \frac{\sqrt{w} - 1}{\sqrt{w} - 1} = \frac{\sqrt{w} - 1 - w\sqrt{w} + w}{w - 1} = \\
&\frac{(w - 1) - \sqrt{w}(w - 1)}{w - 1} = \frac{(w - 1)(1 - \sqrt{w})}{w - 1} = \\
&1 - \sqrt{w}
\end{aligned}
$$

120. $\dfrac{7\sqrt{3}}{39}$

121. $x - 5 = (\sqrt{x})^2 - (\sqrt{5})^2 = (\sqrt{x} + \sqrt{5})(\sqrt{x} - \sqrt{5})$

122. $(\sqrt{y} + \sqrt{7})(\sqrt{y} - \sqrt{7})$

123. $x - a = (\sqrt{x})^2 - (\sqrt{a})^2 = (\sqrt{x} + \sqrt{a})(\sqrt{x} - \sqrt{a})$

124. 6

125.
$$
\begin{aligned}
&(\sqrt{x + 2} - \sqrt{x - 2})^2 = \\
&x + 2 - 2\sqrt{(x + 2)(x - 2)} + x - 2 = \\
&x + 2 - 2\sqrt{x^2 - 4} + x - 2 = 2x - 2\sqrt{x^2 - 4}
\end{aligned}
$$

126. $\dfrac{ab + (a - b)\sqrt{a + b} - a - b}{a + b - b^2}$

127.
$$
\begin{aligned}
\frac{b + \sqrt{b}}{1 + b + \sqrt{b}} &= \frac{b + \sqrt{b}}{(1 + b) + \sqrt{b}} \cdot \frac{(1 + b) - \sqrt{b}}{(1 + b) - \sqrt{b}} \\
&= \frac{(b + \sqrt{b})(1 + b - \sqrt{b})}{(1 + b)^2 - (\sqrt{b})^2} \\
&= \frac{b + b^2 - b\sqrt{b} + \sqrt{b} + b\sqrt{b} - b}{1 + 2b + b^2 - b} \\
&= \frac{b^2 + \sqrt{b}}{1 + b + b^2}
\end{aligned}
$$

128. $\dfrac{1}{\sqrt{y+18}+\sqrt{y}}$

129.
$$\dfrac{\sqrt{x+6}-5}{\sqrt{x+6}+5} = \dfrac{\sqrt{x+6}-5}{\sqrt{x+6}+5} \cdot \dfrac{\sqrt{x+6}+5}{\sqrt{x+6}+5}$$
$$= \dfrac{(x+6)-25}{(x+6)+10\sqrt{x+6}+25}$$
$$= \dfrac{x-19}{x+10\sqrt{x+6}+31}$$

Exercise Set 7.6

1.
$$\sqrt{x+3} = 5$$
$$(\sqrt{x+3})^2 = 5^2 \quad \text{Principle of powers (squaring)}$$
$$x+3 = 25$$
$$x = 22$$

Check:
$$\dfrac{\sqrt{x+3} = 5}{}$$
$$\sqrt{22+3} \ ? \ 5$$
$$\sqrt{25} \ \Big|$$
$$5 \ \Big| \ 5 \quad \text{TRUE}$$

The solution is 22.

2. $\dfrac{63}{5}$

3.
$$\sqrt{2x}-1 = 2$$
$$\sqrt{2x} = 3 \quad \text{Adding to isolate the radical}$$
$$(\sqrt{2x})^2 = 3^2 \quad \text{Principle of powers (squaring)}$$
$$2x = 9$$
$$x = \dfrac{9}{2}$$

Check:
$$\dfrac{\sqrt{2x}-1 = 2}{}$$
$$\sqrt{2 \cdot \dfrac{9}{2} - 1} \ ? \ 2$$
$$\sqrt{9}-1 \ \Big|$$
$$3-1 \ \Big|$$
$$2 \ \Big| \ 2 \quad \text{TRUE}$$

The solution is $\dfrac{9}{2}$.

4. $\dfrac{25}{3}$

5.
$$\sqrt{x-2}-7 = -4$$
$$\sqrt{x-2} = 3 \quad \text{Adding to isolate the radical}$$
$$(\sqrt{x-2})^2 = 3^2 \quad \text{Principle of powers (squaring)}$$
$$x-2 = 9$$
$$x = 11$$

Check:
$$\dfrac{\sqrt{x-2}-7 = -4}{}$$
$$\sqrt{11-2}-7 \ ? \ -4$$
$$\sqrt{9}-7 \ \Big|$$
$$3-7 \ \Big|$$
$$-4 \ \Big| \ -4 \quad \text{TRUE}$$

The solution is 11.

6. 168

7.
$$\sqrt{y+4}+6 = 7$$
$$\sqrt{y+4} = 1 \quad \text{Adding to isolate the radical}$$
$$(\sqrt{y+4})^2 = 1^2 \quad \text{Principle of powers (squaring)}$$
$$y+4 = 1$$
$$y = -3$$

Check:
$$\dfrac{\sqrt{y+4}+6 = 7}{}$$
$$\sqrt{-3+4}+6 \ ? \ 7$$
$$\sqrt{1}+6 \ \Big|$$
$$1+6 \ \Big|$$
$$7 \ \Big| \ 7 \quad \text{TRUE}$$

The solution is -3.

8. 56

9.
$$\sqrt[3]{x-2} = 3$$
$$(\sqrt[3]{x-2})^3 = 3^3$$
$$x-2 = 27$$
$$x = 29$$

Check:
$$\dfrac{\sqrt[3]{x-2} = 3}{}$$
$$\sqrt[3]{29-2} \ ? \ 3$$
$$\sqrt[3]{27} \ \Big|$$
$$3 \ \Big| \ 3 \quad \text{TRUE}$$

The solution is 29.

10. 3

11.
$$\sqrt[4]{x+3} = 2$$
$$(\sqrt[4]{x+3})^4 = 2^4$$
$$x+3 = 16$$
$$x = 13$$

Check:
$$\dfrac{\sqrt[4]{x+3} = 2}{}$$
$$\sqrt[4]{13+3} \ ? \ 2$$
$$\sqrt[4]{16} \ \Big|$$
$$2 \ \Big| \ 2 \quad \text{TRUE}$$

The solution is 13.

12. 82

13.
$$8\sqrt{y} = y$$
$$(8\sqrt{y})^2 = y^2$$
$$64y = y^2$$
$$0 = y^2 - 64y$$
$$0 = y(y - 64)$$
$$y = 0 \quad or \quad y - 64 = 0$$
$$y = 0 \quad or \quad y = 64$$

Check:

For 0: $8\sqrt{y} = y$

$$8\sqrt{0} \; ? \; 0$$
$$8 \cdot 0 \; |$$
$$0 \; | \; 0 \qquad \text{TRUE}$$

For 64: $8\sqrt{y} = y$

$$8\sqrt{64} \; ? \; 64$$
$$8 \cdot 8 \; |$$
$$64 \; | \; 64 \qquad \text{TRUE}$$

The solutions are 0 and 64.

14. 0, 9

15.
$$3x^{1/2} + 12 = 9$$
$$3\sqrt{x} + 12 = 9$$
$$3\sqrt{x} = -3$$
$$\sqrt{x} = -1$$

Since the principal square root is never negative, this equation has no solution.

16. 64

17.
$$\sqrt[3]{y} = -4$$
$$(\sqrt[3]{y})^3 = (-4)^3$$
$$y = -64$$

Check: $\sqrt[3]{y} = -4$

$$\sqrt[3]{-64} \; ? \; -4$$
$$-4 \; | \; -4 \qquad \text{TRUE}$$

The solution is -64.

18. -27

19.
$$x^{1/4} - 2 = 1$$
$$x^{1/4} = 3$$
$$(x^{1/4})^4 = 3^4$$
$$x = 81$$

Check: $x^{1/4} - 2 = 1$

$$81^{1/4} - 2 \; ? \; 1$$
$$3 - 2 \; |$$
$$1 \; | \; 1 \qquad \text{TRUE}$$

The solution is 81.

20. 125

21.
$$(y - 3)^{1/2} = -2$$
$$\sqrt{y - 3} = -2$$

This equation has no solution, since the principal square root is never negative.

22. No solution

23.
$$\sqrt[4]{3x + 1} - 4 = -1$$
$$\sqrt[4]{3x + 1} = 3$$
$$(\sqrt[4]{3x + 1})^4 = 3^4$$
$$3x + 1 = 81$$
$$3x = 80$$
$$x = \frac{80}{3}$$

Check: $\sqrt[4]{3x + 1} - 4 = -1$

$$\sqrt[4]{3 \cdot \frac{80}{3}} - 4 \; ? \; -1$$
$$\sqrt[4]{81} - 4 \; |$$
$$3 - 4 \; |$$
$$-1 \; | \; -1 \qquad \text{TRUE}$$

The solution is $\frac{80}{3}$.

24. 39

25.
$$(x + 7)^{1/3} = 4$$
$$[(x + 7)^{1/3}]^3 = 4^3$$
$$x + 7 = 64$$
$$x = 57$$

Check: $(x + 7)^{1/3} = 4$

$$(57 + 7)^{1/3} \; ? \; 4$$
$$64^{1/3} \; |$$
$$4 \; | \; 4 \qquad \text{TRUE}$$

The solution is 57.

26. 88

27.
$$\sqrt[3]{3y + 6} + 2 = 3$$
$$\sqrt[3]{3y + 6} = 1$$
$$(\sqrt[3]{3y + 6})^3 = 1^3$$
$$3y + 6 = 1$$
$$3y = -5$$
$$y = -\frac{5}{3}$$

Check: $\dfrac{\sqrt[3]{3y+6}+2=3}{}$

$$\sqrt[3]{3\left(-\dfrac{5}{3}\right)+6}+2\ ?\ 3$$
$$\sqrt[3]{1}+2\ \bigg|$$
$$1+2\ \bigg|$$
$$3\ \bigg|\ 3 \qquad \text{TRUE}$$

The solution is $-\dfrac{5}{3}$.

28. -6

29. $\sqrt{3t+4}=\sqrt{4t+3}$
$$(\sqrt{3t+4})^2=(\sqrt{4t+3})^2$$
$$3t+4=4t+3$$
$$4=t+3$$
$$1=t$$

Check: $\dfrac{\sqrt{3t+4}=\sqrt{4t+3}}{}$

$$\sqrt{3\cdot1+4}\ ?\ \sqrt{4\cdot1+3}$$
$$\sqrt{7}\ \bigg|\ \sqrt{7} \qquad \text{TRUE}$$

The solution is 1.

30. 5

31. $3(4-t)^{1/4}=6^{1/4}$
$$[3(4-t)^{1/4}]^4=(6^{1/4})^4$$
$$81(4-t)=6$$
$$324-81t=6$$
$$-81t=-318$$
$$t=\dfrac{106}{27}$$

The number $\dfrac{106}{27}$ checks and is the solution.

32. $\dfrac{1}{2}$

33. $3+\sqrt{5-x}=x$
$$\sqrt{5-x}=x-3$$
$$(\sqrt{5-x})^2=(x-3)^2$$
$$5-x=x^2-6x+9$$
$$0=x^2-5x+4$$
$$0=(x-1)(x-4)$$
$$x-1=0\ or\ x-4=0$$
$$x=1\ or\ \qquad x=4$$

Check:
For 1: $\dfrac{3+\sqrt{5-x}=x}{}$
$$3+\sqrt{5-1}\ ?\ 1$$
$$3+\sqrt{4}\ \bigg|$$
$$3+2\ \bigg|$$
$$5\ \bigg|\ 1 \qquad \text{FALSE}$$

For 4: $\dfrac{3+\sqrt{5-x}=x}{}$
$$3+\sqrt{5-4}\ ?\ 4$$
$$3+\sqrt{1}\ \bigg|$$
$$3+1\ \bigg|$$
$$4\ \bigg|\ 4 \qquad \text{TRUE}$$

Since 4 checks but 1 does not, the solution is 4.

34. 5

35. $\sqrt{4x-3}=2+\sqrt{2x-5}$ One radical is already isolated.

$$(\sqrt{4x-3})^2=(2+\sqrt{2x-5})^2 \qquad \text{Squaring both sides}$$
$$4x-3=4+4\sqrt{2x-5}+2x-5$$
$$2x-2=4\sqrt{2x-5}$$
$$x-1=2\sqrt{2x-5}$$
$$x^2-2x+1=8x-20$$
$$x^2-10x+21=0$$
$$(x-7)(x-3)=0$$
$$x-7=0\ or\ x-3=0$$
$$x=7\ or\ \qquad x=3$$

Both numbers check. The solutions are 7 and 3.

36. 7

37. $\sqrt{20-x}+8=\sqrt{9-x}+11$
$$\sqrt{20-x}=\sqrt{9-x}+3 \qquad \text{Isolating one radical}$$
$$(\sqrt{20-x})^2=(\sqrt{9-x}+3)^2 \qquad \text{Squaring both sides}$$
$$20-x=9-x+6\sqrt{9-x}+9$$
$$2=6\sqrt{9-x} \qquad \text{Isolating the remaining radical}$$
$$1=3\sqrt{9-x} \qquad \text{Multiplying by }\dfrac{1}{2}$$
$$1^2=(3\sqrt{9-x})^2 \qquad \text{Squaring both sides}$$
$$1=9(9-x)$$
$$1=81-9x$$
$$-80=-9x$$
$$\dfrac{80}{9}=x$$

The number $\dfrac{80}{9}$ checks and is the solution.

38. $\dfrac{15}{4}$

39. $\sqrt{x+2} + \sqrt{3x+4} = 2$

$\sqrt{x+2} = 2 - \sqrt{3x+4}$ Isolating one radical

$(\sqrt{x+2})^2 = (2 - \sqrt{3x+4})^2$

$x + 2 = 4 - 4\sqrt{3x+4} + 3x + 4$

$-2x - 6 = -4\sqrt{3x+4}$ Isolating the remaining radical

$x + 3 = 2\sqrt{3x+4}$ Multiplying by $-\dfrac{1}{2}$

$(x+3)^2 = (2\sqrt{3x+4})^2$

$x^2 + 6x + 9 = 4(3x+4)$

$x^2 + 6x + 9 = 12x + 16$

$x^2 - 6x - 7 = 0$

$(x-7)(x+1) = 0$

$x - 7 = 0 \ \text{ or } \ x + 1 = 0$

$x = 7 \ \text{ or } \ x = -1$

Check:

For 7:

$$\dfrac{\sqrt{x+2}+\sqrt{3x+4}=2}{\sqrt{7+2}+\sqrt{3\cdot 7+4} \ ? \ 2}$$
$$\sqrt{9}+\sqrt{25}$$
$$8 \ \bigg| \ 2 \qquad \text{FALSE}$$

For -1:

$$\dfrac{\sqrt{x+2}+\sqrt{3x+4}=2}{\sqrt{-1+2}+\sqrt{3\cdot(-1)+4} \ ? \ 2}$$
$$\sqrt{1}+\sqrt{1}$$
$$2 \ \bigg| \ 2 \qquad \text{TRUE}$$

Since -1 checks but 7 does not, the solution is -1.

40. $-1, \dfrac{1}{3}$

41. We must have $f(x) = 2$, or $\sqrt{x} + \sqrt{x-9} = 1$.

$\sqrt{x} + \sqrt{x-9} = 1$

$\sqrt{x-9} = 1 - \sqrt{x}$ Isolating one radical term

$(\sqrt{x-9})^2 = (1-\sqrt{x})^2$

$x - 9 = 1 - 2\sqrt{x} + x$

$-10 = -2\sqrt{x}$ Isolating the remaining radical term

$5 = \sqrt{x}$

$25 = x$

This value does not check. There is no solution, so there is no value of x for which $f(x) = 1$.

42. 9

43. $\sqrt{a-2} - \sqrt{4a+1} = -3$

$\sqrt{a-2} = \sqrt{4a+1} - 3$

$(\sqrt{a-2})^2 = (\sqrt{4a+1}-3)^2$

$a - 2 = 4a + 1 - 6\sqrt{4a+1} + 9$

$-3a - 12 = -6\sqrt{4a+1}$

$a + 4 = 2\sqrt{4a+1}$

$(a+4)^2 = (2\sqrt{4a+1})^2$

$a^2 + 8a + 16 = 4(4a+1)$

$a^2 + 8a + 16 = 16a + 4$

$a^2 - 8a + 12 = 0$

$(a-2)(a-6) = 0$

$a - 2 = 0 \ \text{ or } \ a - 6 = 0$

$a = 2 \ \text{ or } \qquad a = 6$

Both numbers check, so we have $f(a) = -3$ when $a = 2$ and when $a = 6$.

44. 1

45. We must have $\sqrt{2x-3} = \sqrt{x+7} - 2$.

$\sqrt{2x-3} = \sqrt{x+7} - 2$

$(\sqrt{2x-3})^2 = (\sqrt{x+7}-2)^2$

$2x - 3 = x + 7 - 4\sqrt{x+7} + 4$

$x - 14 = -4\sqrt{x+7}$

$(x-14)^2 = (-4\sqrt{x+7})^2$

$x^2 - 28x + 196 = 16(x+7)$

$x^2 - 28x + 196 = 16x + 112$

$x^2 - 44x + 84 = 0$

$(x-2)(x-42) = 0$

$x = 2 \ \text{ or } \ x = 42$

Since 2 checks but 42 does not, we have $f(x) = g(x)$ when $x = 2$.

46. 10

47. We must have $4 - \sqrt{a-3} = (a+5)^{1/2}$.

$4 - \sqrt{a-3} = (a+5)^{1/2}$

$(4 - \sqrt{a-3})^2 = [(a+5)^{1/2}]^2$

$16 - 8\sqrt{a-3} + a - 3 = a + 5$

$-8\sqrt{a-3} = -8$

$\sqrt{a-3} = 1$

$(\sqrt{a-3})^2 = 1^2$

$a - 3 = 1$

$a = 4$

The number 4 checks, so we have $f(a) = g(a)$ when $a = 4$.

48. 15

49. *Writing Exercise*

50. *Writing Exercise*

51. Familiarize. Let h = the height of the triangle, in inches. Then $h + 2$ = the base. Recall that the formula for the area of a triangle with base b and height h is $A = \frac{1}{2}bh$.

Translate. Substitute in the formula.
$$31\frac{1}{2} = \frac{1}{2}(h+2)(h)$$

Carry out. We solve the equation.
$$31\frac{1}{2} = \frac{1}{2}(h+2)(h)$$
$$\frac{63}{2} = \frac{1}{2}(h+2)(h)$$
$$63 = (h+2)(h) \quad \text{Multiplying by 2}$$
$$63 = h^2 + 2h$$
$$0 = h^2 + 2h - 63$$
$$0 = (h+9)(h-7)$$
$$h + 9 = 0 \quad \text{or} \quad h - 7 = 0$$
$$h = -9 \quad \text{or} \quad h = 7$$

Check. Since the height of the triangle cannot be negative we check only 7. If the height is 7 in., then the base is $7 + 2$, or 9 in., and the area is $\frac{1}{2} \cdot 9 \cdot 7 = \frac{63}{2} = 31\frac{1}{2}$ in². The answer checks.

State. The height of the triangle is 7 in., and the base is 9 in.

52. 8 60-sec commercials

53. Familiarize. Let t = the time, in hours, it takes Gonzalo to sew the quilt. Then $t - 6$ = the time it takes Elaine to sew the quilt. In 4 hours Gonzalo does $\frac{4}{t}$ of the job and Elaine does $\frac{4}{t-6}$ of the job.

Translate. Together, in 4 hr one entire job is done.
$$\frac{4}{t} + \frac{4}{t-6} = 1$$

Carry out. We solve the equation. The LCD is $t(t-6)$.
$$\frac{4}{t} + \frac{4}{t-6} = 1$$
$$t(t-6)\left(\frac{4}{t} + \frac{4}{t-6}\right) = t(t-6) \cdot 1$$
$$4(t-6) + 4t = t^2 - 6t$$
$$4t - 24 + 4t = t^2 - 6t$$
$$8t - 24 = t^2 - 6t$$
$$0 = t^2 - 14t + 24$$
$$0 = (t-2)(t-12)$$

$$t - 2 = 0 \quad \text{or} \quad t - 12 = 0$$
$$t = 2 \quad \text{or} \quad t = 12$$

Check. If $t = 2$, then $t - 6 = 2 - 6 = -4$. Since time cannot be negative in this application, 2 is not a solution. If Gonzalo sews the quilt in 12 hr, then Elaine sews it in $12 - 6$, or 6 hr. In 4 hr Gonzalo does 4/12, or 1/3 of the job and Elaine does 4/6, or 2/3 of the job. Together they do $1/3 + 2/3$, or 1 entire job. The answer checks.

State. It would take Elaine 6 hr and Gonzalo 12 hr to sew the quilt working alone.

54. 3.2 hr

55. Graph $y > 3x + 5$.

First graph the related equation, $y = 3x + 5$. Use a dashed line since the inequality symbol is $>$. Then test a point not on the line to determine if it is a solution of the inequality. We use $(0, 0)$.

$$\begin{array}{c} y > 3x + 5 \\ \hline 0 \ ? \ 3 \cdot 0 + 5 \\ \hline 0 \ | \ 5 \qquad \text{FALSE} \end{array}$$

Since $0 > 5$ is false, we shade the half-plane that does not contain $(0, 0)$.

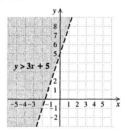

56.

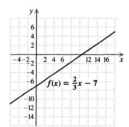

57. *Writing Exercise*

58. *Writing Exercise*

59. $S(t) = 1.087.7\sqrt{\dfrac{9t + 2617}{2457}}$

Substitute 1502.3 for $S(t)$ and solve for t.

$1502.3 = 1087.7\sqrt{\dfrac{9t + 2617}{2457}}$

$1.3812 \approx \sqrt{\dfrac{9t + 2617}{2457}}$ Dividing by 1087.7

$(1.3812)^2 \approx \left(\sqrt{\dfrac{9t + 2617}{2457}}\right)^2$

$1.9077 \approx \dfrac{9t + 2617}{2457}$

$4687.2189 \approx 9t + 2617$

$2070.2189 \approx 9t$

$230.0243 \approx t$

The temperature is about $230.0°C$.

60. $524.8°C$

61.

$S = 1087.7\sqrt{\dfrac{9t + 2617}{2457}}$

$\dfrac{S}{1087.7} = \sqrt{\dfrac{9t + 2617}{2457}}$

$\left(\dfrac{S}{1087.7}\right)^2 = \left(\sqrt{\dfrac{9t + 2617}{2457}}\right)^2$

$\dfrac{S^2}{1087.7^2} = \dfrac{9t + 2617}{2457}$

$\dfrac{2457S^2}{1087.7^2} = 9t + 2617$

$\dfrac{2457S^2}{1087.7^2} - 2617 = 9t$

$\dfrac{1}{9}\left(\dfrac{2457S^2}{1087.7^2} - 2617\right) = t$

62. About 4166 rpm

63. $d(n) = 0.75\sqrt{2.8n}$

Substitute 84 for $d(n)$ and solve for n.

$84 = 0.75\sqrt{2.8n}$

$112 = \sqrt{2.8n}$

$(112)^2 = (\sqrt{2.8n})^2$

$12,544 = 2.8n$

$4480 = n$

About 4480 rpm will produce peak performance.

64. $h = \dfrac{v^2 r}{2gr - v^2}$

65.

$v = \sqrt{2gr}\sqrt{\dfrac{h}{r + h}}$

$v^2 = 2gr \cdot \dfrac{h}{r + h}$ Squaring both sides

$v^2(r + h) = 2grh$ Multiplying by $r+h$

$v^2 r + v^2 h = 2grh$

$v^2 h = 2grh - v^2 r$

$v^2 h = r(2gh - v^2)$

$\dfrac{v^2 h}{2gh - v^2} = r$

66. 22,500 ft

67. $D(h) = 1.2\sqrt{h}$

$10.2 = 1.2\sqrt{h}$

$8.5 = \sqrt{h}$

$(8.5)^2 = (\sqrt{h})^2$

$72.25 = h$

The sailor must climb 72.25 ft above sea level.

68. $-\dfrac{8}{9}$

69.

$\left(\dfrac{z}{4} - 5\right)^{2/3} = \dfrac{1}{25}$

$\left[\left(\dfrac{z}{4} - 5\right)^{2/3}\right]^3 = \left(\dfrac{1}{25}\right)^3$

$\left(\dfrac{z}{4} - 5\right)^2 = \dfrac{1}{15,625}$

$\dfrac{z^2}{16} - \dfrac{5}{2}z + 25 = \dfrac{1}{15,625}$

$15,625z^2 - 625,000z + 6,250,000 = 16$

$15,625z^2 - 625,000z + 6,249,984 = 0$

$(125z - 2504)(125z - 2496) = 0$

$125z - 2504 = 0$ or $125z - 2496 = 0$

$125z = 2504$ or $125z = 2496$

$z = \dfrac{2504}{125}$ or $z = \dfrac{2496}{125}$

Both numbers check. The solutions are $\dfrac{2504}{125}$ and $\dfrac{2496}{125}$.

70. $-8, 8$

71.

$\sqrt{\sqrt{y} + 49} = 7$

$(\sqrt{\sqrt{y} + 49})^2 = 7^2$

$\sqrt{y} + 49 = 49$

$\sqrt{y} = 0$

$(\sqrt{y})^2 = 0^2$

$y = 0$

This number 0 checks and is the solution.

72. $-1, 6$

73.
$$\sqrt{8-b} = b\sqrt{8-b}$$
$$(\sqrt{8-b})^2 = (b\sqrt{8-b})^2$$
$$(8-b) = b^2(8-b)$$
$$0 = b^2(8-b) - (8-b)$$
$$0 = (8-b)(b^2-1)$$
$$0 = (8-b)(b+1)(b-1)$$
$$8-b = 0 \text{ or } b+1 = 0 \text{ or } b-1 = 0$$
$$8 = b \text{ or } \quad b = -1 \text{ or } \quad b = 1$$

Since the numbers 8 and 1 check but -1 does not, 8 and 1 are the solutions.

74. $(2, 0)$

75. We find the values of x for which $g(x) = 0$.
$$6x^{1/2} + 6x^{-1/2} - 37 = 0$$
$$6\sqrt{x} + \frac{6}{\sqrt{x}} = 37$$
$$\left(6\sqrt{x} + \frac{6}{\sqrt{x}}\right)^2 = 37^2$$
$$36x + 72 + \frac{36}{x} = 1369$$
$$36x^2 + 72x + 36 = 1369x \quad \text{Multiplying by } x$$
$$36x^2 - 1297x + 36 = 0$$
$$(36x - 1)(x - 36) = 0$$
$$36x - 1 = 0 \quad or \quad x - 36 = 0$$
$$36x = 1 \quad or \quad \quad x = 36$$
$$x = \frac{1}{36} \quad or \quad \quad x = 36$$

Both numbers check. The x-intercepts are $\left(\frac{1}{36}, 0\right)$ and $(36, 0)$.

76. $(0, 0), \left(\frac{125}{4}, 0\right)$

77. *Writing Exercise*

Exercise Set 7.7

1. $a = 5, \quad b = 3$

Find c.
$$c^2 = a^2 + b^2 \quad \text{Pythagorean equation}$$
$$c^2 = 5^2 + 3^2 \quad \text{Substituting}$$
$$c^2 = 25 + 9$$
$$c^2 = 34$$
$$c = \sqrt{34} \quad \quad \text{Exact answer}$$
$$c \approx 5.831 \quad \quad \text{Approximation}$$

2. $\sqrt{164}; 12.806$

3. $a = 9, \quad b = 9$

Observe that the legs have the same length, so this is an isosceles right triangle. Then we know that the length of the hypotenuse is the length of a leg times $\sqrt{2}$, or $9\sqrt{2}$, or approximately 12.728.

4. $10\sqrt{2}; 14.142$

5. $b = 12, \quad c = 13$

Find a.
$$a^2 + b^2 = c^2 \quad \text{Pythagorean equation}$$
$$a^2 + 12^2 = 13^2 \quad \text{Substituting}$$
$$a^2 + 144 = 169$$
$$a^2 = 25$$
$$a = 5$$

6. $\sqrt{119}; 10.909$

7. $c = 6, \quad a = \sqrt{5}$

Find b.
$$c^2 = a^2 + b^2$$
$$(\sqrt{5})^2 + b^2 = 6^2$$
$$5 + b^2 = 36$$
$$b^2 = 31$$
$$b = \sqrt{31} \quad \quad \text{Exact answer}$$
$$b \approx 5.568 \quad \quad \text{Approximation}$$

8. 4

9. $b = 2, \quad c = \sqrt{15}$

Find a.
$$a^2 + b^2 = c^2 \quad \quad \text{Pythagorean equation}$$
$$a^2 + 2^2 = (\sqrt{15})^2 \quad \text{Substituting}$$
$$a^2 + 4 = 15$$
$$a^2 = 11$$
$$a = \sqrt{11} \quad \quad \text{Exact answer}$$
$$a \approx 3.317 \quad \quad \text{Approximation}$$

10. $\sqrt{19}; 4.359$

11. $a = 1, \quad c = \sqrt{2}$

Observe that the length of the hypotenuse, $\sqrt{2}$, is $\sqrt{2}$ times the length of the given leg, 1. Thus, we have an isosceles right triangle and the length of the other leg is also 1.

12. $\sqrt{3}; 1.732$

13. We make a drawing and let d = the length of the guy wire.

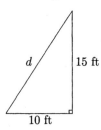

We use the Pythagorean equation to find d.

$d^2 = 10^2 + 15^2$

$d^2 = 100 + 225$

$d^2 = 325$

$d = \sqrt{325}$

$d \approx 18.028$

The wire is $\sqrt{325}$ ft, or about 18.028 ft long.

14. $\sqrt{8450}$ ft; 91.924 ft

15. We first make a drawing and let d = the distance, in feet, to second base. A right triangle is formed in which the length of the leg from second base to third base is 90 ft. The length of the leg from third base to where the catcher fields the ball is $90 - 10$, or 80 ft.

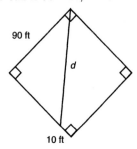

We substitute these values into the Pythagorean equation to find d.

$d^2 = 90^2 + 80^2$

$d^2 = 8100 + 6400$

$d^2 = 14,500$

$d = \sqrt{14,500}$

Exact answer: $d = \sqrt{14,500}$ ft

Approximation: $d \approx 120.416$ ft

16. 12 in.

17. We make a drawing.

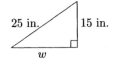

We use the Pythagorean equation to find w.

$w^2 + 15^2 = 25^2$

$w^2 + 225 = 625$

$w^2 = 400$

$w = 20$

The width is 20 in.

18. $(\sqrt{340} + 8)$ ft; 26.439 ft

19.

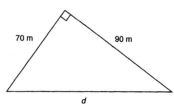

$d^2 = 70^2 + 90^2$

$d^2 = 4900 + 8100$

$d^2 = 13,000$

$d = \sqrt{13,000}$ m   Exact answer

$d \approx 114.018$ m   Approximation

20. 50 ft

21. Since one acute angle is 45°, this is an isosceles right triangle with $b = 5$. Then $a = 5$ also. We substitute to find c.

$c = a\sqrt{2}$

$c = 5\sqrt{2}$

Exact answer: $a = 5$, $c = 5\sqrt{2}$

Approximation: $c \approx 7.071$

22. $a = 14,;\ c = 14\sqrt{2} \approx 19.799$

23. This is a 30-60-90 right triangle with $c = 14$. We substitute to find a and b.

$c = 2a$

$14 = 2a$

$7 = a$

$b = a\sqrt{3}$

$b = 7\sqrt{3}$

Exact answer: $a = 7$, $b = 7\sqrt{3}$

Approximation: $b \approx 12.124$

24. $a = 9; b = 9\sqrt{3} \approx 15.588$

25. This is a 30-60-90 right triangle with $b = 15$. We substitute to find a and c.

$$b = a\sqrt{3}$$
$$15 = a\sqrt{3}$$
$$\frac{15}{\sqrt{3}} = a$$
$$\frac{15\sqrt{3}}{3} = a \qquad \text{Rationalizing the denominator}$$
$$5\sqrt{3} = a \qquad \text{Simplifying}$$
$$c = 2a$$
$$c = 2 \cdot 5\sqrt{3}$$
$$c = 10\sqrt{3}$$

Exact answer: $a = 5\sqrt{3}$, $c = 10\sqrt{3}$

Approximations: $a \approx 8.660$, $c \approx 17.321$

26. $a = 4\sqrt{2} \approx 5.657; b = 4\sqrt{2} \approx 5.657$

27. This is an isosceles right triangle with $c = 13$. We substitute to find a.

$$a = \frac{c\sqrt{2}}{2}$$
$$a = \frac{13\sqrt{2}}{2}$$

Since $a = b$, we have $b = \frac{13\sqrt{2}}{2}$ also.

Exact answer: $a = \frac{13\sqrt{2}}{2}$, $b = \frac{13\sqrt{2}}{2}$

Approximations: $a \approx 9.192$, $b \approx 9.192$

28. $a = \frac{7\sqrt{3}}{3} \approx 4.041; c = \frac{14\sqrt{3}}{3} \approx 8.083$

29. This is a 30-60-90 triangle with $a = 14$. We substitute to find b and c.

$$b = a\sqrt{3} \qquad\qquad c = 2a$$
$$b = 14\sqrt{3} \qquad\qquad c = 2 \cdot 14$$
$$\qquad\qquad\qquad\qquad c = 28$$

Exact answer: $b = 14\sqrt{3}$, $c = 28$

Approximation: $b \approx 24.249$

30. $b = 9\sqrt{3} \approx 15.588; c = 18$

31.

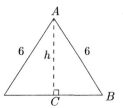

This is an equilateral triangle, so all the angles are 60°. The altitude bisects one angle and one side. Then triangle ABC is a 30-60-90 right triangle with the shorter leg of length 6/2, or 3, and hypotenuse of length 6. We substitute to find the length of the other leg.

$$b = a\sqrt{3}$$
$$h = 3\sqrt{3} \qquad \text{Substituting } h \text{ for } b \text{ and 3 for } a$$

Exact answer: $h = 3\sqrt{3}$

Approximation: $h \approx 5.196$

32. $5\sqrt{3} \approx 8.660$

33.

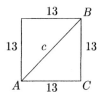

Triangle ABC is an isosceles right triangle with $a = 13$. We substitute to find c.

$$c = a\sqrt{2}$$
$$c = 13\sqrt{2}$$

Exact answer: $c = 13\sqrt{2}$

Approximation: $c \approx 18.385$

34. $7\sqrt{2} \approx 9.899$

35.

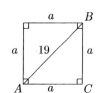

Triangle ABC is an isosceles right triangle with $c = 19$. We substitute to find a.

$$a = \frac{c\sqrt{2}}{2}$$
$$a = \frac{19\sqrt{2}}{2}$$

Exact answer: $a = \frac{19\sqrt{2}}{2}$

Approximation: $a \approx 13.435$

36. $\dfrac{15\sqrt{2}}{2} \approx 10.607$

37. We will express all distances in feet. Recall that 1 mi = 5280 ft.

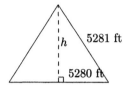

We use the Pythagorean equation to find h.
$$h^2 + (5280)^2 = (5281)^2$$
$$h^2 + 27{,}878{,}400 = 27{,}888{,}961$$
$$h^2 = 10{,}561$$
$$h = \sqrt{10{,}561}$$
$$h \approx 102.767$$

The height of the bulge is $\sqrt{10{,}561}$ ft, or about 102.767 ft.

38. Neither; they have the same area.

39.

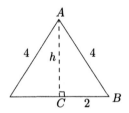

The entrance is an equilateral triangle, so all the angles are 60°. The altitude bisects one angle and one side. Then triangle ABC is a 30-60-90 right triangle with the shorter leg of length 4/2, or 2, and hypotenuse of length 4. We substitute to find h, the height of the tent.

$b = a\sqrt{3}$

$h = 2\sqrt{3}$ Substituting h for b and 2 for a

Exact answer: $h = 2\sqrt{3}$ ft

Approximation: $h \approx 3.464$ ft

40. $d = s + s\sqrt{2}$

41.

Triangle ABC is an isosceles right triangle with $c = 8\sqrt{2}$. We substitute to find a.

$$a = \frac{c\sqrt{2}}{2} = \frac{8\sqrt{2}\cdot\sqrt{2}}{2} = \frac{8\cdot 2}{2} = 8$$

The length of a side of the square is 8 ft.

42. $\sqrt{181}$ cm ≈ 13.454 cm

43.

$$|y|^2 + 3^2 = 5^2$$
$$y^2 + 9 = 25$$
$$y^2 = 16$$
$$y = \pm 4$$

The points are $(0, -4)$ and $(0, 4)$.

44. $(-3, 0)$, $(3, 0)$

45. *Writing Exercise*

46. *Writing Exercise*

47. $47(-1)^{19} = 47(-1) = -47$

48. 5

49. $x^3 - 9x = x \cdot x^2 - 9 \cdot x = x(x^2 - 9) = x(x+3)(x-3)$

50. $7a(a+2)(a-2)$

51. $|3x - 5| = 7$

$3x - 5 = 7$ *or* $3x - 5 = -7$

$3x = 12$ *or* $3x = -2$

$x = 4$ *or* $x = -\dfrac{2}{3}$

The solution set is $\left\{4, -\dfrac{2}{3}\right\}$.

52. $\left\{10, -\dfrac{4}{3}\right\}$

53. *Writing Exercise*

54. *Writing Exercise*

55.

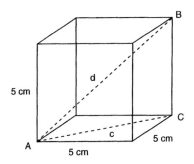

First find the length of a diagonal of the base of the cube. It is the hypotenuse of an isosceles right triangle with $a = 5$ cm. Then $c = a\sqrt{2} = 5\sqrt{2}$ cm.

Triangle ABC is a right triangle with legs of $5\sqrt{2}$ cm and 5 cm and hypotenuse d. Use the Pythagorean equation to find d, the length of the diagonal that connects two opposite corners of the cube.

$$d^2 = (5\sqrt{2})^2 + 5^2$$
$$d^2 = 25 \cdot 2 + 25$$
$$d^2 = 50 + 25$$
$$d^2 = 75$$
$$d = \sqrt{75}$$

Exact answer: $d = \sqrt{75}$ cm

56. 9 packets

57.

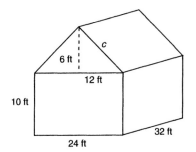

The area to be painted consists of two 10 ft by 24 ft rectangles, two 10 ft by 32 ft rectangles, and two triangles with height 6 ft and base 24 ft. The area of the two 10 ft by 24 ft rectangles is $2 \cdot 10$ ft $\cdot 24$ ft $= 480$ ft^2. The area of the two 10 ft by 32 ft rectangles is $2 \cdot 10$ ft$\cdot 32$ ft $= 640$ ft^2. The area of the two triangles is $2 \cdot \frac{1}{2} \cdot 24$ ft $\cdot 6$ ft $= 144$ ft^2. Thus, the total area to be painted is 480 ft^2 + 640 ft^2 + 144 ft^2 = 1264 ft^2.

One gallon of paint covers 275 ft^2, so we divide to determine how many gallons of paint are required: $\frac{1264}{275} \approx 4.6$. Thus, 4 gallons of paint should be bought to paint the house. This answer assumes that the total area of the doors and windows is 164 ft^2 or more. ($4 \cdot 275 = 1100$ and $1264 = 1100 + 164$)

58. 49.5 ft by 49.5 ft

59. First we find the radius of a circle with an area of 6160 ft^2.
$$A = \pi r^2$$
$$6160 = \pi r^2$$
$$\frac{6160}{\pi} = r^2$$
$$\sqrt{\frac{6160}{\pi}} = r$$
$$44.28 \approx r$$

Now we make a drawing. Let $s =$ the length of a side of the room.

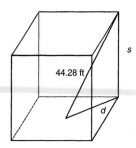

We make a drawing of the floor of the room to help us find d.

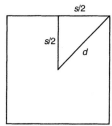

We have an isosceles right triangle, so $d = \frac{s}{2} \cdot \sqrt{2}$, or $\frac{s\sqrt{2}}{2}$.

Now we use the Pythagorean theorem to find s.
$$d^2 + s^2 = (44.28)^2$$
$$\left(\frac{s\sqrt{2}}{2}\right)^2 + s^2 = (44.28)^2 \quad \text{Substituting } \frac{s\sqrt{2}}{2} \text{ for } d.$$
$$\frac{s^2}{2} + s^2 = (44.28)^2$$
$$\frac{3s^2}{2} = (44.28)^2$$
$$s^2 = \frac{2}{3}(44.28)^2$$
$$s = 44.28\sqrt{\frac{2}{3}}$$
$$s \approx 36.15$$

The dimensions of the room are 36.15 ft by 36.15 ft by 36.15 ft.

Exercise Set 7.8

1. $\sqrt{-25} = \sqrt{-1 \cdot 25} = \sqrt{-1} \cdot \sqrt{25} = i \cdot 5 = 5i$

2. $6i$

3. $\sqrt{-13} = \sqrt{-1 \cdot 13} = \sqrt{-1} \cdot \sqrt{13} = i\sqrt{13}$, or $\sqrt{13}i$

4. $i\sqrt{19}$, or $\sqrt{19}i$

5. $\sqrt{-18} = \sqrt{-1} \cdot \sqrt{9} \cdot \sqrt{2} = i \cdot 3 \cdot \sqrt{2} = 3i\sqrt{2}$, or $3\sqrt{2}i$

6. $7i\sqrt{2}$, or $7\sqrt{2}i$

7. $\sqrt{-3} = \sqrt{-1 \cdot 3} = \sqrt{-1} \cdot \sqrt{3} = i\sqrt{3}$, or $\sqrt{3}i$

8. $2i$

9. $\sqrt{-81} = \sqrt{-1 \cdot 81} = \sqrt{-1} \cdot \sqrt{81} = i \cdot 9 = 9i$

10. $3i\sqrt{3}$, or $3\sqrt{3}i$

11. $\sqrt{-300} = \sqrt{-1} \cdot \sqrt{100} \cdot \sqrt{3} = i \cdot 10 \cdot \sqrt{3} = 10i\sqrt{3}$, or $10\sqrt{3}i$

12. $-5i\sqrt{3}$, or $-5\sqrt{3}i$

13. $-\sqrt{-49} = -\sqrt{-1 \cdot 49} = -\sqrt{-1} \cdot \sqrt{49} = -i \cdot 7 = -7i$

14. $-5i\sqrt{5}$, or $-5\sqrt{5}i$

15. $4 - \sqrt{-60} = 4 - \sqrt{-1 \cdot 60} = 4 - \sqrt{-1} \cdot \sqrt{60} = 4 - i \cdot 2\sqrt{15} = 4 - 2\sqrt{15}i$, or $4 - 2i\sqrt{15}$

16. $6 - 2i\sqrt{21}$, or $6 - 2\sqrt{21}i$

17. $\sqrt{-4} + \sqrt{-12} = \sqrt{-1 \cdot 4} + \sqrt{-1 \cdot 12} = \sqrt{-1} \cdot \sqrt{4} + \sqrt{-1} \cdot \sqrt{12} = i \cdot 2 + i \cdot 2\sqrt{3} = (2 + 2\sqrt{3})i$

18. $(-2\sqrt{19} + 5\sqrt{5})i$

19. $\sqrt{-72} - \sqrt{-25} = \sqrt{-1 \cdot 36 \cdot 2} - \sqrt{-1 \cdot 25} = \sqrt{-1}\sqrt{36 \cdot 2} - \sqrt{-1}\sqrt{25} = i \cdot 6\sqrt{2} - i \cdot 5 = (6\sqrt{2} - 5)i$

20. $(3\sqrt{2} - 10)i$

21. $\quad (7 + 8i) + (5 + 3i)$
$\quad = (7 + 5) + (8 + 3)i \quad$ Combining the real and
$\qquad\qquad\qquad\qquad$ the imaginary parts
$\quad = 12 + 11i$

22. $7 + 4i$

23. $(9 + 8i) - (5 + 3i) = (9 - 5) + (8 - 3)i$
$\qquad\qquad\qquad\qquad\quad = 4 + 5i$

24. $7 + 3i$

25. $(5 - 3i) - (9 + 2i) = (5 - 9) + (-3 - 2)i$
$\qquad\qquad\qquad\qquad\quad = -4 - 5i$

26. $2 - i$

27. $(-2 + 6i) - (-7 + i) = -2 - (-7) + (6 - 1)i$
$\qquad\qquad\qquad\qquad\qquad = 5 + 5i$

28. $-12 - 5i$

29. $6i \cdot 9i = 54 \cdot i^2$
$\qquad\quad = 54 \cdot (-1) \qquad\quad i^2 = -1$
$\qquad\quad = -54$

30. -42

31. $7i \cdot (-8i) = -56 \cdot i^2$
$\qquad\qquad = -56 \cdot (-1) \qquad i^2 = -1$
$\qquad\qquad = 56$

32. -24

33. $\sqrt{-49}\sqrt{-25} = \sqrt{-1} \cdot \sqrt{49} \cdot \sqrt{-1} \cdot \sqrt{25}$
$\qquad\qquad\quad = i \cdot 7 \cdot i \cdot 5$
$\qquad\qquad\quad = i^2 \cdot 35$
$\qquad\qquad\quad = -1 \cdot 35$
$\qquad\qquad\quad = -35$

34. -18

35. $\sqrt{-6}\sqrt{-7} = \sqrt{-1} \cdot \sqrt{6} \cdot \sqrt{-1} \cdot \sqrt{7}$
$\qquad\qquad\; = i \cdot \sqrt{6} \cdot i \cdot \sqrt{7}$
$\qquad\qquad\; = i^2 \cdot \sqrt{42}$
$\qquad\qquad\; = -1 \cdot \sqrt{42}$
$\qquad\qquad\; = -\sqrt{42}$

36. $-\sqrt{10}$

37. $\sqrt{-15}\sqrt{-10} = \sqrt{-1} \cdot \sqrt{15} \cdot \sqrt{-1} \cdot \sqrt{10}$
$\qquad\qquad\qquad = i \cdot \sqrt{15} \cdot i \cdot \sqrt{10}$
$\qquad\qquad\qquad = i^2 \cdot \sqrt{150}$
$\qquad\qquad\qquad = -\sqrt{25 \cdot 6}$
$\qquad\qquad\qquad = -5\sqrt{6}$

38. $-3\sqrt{14}$

39. $2i(7 + 3i)$
$= 2i \cdot 7 + 2i \cdot 3i$ Using the distributive law
$= 14i + 6i^2$
$= 14i - 6$ $i^2 = -1$
$= -6 + 14i$

40. $-30 + 10i$

41. $-4i(6 - 5i) = -4i \cdot 6 - 4i(-5i)$
$= -24i + 20i^2$
$= -24i - 20$
$= -20 - 24i$

42. $-28 - 21i$

43. $(1 + 5i)(4 + 3i)$
$= 4 + 3i + 20i + 15i^2$ Using FOIL
$= 4 + 3i + 20i - 15$ $i^2 = -1$
$= -11 + 23i$

44. $1 + 5i$

45. $(5 - 6i)(2 + 5i) = 10 + 25i - 12i - 30i^2$
$= 10 + 25i - 12i + 30$
$= 40 + 13i$

46. $38 + 9i$

47. $(-4 + 5i)(3 - 4i) = -12 + 16i + 15i - 20i^2$
$= -12 + 16i + 15i + 20$
$= 8 + 31i$

48. $2 - 46i$

49. $(7 - 3i)(4 - 7i) = 28 - 49i - 12i + 21i^2 =$
$28 - 49i - 12i - 21 = 7 - 61i$

50. $5 - 37i$

51. $(-3 + 6i)(-3 + 4i) = 9 - 12i - 18i + 24i^2 =$
$9 - 12i - 18i - 24 = -15 - 30i$

52. $-11 - 16i$

53. $(2 + 9i)(-3 - 5i) = -6 - 10i - 27i - 45i^2 =$
$-6 - 10i - 27i + 45 = 39 - 37i$

54. $13 - 47i$

55. $(1 - 2i)^2$
$= 1^2 - 2 \cdot 1 \cdot 2i + (2i)^2$ Squaring a binomial
$= 1 - 4i + 4i^2$
$= 1 - 4i - 4$ $i^2 = -1$
$= -3 - 4i$

56. $12 - 16i$

57. $(3 + 2i)^2$
$= 3^2 + 2 \cdot 3 \cdot 2i + (2i)^2$ Squaring a binomial
$= 9 + 12i + 4i^2$
$= 9 + 12i - 4$ $i^2 = -1$
$= 5 + 12i$

58. $-5 + 12i$

59. $(-5 - 2i)^2 = 25 + 20i + 4i^2 = 25 + 20i - 4 =$
$21 + 20i$

60. $-5 - 12i$

61. $\dfrac{3}{2 - i}$
$= \dfrac{3}{2 - i} \cdot \dfrac{2 + i}{2 + i}$ Multiplying by 1, using the conjugate
$= \dfrac{6 + 3i}{4 - i^2}$ Multiplying
$= \dfrac{6 + 3i}{4 - (-1)}$ $i^2 = -1$
$= \dfrac{6 + 3i}{5}$
$= \dfrac{6}{5} + \dfrac{3}{5}i$

62. $\dfrac{6}{5} - \dfrac{2}{5}i$

63. $\dfrac{3i}{5 + 2i}$
$= \dfrac{3i}{5 + 2i} \cdot \dfrac{5 - 2i}{5 - 2i}$ Multiplying by 1, using the conjugate
$= \dfrac{15i - 6i^2}{25 - 4i^2}$ Multiplying
$= \dfrac{15i + 6}{25 + 4}$
$= \dfrac{15i + 6}{29}$
$= \dfrac{6}{29} + \dfrac{15}{29}i$

64. $-\dfrac{6}{17} + \dfrac{10}{17}i$

65. $\dfrac{7}{9i} = \dfrac{7}{9i} \cdot \dfrac{i}{i} = \dfrac{7i}{9i^2} = \dfrac{7i}{-9} = -\dfrac{7}{9}i$

66. $-\dfrac{5}{8}i$

67. $\dfrac{5-3i}{4i} = \dfrac{5-3i}{4i} \cdot \dfrac{i}{i} = \dfrac{5i - 3i^2}{4i^2} = \dfrac{5i+3}{-4} =$

$-\dfrac{3}{4} - \dfrac{5}{4}i$

68. $\dfrac{7}{5} - \dfrac{2}{5}i$

69. $\dfrac{7i+14}{7i} = \dfrac{7i}{7i} + \dfrac{14}{7i} = 1 + \dfrac{2}{i} = 1 + \dfrac{2}{i} \cdot \dfrac{i}{i} =$

$1 + \dfrac{2i}{i^2} = 1 + \dfrac{2i}{-1} = 1 - 2i$

70. $2 - i$

71. $\dfrac{4+5i}{3-7i} = \dfrac{4+5i}{3-7i} \cdot \dfrac{3+7i}{3+7i} = \dfrac{12 + 28i + 15i + 35i^2}{9 - 49i^2} =$

$\dfrac{12 + 28i + 15i - 35}{9 + 49} = \dfrac{-23 + 43i}{58} = -\dfrac{23}{58} + \dfrac{43}{58}i$

72. $\dfrac{23}{65} + \dfrac{41}{65}i$

73. $\dfrac{3-2i}{4+3i} = \dfrac{3-2i}{4+3i} \cdot \dfrac{4-3i}{4-3i} = \dfrac{12 - 9i - 8i + 6i^2}{16 - 9i^2} =$

$\dfrac{12 - 9i - 8i - 6}{16 + 9} = \dfrac{6 - 17i}{25} = \dfrac{6}{25} - \dfrac{17}{25}i$

74. $\dfrac{1}{15} - \dfrac{4}{5}i$

75. $i^7 = i^6 \cdot i = (i^2)^3 \cdot i = (-1)^3 \cdot i = -1 \cdot i = -i$

76. $-i$

77. $i^{24} = (i^2)^{12} = (-1)^{12} = 1$

78. $-i$

79. $i^{42} = (i^2)^{21} = (-1)^{21} = -1$

80. 1

81. $i^9 = (i^2)^4 \cdot i = (-1)^4 \cdot i = 1 \cdot i = i$

82. i

83. $i^6 = (i^2)^3 = (-1)^3 = -1$

84. 1

85. $(5i)^3 = 5^3 \cdot i^3 = 125 \cdot i^2 \cdot i = 125(-1)(i) = -125i$

86. $-243i$

87. $i^2 + i^4 = -1 + (i^2)^2 = -1 + (-1)^2 = -1 + 1 = 0$

88. i

89. *Writing Exercise*

90. *Writing Exercise*

91. $f(x) = x^2 - 3x,\ g(x) = 2x - 5$

$(f+g)(-2) = f(-2) + g(-2)$

$\qquad = (-2)^2 - 3(-2) + 2(-2) - 5$

$\qquad = 4 + 6 - 4 - 5$

$\qquad = 1$

92. 1

93. $(f \cdot g)(5) = f(5)g(5)$

$\qquad = (5^2 - 3 \cdot 5)(2 \cdot 5 - 5)$

$\qquad = (25 - 15)(10 - 5)$

$\qquad = 10 \cdot 5$

$\qquad = 50$

94. 0

95. $28 = 3x^2 - 17x$

$0 = 3x^2 - 17x - 28$

$0 = (3x + 4)(x - 7)$

$3x + 4 = 0 \quad or \quad x - 7 = 0$

$3x = -4 \quad or \qquad x = 7$

$x = -\dfrac{4}{3} \quad or \qquad x = 7$

Both values check. The solutions are $-\dfrac{4}{3}$ and 7.

96. $\left\{ x \middle| -\dfrac{29}{3} < x < 5 \right\}$, or $\left(-\dfrac{29}{3}, 5 \right)$

97. *Writing Exercise*

98. *Writing Exercise*

99. $g(3i) = \dfrac{(3i)^4 - (3i)^2}{3i - 1} = \dfrac{81i^4 - 9i^2}{-1 + 3i} = \dfrac{81 + 9}{-1 + 3i} =$

$\dfrac{90}{-1 + 3i} = \dfrac{90}{-1 + 3i} \cdot \dfrac{-1 - 3i}{-1 - 3i} = \dfrac{90(-1 - 3i)}{1 - 9i^2} =$

$\dfrac{90(-1 - 3i)}{1 + 9} = \dfrac{90(-1 - 3i)}{10} = \dfrac{9 \cdot \cancel{10}(-1 - 3i)}{\cancel{10}} =$

$9(-1 - 3i) = -9 - 27i$

100. $-2 + 4i$

101. First we simplify $g(z)$.

$g(z) = \dfrac{z^4 - z^2}{z - 1} = \dfrac{z^2(z^2 - 1)}{z - 1} = \dfrac{z^2(z + 1)(z - 1)}{z - 1} =$

$\dfrac{z^2(z + 1)\cancel{(z-1)}}{\cancel{z-1}} = z^2(z + 1)$

Now we substitute.

$g(5i - 1) = (5i - 1)^2(5i - 1 + 1) =$

$(25i^2 - 10i + 1)(5i) =$

$(-25 - 10i + 1)(5i) = (-24 - 10i)(5i) =$

$-120i - 50i^2 = 50 - 120i$

102. $-51 - 21i$

103. $\dfrac{1}{\dfrac{1-i}{10} - \left(\dfrac{1-i}{10}\right)^2} = \dfrac{1}{\dfrac{1-i}{10} - \left(\dfrac{-2i}{100}\right)} =$

$\dfrac{1}{\dfrac{1-i}{10} + \dfrac{i}{50}} = \dfrac{1}{\dfrac{1-i}{10} + \dfrac{i}{50}} \cdot \dfrac{50}{50} = \dfrac{50}{5 - 5i + i} =$

$\dfrac{50}{5 - 4i} = \dfrac{50}{5 - 4i} \cdot \dfrac{5 + 4i}{5 + 4i} = \dfrac{250 + 200i}{41} = \dfrac{250}{41} + \dfrac{200}{41}i$

104. 0

105. $(1 - i)^3(1 + i)^3 =$

$(1 - i)(1 + i) \cdot (1 - i)(1 + i) \cdot (1 - i)(1 + i) =$

$(1 - i^2)(1 - i^2)(1 - i^2) = (1 + 1)(1 + 1)(1 + 1) =$

$2 \cdot 2 \cdot 2 = 8$

106. $-1 - \sqrt{5}i$

107. $\dfrac{6}{1 + \dfrac{3}{i}} = \dfrac{6}{\dfrac{i + 3}{i}} = \dfrac{6i}{i + 3} = \dfrac{6i}{i + 3} \cdot \dfrac{-i + 3}{-i + 3} =$

$\dfrac{-6i^2 + 18i}{-i^2 + 9} = \dfrac{6 + 18i}{10} = \dfrac{6}{10} + \dfrac{18}{10}i = \dfrac{3}{5} + \dfrac{9}{5}i$

108. $-\dfrac{2}{3}i$

109. $\dfrac{i - i^{38}}{1 + i} = \dfrac{i - (i^2)^{19}}{1 + i} = \dfrac{i - (-1)^{19}}{1 + i} = \dfrac{i - (-1)}{1 + i} =$

$\dfrac{i + 1}{1 + i} = 1$

Chapter 8

Quadratic Functions and Equations

Exercise Set 8.1

1. There are 2 x-intercepts, so there are 2 real-number solutions.

2. 0

3. There is 1 x-intercept, so there is 1 real-number solution.

4. 2

5. There are no x-intercepts, so there are no real-number solutions.

6. 1

7.
$$7x^2 = 21$$
$$x^2 = 3 \qquad \text{Multiplying by } \frac{1}{7}$$
$$x = \sqrt{3} \text{ or } x = -\sqrt{3} \qquad \text{Using the principle of square roots}$$
The solutions are $\sqrt{3}$ and $-\sqrt{3}$, or $\pm\sqrt{3}$.

8. $\pm\sqrt{5}$

9.
$$25x^2 + 4 = 0$$
$$x^2 = -\frac{4}{25} \qquad \text{Isolating } x^2$$
$$x = \sqrt{-\frac{4}{25}} \text{ or } x = -\sqrt{-\frac{4}{25}} \qquad \text{Principle of square roots}$$
$$x = \sqrt{\frac{4}{25}}\sqrt{-1} \text{ or } x = -\sqrt{\frac{4}{25}}\sqrt{-1}$$
$$x = \frac{2}{5}i \text{ or } x = -\frac{2}{5}i$$
The solutions are $\frac{2}{5}i$ and $-\frac{2}{5}i$, or $\pm\frac{2}{5}i$.

10. $\pm\frac{4}{3}i$

11.
$$3t^2 - 2 = 0$$
$$3t^2 = 2$$
$$t^2 = \frac{2}{3}$$

$$t = \sqrt{\frac{2}{3}} \quad \text{or} \quad t = -\sqrt{\frac{2}{3}} \qquad \text{Principle of square roots}$$
$$t = \sqrt{\frac{2}{3} \cdot \frac{3}{3}} \quad \text{or} \quad t = -\sqrt{\frac{2}{3} \cdot \frac{3}{3}} \qquad \text{Rationalizing denominators}$$
$$t = \frac{\sqrt{6}}{3} \quad \text{or} \quad t = \frac{-\sqrt{6}}{3}$$

The solutions are $\sqrt{\frac{2}{3}}$ and $-\sqrt{\frac{2}{3}}$. This can also be written as $\pm\sqrt{\frac{2}{3}}$ or, if we rationalize the denominator, $\pm\frac{\sqrt{6}}{3}$.

12. $\pm\frac{\sqrt{35}}{5}$

13.
$$(x + 2)^2 = 25$$
$$x + 2 = 5 \text{ or } x + 2 = -5 \qquad \text{Principle of square roots}$$
$$x = 3 \text{ or } x = -7$$
The solutions are 3 and -7.

14. $-6, 8$

15.
$$(a + 5)^2 = 8$$
$$a + 5 = \sqrt{8} \text{ or } a + 5 = -\sqrt{8} \qquad \text{Principle of square roots}$$
$$a + 5 = 2\sqrt{2} \text{ or } a + 5 = -2\sqrt{2} \qquad (\sqrt{8} = \sqrt{4 \cdot 2} = 2\sqrt{2})$$
$$a = -5 + 2\sqrt{2} \text{ or } a = -5 - 2\sqrt{2}$$
The solutions are $-5 + 2\sqrt{2}$ and $-5 - 2\sqrt{2}$, or $-5 \pm 2\sqrt{2}$.

16. $13 \pm 3\sqrt{2}$

17. $(x - 1)^2 = -49$
$$x - 1 = \sqrt{-49} \quad \text{or} \quad x - 1 = -\sqrt{-49}$$
$$x - 1 = 7i \qquad \text{or} \quad x - 1 = -7i$$
$$x = 1 + 7i \quad \text{or} \qquad x = 1 - 7i$$
The solutions are $1 + 7i$ and $1 - 7i$, or $1 \pm 7i$.

18. $-1 \pm 3i$

19. $\left(t + \dfrac{3}{2}\right)^2 = \dfrac{7}{2}$

$$t + \dfrac{3}{2} = \sqrt{\dfrac{7}{2}} \ \text{or} \ t + \dfrac{3}{2} = -\sqrt{\dfrac{7}{2}}$$

$$t + \dfrac{3}{2} = \sqrt{\dfrac{7}{2} \cdot \dfrac{2}{2}} \ \text{or} \ t + \dfrac{3}{2} = -\sqrt{\dfrac{7}{2} \cdot \dfrac{2}{2}}$$

$$t + \dfrac{3}{2} = \dfrac{\sqrt{14}}{2} \ \text{or} \ t + \dfrac{3}{2} = -\dfrac{\sqrt{14}}{2}$$

$$t = -\dfrac{3}{2} + \dfrac{\sqrt{14}}{2} \ \text{or} \ t = -\dfrac{3}{2} - \dfrac{\sqrt{14}}{2}$$

$$t = \dfrac{-3 + \sqrt{14}}{2} \ \text{or} \ t = \dfrac{-3 - \sqrt{14}}{2}$$

The solutions are $\dfrac{-3 + \sqrt{14}}{2}$ and $\dfrac{-3 - \sqrt{14}}{2}$, or $\dfrac{-3 \pm \sqrt{14}}{2}$.

20. $\dfrac{-3 \pm \sqrt{17}}{4}$

21. $x^2 - 6x + 9 = 100$

$$(x - 3)^2 = 100$$

$$x - 3 = 10 \ \text{or} \ x - 3 = -10$$

$$x = 13 \ \text{or} \ x = -7$$

The solutions are 13 and -7.

22. $-3, 13$

23. $f(x) = 16$

$\quad (x - 5)^2 = 16$ $\qquad$ Substituting

$\quad x - 5 = 4 \ \text{or} \ x - 5 = -4$

$\quad x = 9 \ \text{or} \ \ x = 1$

The solutions are 9 and 1.

24. $-3, 7$

25. $F(t) = 13$

$\quad (t + 4)^2 = 13$ $\quad$ Substituting

$\quad t + 4 = \sqrt{13} \qquad \text{or} \ \ t + 4 = -\sqrt{13}$

$\qquad t = -4 + \sqrt{13} \ \text{or} \qquad t = -4 - \sqrt{13}$

The solutions are $-4 + \sqrt{13}$ and $-4 - \sqrt{13}$, or $-4 \pm \sqrt{13}$.

26. $-6 \pm \sqrt{15}$

27. $g(x) = x^2 + 14x + 49$

Observe first that $g(0) = 49$. Also observe that when $x = -14$, then $x^2 + 14x = (-14)^2 - (14)(14) = (14)^2 - (14)^2 = 0$, so $g(-14) = 49$ as well. Thus, we have $x = 0$ or $x = 14$.

We can also do this problem as follows.

$$g(x) = 49$$

$$x^2 + 14x + 49 = 49 \quad \text{Substituting}$$

$$(x + 7)^2 = 49$$

$$x + 7 = 7 \ \ \text{or} \ \ x + 7 = -7$$

$$x = 0 \ \ \text{or} \qquad x = -14$$

The solutions are 0 and -14.

28. $-7, -1$

29. $x^2 + 8x$

We take half the coefficient of x and square it:

Half of 8 is 4, and $4^2 = 16$. We add 16.

$x^2 + 8x + 16, \ (x + 4)^2$

30. $x^2 + 16x + 64, \ (x + 8)^2$

31. $x^2 - 6x$

We take half the coefficient of x and square it:

Half of -6 is -3, and $(-3)^2 = 9$. We add 9.

$x^2 - 6x + 9, \ (x - 3)^2$

32. $x^2 - 10x + 25, \ (x - 5)^2$

33. $x^2 - 24x$

We take half the coefficient of x and square it:

$\dfrac{1}{2}(-24) = -12$ and $(-12)^2 = 144$. We add 144.

$x^2 - 24x + 144, \ (x - 12)^2$

34. $x^2 - 18x + 81, \ (x - 9)^2$

35. $t^2 + 9t$

$\dfrac{1}{2} \cdot 9 = \dfrac{9}{2}$, and $\left(\dfrac{9}{2}\right)^2 = \dfrac{81}{4}$. We add $\dfrac{81}{4}$.

$t^2 + 9t + \dfrac{81}{4}, \ \left(t + \dfrac{9}{2}\right)^2$

36. $t^2 + 3t + \dfrac{9}{4}, \ \left(t + \dfrac{3}{2}\right)^2$

37. $x^2 - 3x$

We take half the coefficient of x and square it:

$\dfrac{1}{2}(-3) = -\dfrac{3}{2}$ and $\left(-\dfrac{3}{2}\right)^2 = \dfrac{9}{4}$. We add $\dfrac{9}{4}$.

$x^2 - 3x + \dfrac{9}{4}, \ \left(x - \dfrac{3}{2}\right)^2$

38. $x^2 - 7x + \dfrac{49}{4}, \ \left(x - \dfrac{7}{2}\right)^2$

39. $x^2 + \frac{2}{3}x$

$\frac{1}{2} \cdot \frac{2}{3} = \frac{1}{3}$, and $\left(\frac{1}{3}\right)^2 = \frac{1}{9}$. We add $\frac{1}{9}$.

$x^2 + \frac{2}{3}x + \frac{1}{9}, \ \left(x + \frac{1}{3}\right)^2$

40. $x^2 + \frac{2}{5}x + \frac{1}{25}, \ \left(x + \frac{1}{5}\right)^2$

41. $t^2 - \frac{5}{3}t$

$\frac{1}{2}\left(-\frac{5}{3}\right) = -\frac{5}{6}$, and $\left(-\frac{5}{6}\right)^2 = \frac{25}{36}$. We add $\frac{25}{36}$.

$t^2 - \frac{5}{3}t + \frac{25}{36}, \ \left(t - \frac{5}{6}\right)^2$

42. $t^2 - \frac{5}{6}t + \frac{25}{144}, \ \left(t - \frac{5}{12}\right)^2$

43. $x^2 + \frac{9}{5}x$

$\frac{1}{2} \cdot \frac{9}{5} = \frac{9}{10}$, and $\left(\frac{9}{10}\right)^2 = \frac{81}{100}$. We add $\frac{81}{100}$.

$x^2 + \frac{9}{5}x + \frac{81}{100}, \ \left(x + \frac{9}{10}\right)^2$

44. $x^2 + \frac{9}{4}x + \frac{81}{64}, \ \left(x + \frac{9}{8}\right)^2$

45.
$$x^2 + 6x = 7$$
$$x^2 + 6x + 9 = 7 + 9 \quad \text{Adding 9 to both sides to complete the square}$$
$$(x + 3)^2 = 16 \quad \text{Factoring}$$
$$x + 3 = \pm 4 \quad \text{Principle of square roots}$$
$$x = -3 \pm 4$$
$$x = -3 + 4 \ \text{or} \ x = -3 - 4$$
$$x = 1 \qquad \text{or} \ x = -7$$
The solutions are 1 and -7.

46. $-9, 1$

47.
$$x^2 - 10x = 22$$
$$x^2 - 10x + 25 = 22 + 25 \quad \text{Adding 25 to both sides to complete the square}$$
$$(x - 5)^2 = 47$$
$$x - 5 = \pm\sqrt{47} \quad \text{Principle of square roots}$$
$$x = 5 \pm \sqrt{47}$$
The solutions are $5 \pm \sqrt{47}$.

48. $2 \pm i\sqrt{5}$

49.
$$x^2 + 8x + 7 = 0$$
$$x^2 + 8x = -7 \quad \text{Adding } -7 \text{ to both sides}$$
$$x^2 + 8x + 16 = -7 + 16 \quad \text{Completing the square}$$
$$(x + 4)^2 = 9$$
$$x + 4 = \pm 3$$
$$x = -4 \pm 3$$
$$x = -4 - 3 \ \text{or} \ x = -4 + 3$$
$$x = -7 \qquad \text{or} \ x = -1$$
The solutions are -7 and -1.

50. $-9, -1$

51.
$$x^2 - 10x + 21 = 0$$
$$x^2 - 10x = -21$$
$$x^2 - 10x + 25 = -21 + 25$$
$$(x - 5)^2 = 4$$
$$x - 5 = \pm 2$$
$$x = 5 \pm 2$$
$$x = 5 - 2 \ \text{or} \ x = 5 + 2$$
$$x = 3 \qquad \text{or} \ x = 7$$
The solutions are 3 and 7.

52. $4, 6$

53.
$$t^2 + 5t + 3 = 0$$
$$t^2 + 5t = -3$$
$$t^2 + 5t + \frac{25}{4} = -3 + \frac{25}{4}$$
$$\left(t + \frac{5}{2}\right)^2 = \frac{13}{4}$$
$$t + \frac{5}{2} = \pm\frac{\sqrt{13}}{2}$$
$$t = -\frac{5}{2} \pm \frac{\sqrt{13}}{2}$$
$$t = \frac{-5 \pm \sqrt{13}}{2}$$
The solutions are $\dfrac{-5 \pm \sqrt{13}}{2}$.

54. $-3 \pm \sqrt{2}$

55.
$$x^2 + 10 = 6x$$
$$x^2 - 6x = -10$$
$$x^2 - 6x + 9 = -10 + 9$$
$$(x-3)^2 = -1$$
$$x - 3 = \pm\sqrt{-1}$$
$$x - 3 = \pm i$$
$$x = 3 \pm i$$

The solutions are $3 \pm i$.

56. $5 \pm \sqrt{2}$

57.
$$s^2 + 4s + 13 = 0$$
$$s^2 + 4s = -13$$
$$s^2 + 4s + 4 = -13 + 4$$
$$(s+2)^2 = -9$$
$$s + 2 = \pm\sqrt{-9}$$
$$s + 2 = \pm 3i$$
$$s = -2 \pm 3i$$

The solutions are $-2 \pm 3i$.

58. $-6 \pm \sqrt{11}$

59.
$$2x^2 - 5x - 3 = 0$$
$$2x^2 - 5x = 3$$
$$x^2 - \frac{5}{2}x = \frac{3}{2} \quad \text{Dividing both sides by 2}$$
$$x^2 - \frac{5}{2}x + \frac{25}{16} = \frac{3}{2} + \frac{25}{16}$$
$$\left(x - \frac{5}{4}\right)^2 = \frac{49}{16}$$
$$x - \frac{5}{4} = \pm\frac{7}{4}$$
$$x = \frac{5}{4} \pm \frac{7}{4}$$
$$x = \frac{5}{4} - \frac{7}{4} \quad or \quad x = \frac{5}{4} + \frac{7}{4}$$
$$x = -\frac{1}{2} \quad or \quad x = 3$$

The solutions are $-\frac{1}{2}$ and 3.

60. $-2, \frac{1}{3}$

61.
$$4x^2 + 8x + 3 = 0$$
$$4x^2 + 8x = -3$$
$$x^2 + 2x = -\frac{3}{4}$$
$$x^2 + 2x + 1 = -\frac{3}{4} + 1$$
$$(x+1)^2 = \frac{1}{4}$$
$$x + 1 = \pm\frac{1}{2}$$
$$x = -1 \pm \frac{1}{2}$$
$$x = -1 - \frac{1}{2} \quad or \quad x = -1 + \frac{1}{2}$$
$$x = -\frac{3}{2} \qquad or \quad x = -\frac{1}{2}$$

The solutions are $-\frac{3}{2}$ and $-\frac{1}{2}$.

62. $-\frac{4}{3}, -\frac{2}{3}$

63.
$$6x^2 - x = 15$$
$$x^2 - \frac{1}{6}x = \frac{5}{2}$$
$$x^2 - \frac{1}{6}x + \frac{1}{144} = \frac{5}{2} + \frac{1}{144}$$
$$\left(x - \frac{1}{12}\right)^2 = \frac{361}{144}$$
$$x - \frac{1}{12} = \pm\frac{19}{12}$$
$$x = \frac{1}{12} \pm \frac{19}{12}$$
$$x = \frac{1}{12} + \frac{19}{12} \quad or \quad x = \frac{1}{12} - \frac{19}{12}$$
$$x = \frac{20}{12} \qquad or \quad x = -\frac{18}{12}$$
$$x = \frac{5}{3} \qquad or \quad x = -\frac{3}{2}$$

The solutions are $\frac{5}{3}$ and $-\frac{3}{2}$.

64. $-\frac{1}{2}, \frac{2}{3}$

65. $2x^2 + 4x + 1 = 0$

$$2x^2 + 4x = -1$$

$$x^2 + 2x = -\frac{1}{2}$$

$$x^2 + 2x + 1 = -\frac{1}{2} + 1$$

$$(x+1)^2 = \frac{1}{2}$$

$$x + 1 = \pm\sqrt{\frac{1}{2}}$$

$$x + 1 = \pm\frac{\sqrt{2}}{2} \qquad \text{Rationalizing the denominator}$$

$$x = -1 \pm \frac{\sqrt{2}}{2}$$

The solutions are $-1 \pm \dfrac{\sqrt{2}}{2}$, or $\dfrac{-2 \pm \sqrt{2}}{2}$.

66. $-2, -\dfrac{1}{2}$

67. $3x^2 - 5x - 3 = 0$

$$3x^2 - 5x = 3$$

$$x^2 - \frac{5}{3}x = 1$$

$$x^2 - \frac{5}{3}x + \frac{25}{36} = 1 + \frac{25}{36}$$

$$\left(x - \frac{5}{6}\right)^2 = \frac{61}{36}$$

$$x - \frac{5}{6} = \pm\frac{\sqrt{61}}{6}$$

$$x = \frac{5 \pm \sqrt{61}}{6}$$

The solutions are $\dfrac{5 \pm \sqrt{61}}{6}$.

68. $\dfrac{3 \pm \sqrt{13}}{4}$

69. Familiarize. We are already familiar with the compound-interest formula.

Translate. We substitute into the formula.

$$A = P(1 + r)^t$$

$$2420 = 2000(1 + r)^2$$

Carry out. We solve for r.

$$2420 = 2000(1 + r)^2$$

$$\frac{2420}{2000} = (1 + r)^2$$

$$\frac{121}{100} = (1 + r)^2$$

$$\pm\sqrt{\frac{121}{100}} = 1 + r$$

$$\pm\frac{11}{10} = 1 + r$$

$$-\frac{10}{10} + \frac{11}{10} = r$$

$$\frac{1}{10} = r \text{ or } -\frac{21}{10} = r$$

Check. Since the interest rate cannot be negative, we need only check $\dfrac{1}{10}$, or 10%. If $2000 were invested at 10% interest, compounded annually, then in 2 years it would grow to $2000(1.1)^2$, or $2420. The number 10% checks.

State. The interest rate is 10%.

70. 6.25%

71. Familiarize. We are already familiar with the compound-interest formula.

Translate. We substitute into the formula.

$$A = P(1 + r)^t$$

$$1805 = 1280(1 + r)^2$$

Carry out. We solve for r.

$$1805 = 1280(1 + r)^2$$

$$\frac{1805}{1280} = (1 + r)^2$$

$$\frac{361}{256} = (1 + r)^2$$

$$\pm\frac{19}{16} = 1 + r$$

$$-\frac{16}{16} \pm \frac{19}{16} = r$$

$$\frac{3}{16} = r \text{ or } -\frac{35}{16} = r$$

Check. Since the interest rate cannot be negative, we need only check $\dfrac{3}{16}$ or 18.75%. If $1280 were invested at 18.75% interest, compounded annually, then in 2 years it would grow to $1280(1.1875)^2$, or $1805. The number 18.75% checks.

State. The interest rate is 18.75%.

72. 20%

73. *Familiarize*. We are already familiar with the compound-interest formula.

***Translate*.** We substitute into the formula.
$$A = P(1 + r)^t$$
$$6760 = 6250(1 + r)^2$$

***Carry out*.** We solve for r.
$$\frac{6760}{6250} = (1 + r)^2$$
$$\frac{676}{625} = (1 + r)^2$$
$$\pm\frac{26}{25} = 1 + r$$
$$-\frac{25}{25} \pm \frac{26}{25} = r$$
$$\frac{1}{25} = r \ or \ -\frac{51}{25} = r$$

***Check*.** Since the interest rate cannot be negative, we need only check $\frac{1}{25}$, or 4%. If \$6250 were invested at 4% interest, compounded annually, then in 2 years it would grow to \$6250(1.04)^2, or \$6760. The number 4% checks.

***State*.** The interest rate is 4%.

74. 8%

75. *Familiarize*. We will use the formula $s = 16t^2$.

***Translate*.** We substitute into the formula.
$$s = 16t^2$$
$$1815 = 16t^2$$

***Carry out*.** We solve for t.
$$1815 = 16t^2$$
$$\frac{1815}{16} = t^2$$
$$\sqrt{\frac{1815}{16}} = t \quad \text{Principle of square roots;}$$
$$\text{rejecting the negative}$$
$$\text{square root}$$
$$10.7 \approx t$$

***Check*.** Since $16(10.7)^2 = 1831.84 \approx 1815$, our answer checks.

***State*.** It would take an object about 10.7 sec to fall freely from the top of the CN Tower.

76. About 6.8 sec

77. *Familiarize*. We will use the formula $s = 16t^2$.

***Translate*.** We substitute into the formula.
$$s = 16t^2$$
$$640 = 16t^2$$

***Carry out*.** We solve for t.
$$640 = 16t^2$$
$$40 = t^2$$
$$\sqrt{40} = t \quad \text{Principle of square roots;}$$
$$\text{rejecting the negative square}$$
$$\text{root}$$
$$6.3 \approx t$$

***Check*.** Since $16(6.3)^2 = 635.04 \approx 640$, our answer checks.

***State*.** It would take an object about 6.3 sec to fall freely from the top of the Gateway Arch.

78. About 9.5 sec

79. *Writing Exercise*

80. *Writing Exercise*

81. $at^2 - bt = 3 \cdot 4^2 - 5 \cdot 4$
$$= 3 \cdot 16 - 5 \cdot 4$$
$$= 48 - 20$$
$$= 28$$

82. -92

83. $\sqrt[3]{270} = \sqrt[3]{27 \cdot 10} = \sqrt[3]{27}\sqrt[3]{10} = 3\sqrt[3]{10}$

84. $4\sqrt{5}$

85. $f(x) = \sqrt{3x - 5}$
$$f(10) = \sqrt{3 \cdot 10 - 5} = \sqrt{30 - 5} = \sqrt{25} = 5$$

86. 7

87. *Writing Exercise*

88. *Writing Exercise*

89. In order for $x^2 + bx + 81$ to be a square, the following must be true:
$$\left(\frac{b}{2}\right)^2 = 81$$
$$\frac{b^2}{4} = 81$$
$$b^2 = 324$$
$$b = 18 \ or \ b = -18$$

90. ± 14

91. We see that x is a factor of each term, so x is also a factor of $f(x)$. We have $f(x) = x(2x^4 - 9x^3 - 66x^2 + 45x + 280)$. Since $x^2 - 5$ is a factor of $f(x)$ it is also a factor of $2x^4 - 9x^3 - 66x^2 + 45x + 280$. We divide to find another factor.

$$
\begin{array}{r}
2x^2 - 9x - 56 \\
x^2 - 5 \overline{\smash{\big)}\ 2x^4 - 9x^3 - 66x^2 + 45x + 280} \\
\underline{2x^4 \phantom{{}-9x^3} - 10x^2 } \\
-9x^3 - 56x^2 + 45x \\
\underline{-9x^3 + 45x } \\
-56x^2 + 280 \\
\underline{-56x^2 + 280} \\
0
\end{array}
$$

Then we have $f(x) = x(x^2 - 5)(2x^2 - 9x - 56)$, or $f(x) = x(x^2 - 5)(2x + 7)(x - 8)$. Now we find the values of a for which $f(a) = 0$.

$$f(a) = 0$$
$$a(a^2 - 5)(2a + 7)(a - 8) = 0$$

$a=0 \ \ or \ \ a^2-5=0 \ \ \ or \ \ 2a+7=0 \ \ \ or \ \ a-8=0$

$a=0 \ \ or \ \ \ \ \ a^2=5 \ \ \ \ or \ \ \ \ \ 2a=-7 \ \ or \ \ \ \ \ a=8$

$a=0 \ \ or \ \ \ \ \ a=\pm\sqrt{5} \ \ or \ \ \ \ \ a=-\dfrac{7}{2} \ \ or \ \ \ \ \ a=8$

The solutions are 0, $\sqrt{5}$, $-\sqrt{5}$, $-\dfrac{7}{2}$, and 8.

92. $\dfrac{1}{3}, \pm\dfrac{2\sqrt{6}}{3}i$

93. *Familiarize.* It is helpful to list information in a chart and make a drawing. Let r represent the speed of the fishing boat. Then $r - 7$ represents the speed of the barge.

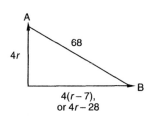

Boat	r	t	d
Fishing	r	4	$4r$
Barge	$r - 7$	4	$4(r - 7)$

Translate. We use the Pythagorean equation:
$$a^2 + b^2 = c^2$$
$$(4r - 28)^2 + (4r)^2 = 68^2$$

Carry out.
$$(4r - 28)^2 + (4r)^2 = 68^2$$
$$16r^2 - 224r + 784 + 16r^2 = 4624$$
$$32r^2 - 224r - 3840 = 0$$
$$r^2 - 7r - 120 = 0$$
$$(r + 8)(r - 15) = 0$$

$r + 8 = 0 \ \ \ or \ \ r - 15 = 0$
$r = -8 \ \ or \ \ r = 15$

Check. We check only 15 since the speeds of the boats cannot be negative. If the speed of the fishing boat is 15 km/h, then the speed of the barge is $15 - 7$, or 8 km/h, and the distances they travel are $4 \cdot 15$ (or 60) and $4 \cdot 8$ (or 32).

$$60^2 + 32^2 = 3600 + 1024 = 4624 = 68^2$$

The values check.

State. The speed of the fishing boat is 15 km/h, and the speed of the barge is 8 km/h.

94. 5, 6, 7

Exercise Set 8.2

---

1. $x^2 + 7x - 3 = 0$

$a = 1, \ b = 7, \ c = -3$

$$x = \frac{-b \pm \sqrt{b^2 - 4ac}}{2a}$$

$$x = \frac{-7 \pm \sqrt{7^2 - 4 \cdot 1 \cdot (-3)}}{2 \cdot 1} = \frac{-7 \pm \sqrt{49 + 12}}{2}$$

$$x = \frac{-7 \pm \sqrt{61}}{2}$$

The solutions are $\dfrac{7 + \sqrt{61}}{2}$ and $\dfrac{7 - \sqrt{61}}{2}$.

2. $\dfrac{7 \pm \sqrt{33}}{2}$

3. $ 3p^2 = 18p - 6$

$$3p^2 - 18p + 6 = 0$$
$$p^2 - 6p + 2 = 0 \text{Dividing by 3}$$
$$a = 1, \ b = -6, \ c = 2$$
$$p = \frac{-b \pm \sqrt{b^2 - 4ac}}{2a}$$
$$p = \frac{-(-6) \pm \sqrt{(-6)^2 - 4 \cdot 1 \cdot 2}}{2 \cdot 1} = \frac{6 \pm \sqrt{36 - 8}}{2}$$
$$p = \frac{6 \pm \sqrt{28}}{2} = \frac{6 \pm 2\sqrt{7}}{2}$$
$$p = \frac{2(3 \pm \sqrt{7})}{2} = 3 \pm \sqrt{7}$$

The solutions are $3 + \sqrt{7}$ and $3 - \sqrt{7}$.

4. $1, \dfrac{5}{3}$

5. $x^2 - x + 2 = 0$

$a = 1, b = -1, c = 2$

$x = \dfrac{-b \pm \sqrt{b^2 - 4ac}}{2a}$

$x = \dfrac{-(-1) \pm \sqrt{(-1)^2 - 4 \cdot 1 \cdot 2}}{2 \cdot 1} = \dfrac{1 \pm \sqrt{1 - 8}}{2}$

$x = \dfrac{1 \pm \sqrt{-7}}{2} = \dfrac{1 \pm i\sqrt{7}}{2}$

The solutions are $\dfrac{1 + i\sqrt{7}}{2}$ and $\dfrac{1 - i\sqrt{7}}{2}$, or $\dfrac{1}{2} + \dfrac{\sqrt{7}}{2}i$ and $\dfrac{1}{2} - \dfrac{\sqrt{7}}{2}i$.

6. $-\dfrac{1}{2} \pm \dfrac{\sqrt{3}}{2}i$

7. $x^2 + 13 = 4x$

$x^2 - 4x + 13 = 0$

$a = 1, b = -4, c = 13$

$x = \dfrac{-b \pm \sqrt{b^2 - 4ac}}{2a}$

$x = \dfrac{-(-4) \pm \sqrt{(-4)^2 - 4 \cdot 1 \cdot 13}}{2 \cdot 1} = \dfrac{4 \pm \sqrt{16 - 52}}{2}$

$x = \dfrac{4 \pm \sqrt{-36}}{2} = \dfrac{4 \pm 6i}{2}$

$x = \dfrac{2(2 \pm 3i)}{2} = 2 \pm 3i$

The solutions are $2 + 3i$ and $2 - 3i$.

8. $3 \pm 2i$

The solutions are $3 + 2i$ and $3 - 2i$.

9. $h^2 + 4 = 6h$

$h^2 - 6h + 4 = 0$

$a = 1, b = -6, c = 4$

$x = \dfrac{-(-6) \pm \sqrt{(-6)^2 - 4 \cdot 1 \cdot 4}}{2 \cdot 1} = \dfrac{6 \pm \sqrt{36 - 16}}{2}$

$x = \dfrac{6 \pm \sqrt{20}}{2} = \dfrac{6 \pm \sqrt{4 \cdot 5}}{2} = \dfrac{6 \pm 2\sqrt{5}}{2}$

$x = 3 \pm \sqrt{5}$

The solutions are $3 + \sqrt{5}$ and $3 - \sqrt{5}$.

10. $\dfrac{-3 \pm \sqrt{41}}{2}$

11. $3 + \dfrac{8}{x} = \dfrac{1}{x^2}$, LCD is x^2

$x^2\left(3 + \dfrac{8}{x}\right) = x^2 \cdot \dfrac{1}{x^2}$

$3x^2 + 8x = 1$

$3x^2 + 8x - 1 = 0$

$a = 3, b = 8, c = -1$

$x = \dfrac{-8 \pm \sqrt{8^2 - 4 \cdot 3 \cdot (-1)}}{2 \cdot 3} = \dfrac{-8 \pm \sqrt{64 + 12}}{6}$

$x = \dfrac{-8 \pm \sqrt{76}}{6} = \dfrac{-8 \pm \sqrt{4 \cdot 19}}{6} = \dfrac{-8 \pm 2\sqrt{19}}{6}$

$x = \dfrac{-4 \pm \sqrt{19}}{3}$

The solutions are $\dfrac{-4 + \sqrt{19}}{3}$ and $\dfrac{-4 - \sqrt{19}}{3}$.

12. $\dfrac{9 \pm \sqrt{41}}{4}$

13. $3x + x(x - 2) = 4$

$3x + x^2 - 2x = 4$

$x^2 + x = 4$

$x^2 + x - 4 = 0$

$a = 1, b = 1, c = -4$

$x = \dfrac{-1 \pm \sqrt{1^2 - 4 \cdot 1 \cdot (-4)}}{2 \cdot 1} = \dfrac{-1 \pm \sqrt{1 + 16}}{2}$

$x = \dfrac{-1 \pm \sqrt{17}}{2}$

The solutions are $\dfrac{-1 + \sqrt{17}}{2}$ and $\dfrac{-1 - \sqrt{17}}{2}$.

14. $\dfrac{-1 \pm \sqrt{21}}{2}$

15. $12x^2 + 9t = 1$

$12t^2 + 9t - 1 = 0$

$a = 12, b = 9, c = -1$

$t = \dfrac{-9 \pm \sqrt{9^2 - 4 \cdot 12 \cdot (-1)}}{2 \cdot 12} = \dfrac{-9 \pm \sqrt{81 + 48}}{24}$

$t = \dfrac{-9 \pm \sqrt{129}}{24}$

The solutions are $\dfrac{-9 + \sqrt{129}}{24}$ and $\dfrac{-9 - \sqrt{129}}{24}$.

16. $-\dfrac{2}{3}, \dfrac{1}{5}$

17. $25x^2 - 20x + 4 = 0$

$(5x - 2)(5x - 2) = 0$

$5x - 2 = 0 \quad or \quad 5x - 2 = 0$

$5x = 2 \quad or \qquad 5x = 2$

$x = \dfrac{2}{5} \quad or \qquad x = \dfrac{2}{5}$

The solution is $\dfrac{2}{5}$.

18. $-\dfrac{7}{6}$

19. $7x(x+2) + 5 = 3x(x+1)$

$7x^2 + 14x + 5 = 3x^2 + 3x$

$4x^2 + 11x + 5 = 0$

$a = 4,\ b = 11,\ c = 5$

$x = \dfrac{-11 \pm \sqrt{11^2 - 4 \cdot 4 \cdot 5}}{2 \cdot 4} = \dfrac{-11 \pm \sqrt{121 - 80}}{8}$

$x = \dfrac{-11 \pm \sqrt{41}}{8}$

The solutions are $\dfrac{-11 + \sqrt{41}}{8}$ and $\dfrac{-11 - \sqrt{41}}{8}$.

20. $\dfrac{-3 \pm \sqrt{37}}{2}$

21. $14(x-4) - (x+2) = (x+2)(x-4)$

$14x - 56 - x - 2 = x^2 - 2x - 8$ Removing parentheses

$13x - 58 = x^2 - 2x - 8$

$0 = x^2 - 15x + 50$

$0 = (x-10)(x-5)$

$x - 10 = 0 \quad or \quad x - 5 = 0$

$x = 10 \quad or \qquad x = 5$

The solutions are 10 and 5.

22. 1, 15

23. $5x^2 = 13x + 17$

$5x^2 - 13x - 17 = 0$

$a = 5,\ b = -13,\ c = -17$

$x = \dfrac{-(-13) \pm \sqrt{(-13)^2 - 4(5)(-17)}}{2 \cdot 5}$

$x = \dfrac{13 \pm \sqrt{169 + 340}}{10} = \dfrac{13 \pm \sqrt{509}}{10}$

The solutions are $\dfrac{13 + \sqrt{509}}{10}$ and $\dfrac{13 - \sqrt{509}}{10}$.

24. $\dfrac{4}{3}, 7$

25. $x^2 + 9 = 4x$

$x^2 - 4x + 9 = 0$

$a = 1,\ b = -4,\ c = 9$

$x = \dfrac{-(-4) \pm \sqrt{(-4)^2 - 4 \cdot 1 \cdot 9}}{2 \cdot 1} = \dfrac{4 \pm \sqrt{16 - 36}}{2}$

$x = \dfrac{4 \pm \sqrt{-20}}{2} = \dfrac{4 \pm \sqrt{-4 \cdot 5}}{2}$

$x = \dfrac{4 \pm 2i\sqrt{5}}{2} = 2 \pm i\sqrt{5}$

The solutions are $2 + i\sqrt{5}$ and $2 - i\sqrt{5}$.

26. $\dfrac{3}{2} \pm \dfrac{\sqrt{19}}{2} i$

27. $x^3 - 8 = 0$

$x^3 - 2^3 = 0$

$(x - 2)(x^2 + 2x + 4) = 0$

$x - 2 = 0 \quad or \quad x^2 + 2x + 4 = 0$

$x = 2 \quad or \quad x = \dfrac{-2 \pm \sqrt{2^2 - 4 \cdot 1 \cdot 4}}{2 \cdot 1}$

$x = 2 \quad or \quad x = \dfrac{-2 \pm \sqrt{-12}}{2} = \dfrac{-2 \pm 2i\sqrt{3}}{2}$

$x = 2 \quad or \quad x = -1 \pm i\sqrt{3}$

The solutions are 2, $-1 + i\sqrt{3}$, and $-1 - i\sqrt{3}$.

28. $\dfrac{1}{2} \pm \dfrac{\sqrt{3}}{2} i$

29. $f(x) = 0$

$3x^2 - 5x - 1 = 0$ Substituting

$a = 3,\ b = -5,\ c = -1$

$x = \dfrac{-(-5) \pm \sqrt{(-5)^2 - 4 \cdot 3 \cdot (-1)}}{2 \cdot 3}$

$x = \dfrac{5 \pm \sqrt{25 + 12}}{6} = \dfrac{5 \pm \sqrt{37}}{6}$

The solutions are $\dfrac{5 + \sqrt{37}}{6}$ and $\dfrac{5 - \sqrt{37}}{6}$.

30. $\dfrac{1 \pm \sqrt{13}}{4}$

31. $f(x) = 1$

$\dfrac{7}{x} + \dfrac{7}{x+4} = 1$ Substituting

$x(x+4)\left(\dfrac{7}{x} + \dfrac{7}{x+4}\right) = x(x+4) \cdot 1$

Multiplying by the LCD

$7(x+4) + 7x = x^2 + 4x$

$7x + 28 + 7x = x^2 + 4x$

$14x + 28 = x^2 + 4x$

$0 = x^2 - 10x - 28$

$a = 1,\ b = -10,\ c = -28$

$x = \dfrac{-(-10) \pm \sqrt{(-10)^2 - 4 \cdot 1 \cdot (-28)}}{2 \cdot 1}$

$x = \dfrac{10 \pm \sqrt{100 + 112}}{2} = \dfrac{10 \pm \sqrt{212}}{2}$

$x = \dfrac{10 \pm \sqrt{4 \cdot 53}}{2} = \dfrac{10 \pm 2\sqrt{53}}{2}$

$x = 5 \pm \sqrt{53}$

The solutions are $5 + \sqrt{53}$ and $5 - \sqrt{53}$.

32. $-2, 3$

33.
$$F(x) = G(x)$$
$$\frac{x+3}{x} = \frac{x-4}{3} \qquad \text{Substituting}$$
$$3x\left(\frac{x+3}{x}\right) = 3x\left(\frac{x-4}{3}\right) \qquad \begin{array}{l}\text{Multiplying}\\ \text{by the LCD}\end{array}$$
$$3x + 9 = x^2 - 4x$$
$$0 = x^2 - 7x - 9$$
$$a = 1,\, b = -7,\, c = -9$$
$$x = \frac{-(-7) \pm \sqrt{(-7)^2 - 4 \cdot 1 \cdot (-9)}}{2 \cdot 1}$$
$$x = \frac{7 \pm \sqrt{49 + 36}}{2} = \frac{7 \pm \sqrt{85}}{2}$$
The solutions are $\dfrac{7 + \sqrt{85}}{2}$ and $\dfrac{7 - \sqrt{85}}{2}$.

34. $\dfrac{3 \pm \sqrt{5}}{2}$

35.
$$f(x) = g(x)$$
$$\frac{15 - 2x}{6} = \frac{3}{x}, \text{ LCD is } 6x$$
$$6x \cdot \frac{15 - 2x}{6} = 6x \cdot \frac{3}{x}$$
$$x(15 - 2x) = 6 \cdot 3$$
$$15x - 2x^2 = 18$$
$$0 = 2x^2 - 15x + 18$$
$$0 = (2x - 3)(x - 6)$$
$$2x - 3 = 0 \quad or \quad x - 6 = 0$$
$$2x = 3 \quad or \qquad x = 6$$
$$x = \frac{3}{2} \quad or \qquad x = 6$$
The solutions are $\dfrac{3}{2}$ and 6.

36. $\pm 2\sqrt{7}$

37. $x^2 + 4x - 7 = 0$
$$a = 1,\, b = 4,\, c = -7$$
$$x = \frac{-4 \pm \sqrt{4^2 - 4 \cdot 1 \cdot (-7)}}{2 \cdot 1} = \frac{-4 \pm \sqrt{16 + 28}}{2}$$
$$x = \frac{-4 \pm \sqrt{44}}{2}$$
Using a calculator we find that $\dfrac{-4 + \sqrt{44}}{2} \approx 1.3166$ and
$\dfrac{-4 - \sqrt{44}}{2} \approx -5.3166$.
The solutions are approximately 1.3166 and -5.3166.

38. $-5.2361, -0.7639$

39. $x^2 - 6x + 4 = 0$
$$a = 1,\, b = -6,\, c = 4$$
$$x = \frac{-(-6) \pm \sqrt{(-6)^2 - 4 \cdot 1 \cdot 4}}{2 \cdot 1} = \frac{6 \pm \sqrt{36 - 16}}{2}$$
$$x = \frac{6 \pm \sqrt{20}}{2}$$
Using a calculator we find that $\dfrac{6 + \sqrt{20}}{2} \approx 5.2361$ and $\dfrac{6 - \sqrt{20}}{2} \approx 0.7639$.
The solutions are approximately 5.2361 and 0.7639.

40. $0.2679, 3.7321$

41. $2x^2 - 3x - 7 = 0$
$$a = 2,\, b = -3,\, c = -7$$
$$x = \frac{-(-3) \pm \sqrt{(-3)^2 - 4 \cdot 2 \cdot (-7)}}{2 \cdot 2}$$
$$x = \frac{3 \pm \sqrt{9 + 56}}{4} = \frac{3 \pm \sqrt{65}}{4}$$
Using a calculator we find that $\dfrac{3 + \sqrt{65}}{4} \approx 2.7656$ and $\dfrac{3 - \sqrt{65}}{4} \approx -1.2656$.
The solutions are approximately 2.7656 and -1.2656.

42. $-0.4574, 1.4574$

43. *Writing Exercise*

44. *Writing Exercise*

45. *Familiarize.* Let $x =$ the number of pounds of Kenyan coffee and $y =$ the number of pounds of Peruvian coffee in the mixture. We organize the information in a table.

Type of Coffee	Kenyan	Peruvian	Mixture
Price per pound	$6.75	$11.25	$8.55
Number of pounds	x	y	50
Total cost	$6.75x$	$11.25y$	8.55×50, or $427.50

Translate. From the last two rows of the table we get a system of equations.
$$x + y = 50,$$
$$6.75x + 11.25y = 427.50$$

Solve. Solving the system of equations, we get $(30, 20)$.

Check. The total number of pounds in the mixture is $30 + 20$, or 50. The total cost of the mixture is $6.75(30) + 11.25(20) = \$427.50$. The values check.

State. The mixture should consist of 30 lb of Kenyan coffee and 20 lb of Peruvian coffee.

46. 46 cream-filled; 44 glazed

47. $\sqrt{27a^2b^5} \cdot \sqrt{6a^3b} = \sqrt{27a^2b^5 \cdot 6a^3b} =$
$\sqrt{162a^5b^6} = \sqrt{81a^4b^6 \cdot 2a} = \sqrt{81a^4b^6}\sqrt{2a} =$
$9a^2b^3\sqrt{2a}$

48. $4a^2b^3\sqrt{6}$

49.
$$\dfrac{\dfrac{3}{x-1}}{\dfrac{1}{x+1}+\dfrac{2}{x-1}}$$

$$= \dfrac{\dfrac{3}{x-1}}{\dfrac{1}{x+1}+\dfrac{2}{x-1}} \cdot \dfrac{(x-1)(x+1)}{(x-1)(x+1)}$$

$$= \dfrac{3(x+1)}{x-1+2(x+1)}$$

$$= \dfrac{3x+3}{x-1+2x+2}$$

$$= \dfrac{3x+3}{3x+1}, \text{ or } \dfrac{3(x+1)}{3x+1}$$

50. $\dfrac{4b}{3ab^2-4a^2}$

51. *Writing Exercise*

52. *Writing Exercise*

53. $f(x) = \dfrac{x^2}{x-2} + 1$

To find the x-coordinates of the x-intercepts of the graph of f, we solve $f(x) = 0$.

$$\dfrac{x^2}{x-2} + 1 = 0$$
$$x^2 + x - 2 = 0 \quad \text{Multiplying by } x-2$$
$$(x+2)(x-1) = 0$$
$$x = -2 \ \text{ or } \ x = 1$$

The x-intercepts are $(-2, 0)$ and $(1, 0)$.

54. $(-5 - \sqrt{37}, 0), (-5 + \sqrt{37}, 0)$

55.
$$f(x) = g(x)$$
$$\dfrac{x^2}{x-2} + 1 = \dfrac{4x-2}{x-2} + \dfrac{x+4}{2}$$
Substituting
$$2(x-2)\left(\dfrac{x^2}{x-2}+1\right) = 2(x-2)\left(\dfrac{4x-2}{x-2}+\dfrac{x+4}{2}\right)$$
Multiplying by the LCD
$$2x^2 + 2(x-2) = 2(4x-2) + (x-2)(x+4)$$
$$2x^2 + 2x - 4 = 8x - 4 + x^2 + 2x - 8$$
$$2x^2 + 2x - 4 = x^2 + 10x - 12$$
$$x^2 - 8x + 8 = 0$$
$$a = 1, \, b = -8, \, c = 8$$
$$x = \dfrac{-(-8) \pm \sqrt{(-8)^2 - 4 \cdot 1 \cdot 8}}{2 \cdot 1} = \dfrac{8 \pm \sqrt{64-32}}{2}$$
$$x = \dfrac{8 \pm \sqrt{32}}{2} = \dfrac{8 \pm \sqrt{16 \cdot 2}}{2} = \dfrac{8 \pm 4\sqrt{2}}{2}$$
$$x = 4 \pm 2\sqrt{2}$$

The solutions are $4 + 2\sqrt{2}$ and $4 - 2\sqrt{2}$.

56. $-0.4253905297, 1.17539053$

57. $z^2 + 0.84z - 0.4 = 0$
$$a = 1, \, b = 0.84, \, c = -0.4$$
$$z = \dfrac{-0.84 \pm \sqrt{(0.84)^2 - 4 \cdot 1 \cdot (-0.4)}}{2 \cdot 1}$$
$$z = \dfrac{-0.84 \pm \sqrt{2.3056}}{2}$$
$$z = \dfrac{-0.84 + \sqrt{2.3056}}{2} \approx 0.3392101158$$
$$z = \dfrac{-0.84 - \sqrt{2.3056}}{2} \approx -1.179210116$$

The solutions are approximately 0.3392101158 and -1.179210116.

58. $\sqrt{3}, \dfrac{3-\sqrt{3}}{2}$

59. $\sqrt{2}x^2 + 5x + \sqrt{2} = 0$
$$x = \dfrac{-5 \pm \sqrt{5^2 - 4 \cdot \sqrt{2} \cdot \sqrt{2}}}{2\sqrt{2}} = \dfrac{-5 \pm \sqrt{17}}{2\sqrt{2}}, \text{ or}$$
$$x = \dfrac{-5 \pm \sqrt{17}}{2\sqrt{2}} \cdot \dfrac{\sqrt{2}}{\sqrt{2}} = \dfrac{-5\sqrt{2} \pm \sqrt{34}}{4}$$

The solutions are $\dfrac{-5\sqrt{2} \pm \sqrt{34}}{4}$.

60. $-i \pm i\sqrt{1-i}$

61.
$$kx^2 + 3x - k = 0$$
$$k(-2)^2 + 3(-2) - k = 0 \quad \text{Substituting } -2 \text{ for } x$$
$$4k - 6 - k = 0$$
$$3k = 6$$
$$k = 2$$
$$2x^2 + 3x - 2 = 0 \quad \text{Substituting 2 for } k$$
$$(2x - 1)(x + 2) = 0$$
$$2x - 1 = 0 \quad or \quad x + 2 = 0$$
$$x = \frac{1}{2} \quad or \quad x = -2$$

The other solution is $\dfrac{1}{2}$.

62. *Writing Exercise*

Exercise Set 8.3

1. *Familiarize.* We first make a drawing, labeling it with the known and unknown information. We can also organize the information in a table. We let r represent the speed and t the time for the first part of the trip.

$$\underset{60 \text{ km}}{r \text{ km/h} \quad t \text{ hr}} \bullet \underset{24 \text{ km}}{r - 4 \text{ km/h} \quad 8 - t \text{ hr}}$$

Canoe trip	Distance	Speed	Time
1st part	60	r	t
2nd part	24	$r - 4$	$8 - t$

Translate. Using $r = \dfrac{d}{t}$, we get two equations from the table, $r = \dfrac{60}{t}$ and $r - 4 = \dfrac{24}{8 - t}$.

Carry out. We substitute $\dfrac{60}{t}$ for r in the second equation and solve for t.
$$\frac{60}{t} - 4 = \frac{24}{8 - t}, \quad \text{LCD is } t(8 - t)$$
$$t(8 - t)\left(\frac{60}{t} - 4\right) = t(8 - t) \cdot \frac{24}{8 - t}$$
$$60(8 - t) - 4t(8 - t) = 24t$$
$$480 - 60t - 32t + 4t^2 = 24t$$
$$4t^2 - 116t + 480 = 0 \quad \text{Standard form}$$
$$t^2 - 29t + 120 = 0 \quad \text{Multiplying by } \frac{1}{4}$$
$$(t - 24)(t - 5) = 0$$
$$t = 24 \quad or \quad t = 5$$

Check. Since the time cannot be negative (If $t = 24$, $8 - t = -16$.), we check only 5 hr. If $t = 5$, then $8 - t = 3$. The speed of the first part is $\dfrac{60}{5}$, or 12 km/h. The speed of

the second part is $\dfrac{24}{3}$, or 8 km/h. The speed of the second part is 4 km/h slower than the first part. The value checks.

State. The speed of the first part was 12 km/h, and the speed of the second part was 8 km/h.

2. First part: 60 mph; second part: 50 mph

3. *Familiarize.* We first make a drawing. We also organize the information in a table. We let $r =$ the speed and $t =$ the time of the slower trip.

$$\underset{280 \text{ mi}}{280 \text{ mi} \quad r \text{ mph} \quad t \text{ hr}} \bullet \underset{}{r + 5 \text{ mph} \quad t - 1 \text{ hr}}$$

Trip	Distance	Speed	Time
Slower	280	r	t
Faster	280	$r + 5$	$t - 1$

Translate. Using $t = \dfrac{d}{r}$, we get two equations from the table, $t = \dfrac{280}{r}$, and $t - 1 = \dfrac{280}{r + 5}$.

Carry out. We substitute $\dfrac{280}{r}$ for t in the second equation and solve for r.
$$\frac{280}{r} - 1 = \frac{280}{r + 5}, \quad \text{LCD is } r(r + 5)$$
$$r(r + 5)\left(\frac{280}{r} - 1\right) = r(r + 5) \cdot \frac{280}{r + 5}$$
$$280(r + 5) - r(r + 5) = 280r$$
$$280r + 1400 - r^2 - 5r = 280r$$
$$0 = r^2 + 5r - 1400$$
$$0 = (r - 35)(r + 40)$$
$$r = 35 \quad or \quad r = -40$$

Check. Since negative speed has no meaning in this problem, we check only 35. If $r = 35$, then the time for the slow trip is $\dfrac{280}{35}$, or 8 hours. If $r = 35$ then $r + 5 = 40$ and the time for the fast trip is $\dfrac{280}{40}$, or 7 hours. This is 1 hour less time than the slow trip took, so we have an answer to the problem.

State. The speed is 35 mph.

4. 40 mph

5. *Familiarize.* We make a drawing and then organize the information in a table. We let $r =$ the speed and $t =$ the time of the Cessna.

$$\underset{1000 \text{ mi}}{600 \text{ mi} \quad r \text{ mph} \quad t \text{ hr}} \bullet \underset{}{r + 50 \text{ mph} \quad t + 1 \text{ hr}}$$

Plane	Distance	Speed	Time
Cessna	600	r	t
Beechcraft	1000	$r + 50$	$t + 1$

Translate. Using $t = d/r$, we get two equations from the table:

$$t = \frac{600}{r} \quad \text{and} \quad t + 1 = \frac{1000}{r + 50}$$

Carry out. We substitute $\frac{600}{r}$ for t in the second equation and solve for r.

$$\frac{600}{r} + 1 = \frac{1000}{r + 50},$$
$$\text{LCD is } r(r + 50)$$
$$r(r + 50)\left(\frac{600}{r} + 1\right) = r(r + 50) \cdot \frac{1000}{r + 50}$$
$$600(r + 50) + r(r + 50) = 1000r$$
$$600r + 30,000 + r^2 + 50r = 1000r$$
$$r^2 - 350r + 30,000 = 0$$
$$(r - 150)(r - 200) = 0$$
$$r = 150 \quad or \quad r = 200$$

Check. If $r = 150$, then the Cessna's time is $\frac{600}{150}$, or 4 hr and the Beechcraft's time is $\frac{1000}{150 + 50}$, or $\frac{1000}{200}$, or 5 hr. If $r = 200$, then the Cessna's time is $\frac{600}{200}$, or 3 hr and the Beechcraft's time is $\frac{1000}{200 + 50}$, or $\frac{1000}{250}$, or 4 hr. Since the Beechcraft's time is 1 hr longer in each case, both values check. There are two solutions.

State. The speed of the Cessna is 150 mph and the speed of the Beechcraft is 200 mph; or the speed of the Cessna is 200 mph and the speed of the Beechcraft is 250 mph.

6. Super-prop: 350 mph; turbo-jet: 400 mph

7. *Familiarize.* We make a drawing and then organize the information in a table. We let r represent the speed and t the time of the trip to Hillsboro.

Hillsboro

40 mi $\quad r$ mph $\quad t$ hr

40 mi $\quad r - 6$ mph $\quad 14 - t$ hr

Trip	Distance	Speed	Time
To Hillsboro	40	r	t
Return	40	$r - 6$	$14 - t$

Translate. Using $t = \frac{d}{r}$, we get two equations from the table,

$$t = \frac{40}{r} \quad \text{and} \quad 14 - t = \frac{40}{r - 6}.$$

Carry out. We substitute $\frac{40}{r}$ for t in the second equation and solve for r.

$$14 - \frac{40}{r} = \frac{40}{r - 6},$$
$$\text{LCD is } r(r - 6)$$
$$r(r - 6)\left(14 - \frac{40}{r}\right) = r(r - 6) \cdot \frac{40}{r - 6}$$
$$14r(r - 6) - 40(r - 6) = 40r$$
$$14r^2 - 84r - 40r + 240 = 40r$$
$$14r^2 - 164r + 240 = 0$$
$$7r^2 - 82r + 120 = 0$$
$$(7r - 12)(r - 10) = 0$$
$$r = \frac{12}{7} \quad or \quad r = 10$$

Check. Since negative speed has no meaning in this problem (If $r = \frac{12}{7}$, then $r - 6 = -\frac{30}{7}$.), we check only 10 mph. If $r = 10$, then the time of the trip to Hillsboro is $\frac{40}{10}$, or 4 hr. The speed of the return trip is $10 - 6$, or 4 mph, and the time is $\frac{40}{4}$, or 10 hr. The total time for the round trip is 4 hr + 10 hr, or 14 hr. The value checks.

State. Naoki's speed on the trip to Hillsboro was 10 mph and it was 4 mph on the return trip.

8. Average speed to Richmond: 60 mph; average speed returning: 50 mph

9. *Familiarize.* We make a drawing and organize the information in a table. Let r represent the speed of the barge in still water, and let t represent the time of the trip upriver.

24 mi $\quad r - 4$ mph $\quad t$ hr
$\longrightarrow$ Upriver

Downriver $\longleftarrow$ 24 mi $\quad r + 4$ mph $\quad 5 - t$ hr

Trip	Distance	Speed	Time
Upriver	24	$r - 4$	t
Downriver	24	$r + 4$	$5 - t$

Translate. Using $t = \frac{d}{r}$, we get two equations from the table,

$$t = \frac{24}{r - 4} \quad \text{and} \quad 5 - t = \frac{24}{r + 4}.$$

Carry out. We substitute $\dfrac{24}{r-4}$ for t in the second equation and solve for r.

$$5 - \frac{24}{r-4} = \frac{24}{r+4},$$

LCD is $(r-4)(r+4)$

$$(r-4)(r+4)\left(5 - \frac{24}{r-4}\right) = (r-4)(r+4)\cdot\frac{24}{r+4}$$

$$5(r-4)(r+4) - 24(r+4) = 24(r-4)$$

$$5r^2 - 80 - 24r - 96 = 24r - 96$$

$$5r^2 - 48r - 80 = 0$$

We use the quadratic formula.

$$r = \frac{-(-48) \pm \sqrt{(-48)^2 - 4\cdot 5\cdot(-80)}}{2\cdot 5}$$

$$r = \frac{48 \pm \sqrt{3904}}{10}$$

$$r \approx 11 \quad or \quad r \approx -1.5$$

Check. Since negative speed has no meaning in this problem, we check only 11 mph. If $r \approx 11$, then the speed upriver is about $11 - 4$, or 7 mph, and the time is about $\dfrac{24}{7}$, or 3.4 hr. The speed downriver is about $11 + 4$, or 15 mph, and the time is about $\dfrac{24}{15}$, or 1.6 hr. The total time of the round trip is $3.4 + 1.6$, or 5 hr. The value checks.

State. The barge must be able to travel about 11 mph in still water.

10. About 14 mph

11. *Familiarize*. Let x represent the time it takes one well to fill the pool. Then $x - 6$ represents the time it takes the other well to fill the pool. It takes them 4 hr to fill the pool when both wells are working together, so they can fill $\dfrac{1}{4}$ of the pool in 1 hr. The first well will fill $\dfrac{1}{x}$ of the pool in 1 hr, and the other well will fill $\dfrac{1}{x-6}$ of the pool in 1 hr.

Translate. We have an equation.

$$\frac{1}{x} + \frac{1}{x-6} = \frac{1}{4}$$

Carry out. We solve the equation.

We multiply by the LCD, $4x(x-6)$.

$$4x(x-6)\left(\frac{1}{x} + \frac{1}{x-6}\right) = 4x(x-6)\cdot\frac{1}{4}$$

$$4(x-6) + 4x = x(x-6)$$

$$4x - 24 + 4x = x^2 - 6x$$

$$0 = x^2 - 14x + 24$$

$$0 = (x-2)(x-12)$$

$$x = 2 \quad or \quad x = 12$$

Check. Since negative time has no meaning in this problem, 2 is not a solution $(2 - 6 = -4)$. We check only 12 hr.

This is the time it would take the first well working alone. Then the other well would take $12 - 6$, or 6 hr working alone. The second well would fill $4\left(\dfrac{1}{6}\right)$, or $\dfrac{2}{3}$, of the pool in 4 hr, and the first well would fill $4\left(\dfrac{1}{12}\right)$, or $\dfrac{1}{3}$, of the pool in 4 hr. Thus in 4 hr they would fill $\dfrac{2}{3} + \dfrac{1}{3}$ of the pool. This is all of it, so the numbers check.

State. It takes the first well, working alone, 12 hr to fill the pool.

12. 6 hr

13. We make a drawing and then organize the information in a table. We let r represent Ellen's speed in still water. Then $r - 2$ is the speed upstream and $r + 2$ is the speed downstream. Using $t = \dfrac{d}{r}$, we let $\dfrac{1}{r-2}$ represent the time upstream and $\dfrac{1}{r+2}$ represent the time downstream.

	1 mi	$r - 2$ mph	
			Upstream
Downstream	1 mi	$r + 2$ mph	

Trip	Distance	Speed	Time
Upstream	1	$r - 2$	$\dfrac{1}{r-2}$
Downstream	1	$r + 2$	$\dfrac{1}{r+2}$

Translate. The time for the round trip is 1 hour. We now have an equation.

$$\frac{1}{r-2} + \frac{1}{r+2} = 1$$

Carry out. We solve the equation. We multiply by the LCD, $(r-2)(r+2)$.

$$(r-2)(r+2)\left(\frac{1}{r-2} + \frac{1}{r+2}\right) = (r-2)(r+2)\cdot 1$$

$$(r+2) + (r-2) = (r-2)(r+2)$$

$$2r = r^2 - 4$$

$$0 = r^2 - 2r - 4$$

$a = 1,\ b = -2,\ c = -4$

$$r = \frac{-(-2) \pm \sqrt{(-2)^2 - 4\cdot 1(-4)}}{2\cdot 1}$$

$$r = \frac{2 \pm \sqrt{4 + 16}}{2} = \frac{2 \pm \sqrt{20}}{2}$$

$$r = \frac{2 \pm 2\sqrt{5}}{2} = 1 \pm \sqrt{5}$$

$$1 + \sqrt{5} \approx 1 + 2.236 \approx 3.24$$

$$1 - \sqrt{5} \approx 1 - 2.236 \approx -1.24$$

Check. Since negative speed has no meaning in this problem, we check only 3.24 mph. If $r \approx 3.24$, then $r - 2 \approx 1.24$ and $r + 2 \approx 5.24$. The time it takes to travel upstream is approximately $\dfrac{1}{1.24}$, or 0.806 hr, and the time it takes to travel downstream is approximately $\dfrac{1}{5.24}$, or 0.191 hr. The total time is 0.997 which is approximately 1 hour. The value checks.

State. Ellen's speed in still water is approximately 3.24 mph.

14. About 9.34 km/h

15. $\qquad A = 4\pi r^2$

$\qquad \dfrac{A}{4\pi} = r^2 \qquad$ Dividing by 4π

$\qquad \dfrac{1}{2}\sqrt{\dfrac{A}{\pi}} = r \qquad$ Taking the positive square root

16. $s = \sqrt{\dfrac{A}{6}}$

17. $\quad A = 2\pi r^2 + 2\pi rh$

$\quad 0 = 2\pi r^2 + 2\pi rh - A \qquad$ Standard form

$a = 2\pi, \; b = 2\pi h, \; c = -A$

$r = \dfrac{-2\pi h \pm \sqrt{(2\pi h)^2 - 4 \cdot 2\pi \cdot (-A)}}{2 \cdot 2\pi} \qquad$ Using the quadratic formula

$r = \dfrac{-2\pi h \pm \sqrt{4\pi^2 h^2 + 8\pi A}}{4\pi}$

$r = \dfrac{-2\pi h \pm 2\sqrt{\pi^2 h^2 + 2\pi A}}{4\pi}$

$r = \dfrac{-\pi h \pm \sqrt{\pi^2 h^2 + 2\pi A}}{2\pi}$

Since taking the negative square root would result in a negative answer, we take the positive one.

$r = \dfrac{-\pi h + \sqrt{\pi^2 h^2 + 2\pi A}}{2\pi}$

18. $r = \sqrt{\dfrac{Gm_1 m_2}{F}}$

19. $\qquad N = \dfrac{kQ_1 Q_2}{s^2}$

$\qquad Ns^2 = kQ_1 Q_2 \qquad$ Multiplying by s^2

$\qquad s^2 = \dfrac{kQ_1 Q_2}{N} \qquad$ Dividing by N

$\qquad s = \sqrt{\dfrac{kQ_1 Q_2}{N}} \qquad$ Taking the positive square root

20. $r = \sqrt{\dfrac{A}{\pi}}$

21. $\qquad T = 2\pi\sqrt{\dfrac{l}{g}}$

$\qquad \dfrac{T}{2\pi} = \sqrt{\dfrac{l}{g}} \qquad$ Multiplying by $\dfrac{1}{2\pi}$

$\qquad \dfrac{T^2}{4\pi^2} = \dfrac{l}{g} \qquad$ Squaring

$\qquad gT^2 = 4\pi^2 l \qquad$ Multiplying by $4\pi^2 g$

$\qquad g = \dfrac{4\pi^2 l}{T^2} \qquad$ Multiplying by $\dfrac{1}{T^2}$

22. $b = \sqrt{c^2 - a^2}$

23. $\quad a^2 + b^2 + c^2 = d^2$

$\qquad c^2 = d^2 - a^2 - b^2 \qquad$ Subtracting a^2 and b^2

$\qquad c = \sqrt{d^2 - a^2 - b^2} \qquad$ Taking the positive square root

24. $k = \dfrac{3 + \sqrt{9 + 8N}}{2}$

25. $\quad s = v_0 t + \dfrac{gt^2}{2}$

$\quad 0 = \dfrac{gt^2}{2} + v_0 t - s \qquad$ Standard form

$a = \dfrac{g}{2}, \; b = v_0, \; c = -s$

$t = \dfrac{-v_0 \pm \sqrt{v_0^2 - 4\left(\dfrac{g}{2}\right)(-s)}}{2\left(\dfrac{g}{2}\right)}$

$t = \dfrac{-v_0 \pm \sqrt{v_0^2 + 2gs}}{g}$

Since taking the negative square root would result in a negative answer, we take the positive one.

$t = \dfrac{-v_0 + \sqrt{v_0^2 + 2gs}}{g}$

26. $r = \dfrac{-\pi s + \sqrt{\pi^2 s^2 + 4\pi A}}{2\pi}$

27. $\quad N = \dfrac{1}{2}(n^2 - n)$

$\quad N = \dfrac{1}{2}n^2 - \dfrac{1}{2}n$

$\quad 0 = \dfrac{1}{2}n^2 - \dfrac{1}{2}n - N$

$\quad a = \dfrac{1}{2}, \; b = -\dfrac{1}{2}, \; c = -N$

$$n = \frac{-\left(-\frac{1}{2}\right) \pm \sqrt{\left(-\frac{1}{2}\right)^2 - 4 \cdot \frac{1}{2} \cdot (-N)}}{2\left(\frac{1}{2}\right)}$$

$$n = \frac{1}{2} \pm \sqrt{\frac{1}{4} + 2N}$$

$$n = \frac{1}{2} \pm \sqrt{\frac{1 + 8N}{4}}$$

$$n = \frac{1}{2} \pm \frac{1}{2}\sqrt{1 + 8N}$$

Since taking the negative square root would result in a negative answer, we take the positive one.

$$n = \frac{1}{2} + \frac{1}{2}\sqrt{1 + 8N}, \text{ or } \frac{1 + \sqrt{1 + 8N}}{2}$$

28. $r = 1 - \sqrt{\dfrac{A}{A_0}}$

29. $\quad V = 3.5\sqrt{h}$

$\quad V = 12.25h \quad$ Squaring

$\quad \dfrac{V^2}{12.25} = h$

30. $L = \dfrac{1}{W^2 C}$

31. $at^2 + bt + c = 0$

The quadratic formula gives the result.

$$t = \frac{-b \pm \sqrt{b^2 - 4ac}}{2a}$$

32. $r = -1 + \dfrac{-P_2 + \sqrt{P_2^2 + 4AP_1}}{2P_1}$

33. a) *Familiarize and Translate.* From Example 4, we know

$$t = \frac{-v_0 + \sqrt{v_0{}^2 + 19.6s}}{9.8}.$$

Carry out. Substituting 500 for s and 0 for v_0, we have

$$t = \frac{0 + \sqrt{0^2 + 19.6(500)}}{9.8}$$

$$t \approx 10.1$$

Check. Substitute 10.1 for t and 0 for v_0 in the original formula. (See Example 4.)

$$s = 4.9t^2 + v_0 t = 4.9(10.1)^2 + 0 \cdot (10.1)^2$$

$$\approx 500$$

The answer checks.

State. It takes about 10.1 sec to reach the ground.

b) *Familiarize and Translate.* From Example 4, we know

$$t = \frac{-v_0 + \sqrt{v_0^2 + 19.6s}}{9.8}.$$

Carry out. Substitute 500 for s and 30 for v_0.

$$t = \frac{-30 + \sqrt{30^2 + 19.6(500)}}{9.8}$$

$$t \approx 7.49$$

Check. Substitute 30 for v_0 and 7.49 for t in the original formula. (See Example 4.)

$$s = 4.9t^2 + v_0 t = 4.9(7.49)^2 + (30)(7.49)$$

$$\approx 500$$

The answer checks.

State. It takes about 7.49 sec to reach the ground.

c) *Familiarize and Translate.* We will use the formula in Example 4, $s = 4.9t^2 + v_0 t$.

Carry out. Substitute 5 for t and 30 for v_0.

$$s = 4.9(5)^2 + 30(5) = 272.5$$

Check. We can substitute 30 for v_0 and 272.5 for s in the form of the formula we used in part (b).

$$t = \frac{-v_0 + \sqrt{v_0^2 + 19.6s}}{9.8}$$

$$= \frac{-30 + \sqrt{(30)^2 + 19.6(272.5)}}{9.8} = 5$$

The answer checks.

State. The object will fall 272.5 m.

34. a) 3.9 sec

b) 1.9 sec

c) 79.6 m

35. *Familiarize and Translate.* From Example 4, we know

$$t = \frac{-v_0 + \sqrt{v_0^2 + 19.6s}}{9.8}.$$

Carry out. Substituting 40 for s and 0 for v_0 we have

$$t = \frac{0 + \sqrt{0^2 + 19.6(40)}}{9.8}$$

$$t \approx 2.9$$

Check. Substitute 2.9 for t and 0 for v_0 in the original formula. (See Example 4.)

$$s = 4.9t^2 + v_0 t = 4.9(2.9)^2 + 0(2.9)$$

$$\approx 40$$

The answer checks.

State. He will be falling for about 2.9 sec.

36. 30.625 m

37. *Familiarize and Translate.* From Example 3, we know

$$T = \frac{\sqrt{3V}}{12}.$$

Carry out. Substituting 36 for V, we have

$$T = \frac{\sqrt{3 \cdot 36}}{12}$$

$$T \approx 0.87$$

Check. Substitute 0.87 for T in the original formula. (See Example 3.)

$$48T^2 = V$$

$$48(0.87)^2 = V$$

$$36 \approx V$$

The answer checks.

State. Vince Carter's hang time is about 0.87 sec.

38. 12

39. Familiarize and Translate. We will use the formula in Example 4, $s = 4.9t^2 + v_0t$.

Carry out. Solve the formula for v_0.

$$s - 4.9t^2 = v_0t$$

$$\frac{s - 4.9t^2}{t} = v_0$$

Now substitute 51.6 for s and 3 for t.

$$\frac{51.6 - 4.9(3)^2}{3} = v_0$$

$$2.5 = v_0$$

Check. Substitute 3 for t and 2.5 for v_0 in the original formula.

$$s = 4.9(3)^2 + 2.5(3) = 51.6$$

The solution checks.

State. The initial velocity is 2.5 m/sec.

40. 3.2 m/sec

41. Familiarize and Translate. From Exercise 32 we know that

$$r = -1 + \frac{-P_2 + \sqrt{P_2^2 + 4P_1A}}{2P_1},$$

where A is the total amount in the account after two years, P_1 is the amount of the original deposit, P_2 is deposited at the beginning of the second year, and r is the annual interest rate.

Carry out. Substitute 3000 for P_1, 1700 for P_2, and 5253.70 for A.

$$r = -1 + \frac{-1700 + \sqrt{(1700)^2 + 4(3000)(5253.70)}}{2(3000)}$$

Using a calculator, we have $r = 0.07$.

Check. Substitute in the original formula in Exercise 32.

$$P_1(1 + r)^2 + P_2(1 + r) = A$$

$$3000(1.07)^2 + 1700(1.07) = A$$

$$5253.70 = A$$

The answer checks.

State. The annual interest rate is 0.07, or 7%.

42. 8.5%

43. *Writing Exercise*

44. *Writing Exercise*

45. $b^2 - 4ac = 6^2 - 4 \cdot 5 \cdot 7$
$$= 36 - 4 \cdot 5 \cdot 7$$
$$= 36 - 140$$
$$= -104$$

46. $2i\sqrt{11}$

47. $\dfrac{x^2 + xy}{2x} = \dfrac{x(x + y)}{2x}$

$$= \frac{x(x + y)}{2 \cdot x}$$

$$= \frac{\cancel{x}(x + y)}{2 \cdot \cancel{x}}$$

$$= \frac{x + y}{2}$$

48. $\dfrac{a^2 - b^2}{b}$

49. $\dfrac{3 + \sqrt{45}}{6} = \dfrac{3 + \sqrt{9 \cdot 5}}{6} = \dfrac{3 + 3\sqrt{5}}{6} = \dfrac{\cancel{3}(1 + \sqrt{5})}{\cancel{3} \cdot 2} = $
$$\frac{1 + \sqrt{5}}{2}$$

50. $\dfrac{1 - \sqrt{7}}{5}$

51. *Writing Exercise*

52. *Writing Exercise*

53. $$A = 6.5 - \frac{20.4t}{t^2 + 36}$$

$$(t^2 + 36)A = (t^2 + 36)\left(6.5 - \frac{20.4t}{t^2 + 36}\right)$$

$$At^2 + 36A = (t^2 + 36)(6.5) - (t^2 + 36)\left(\frac{20.4t}{t^2 + 36}\right)$$

$$At^2 + 36A = 6.5t^2 + 234 - 20.4t$$

$$At^2 - 6.5t^2 + 20.4 + 36A - 234 = 0$$

$$(A - 6.5)t^2 + 20.4t + (36A - 234) = 0$$

$$a = A - 6.5, \ b = 20.4, \ c = 36A - 234$$

$$t = \frac{-20.4 \pm \sqrt{(20.4)^2 - 4(A - 6.5)(36A - 234)}}{2(A - 6.5)}$$

$$t = \frac{-20.4 \pm \sqrt{416.16 - 144A^2 + 1872A - 6084}}{2(A - 6.5)}$$

$$t = \frac{-20.4 \pm \sqrt{-144A^2 + 1872A - 5667.84}}{2(A - 6.5)}$$

$$t = \frac{-20.4 \pm \sqrt{144(-A^2 + 13A - 39.36)}}{2(A - 6.5)}$$

$$t = \frac{-20.4 \pm 12\sqrt{-A^2 + 13A - 39.36}}{2(A - 6.5)}$$

$$t = \frac{2(-10.2 \pm 6\sqrt{-A^2 + 13A - 39.36})}{2(A - 6.5)}$$

$$t = \frac{-10.2 \pm 6\sqrt{-A^2 + 13A - 39.36}}{A - 6.5}$$

54. $c = \dfrac{mv}{\sqrt{m^2 - m_0^2}}$

55.
$$\frac{w}{l} = \frac{l}{w + l}$$

$$l(w + l) \cdot \frac{w}{l} = l(w + l) \cdot \frac{l}{w + l}$$

$$w(w + l) = l^2$$

$$w^2 + lw = l^2$$

$$0 = l^2 - lw - w^2$$

Use the quadratic formula with $a = 1$, $b = -w$, and $c = -w^2$.

$$l = \frac{-(-w) \pm \sqrt{(-w)^2 - 4 \cdot 1(-w^2)}}{2 \cdot 1}$$

$$l = \frac{w \pm \sqrt{w^2 + 4w^2}}{2} = \frac{w \pm \sqrt{5w^2}}{2}$$

$$l = \frac{w \pm w\sqrt{5}}{2}$$

Since $\dfrac{w - w\sqrt{5}}{2}$ is negative we use the positive square root:

$$l = \frac{w + w\sqrt{5}}{2}$$

56. $L(A) = \sqrt{\dfrac{A}{2}}$

57. Familiarize. Let a = the number. Then $a - 1$ is 1 less than a and the reciprocal of that number is $\dfrac{1}{a - 1}$. Also, 1 more than the number is $a + 1$.

Translate.

The reciprocal of 1 less than a number	is	1 more than the number.
$\downarrow$	$\downarrow$	$\downarrow$
$\dfrac{1}{(a - 1)}$	$=$	$a + 1$

Carry out. We solve the equation.

$$\frac{1}{a - 1} = a + 1, \text{ LCD is } a - 1$$

$$(a - 1) \cdot \frac{1}{a - 1} = (a - 1)(a + 1)$$

$$1 = a^2 - 1$$

$$2 = a^2$$

$$\pm\sqrt{2} = a$$

Check. $\dfrac{1}{\sqrt{2} - 1} \approx 2.4142 \approx \sqrt{2} + 1$ and $\dfrac{1}{-\sqrt{2} - 1} \approx -0.4142 \approx -\sqrt{2} + 1$. The answers check.

State. The numbers are $\sqrt{2}$ and $-\sqrt{2}$, or $\pm\sqrt{2}$.

58. \$2.50

59. $mn^4 - r^2pm^3 - r^2n^2 + p = 0$

Let $u = n^2$. Substitute and rearrange.

$$mu^2 - r^2u - r^2pm^3 + p = 0$$

$$a = m, \ b = -r^2, \ c = -r^2pm^3 + p$$

$$u = \frac{-(-r^2) \pm \sqrt{(-r^2)^2 - 4 \cdot m(-r^2pm^3 + p)}}{2 \cdot m}$$

$$u = \frac{r^2 \pm \sqrt{r^4 + 4m^4r^2p - 4mp}}{2m}$$

$$n^2 = \frac{r^2 \pm \sqrt{r^4 + 4m^4r^2p - 4mp}}{2m}$$

$$n = \pm\sqrt{\frac{r^2 \pm \sqrt{r^4 + 4m^4r^2p - 4mp}}{2m}}$$

60. $d = \dfrac{-\pi h + \sqrt{\pi^2 h^2 + 2\pi A}}{\pi}$

61. Let s represent a length of a side of the cube, let S represent the surface area of the cube, and let A represent the surface area of the sphere. Then the diameter of the sphere is s, so the radius r is $s/2$. From Exercise 15, we know, $A = 4\pi r^2$, so when $r = s/2$ we have $A = 4\pi \left(\dfrac{s}{2}\right)^2 = 4\pi \cdot \dfrac{s^2}{4} = \pi s^2$. From the formula for the surface area of a cube (See Exercise 16.) we know that $S = 6s^2$, so $\dfrac{S}{6} = s^2$ and then $A = \pi \cdot \dfrac{S}{6}$, or $A(S) = \dfrac{\pi S}{6}$.

62. *Writing Exercise*

Exercise Set 8.4

1. $x^2 - 5x + 3 = 0$

$a = 1, \ b = -5, \ c = 3$

We substitute and compute the discriminant.

$$b^2 - 4ac = (-5)^2 - 4 \cdot 1 \cdot 3$$
$$= 25 - 12$$
$$= 13$$

Since the discriminant is a positive number that is not a perfect square, there are two irrational solutions.

2. Two irrational

3. $x^2 + 5 = 0$

$a = 1, b = 0, c = 5$

We substitute and compute the discriminant.

$$b^2 - 4ac = 0^2 - 4 \cdot 1 \cdot 5$$
$$= -20$$

Since the discriminant is negative, there are two imaginary-number solutions.

4. Two imaginary

5. $x^2 - 3 = 0$

$a = 1, b = 0, c = -3$

We substitute and compute the discriminant.

$$b^2 - 4ac = 0^2 - 4 \cdot 1 \cdot (-3)$$
$$= 12$$

Since the discriminant is a positive number that is not a perfect square, there are two irrational solutions.

6. Two irrational

7. $4x^2 - 12x + 9 = 0$

$a = 4, \; b = -12, \; c = 9$

We substitute and compute the discriminant.

$$b^2 - 4ac = (-12)^2 - 4 \cdot 4 \cdot 9$$
$$= 144 - 144$$
$$= 0$$

Since the discriminant is 0, there is just one solution, and it is a rational number.

8. Two rational

9. $x^2 - 2x + 4 = 0$

$a = 1, \; b = -2, \; c = 4$

We substitute and compute the discriminant.

$$b^2 - 4ac = (-2)^2 - 4 \cdot 1 \cdot 4$$
$$= 4 - 16$$
$$= -12$$

Since the discriminant is negative, there are two imaginary-number solutions.

10. Two imaginary

11. $6t^2 - 19t - 20 = 0$

$a = 6, b = -19, c = -20$

We substitute and compute the discriminant.

$$b^2 - 4ac = (-19)^2 - 4 \cdot 6 \cdot (-20)$$
$$= 361 + 480$$
$$= 841$$

Since the discriminant is a positive number and a perfect square, there are two rational solutions.

12. One rational

13. $6x^2 + 5x - 4 = 0$

$a = 6, b = 5, c = -4$

We substitute and compute the discriminant.

$$b^2 - 4ac = 5^2 - 4 \cdot 6 \cdot (-4)$$
$$= 25 + 96 = 121$$

Since the discriminant is a positive number and a perfect square, there are two rational solutions.

14. Two rational

15. $9t^2 - 3t = 0$

Observe that we can factor $9t^2 - 3t$. This tells us that there are two rational solutions. We could also do this problem as follows.

$a = 9, \; b = -3, \; c = 0$

We substitute and compute the discriminant.

$$b^2 - 4ac = (-3)^2 - 4 \cdot 9 \cdot 0$$
$$= 9 - 0$$
$$= 9$$

Since the discriminant is a positive number and a perfect square, there are two rational solutions.

16. Two rational

17. $x^2 + 4x = 8$

$\quad x^2 + 4x - 8 = 0 \quad$ Standard form

$a = 1, b = 4, c = -8$

We substitute and compute the discriminant.

$$b^2 - 4ac = 4^2 - 4 \cdot 1 \cdot (-8)$$
$$= 16 + 32 = 48$$

Since the discriminant is a positive number that is not a perfect square, there are two irrational solutions.

18. Two irrational

19. $\quad 2a^2 - 3a = -5$

$\quad 2a^2 - 3a + 5 = 0 \quad$ Standard form

$a = 2, b = -3, c = 5$

We substitute and compute the discriminant.

$b^2 - 4ac = (-3)^2 - 4 \cdot 2 \cdot 5$

$= 9 - 40$

$= -31$

Since the discriminant is negative, there are two imaginary-number solutions.

20. Two imaginary

21.
$$y^2 + \frac{9}{4} = 4y$$

$y^2 - 4y + \frac{9}{4} = 0 \quad$ Standard form

$a = 1, b = -4, c = \frac{9}{4}$

We substitute and compute the discriminant.

$b^2 - 4ac = (-4)^2 - 4 \cdot 1 \cdot \frac{9}{4}$

$= 16 - 9$

$= 7$

The discriminant is a positive number that is not a perfect square. There are two irrational solutions.

22. Two imaginary

23. The solutions are -7 and 3.

$x = -7 \quad or \quad\quad x = 3$

$x + 7 = 0 \quad or \quad x - 3 = 0$

$(x + 7)(x - 3) = 0 \quad$ Principle of zero products

$x^2 + 4x - 21 = 0 \quad$ FOIL

24. $x^2 + 2x - 24 = 0$

25. The only solution is 3. It must be a repeated solution.

$x = 3 \quad or \quad\quad x = 3$

$x - 3 = 0 \quad or \quad x - 3 = 0$

$(x - 3)(x - 3) = 0 \quad$ Principle of zero products

$x^2 - 6x + 9 = 0 \quad$ FOIL

26. $x^2 + 10x + 25 = 0$

27. The solutions are -2 and -5.

$x = -2 \quad or \quad\quad x = -5$

$x + 2 = 0 \quad or \quad x + 5 = 0$

$(x + 2)(x + 5) = 0$

$x^2 + 7x + 10 = 0$

28. $x^2 + 4x + 3 = 0$

29. The solutions are 4 and $\frac{2}{3}$.

$x = 4 \quad or \quad\quad x = \frac{2}{3}$

$x - 4 = 0 \quad or \quad x - \frac{2}{3} = 0$

$(x - 4)\left(x - \frac{2}{3}\right) = 0$

$x^2 - \frac{2}{3}x - 4x + \frac{8}{3} = 0$

$x^2 - \frac{14}{3}x + \frac{8}{3} = 0$

$3x^2 - 14x + 8 = 0 \quad$ Multiplying by 3

30. $4x^2 - 23x + 15 = 0$

31. The solutions are $\frac{1}{2}$ and $\frac{1}{3}$.

$x = \frac{1}{2} \quad or \quad\quad x = \frac{1}{3}$

$x - \frac{1}{2} = 0 \quad or \quad x - \frac{1}{3} = 0$

$\left(x - \frac{1}{2}\right)\left(x - \frac{1}{3}\right) = 0$

$x^2 - \frac{1}{3}x - \frac{1}{2}x + \frac{1}{6} = 0$

$x^2 - \frac{5}{6}x + \frac{1}{6} = 0$

$6x^2 - 5x + 1 = 0 \quad$ Multiplying by 6

32. $8x^2 + 6x + 1 = 0$

33. The solutions are -0.6 and 1.4.

$x = -0.6 \quad or \quad\quad x = 1.4$

$x + 0.6 = 0 \quad or \quad x - 1.4 = 0$

$(x + 0.6)(x - 1.4) = 0$

$x^2 - 1.4x + 0.6x - 0.84 = 0$

$x^2 - 0.8x - 0.84 = 0$

34. $x^2 - 2x - 0.96 = 0$

35. The solutions are $-\sqrt{7}$ and $\sqrt{7}$.

$x = -\sqrt{7} \quad or \quad\quad x = \sqrt{7}$

$x + \sqrt{7} = 0 \quad or \quad x - \sqrt{7} = 0$

$(x + \sqrt{7})(x - \sqrt{7}) = 0$

$x^2 - 7 = 0$

36. $x^2 - 3 = 0$

37. The solutions are $3\sqrt{2}$ and $-3\sqrt{2}$.

$x = 3\sqrt{2} \quad or \quad\quad x = -3\sqrt{2}$

$x - 3\sqrt{2} = 0 \quad or \quad x + 3\sqrt{2} = 0$

$$(x - 3\sqrt{2})(x + 3\sqrt{2}) = 0$$
$$x^2 - (3\sqrt{2})^2 = 0$$
$$x^2 - 9 \cdot 2 = 0$$
$$x^2 - 18 = 0$$

38. $x^2 - 20 = 0$

39. The solutions are $3i$ and $-3i$.
$$x = 3i \quad or \qquad x = -3i$$
$$x - 3i = 0 \quad or \quad x + 3i = 0$$
$$(x - 3i)(x + 3i) = 0$$
$$x^2 - (3i)^2 = 0$$
$$x^2 + 9 = 0$$

40. $x^2 + 16 = 0$

41. The solutions are $5 - 2i$ and $5 + 2i$.
$$x = 5 - 2i \quad or \qquad x = 5 + 2i$$
$$x - 5 + 2i = 0 \quad or \quad x - 5 - 2i = 0$$
$$[x + (-5 + 2i)][x + (-5 - 2i)] = 0$$
$$x^2 + x(-5 - 2i) + x(-5 + 2i) + (-5 + 2i)(-5 - 2i) = 0$$
$$x^2 - 5x - 2xi - 5x + 2xi + 25 - 4i^2 = 0$$
$$x^2 - 10x + 29 = 0$$
$$(i^2 = -1)$$

42. $x^2 - 4x + 53 = 0$

43. The solutions are $2 - \sqrt{10}$ and $2 + \sqrt{10}$.
$$x = 2 - \sqrt{10} \quad or \qquad x = 2 + \sqrt{10}$$
$$x - (2 - \sqrt{10}) = 0 \quad or \quad x - (2 + \sqrt{10}) = 0$$
$$[x - (2 - \sqrt{10})][x - (2 + \sqrt{10})] = 0$$
$$x^2 - x(2 + \sqrt{10}) - x(2 - \sqrt{10}) + (2 - \sqrt{10})(2 + \sqrt{10}) = 0$$
$$x^2 - 2x - x\sqrt{10} - 2x + x\sqrt{10} + 4 - 10 = 0$$
$$x^2 - 4x - 6 = 0$$

44. $x^2 - 6x - 5 = 0$

45. The solutions are -2, 1, and 5.
$$x = -2 \quad or \quad x = 1 \quad or \qquad x = 5$$
$$x + 2 = 0 \quad or \quad x - 1 = 0 \quad or \quad x - 5 = 0$$
$$(x + 2)(x - 1)(x - 5) = 0$$
$$(x^2 + x - 2)(x - 5) = 0$$
$$x^3 + x^2 - 2x - 5x^2 - 5x + 10 = 0$$
$$x^3 - 4x^2 - 7x + 10 = 0$$

46. $x^3 + 3x^2 - 10x = 0$

47. The solutions are -1, 0, and 3.
$$x = -1 \quad or \quad x = 0 \quad or \qquad x = 3$$
$$x + 1 = 0 \quad or \quad x = 0 \quad or \quad x - 3 = 0$$
$$(x + 1)(x)(x - 3) = 0$$
$$(x^2 + x)(x - 3) = 0$$
$$x^3 - 3x^2 + x^2 - 3x = 0$$
$$x^3 - 2x^2 - 3x = 0$$

48. $x^3 - 3x^2 - 4x + 12 = 0$

49. *Writing Exercise*

50. *Writing Exercise*

51. $(3a^2)^4 = 3^4(a^2)^4 = 81a^{2 \cdot 4} = 81a^8$

52. $16x^6$

53. $f(x) = x^2 - 7x - 8$
We find the values of x for which $f(x) = 0$.
$$x^2 - 7x - 8 = 0$$
$$(x - 8)(x + 1) = 0$$
$$x - 8 = 0 \quad or \quad x + 1 = 0$$
$$x = 8 \quad or \qquad x = -1$$
The x-intercepts are $(8, 0)$ and $(-1, 0)$.

54. $(2, 0)$, $(4, 0)$

55. *Familiarize*. Let x and y represent the number of 30-sec and 60-sec commercials, respectively. Then the amount of time for the 30-sec commercials was $30x$ sec, or $\frac{30x}{60} = \frac{x}{2}$ min. The amount of time for the 60-sec commercials was $60x$ sec, or $\frac{60x}{60} = x$ min.

Translate. Rewording, we write two equations. We will express time in minutes.

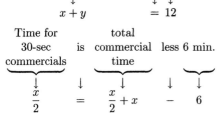

Carry out. Solving the system of equations we get $(6, 6)$.

Check. If there are six 30-sec and six 60-sec commercials, the total number of commercials is 12. The amount of time for six 30-sec commercials is 180 sec, or 3 min, and for six 60-sec commercials is 360 sec, or 6 min. The total commercial time is 9 min, and the amount of time for 30-sec commercials is 6 min less than this. The numbers check.

State. There were six 30-sec commercials.

56.

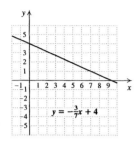

$$y = -\frac{3}{7}x + 4$$

57. *Writing Exercise*

58. *Writing Exercise*

59. The graph includes the points $(-3,0)$, $(0,-3)$, and $(1,0)$. Substituting in $y = ax^2 + bx + c$, we have three equations.
$$0 = 9a - 3b + c,$$
$$-3 = \qquad\qquad c,$$
$$0 = a + b + c$$
The solution of this system of equations is $a = 1$, $b = 2$, $c = -3$.

60. Consider a quadratic equation in standard form, $ax^2 + bx + c = 0$. The solutions are
$$\frac{-b \pm \sqrt{b^2 - 4ac}}{2a}.$$
The product of the solutions is
$$\left(\frac{-b + \sqrt{b^2 - 4ac}}{2a}\right)\left(\frac{-b - \sqrt{b^2 - 4ac}}{2a}\right) =$$
$$\frac{(-b)^2 - (\sqrt{b^2 - 4ac})^2}{(2a)^2} = \frac{b^2 - (b^2 - 4ac)}{4a^2} = \frac{4ac}{4a^2} = \frac{c}{a}.$$

61. a) $kx^2 - 2x + k = 0$; one solution is -3
We first find k by substituting -3 for x.
$$k(-3)^2 - 2(-3) + k = 0$$
$$9k + 6 + k = 0$$
$$10k = -6$$
$$k = -\frac{6}{10}$$
$$k = -\frac{3}{5}$$

b) Now substitute $-\frac{3}{5}$ for k in the original equation.
$$-\frac{3}{5}x^2 - 2x + \left(-\frac{3}{5}\right) = 0$$
$$3x^2 + 10x + 3 = 0 \quad \text{Multiplying by } -5$$
$$(3x + 1)(x + 3) = 0$$

$$x = -\frac{1}{3} \text{ or } x = -3$$
The other solution is $-\frac{1}{3}$.

62. a) 2

b) $1 - i$

63. a) $x^2 - (6 + 3i)x + k = 0$; one solution is 3.
We first find k by substituting 3 for x.
$$3^2 - (6 + 3i)3 + k = 0$$
$$9 - 18 - 9i + k = 0$$
$$-9 - 9i + k = 0$$
$$k = 9 + 9i$$

b) Now we substitute $9 + 9i$ for k in the original equation.
$$x^2 - (6 + 3i)x + (9 + 9i) = 0$$
$$x^2 - (6 + 3i)x + 3(3 + 3i) = 0$$
$$[x - (3 + 3i)][x - 3] = 0$$
$$x = 3 + 3i \text{ or } x = 3$$
The other solution is $3 + 3i$.

64. Consider a quadratic equation in standard form, $ax^2 + bx + c = 0$. The solutions are
$$\frac{-b \pm \sqrt{b^2 - 4ac}}{2a}.$$
The sum of the solutions is
$$\frac{-b + \sqrt{b^2 - 4ac}}{2a} + \frac{-b - \sqrt{b^2 - 4ac}}{2a} = \frac{-2b}{2a} = -\frac{b}{a}.$$

65. The solutions of $ax^2 + bx + c = 0$ are $x = \frac{-b \pm \sqrt{b^2 - 4ac}}{2a}$.
When there is just one solution, $b^2 - 4ac = 0$, so
$$x = \frac{-b \pm 0}{2a} = -\frac{b}{2a}.$$

66. $h = -36$, $k = 15$

67. We substitute $(-3, 0)$, $\left(\frac{1}{2}, 0\right)$, and $(0, -12)$ in $f(x) = ax^2 + bx + c$ and get three equations.
$$0 = 9a - 3b + c,$$
$$0 = \frac{1}{4}a + \frac{1}{2}b + c,$$
$$-12 = c$$
The solution of this system of equations is $a = 8$, $b = 20$, $c = -12$.

68. $x^4 - 14x^3 + 70x^2 - 126x + 29 = 0$

69. If $1 - \sqrt{5}$ and $3 + 2i$ are two solutions, then $1 + \sqrt{5}$ and $3 - 2i$ are also solutions. The equation of lowest degree that has these solutions is found as follows.

$$[x-(1-\sqrt{5})][x-(1+\sqrt{5})][x-(3+2i)][x-(3-2i)] = 0$$
$$(x^2 - 2x - 4)(x^2 - 6x + 13) = 0$$
$$x^4 - 8x^3 + 21x^2 - 2x - 52 = 0$$

70. *Writing Exercise*

Exercise Set 8.5

1. $x^4 - 10x^2 + 9 = 0$

Let $u = x^2$ and $u^2 = x^4$.
$$u^2 - 10u + 9 = 0 \quad \text{Substituting}$$
$$(u - 1)(u - 9) = 0$$
$$u - 1 = 0 \quad or \quad u - 9 = 0$$
$$u = 1 \quad or \quad u = 9$$

Now replace u with x^2 and solve these equations:
$$x^2 = 1 \quad or \quad x^2 = 9$$
$$x = \pm 1 \quad or \quad x = \pm 3$$

The numbers 1, -1, 3, and -3 check. They are the solutions.

2. $\pm 1, \pm 2$

3. $x^4 - 12x^2 + 27 = 0$

Let $u = x^2$ and $u^2 = x^4$.
$$u^2 - 12u + 27 = 0 \quad \text{Substituting } u \text{ for } x^2$$
$$(u - 9)(u - 3) = 0$$
$$u = 9 \quad or \quad u = 3$$

Now replace u with x^2 and solve these equations:
$$x^2 = 9 \quad or \quad x^2 = 3$$
$$x = \pm 3 \quad or \quad x = \pm\sqrt{3}$$

The numbers 3, -3, $\sqrt{3}$, and $-\sqrt{3}$ check. They are the solutions.

4. $\pm\sqrt{5}, \pm 2$

5. $9x^4 - 14x^2 + 5 = 0$

Let $u = x^2$ and $u^2 = x^4$.
$$9u^2 - 14u + 5 = 0 \quad \text{Substituting}$$
$$(9u - 5)(u - 1) = 0$$
$$9u - 5 = 0 \quad or \quad u - 1 = 0$$
$$9u = 5 \quad or \quad u = 1$$
$$u = \frac{5}{9} \quad or \quad u = 1$$

Now replace u with x^2 and solve these equations:
$$x^2 = \frac{5}{9} \quad or \quad x^2 = 1$$
$$x = \pm\frac{\sqrt{5}}{3} \quad or \quad x = \pm 1$$

The numbers $\frac{\sqrt{5}}{3}$, $-\frac{\sqrt{5}}{3}$, 1, and -1 check. They are the solutions.

6. $\pm\dfrac{\sqrt{3}}{2}, \pm 2$

7. $x - 4\sqrt{x} - 1 = 0$

Let $u = \sqrt{x}$ and $u^2 = x$.
$$u^2 - 4u - 1 = 0 \quad \text{Substituting}$$
$$u = \frac{-(-4) \pm \sqrt{(-4)^2 - 4 \cdot 1 \cdot (-1)}}{2 \cdot 1}$$
$$u = \frac{4 \pm \sqrt{20}}{2} = \frac{2 \cdot 2 \pm 2\sqrt{5}}{2}$$
$$u = 2 \pm \sqrt{5}$$
$$u = 2 + \sqrt{5} \quad or \quad u = 2 - \sqrt{5}$$

Replace u with $\sqrt{x}$ and solve these equations.
$$\sqrt{x} = 2 + \sqrt{5} \quad or \quad \sqrt{x} = 2 - \sqrt{5}$$
$$(\sqrt{x})^2 = (2 + \sqrt{5})^2 \quad \text{No solution:}$$
$$2 - \sqrt{5} \text{ is negative}$$
$$x = 4 + 4\sqrt{5} + 5$$
$$x = 9 + 4\sqrt{5}$$

The number $9 + 4\sqrt{5}$ checks. It is the solution.

8. $8 + 2\sqrt{7}$

9. $(x^2 - 7)^2 - 3(x^2 - 7) + 2 = 0$

Let $u = x^2 - 7$ and $u^2 = (x^2 - 7)^2$.
$$u^2 - 3u + 2 = 0 \quad \text{Substituting}$$
$$(u - 1)(u - 2) = 0$$
$$u = 1 \quad or \quad u = 2$$
$$x^2 - 7 = 1 \quad or \quad x^2 - 7 = 2 \quad \text{Replacing } u$$
$$\text{with } x^2 - 7$$
$$x^2 = 8 \quad or \quad x^2 = 9$$
$$x = \pm\sqrt{8} \quad or \quad x = \pm 3$$
$$x = \pm 2\sqrt{2} \quad or \quad x = \pm 3$$

The numbers $2\sqrt{2}$, $-2\sqrt{3}$, 3, and -3 check. They are the solutions.

10. $\pm\sqrt{3}, \pm 2$

11. $(1 + \sqrt{x})^2 + 5(1 + \sqrt{x}) + 6 = 0$

Let $u = 1 + \sqrt{x}$ and $u^2 = (1 + \sqrt{x})^2$.
$$u^2 + 5u + 6 = 0 \quad \text{Substituting}$$
$$(u + 3)(u + 2) = 0$$
$$u = -3 \quad or \quad u = -2$$
$$1 + \sqrt{x} = -3 \quad or \quad 1 + \sqrt{x} = -2 \quad \text{Replacing } u$$
$$\text{with } 1 + \sqrt{x}$$
$$\sqrt{x} = -4 \quad or \quad \sqrt{x} = -3$$

Since the principal square root cannot be negative, this equation has no solution.

12. No solution

13. $x^{-2} - x^{-1} - 6 = 0$

Let $u = x^{-1}$ and $u^2 = x^{-2}$.

$u^2 - u - 6 = 0$ Substituting

$(u - 3)(u + 2) = 0$

$u = 3$ or $u = -2$

Now we replace u with x^{-1} and solve these equations:

$x^{-1} = 3$ or $x^{-1} = -2$

$\dfrac{1}{x} = 3$ or $\dfrac{1}{x} = -2$

$\dfrac{1}{3} = x$ or $-\dfrac{1}{2} = x$

Both $\dfrac{1}{3}$ and $-\dfrac{1}{2}$ check. They are the solutions.

14. $-2, 1$

15. $4x^{-2} + x^{-1} - 5 = 0$

Let $u = x^{-1}$ and $u^2 = x^{-2}$.

$4u^2 + u - 5 = 0$ Substituting

$(4u + 5)(u - 1) = 0$

$u = -\dfrac{5}{4}$ or $u = 1$

Now we replace u with x^{-1} and solve these equations:

$x^{-1} = -\dfrac{5}{4}$ or $x^{-1} = 1$

$\dfrac{1}{x} = -\dfrac{5}{4}$ or $\dfrac{1}{x} = 1$

$4 = -5x$ or $1 = x$

$-\dfrac{4}{5} = x$ or $1 = x$

The numbers $-\dfrac{4}{5}$ and 1 check. They are the solutions.

16. $-\dfrac{1}{10}, 1$

17. $t^{2/3} + t^{1/3} - 6 = 0$

Let $u = t^{1/3}$ and $u^2 = t^{2/3}$.

$u^2 + u - 6 = 0$ Substituting

$(u + 3)(u - 2) = 0$

$u = -3$ or $u = 2$

Now we replace u with $t^{1/3}$ and solve these equations:

$t^{1/3} = -3$ or $t^{1/3} = 2$

$t = (-3)^3$ or $t = 2^3$ Raising to the
                                  third power

$t = -27$ or $t = 8$

Both -27 and 8 check. They are the solutions.

18. $-8, 64$

19. $y^{1/3} - y^{1/6} - 6 = 0$

Let $u = y^{1/6}$ and $u^2 = y^{2/3}$.

$u^2 - u - 6 = 0$ Substituting

$(u - 3)(u + 2) = 0$

$u = 3$ or $u = -2$

Now we replace u with $y^{1/6}$ and solve these equations:

$y^{1/6} = 3$ or $y^{1/6} = -2$

$\sqrt[6]{y} = 3$ or $\sqrt[6]{y} = -2$

$y = 3^6$ This equation has no

$y = 729$ solution since principal
                  sixth roots are never negative.

The number 729 checks and is the solution.

20. No solution

21. $t^{1/3} + 2t^{1/6} = 3$

$t^{1/3} + 2t^{1/6} - 3 = 0$

Let $u = t^{1/6}$ and $u^2 = t^{2/6} = t^{1/3}$.

$u^2 + 2u - 3 = 0$ Substituting

$(u + 3)(u - 1) = 0$

$u = -3$ or $u = 1$

$t^{1/6} = -3$ or $t^{1/6} = 1$ Substituting $t^{1/6}$ for u

No solution $t = 1$

The number 1 checks and is the solution.

22. $16, 81$

23. $(3 - \sqrt{x})^2 - 10(3 - \sqrt{x}) + 23 = 0$

Let $u = 3 - \sqrt{x}$ and $u^2 = (3 - \sqrt{x})^2$.

$u^2 - 10u + 23 = 0$ Substituting

$u = \dfrac{-(-10) \pm \sqrt{(-10)^2 - 4 \cdot 1 \cdot 23}}{2 \cdot 1}$

$u = \dfrac{10 \pm \sqrt{8}}{2} = \dfrac{2 \cdot 5 \pm 2\sqrt{2}}{2}$

$u = 5 \pm \sqrt{2}$

$u = 5 + \sqrt{2}$ or $u = 5 - \sqrt{2}$

Now we replace u with $3 - \sqrt{x}$ and solve these equations:

$3 - \sqrt{x} = 5 + \sqrt{2}$ or $3 - \sqrt{x} = 5 - \sqrt{2}$

$-\sqrt{x} = 2 + \sqrt{2}$ or $-\sqrt{x} = 2 - \sqrt{2}$

$\sqrt{x} = -2 - \sqrt{2}$ or $\sqrt{x} = -2 + \sqrt{2}$

Since both $-2 - \sqrt{2}$ and $-2 + \sqrt{2}$ are negative and principal square roots are never negative, the equation has no solution.

24. $4 + 2\sqrt{3}$

25. $16\left(\dfrac{x-1}{x-8}\right)^2 + 8\left(\dfrac{x-1}{x-8}\right) + 1 = 0$

Let $u = \dfrac{x-1}{x-8}$ and $u^2 = \left(\dfrac{x-1}{x-8}\right)^2$.

$16u^2 + 8u + 1 = 0$ Substituting

$(4u+1)(4u+1) = 0$

$u = -\dfrac{1}{4}$

Now we replace u with $\dfrac{x-1}{x-8}$ and solve this equation:

$\dfrac{x-1}{x-8} = -\dfrac{1}{4}$

$4x - 4 = -x + 8$ Multiplying by $4(x-8)$

$5x = 12$

$x = \dfrac{12}{5}$

The number $\dfrac{12}{5}$ checks and is the solution.

26. $-\dfrac{3}{2}$

27. The x-intercepts occur where $f(x) = 0$. Thus, we must have $5x + 13\sqrt{x} - 6 = 0$.

Let $u = \sqrt{x}$ and $u^2 = x$.

$5u^2 + 13u - 6 = 0$ Substituting

$(5u - 2)(u + 3) = 0$

$u = \dfrac{2}{5}$ or $u = -3$

Now replace u with $\sqrt{x}$ and solve these equations:

$\sqrt{x} = \dfrac{2}{5}$ or $\sqrt{x} = -3$

$x = \dfrac{4}{25}$ No solution

The number $\dfrac{4}{25}$ checks. Thus, the x-intercept is $\left(\dfrac{4}{25}, 0\right)$.

28. $\left(\dfrac{4}{9}, 0\right)$

29. The x-intercepts occur where $f(x) = 0$. Thus, we must have $(x^2 - 3x)^2 - 10(x^2 - 3x) + 24 = 0$.

Let $u = x^2 - 3x$ and $u^2 = (x^2 - 3x)^2$.

$u^2 - 10u + 24 = 0$ Substituting

$(u - 6)(u - 4) = 0$

$u = 6$ or $u = 4$

Now replace u with $x^2 - 3x$ and solve these equations:

$x^2 - 3x = 6$ or $x^2 - 3x = 4$

$x^2 - 3x - 6 = 0$ or $x^2 - 3x - 4 = 0$

$x = \dfrac{-(-3) \pm \sqrt{(-3)^2 - 4(1)(-6)}}{2 \cdot 1}$ or

$(x - 4)(x + 1) = 0$

$x = \dfrac{3 \pm \sqrt{33}}{2}$ or $x = 4$ or $x = -1$

All four numbers check. Thus, the x-intercepts are $\left(\dfrac{3 + \sqrt{33}}{2}, 0\right)$, $\left(\dfrac{3 - \sqrt{33}}{2}, 0\right)$, $(4, 0)$, and $(-1, 0)$.

30. $(-1, 0)$, $(1, 0)$, $(5, 0)$, $(7, 0)$

31. The x-intercepts occur where $f(x) = 0$. Thus, we must have $x^{2/5} + x^{1/5} - 6 = 0$.

Let $u = x^{1/5}$ and $u^2 = x^{2/5}$.

$u^2 + u - 6 = 0$ Substituting

$(u + 3)(u - 2) = 0$

$u = -3$ or $u = 2$

$x^{1/5} = -3$ or $x^{1/5} = 2$ Replacing u with $x^{1/5}$

$x = -243$ or $x = 32$ Raising to the fifth power

Both -243 and 32 check. Thus, the x-intercepts are $(-243, 0)$ and $(32, 0)$.

32. $(81, 0)$

33. $f(x) = \left(\dfrac{x^2 + 2}{x}\right)^4 + 7\left(\dfrac{x^2 + 2}{x}\right)^2 + 5$

Observe that, for all real numbers x, each term is positive. Thus, there are no real-number values of x for which $f(x) = 0$ and hence no x-intercepts.

34. No x-intercepts

35. *Writing Exercise*

36. *Writing Exercise*

37. Graph $f(x) = \dfrac{3}{2}x$.

We find some ordered pairs, plot points, and draw the graph.

x	y
-4	-6
-2	-3
0	0
2	3
4	6

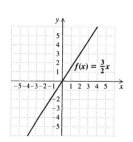

38.

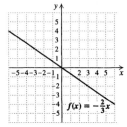

39. Graph $g(x) = \dfrac{2}{x}$.

We find some ordered pairs, plot points, and draw the graph. Note that we cannot use 0 as a first coordinate since division by 0 is undefined.

x	y
-4	$-\dfrac{1}{2}$
-2	-1
$-\dfrac{1}{2}$	-4
$\dfrac{1}{2}$	4
2	1
4	$\dfrac{1}{2}$

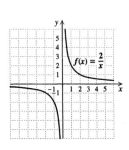

40.

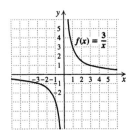

41. Familiarize. Let $a =$ the number of liters of solution A in the mixture and $b =$ the number of liters of solution B. We organize the information in a table.

Solution	A	B	Mixture
Number of liters	a	b	12
Percent of alcohol	18%	45%	36%
Amount of alcohol	$0.18a$	$0.45b$	0.36(12), or 4.32 L

From the first row of the table we get one equation:

$$a + b = 12$$

We get a second equation from the last row of the table:

$$0.18a + 0.45b = 4.32$$

After clearing decimals, we have the following system of equations:

$$a + \quad b = 12, \quad (1)$$
$$18a + 45b = 432 \quad (2)$$

Carry out. We use the elimination method. First we multiply equation (1) by -18 and then add.

$$
\begin{array}{r}
-18a - 18b = -216 \\
18a + 45b = 432 \\
\hline
27b = 216 \\
b = 8
\end{array}
$$

Now we substitute 8 for b in one of the original equations and solve for a.

$$a + b = 12 \quad (1)$$
$$a + 8 = 12$$
$$a = 4$$

Check. If 4 L of solution A and 8 L of solution B are used, the mixture has $4 + 8$, or 12 L. The amount of alcohol in 4 L of solution A is 0.18(4), or 0.72 L. The amount of alcohol in 8 L of solution B is 0.45(8), or 3.6 L. Then the amount of alcohol in the mixture is $0.72 + 3.6$, or 4.32 L. The answer checks.

State. The mixture should contain 4 L of solution A and 8 L of solution B.

42. $a^2 + a$

43. *Writing Exercise*

44. *Writing Exercise*

45. $5x^4 - 7x^2 + 1 = 0$

Let $u = x^2$ and $u^2 = x^4$.

$5u^2 - 7u + 1 = 0$ Substituting

$$u = \frac{-(-7) \pm \sqrt{(-7)^2 - 4 \cdot 5 \cdot 1}}{2 \cdot 5}$$

$$u = \frac{7 \pm \sqrt{29}}{10}$$

$$x^2 = \frac{7 \pm \sqrt{29}}{10} \qquad \text{Replacing } u \text{ with } x^2$$

$$x = \pm\sqrt{\frac{7 \pm \sqrt{29}}{10}}$$

All four numbers check and are the solutions.

46. $\pm\sqrt{\dfrac{-5 \pm \sqrt{37}}{6}}$

47. $(x^2 - 4x - 2)^2 - 13(x^2 - 4x - 2) + 30 = 0$

Let $u = x^2 - 4x - 2$ and $u^2 = (x^2 - 4x - 2)^2$.

$u^2 - 13u + 30 = 0$ Substituting

$(u - 3)(u - 10) = 0$

$\qquad u = 3 \quad or \qquad\qquad u = 10$

$x^2 - 4x - 2 = 3 \quad or \qquad x^2 - 4x - 2 = 10$

$\qquad\qquad\qquad$ Replacing u with $x^2 - 4x - 2$

$x^2 - 4x - 5 = 0 \quad or \quad x^2 - 4x - 12 = 0$

$(x - 5)(x + 1) = 0 \quad or \quad (x - 6)(x + 2) = 0$

$x = 5 \; or \; x = -1 \; or \; x = 6 \; or \; x = -2$

All four numbers check and are the solutions.

48. $-2, -1, 6, 7$

49. $\dfrac{x}{x - 1} - 6\sqrt{\dfrac{x}{x - 1}} - 40 = 0$

$\quad$ Let $u = \sqrt{\dfrac{x}{x - 1}}$ and $u^2 = \dfrac{x}{x - 1}$.

$\qquad u^2 - 6u - 40 = 0 \quad$ Substituting

$\quad (u - 10)(u + 4) = 0$

$\qquad\qquad u = 10 \qquad\quad or \qquad\quad u = -4$

$\quad \sqrt{\dfrac{x}{x - 1}} = 10 \qquad or \quad \sqrt{\dfrac{x}{x - 1}} = -4$

$\qquad \dfrac{x}{x - 1} = 100 \qquad or \qquad$ No solution

$\qquad\qquad x = 100x - 100$ Multiplying by $(x - 1)$

$\qquad\quad 100 = 99x$

$\qquad\quad \dfrac{100}{99} = x$

The number $\dfrac{100}{99}$ checks. It is the solution.

50. $\dfrac{432}{143}$

51. $\quad a^5(a^2 - 25) + 13a^3(25 - a^2) + 36a(a^2 - 25) = 0$

$\quad a^5(a^2 - 25) - 13a^3(a^2 - 25) + 36a(a^2 - 25) = 0$

$\quad a(a^2 - 25)(a^4 - 13a^2 + 36) = 0$

$\quad a(a^2 - 25)(a^2 - 4)(a^2 - 9) = 0$

$\quad a{=}0 \; or \; a^2 - 25{=}0 \quad or \; a^2 - 4{=}0 \quad or \; a^2 - 9 = 0$

$\quad a{=}0 \; or \qquad a^2{=}25 \; or \qquad a^2{=}4 \quad or \qquad a^2 = 9$

$\quad a{=}0 \; or \qquad a{=}\pm5 \; or \qquad a{=}\pm2 \; or \qquad a = \pm3$

All seven numbers check. The solutions are 0, 5, -5, 2, -2, 3, and -3.

52. 9

53. $x^6 - 28x^3 + 27 = 0$

$\quad$ Let $u = x^3$.

$\qquad u^2 - 28u + 27 = 0$

$\quad (u - 27)(u - 1) = 0$

$\qquad u = 27 \quad or \quad u = 1$

$\qquad x^3 = 27 \quad or \quad x^3 = 1$

$\qquad x = 3 \quad or \quad x = 1$

Both 3 and 1 check. They are the solutions.

54. $-2, 1$

Exercise Set 8.6

1. a) The parabola opens upward, so a is positive.

$\quad$ b) The vertex is $(3, 1)$.

$\quad$ c) The axis of symmetry is $x = 3$.

$\quad$ d) The range is $[1, \infty)$.

2. a) Negative

$\quad$ b) $(-1, 2)$

$\quad$ c) $x = -1$

$\quad$ d) $(-\infty, 2]$

3. a) The parabola opens downward, so a is negative.

$\quad$ b) The vertex is $(-2, -3)$.

$\quad$ c) The axis of symmetry is $x = -2$.

$\quad$ d) The range is $(-\infty, -3]$.

4. a) Positive

$\quad$ b) $(2, 0)$

$\quad$ c) $x = 2$

$\quad$ d) $[0, \infty)$

5. a) The parabola opens upward, so a is positive.

$\quad$ b) The vertex is $(-3, 0)$.

$\quad$ c) The axis of symmetry is $x = -3$.

$\quad$ d) The range is $[0, \infty)$.

6. a) Negative

$\quad$ b) $(1, -2)$

$\quad$ c) $x = 1$

$\quad$ d) $(-\infty, -2]$

7. $a = 3$ and $3 > 0$, so the graph opens up; the vertex is $(0, 0)$. Graph (f) matches this function.

8. (c)

9. $a = -1$ and $-1 < 0$, so the graph opens down; the vertex is $(2, 0)$. Graph (e) matches this function.

10. (b)

11. $a = \dfrac{2}{3}$ and $\dfrac{2}{3} > 0$, so the graph opens up; the vertex is $(-3, 1)$. Graph (d) matches this function.

12. (a)

13. $f(x) = x^2$

See Example 1 in the text.

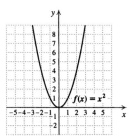

14.

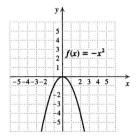

15. $f(x) = -2x^2$

We choose some numbers for x and compute $f(x)$ for each one. Then we plot the ordered pairs $(x, f(x))$ and connect them with a smooth curve.

x	$f(x) = -4x^2$
0	0
1	-2
2	-8
-1	-2
-2	-8

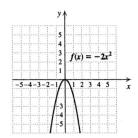

16.

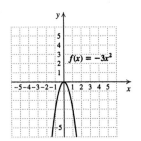

17. $g(x) = \dfrac{1}{3}x^2$

x	$g(x) = \dfrac{1}{3}x^2$
0	0
1	$\dfrac{1}{3}$
2	$\dfrac{4}{3}$
3	3
-1	$\dfrac{1}{3}$
-2	$\dfrac{4}{3}$
-3	3

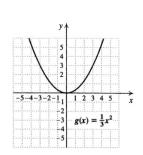

18.

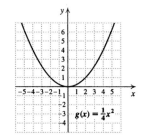

19. $h(x) = -\dfrac{1}{3}x^2$

x	$h(x) = -\dfrac{1}{3}x^2$
0	0
1	$-\dfrac{1}{3}$
2	$-\dfrac{4}{3}$
3	-3
-1	$-\dfrac{1}{3}$
-2	$-\dfrac{4}{3}$
-3	-3

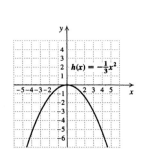

20.

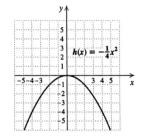

21. $f(x) = \dfrac{3}{2}x^2$

x	$f(x) = \dfrac{3}{2}x^2$
0	0
1	$\dfrac{3}{2}$
2	6
−1	$\dfrac{3}{2}$
−2	6

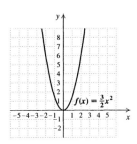

22.

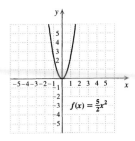

23. $g(x) = (x+1)^2 = [x - (-1)]^2$

We know that the graph of $g(x) = (x+1)^2$ looks like the graph of $f(x) = x^2$ (see Exercise 13) but moved to the left 1 unit.

Vertex: $(-1, 0)$, axis of symmetry: $x = -1$

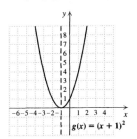

24.

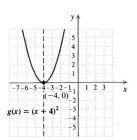

25. $f(x) = (x-2)^2$

The graph of $f(x) = (x-2)^2$ looks like the graph of $f(x) = x^2$ (see Exercise 13) but moved to the right 2 units.

Vertex: $(2, 0)$, axis of symmetry: $x = 2$

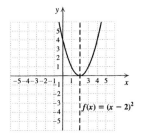

26.

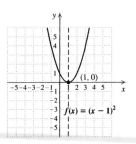

27. $f(x) = -(x+4)^2 = -[x - (-4)]^2$

The graph of $f(x) = -(x+4)^2$ looks like the graph of $f(x) = x^2$ (see Exercise 13) but moved to the left 4 units. It will also open downward because of the negative coefficient, -1.

Vertex: $(-4, 0)$, axis of symmetry: $x = -4$

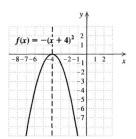

28.

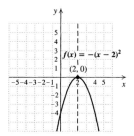

29. $f(x) = 2(x+1)^2$

The graph of $f(x) = 2(x+1)^2$ looks like the graph of $h(x) = 2x^2$ (see graph following Example 1) but moved to the left 1 unit.

Vertex: $(-1, 0)$, axis of symmetry: $x = -1$

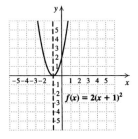

30.

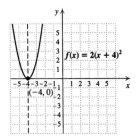

31. $h(x) = -\dfrac{1}{2}(x-3)^2$

The graph of $h(x) = -\dfrac{1}{2}(x-3)^2$ looks like the graph of $g(x) = \dfrac{1}{2}x^2$ (see graph following Example 1) but moved to the right 3 units. It will also open downward because of the negative coefficient, $-\dfrac{1}{2}$.

Vertex: $(3,0)$, axis of symmetry: $x = 3$

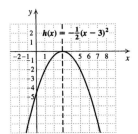

32.

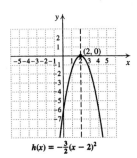

33. $f(x) = \dfrac{1}{2}(x-1)^2$

The graph of $f(x) = \dfrac{1}{2}(x-1)^2$ looks like the graph of $g(x) = \dfrac{1}{2}x^2$ (see graph following Example 1) but moved

to the right 1 unit.

Vertex: $(1,0)$, axis of symmetry: $x = 1$

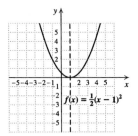

34.

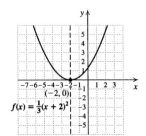

35. $f(x) = (x-5)^2 + 1$

We know that the graph looks like the graph of $f(x) = x^2$ (see Example 1) but moved to the right 5 units and up 1 unit. The vertex is $(5,1)$, and the axis of symmetry is $x = 5$. Since the coefficient of $(x-5)^2$ is positive ($1 > 0$), there is a minimum function value, 1.

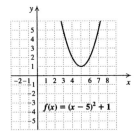

36. $f(x) = (x+3)^2 - 2$

Vertex: $(-3,-2)$, axis of symmetry: $x = -3$

Minimum: -2

37. $f(x) = (x+1)^2 - 2$

We know that the graph looks like the graph of $f(x) = x^2$ (see Example 1) but moved to the left 1 unit and down 2 units. The vertex is $(-1, -2)$, and the axis of symmetry is $x = -1$. Since the coefficient of $(x+1)^2$ is positive $(1 > 0)$, there is a minimum function value, -2.

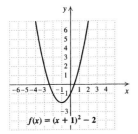

$f(x) = (x + 1)^2 - 2$

38. $g(x) = -(x - 2)^2 - 4$

Vertex: $(2, -4)$, axis of symmetry: $x = 2$

Maximum: -4

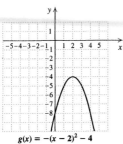

$g(x) = -(x - 2)^2 - 4$

39. $h(x) = -2(x - 1)^2 - 3$

We know that the graph looks like the graph of $h(x) = x^2$ (see Example 1) but it is more narrow, moved to the right 1 unit and down 3 units, and turned upside down. The vertex is $(1, -3)$, and the axis of symmetry is $x = 1$. The maximum function value is -3.

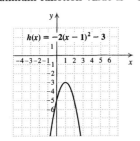

$h(x) = -2(x - 1)^2 - 3$

40. $h(x) = -2(x + 1)^2 + 4$

Vertex: $(-1, 4)$, axis of symmetry: $x = -1$

Maximum: 4

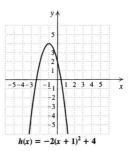

$h(x) = -2(x + 1)^2 + 4$

41. $f(x) = 2(x + 4)^2 + 1$

We know that the graph looks like the graph of $f(x) = x^2$ (see Example 1) but it is more narrow and is moved to the left 4 units and up 1 unit. The vertex is $(-4, 1)$, the axis of symmetry is $x = -4$, and the minimum function value is 1.

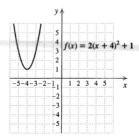

$f(x) = 2(x + 4)^2 + 1$

42. $f(x) = 2(x - 5)^2 - 3$

Vertex: $(5, -3)$, axis of symmetry: $x = 5$

Minimum: -3

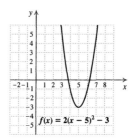

$f(x) = 2(x - 5)^2 - 3$

43. $g(x) = -\dfrac{3}{2}(x - 1)^2 + 2$

We know that the graph looks like the graph of $f(x) = \dfrac{3}{2}x^2$ (see Exercise 21) but moved to the right 1 unit and up 2 units and turned upside down. The vertex is $(1, 2)$, the axis of symmetry is $x = 1$, and the maximum function value is 2.

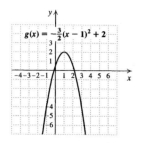

44. $g(x) = \dfrac{3}{2}(x+2)^2 - 1$

Vertex: $(-2, -1)$, axis of symmetry: $x = -2$

Minimum: -1

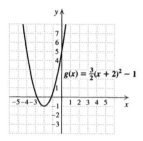

45. $f(x) = 8(x-9)^2 + 5$

This function is of the form $f(x) = a(x-h)^2 + k$ with $a = 8$, $h = 9$, and $k = 5$. The vertex is (h, k), or $(9, 5)$. The axis of symmetry is $x = h$, or $x = 9$. Since $a > 0$, then k, or 5, is the minimum function value. The range is $[5, \infty)$.

46. Vertex: $(-5, -8)$

Axis of symmetry: $x = -5$

Minimum: -8

Range: $[-8, \infty)$

47. $h(x) = -\dfrac{2}{7}(x+6)^2 + 11$

This function is of the form $f(x) = a(x-h)^2 + k$ with $a = -\dfrac{2}{7}$, $h = -6$, and $k = 11$. The vertex is (h, k), or $(-6, 11)$. The axis of symmetry is $x = h$, or $x = -6$. Since $a < 0$, then k, or 11, is the maximum function value. The range is $(-\infty, 11]$.

48. Vertex: $(7, -9)$

Axis of symmetry: $x = 7$

Maximum: -9

Range: $(-\infty, -9]$

49. $f(x) = 5\left(x + \dfrac{1}{4}\right)^2 - 13$

This function is of the form $f(x) = a(x-h)^2 + k$ with $a = 5$, $h = -\dfrac{1}{4}$, and $k = -13$. The vertex is (h, k), or $\left(-\dfrac{1}{4}, -13\right)$.

The axis of symmetry is $x = h$, or $x = -\dfrac{1}{4}$. Since $a > 0$, then k, or -13, is the minimum function value. The range is $[-13, \infty)$.

50. Vertex: $\left(\dfrac{1}{4}, 19\right)$

Axis of symmetry: $x = \dfrac{1}{4}$

Minimum: 19

Range: $[19, \infty)$

51. $f(x) = \sqrt{2}(x + 4.58)^2 + 65\pi$

This function is of the form $f(x) = a(x-h)^2 + k$ with $a = \sqrt{2}$, $h = -4.58$, and $k = 65\pi$. The vertex is (h, k), or $(-4.58, 65\pi)$. The axis of symmetry is $x = h$, or $x = -4.58$. Since $a > 0$, then k, or 65π, is the minimum function value. The range is $[65\pi, \infty)$.

52. Vertex: $(38.2, -\sqrt{34})$

Axis of symmetry: $x = 38.2$

Minimum: $-\sqrt{34}$

Range: $[-\sqrt{34}, \infty)$

53. *Writing Exercise*

54. *Writing Exercise*

55. Graph $2x - 7y = 28$.

Find the x-intercept.
$$2x - 7 \cdot 0 = 28$$
$$2x = 28$$
$$x = 14$$
The x-intercept is $(14, 0)$.

Find the y-intercept.
$$2 \cdot 0 - 7y = 28$$
$$-7y = 28$$
$$y = -4$$
The y-intercept is $(0, -4)$.

Plot the intercepts and draw a line through them. A third point can be plotted as a check.

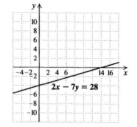

56.

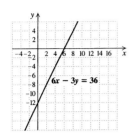

$6x - 3y = 36$

57. $3x + 4y = -19,$ (1)

$7x - 6y = -29$ (2)

Multiply Equation (1) by 3 and multiply Equation (2) by 2. Then add the equations to eliminate the y-term.

$$\begin{array}{r} 9x + 12y = -57 \\ 14x - 12y = -58 \\ \hline 23x \qquad = -115 \\ x = -5 \end{array}$$

Now substitute -5 for x in one of the original equations and solve for y. We use Equation (1).

$$3(-5) + 4y = -19$$
$$-15 + 4y = -19$$
$$4y = -4$$
$$y = -1$$

The pair $(-5, -1)$ checks and it is the solution.

58. $(-1, 2)$

59. $x^2 + 5x$

We take half the coefficient of x and square it.

$$\frac{1}{2} \cdot 5 = \frac{5}{2}, \quad \left(\frac{5}{2}\right)^2 = \frac{25}{4}$$

Then we have $x^2 + 5x + \frac{25}{4}$.

60. $x^2 - 9x + \frac{81}{4}$

61. *Writing Exercise*

62. *Writing Exercise*

63. The equation will be of the form $f(x) = \frac{3}{5}(x - h)^2 + k$ with $h = 4$ and $k = 1$:

$$f(x) = \frac{3}{5}(x - 4)^2 + 1$$

64. $f(x) = \frac{3}{5}(x - 2)^2 + 6$

65. The equation will be of the form $f(x) = \frac{3}{5}(x - h)^2 + k$ with $h = 3$ and $k = -1$:

$$f(x) = \frac{3}{5}(x - 3)^2 + (-1), \text{ or}$$
$$f(x) = \frac{3}{5}(x - 3)^2 - 1$$

66. $f(x) = \frac{3}{5}(x - 5)^2 - 6$

67. The equation will be of the form $f(x) = \frac{3}{5}(x - h)^2 + k$ with $h = -2$ and $k = -5$:

$$f(x) = \frac{3}{5}[x - (-2)]^2 + (-5), \text{ or}$$
$$f(x) = \frac{3}{5}(x + 2)^2 - 5$$

68. $f(x) = \frac{3}{5}(x + 4)^2 - 2$

69. Since there is a maximum at $(5, 0)$, the parabola will have the same shape as $g(x) = -2x^2$. It will be of the form $g(x) = -2(x - h)^2 + k$ with $h = 5$ and $k = 0$: $g(x) = -2(x - 5)^2$

70. $f(x) = 2(x - 2)^2$

71. Since there is a minimum at $(-4, 0)$, the parabola will have the same shape as $f(x) = 2x^2$. It will be of the form $f(x) = 2(x - h)^2 + k$ with $h = -4$ and $k = 0$: $f(x) = 2[x - (-4)]^2$, or $f(x) = 2(x + 4)^2$

72. $g(x) = -2x^2 + 3$

73. Since there is a maximum at $(3, 8)$, the parabola will have the same shape as $g(x) = -2x^2$. It will be of the form $g(x) = -2(x - h)^2 + k$ with $h = 3$ and $k = 8$: $g(x) = -2(x - 3)^2 + 8$

74. $f(x) = 2(x + 2)^2 + 3$

75. The maximum value of $g(x)$ is 1 and occurs at the point $(5, 1)$, so for $F(x)$ we have $h = 5$ and $k = 1$. $F(x)$ has the same shape as $f(x)$ and has a minimum, so $a = 3$. Thus, $F(x) = 3(x - 5)^2 + 1$.

76. $F(x) = -\frac{1}{3}(x + 4)^2 - 6$

77. The function is of the form $F(x) = a(x - h)^2 + k$. Substitute 2 for h, -3 for k, 1 for x, and -5 for $F(x)$ and find a.

$$F(x) = a(x - h)^2 + k$$
$$-5 = a(1 - 2)^2 + (-3)$$
$$-5 = a(-1)^2 - 3$$
$$-5 = a - 3$$
$$-2 = a$$

Keeping in mind that $h = 2$ and $k = -3$, we have $F(x) = -2(x - 2)^2 - 3$.

78. $F(x) = -\dfrac{5}{4}(x-3)^2 - 1$

79. The graph of $y = f(x-1)$ looks like the graph of $y = f(x)$ moved 1 unit to the right.

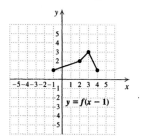

80.

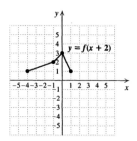

81. The graph of $y = f(x) + 2$ looks like the graph of $y = f(x)$ moved up 2 units.

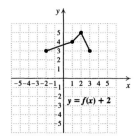

82.

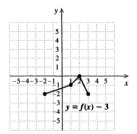

83. The graph of $y = f(x+3) - 2$ looks like the graph of $y = f(x)$ moved 3 units to the left and also moved down 2 units.

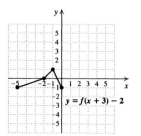

84.

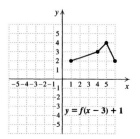

Exercise Set 8.7

1. a) $f(x) = x^2 - 4x + 5$
$\qquad = (x^2 - 4x + 4 - 4) + 5 \quad$ Adding $4 - 4$
$\qquad = (x^2 - 4x + 4) - 4 + 5 \quad$ Regrouping
$\qquad = (x-2)^2 + 1$

b) The vertex is $(2, 1)$; the axis of symmetry is $x = 2$.

2. a) $f(x) = (x+3)^2 + 4$

b) Vertex: $(-3, 4)$; line of symmetry: $x = -3$

3. a) $f(x) = -x^2 + 3x - 10$
$\qquad = -(x^2 - 3x) - 10$
$\qquad = -\left(x^2 - 3x + \dfrac{9}{4} - \dfrac{9}{4}\right) - 10$
$\qquad = -\left(x^2 - 3x + \dfrac{9}{4}\right) + \left[-\left(-\dfrac{9}{4}\right)\right] - 10$
$\qquad = -\left(x - \dfrac{3}{2}\right)^2 + \dfrac{9}{4} - 10$
$\qquad = -\left(x - \dfrac{3}{2}\right)^2 - \dfrac{31}{4}$

b) The vertex is $\left(\dfrac{3}{2}, -\dfrac{31}{4}\right)$; the axis of symmetry is $x = \dfrac{3}{2}$.

4. a) $f(x) = \left(x + \dfrac{5}{2}\right)^2 - \dfrac{9}{4}$

b) Vertex: $\left(-\dfrac{5}{2}, -\dfrac{9}{4}\right)$; axis of symmetry: $x = -\dfrac{5}{2}$

5. a) $f(x) = 2x^2 - 7x + 1$

$$= 2\left(x^2 - \frac{7}{2}x\right) + 1$$

$$= 2\left(x^2 - \frac{7}{2}x + \frac{49}{16} - \frac{49}{16}\right) + 1$$

$$= 2\left(x^2 - \frac{7}{2}x + \frac{49}{16}\right) + 2\left(-\frac{49}{16}\right) + 1$$

$$= 2\left(x - \frac{7}{4}\right)^2 - \frac{49}{8} + 1$$

$$= 2\left(x - \frac{7}{4}\right)^2 - \frac{41}{8}$$

b) Vertex: $\left(\frac{7}{4}, -\frac{41}{8}\right)$; axis of symmetry: $x = \frac{7}{4}$

6. a) $f(x) = -2\left(x - \frac{5}{4}\right)^2 + \frac{17}{8}$

b) Vertex: $\left(\frac{5}{4}, \frac{17}{8}\right)$; axis of symmetry: $x = \frac{5}{4}$

7. $f(x) = x^2 + 4x + 5$

$$= (x^2 + 4x + 4 - 4) + 5 \quad \text{Adding } 4 - 4$$

$$= (x^2 + 4x + 4) - 4 + 5 \quad \text{Regrouping}$$

$$= (x + 2)^2 + 1$$

The vertex is $(-2, 1)$, the axis of symmetry is $x = -2$, and the graph opens upward since the coefficient 1 is positive. We plot a few points as a check and draw the curve.

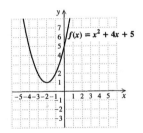

8. $f(x) = (x + 1)^2 - 6$

Vertex: $(-1, -6)$, axis of symmetry: $x = -1$

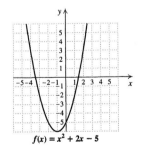

9. $f(x) = x^2 + 8x + 20$

$$= (x^2 + 8x + 16 - 16) + 20 \quad \text{Adding } 16 - 16$$

$$= (x^2 + 8x + 16) - 16 + 20 \quad \text{Regrouping}$$

$$= (x + 4)^2 + 4$$

The vertex is $(-4, 4)$, the axis of symmetry is $x = -4$, and the graph opens upward since the coefficient 1 is positive.

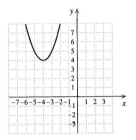

10. $f(x) = (x - 5)^2 - 4$

Vertex: $(5, -4)$, axis of symmetry: $x = 5$

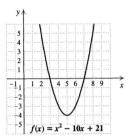

11. $h(x) = 2x^2 - 16x + 25$

$$= 2(x^2 - 8x) + 25 \quad \text{Factoring 2 from the first two terms}$$

$$= 2(x^2 - 8x + 16 - 16) + 25 \quad \text{Adding } 16 - 16 \text{ inside the parentheses}$$

$$= 2(x^2 - 8x + 16) + 2(-16) + 25 \quad \text{Distributing to obtain a trinomial square}$$

$$= 2(x - 4)^2 - 7$$

The vertex is $(4, -7)$, the axis of symmetry is $x = 4$, and the graph opens upward since the coefficient 2 is positive.

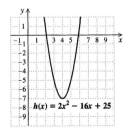

12. $h(x) = 2(x+4)^2 - 9$

Vertex: $(-4, -9)$, axis of symmetry: $x = -4$

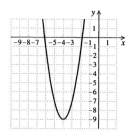

$$h(x) = 2x^2 + 16x + 23$$

13. $f(x) = -x^2 + 2x + 5$

$\qquad = -(x^2 - 2x) + 5 \qquad$ Factoring -1 from the first two terms

$\qquad = -(x^2 - 2x + 1 - 1) + 5$

$\qquad\qquad\qquad\qquad\qquad$ Adding $1 - 1$ inside the parentheses

$\qquad = -(x^2 - 2x + 1) - (-1) + 5$

$\qquad = -(x-1)^2 + 6$

The vertex is $(1, 6)$, the axis of symmetry is $x = 1$, and the graph opens downward since the coefficient -1 is negative.

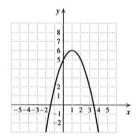

$$f(x) = -x^2 - 2x + 7$$

14. $f(x) = -(x+1)^2 + 8$

Vertex: $(-1, 8)$, axis of symmetry: $x = -1$

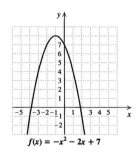

$$f(x) = -x^2 - 2x + 7$$

15. $g(x) = x^2 + 3x - 10$

$\qquad = \left(x^2 + 3x + \dfrac{9}{4} - \dfrac{9}{4}\right) - 10$

$\qquad = \left(x^2 + 3x + \dfrac{9}{4}\right) - \dfrac{9}{4} - 10$

$\qquad = \left(x + \dfrac{3}{2}\right)^2 - \dfrac{49}{4}$

The vertex is $\left(-\dfrac{3}{2}, -\dfrac{49}{4}\right)$, the axis of symmetry is $x = -\dfrac{3}{2}$, and the graph opens upward since the coefficient 1 is positive.

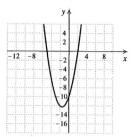

16. $g(x) = \left(x + \dfrac{5}{2}\right)^2 - \dfrac{9}{4}$

Vertex: $\left(-\dfrac{5}{2}, -\dfrac{9}{4}\right)$, axis of symmetry: $x = -\dfrac{5}{2}$

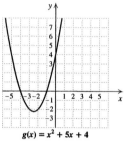

$$g(x) = x^2 + 5x + 4$$

17. $h(x) = x^2 + 7x$

$\qquad = \left(x^2 + 7x + \dfrac{49}{4}\right) - \dfrac{49}{4}$

$\qquad = \left(x + \dfrac{7}{2}\right)^2 - \dfrac{49}{4}$

The vertex is $\left(-\dfrac{7}{2}, -\dfrac{49}{4}\right)$, the axis of symmetry is $x = -\dfrac{7}{2}$, and the graph opens upward since the coefficient 1 is positive.

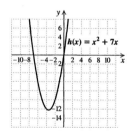

18. $h(x) = \left(x - \dfrac{5}{2}\right)^2 - \dfrac{25}{4}$

Vertex: $\left(\dfrac{5}{2}, -\dfrac{25}{4}\right)$, axis of symmetry: $x = \dfrac{5}{2}$

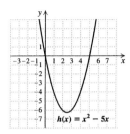

19. $f(x) = -2x^2 - 4x - 6$

$\quad = -2(x^2 + 2x) - 6 \quad$ Factoring

$\quad = -2(x^2 + 2x + 1 - 1) - 6$

$\qquad\qquad$ Adding $1 - 1$ inside
$\qquad\qquad$ the parentheses

$\quad = -2(x^2 + 2x + 1) - 2(-1) - 6$

$\quad = -2(x + 1)^2 - 4$

The vertex is $(-1, -4)$, the axis of symmetry is $x = -1$, and the graph opens downward since the coefficient -2 is negative.

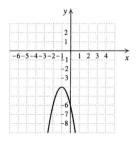

20. $f(x) = -3(x - 1)^2 + 5$

Vertex: $(1, 5)$, axis of symmetry: $x = 1$

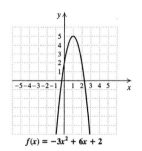

$f(x) = -3x^2 + 6x + 2$

21. $f(x) = -3x^2 + 5x - 2$

$\quad = -3\left(x^2 - \dfrac{5}{3}x\right) - 2 \qquad$ Factoring

$\quad = -3\left(x^2 - \dfrac{5}{3}x + \dfrac{25}{36} - \dfrac{25}{36}\right) - 2$

$\qquad\qquad$ Adding $\dfrac{25}{36} - \dfrac{25}{36}$ inside
$\qquad\qquad$ the parentheses

$\quad = -3\left(x^2 - \dfrac{5}{3}x + \dfrac{25}{36}\right) - 3\left(-\dfrac{25}{36}\right) - 2$

$\quad = -3\left(x - \dfrac{5}{6}\right)^2 + \dfrac{1}{12}$

The vertex is $\left(\dfrac{5}{6}, \dfrac{1}{12}\right)$, the axis of symmetry is $x = \dfrac{5}{6}$, and the graph opens downward since the coefficient -3 is negative.

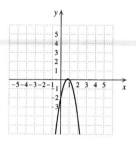

22. $f(x) = -3\left(x + \dfrac{7}{6}\right)^2 + \dfrac{73}{12}$

Vertex: $\left(-\dfrac{7}{6}, \dfrac{73}{12}\right)$, axis of symmetry: $x = -\dfrac{7}{6}$

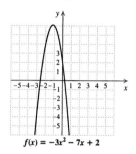

$f(x) = -3x^2 - 7x + 2$

23. $h(x) = \dfrac{1}{2}x^2 + 4x + \dfrac{19}{3}$

$\quad = \dfrac{1}{2}(x^2 + 8x) + \dfrac{19}{3} \qquad$ Factoring

$\quad = \dfrac{1}{2}(x^2 + 8x + 16 - 16) + \dfrac{19}{3}$

$\qquad\qquad$ Adding $16 - 16$ inside
$\qquad\qquad$ the parentheses

$\quad = \dfrac{1}{2}(x^2 + 8x + 16) + \dfrac{1}{2}(-16) + \dfrac{19}{3}$

$\quad = \dfrac{1}{2}(x + 4)^2 - \dfrac{5}{3}$

The vertex is $\left(-4, -\dfrac{5}{3}\right)$, the axis of symmetry is $x = -4$, and the graph opens upward since the coefficient $\dfrac{1}{2}$ is positive.

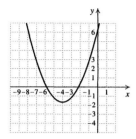

24. $h(x) = \dfrac{1}{2}(x - 3)^2 - \dfrac{5}{2}$

Vertex: $\left(3, -\dfrac{5}{2}\right)$, axis of symmetry: $x = 3$

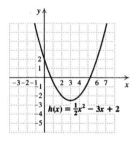

25. $f(x) = x^2 + x - 6$

The coefficient of x^2 is positive so the graph opens upward and the function has a minimum value. Graph the function in a window that shows the vertex. The standard window is one good choice. Then use the Minimum feature from the CALC menu to find that the vertex is $(-0.5, -6.25)$.

26. $(-1, -6)$

27. $f(x) = 5x^2 - x + 1$

The coefficient of x^2 is positive so the graph opens upward and the function has a minimum value. Graph the function in a window that shows the vertex. The standard window is one good choice. Then use the Minimum feature from the CALC menu to find that the vertex is $(0.1, 0.95)$.

28. $(-0.375, 7.5625)$

29. $f(x) = -0.2x^2 + 1.4x - 6.7$

The coefficient of x^2 is negative so the graph opens downward and the function has a maximum value. Graph the function in a window that shows the vertex. The standard window is one good choice. Then use the Maximum feature from the CALC menu to find that the vertex is $(3.5, -4.25)$.

30. $(-2.4, 0.32)$

31. $f(x) = x^2 - 6x + 3$

To find the x-intercepts, solve the equation $0 = x^2 - 6x + 3$. Use the quadratic formula.

$$x = \frac{-(-6) \pm \sqrt{(-6)^2 - 4 \cdot 1 \cdot 3}}{2 \cdot 1}$$

$$x = \frac{6 \pm \sqrt{24}}{2} = \frac{6 \pm 2\sqrt{6}}{2} = 3 \pm \sqrt{6}$$

The x-intercepts are $(3 - \sqrt{6}, 0)$ and $(3 + \sqrt{6}, 0)$.

The y-intercept is $(0, f(0))$, or $(0, 3)$.

32. x-intercepts: $\left(\dfrac{-5 - \sqrt{17}}{2}, 0\right)$, $\left(\dfrac{-5 + \sqrt{17}}{2}, 0\right)$;

y-intercept: $(0, 2)$

33. $g(x) = -x^2 + 2x + 3$

To find the x-intercepts, solve the equation $0 = -x^2 + 2x + 3$. We factor.

$$0 = -x^2 + 2x + 3$$
$$0 = x^2 - 2x - 3 \quad \text{Multiplying by } -1$$
$$0 = (x - 3)(x + 1)$$
$$x = 3 \ or \ x = -1$$

The x-intercepts are $(-1, 0)$ and $(3, 0)$.

The y-intercept is $(0, g(0))$, or $(0, 3)$.

34. x-intercept: $(3, 0)$; y-intercept: $(0, 9)$

35. $f(x) = x^2 - 9x$

To find the x-intercepts, solve the equation $0 = x^2 - 9x$. We factor.

$$0 = x^2 - 9x$$
$$0 = x(x - 9)$$
$$x = 0 \ or \ x = 9$$

The x-intercepts are $(0, 0)$ and $(9, 0)$.

Since $(0, 0)$ is an x-intercept, we observe that $(0, 0)$ is also the y-intercept.

36. x-intercepts: $(0, 0)$, $(7, 0)$; y-intercept: $(0, 0)$

37. $h(x) = -x^2 + 4x - 4$

To find the x-intercepts, solve the equation $0 = -x^2 + 4x - 4$. We factor.

$$0 = -x^2 + 4x - 4$$
$$0 = x^2 - 4x + 4 \quad \text{Multiplying by } -1$$
$$0 = (x - 2)(x - 2)$$
$$x = 2 \ or \ x = 2$$

The x-intercept is $(2, 0)$.

The y-intercept is $(0, h(0))$, or $(0, -4)$.

38. x-intercepts: $\left(\dfrac{3-\sqrt{6}}{2},0\right)$, $\left(\dfrac{3+\sqrt{6}}{2},0\right)$;

y-intercept: $(0,3)$

39. $f(x) = 2x^2 - 4x + 6$

To find the x-intercepts, solve the equation $0 = 2x^2 - 4x + 6$. We use the quadratic formula.

$$x = \frac{-(-4) \pm \sqrt{(-4)^2 - 4 \cdot 2 \cdot 6}}{2 \cdot 2}$$

$$x = \frac{4 \pm \sqrt{-32}}{4} = \frac{4 \pm 4i\sqrt{2}}{2} = 2 \pm 2i\sqrt{2}$$

There are no real-number solutions, so there is no x-intercept.

The y-intercept is $(0, f(0))$, or $(0, 6)$.

40. No x-intercept; y-intercept: $(0, 2)$

41. $f(x) = 2.31x^2 - 3.135x - 5.89$

a) The coefficient of x^2 is positive so the graph opens upward and the function has a minimum value. Graph the function in a window that shows the vertex. The standard window is one good choice. Then use the Minimum feature from the CALC menu to find that the minimum value is about -6.95.

b) To find the first coordinates of the x-intercepts we use the Zero feature from the CALC menu to find the zeros of the function. They are about -1.06 and 2.41, so the x-intercepts are $(-1.06, 0)$ and $(2.41, 0)$.

The y-intercept is $(0, f(0))$, or $(0, -5.89)$.

42. a) About 7.01

b) x-intercepts: $(-0.40, 0)$, $(0.82, 0)$; y-intercept: $(0, 6.18)$

43. $g(x) = -1.25x^2 + 3.42x - 2.79$

a) The coefficient of x^2 is negative so the graph opens downward and the function has a maximum value. Graph the function in a window that shows the vertex. The standard window is one good choice. Then use the Maximum feature from the CALC menu to find that the maximum value is about -0.45.

b) The graph has no x-intercepts. The y-intercept is $(0, f(0))$, or $(0, -2.79)$.

44. a) About 11.28

b) No x-intercepts; y-intercept: $(0, 12.92)$

45. *Writing Exercise*

46. *Writing Exercise*

47. $5x - 3y = 16$,   (1)

    $4x + 2y = 4$ (2)

Multiply equation (1) by 2 and equation (2) by 3 and add.

$$\begin{aligned} 10x - 6y &= 32 \\ \underline{12x + 6y} &= \underline{12} \\ 22x &= 44 \\ x &= 2 \end{aligned}$$

Substitute 2 for x in one of the original equations and solve for y.

$$\begin{aligned} 4x + 2y &= 4 \quad (1) \\ 4 \cdot 2 + 2y &= 4 \\ 8 + 2y &= 4 \\ 2y &= -4 \\ y &= -2 \end{aligned}$$

The solution is $(2, -2)$.

48. $(7, 1)$

49. $4a - 5b + c = 3$,   (1)

   $3a - 4b + 2c = 3$,   (2)

    $a + b - 7c = -2$ (3)

First multiply equation (1) by -2 and add it to equation (2).

$$\begin{aligned} -8a + 10b - 2c &= -6 \\ \underline{3a - 4b + 2c} &= \underline{3} \\ -5a + 6b &= -3 \quad (4) \end{aligned}$$

Next multiply equation (1) by 7 and add it to equation (3).

$$\begin{aligned} 28a - 35b + 7c &= 21 \\ \underline{a + b - 7c} &= \underline{-2} \\ 29a - 34b &= 19 \quad (5) \end{aligned}$$

Now we solve the system of equations (4) and (5). Multiply equation (4) by 29 and equation (5) by 5 and add.

$$\begin{aligned} -145a + 174b &= -87 \\ \underline{145a - 170b} &= \underline{95} \\ 4b &= 8 \\ b &= 2 \end{aligned}$$

Substitute 2 for b in equation (4) and solve for a.

$$\begin{aligned} -5a + 6 \cdot 2 &= -3 \\ -5a + 12 &= -3 \\ -5a &= -15 \\ a &= 3 \end{aligned}$$

Now substitute 3 for a and 2 for b in equation (1) and solve for c.

$$4 \cdot 3 - 5 \cdot 2 + c = 3$$
$$12 - 10 + c = 3$$
$$2 + c = 3$$
$$c = 1$$

The solution is $(3, 2, 1)$.

50. $(1, -3, 2)$

51.

$$\sqrt{4x - 4} = \sqrt{x + 4} + 1$$
$$4x - 4 = x + 4 + 2\sqrt{x + 4} + 1 \quad \text{Squaring both sides}$$
$$3x - 9 = 2\sqrt{x + 4}$$
$$9x^2 - 54x + 81 = 4(x + 4) \quad \text{Squaring both sides again}$$
$$9x^2 - 54x + 81 = 4x + 16$$
$$9x^2 - 58x + 65 = 0$$
$$(9x - 13)(x - 5) = 0$$
$$x = \frac{13}{9} \quad \text{or} \quad x = 5$$

Check: For $x = \frac{13}{9}$:

$$\sqrt{4x - 4} = \sqrt{x + 4} + 1$$

$$\begin{array}{c|c} \sqrt{4\left(\frac{13}{9}\right) - 4} \ ? & \sqrt{\frac{13}{9} + 4} + 1 \\ \sqrt{\frac{16}{9}} & \sqrt{\frac{49}{9}} + 1 \\ \frac{4}{3} & \frac{7}{3} + 1 \\ \frac{4}{3} & \frac{10}{3} \quad \text{FALSE} \end{array}$$

For $x = 5$:

$$\sqrt{4x - 4} = \sqrt{x + 4} + 1$$

$$\begin{array}{c|c} \sqrt{4 \cdot 5 - 4} \ ? & \sqrt{5 + 4} + 1 \\ \sqrt{16} & \sqrt{9} + 1 \\ 4 & 3 + 1 \\ 4 & 4 \quad \text{TRUE} \end{array}$$

5 checks, but $\frac{13}{9}$ does not. The solution is 5.

52. 4

53. *Writing Exercise*

54. *Writing Exercise*

55. $f(x) = x^2 - x - 6$

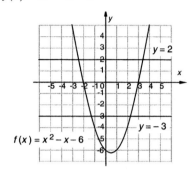

$f(x) = x^2 - x - 6$

a) The solutions of $x^2 - x - 6 = 2$ are the first coordinates of the points of intersection of the graphs of $f(x) = x^2 - x - 6$ and $y = 2$. From the graph we see that the solutions are approximately -2.4 and 3.4.

b) The solutions of $x^2 - x - 6 = -3$ are the first coordinates of the points of intersection of the graphs of $f(x) = x^2 - x - 6$ and $y = -3$. From the graph we see that the solutions are approximately -1.3 and 2.3.

56. a) $-3, 1$

     b) $-3.4, 1.4$

     c) $-3.8, 1.8$

57.
$$f(x) = mx^2 - nx + p$$
$$= m\left(x^2 - \frac{n}{m}x\right) + p$$
$$= m\left(x^2 - \frac{n}{m}x + \frac{n^2}{4m^2} - \frac{n^2}{4m^2}\right) + p$$
$$= m\left(x - \frac{n}{2m}\right)^2 - \frac{n^2}{4m} + p$$
$$= m\left(x - \frac{n}{2m}\right)^2 + \frac{-n^2 + 4mp}{4m}, \quad \text{or}$$
$$m\left(x - \frac{n}{2m}\right)^2 + \frac{4mp - n^2}{4m}$$

58. $f(x) = 3\left[x - \left(-\frac{m}{6}\right)\right]^2 + \frac{11m^2}{12}$

59. The horizontal distance from $(-1, 0)$ to $(3, -5)$ is $|3 - (-1)|$, or 4, so by symmetry the other x-intercept is $(3 + 4, 0)$, or $(7, 0)$. Substituting the three ordered pairs $(-1, 0)$, $(3, -5)$, and $(7, 0)$ in the equation $f(x) = ax^2 + bx + c$ yields a system of equations:

$$0 = a - b + c,$$
$$-5 = 9a + 3b + c,$$
$$0 = 49a + 7b + c$$

The solution of this system of equations is

$\left(\dfrac{5}{16}, -\dfrac{15}{8}, -\dfrac{35}{16}\right)$, so $f(x) = \dfrac{5}{16}x^2 - \dfrac{15}{8}x - \dfrac{35}{16}$.

60. $f(x) = -0.28x^2 - 0.56x + 6.72.$, or $f(x) = -\dfrac{7}{25}(x+1)^2 + 7$

61. $f(x) = |x^2 - 1|$

We plot some points and draw the curve. Note that it will lie entirely on or above the $x-$axis since absolute value is never negative.

x	$f(x)$
-3	8
-2	3
-1	0
0	1
1	0
2	3
3	8

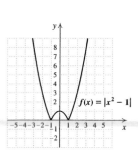

62.

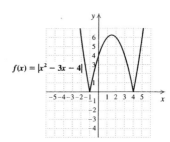

63. $f(x) = |2(x-3)^2 - 5|$

We plot some points and draw the curve. Note that it will lie entirely on or above the $x-$axis since absolute value is never negative.

x	$f(x)$
-1	27
0	13
1	3
2	3
3	5
4	3
5	3
6	13

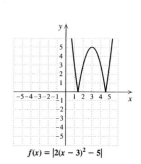

Exercise Set 8.8

1. ***Familiarize and Translate***. We are given the function $V(x) = x^2 - 6x + 13$.

Carry out. To find the value of x for which $V(x)$ is a minimum, we first find $-\dfrac{b}{2a}$:

$$-\dfrac{b}{2a} = -\dfrac{-6}{2 \cdot 1} = 3$$

Now we find the minimum value of the function, $V(3)$:

$$V(3) = 3^2 - 6 \cdot 3 + 13 = 9 - 18 + 13 = 4$$

Check. We can go over the calculations again. We could also solve the problem again by completing the square. The answer checks.

State. The lowest value $V(x)$ will reach is \$4. This occurs 3 months after January 2001.

2. \$120/bicycle; 350 bicycles

3. ***Familiarize and Translate***. We are given the function $N(x) = -0.4x^2 + 9x + 11$.

Carry out. To find the value of x for which $N(x)$ is a maximum, we first find $-\dfrac{b}{2a}$:

$$-\dfrac{b}{2a} = -\dfrac{9}{2(-0.4)} = 11.25$$

Now we find the maximum value of the function $N(11.25)$:

$$N(11.25) = -0.4(11.25)^2 + 9(11.25) + 11 = 61.625$$

Check. We can go over the calculations again. We could also solve the problem again by completing the square. The answer checks.

State. Daily ticket sales will peak 11 days after the concert was announced. About 62 tickets will be sold that day.

4. $P(x) = -x^2 + 980x - 3000$; \$237,100 at $x = 490$

5. ***Familiarize***. We make a drawing and label it.

Perimeter: $2l + 2w = 720$ ft

Area: $A = l \cdot w$

Translate. We have a system of equations.

$$2l + 2w = 720,$$
$$A = lw$$

Carry out. Solving the first equation for l, we get $l = 360 - w$. Substituting for l in the second equation we get a quadratic function A:

$$A = (360 - w)w$$
$$A = -w^2 + 360w$$

Completing the square, we get

$$A = -(w - 180)^2 + 32,400.$$

The maximum function value is 32,400. It occurs when w is 180. When $w = 180$, $l = 360 - 180$, or 180.

Check. We check a function value for w less than 180 and for w greater than 180.

$$A(179) = -179^2 + 360(179) = 32,399$$
$$A(181) = -181^2 + 360(181) = 32,399$$

Since 32,400 is greater than these numbers, it looks as though we have a maximum.

State. The maximum area occurs when the dimensions are 180 ft by 180 ft.

6. 21 in. by 21 in.

7. ***Familiarize***. We make a drawing and label it.

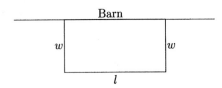

Translate. We have two equations.

$$l + 2w = 40,$$
$$A = lw$$

Carry out. Solve the first equation for l.

$$l = 40 - 2w$$

Substitute for l in the second equation.

$$A = (40 - 2w)w$$
$$A = -2w^2 + 40w$$

Completing the square, we get

$$A = -2(w - 10)^2 + 200.$$

The maximum function value of 200 occurs when $w = 10$. When $w = 10$, $l = 40 - 2 \cdot 10 = 20$.

Check. Check a function value for w less than 10 and for w greater than 10.

$$A(9) = -2 \cdot 9^2 + 40 \cdot 9 = 198$$
$$A(11) = -2 \cdot 11^2 + 40 \cdot 11 = 198$$

Since 200 is greater than these numbers, it looks as though we have a maximum.

State. The maximum area of 200 ft^2 will occur when the dimensions are 10 ft by 20 ft.

8. 450 ft^2; 15 ft by 30 ft (The house serves as the 30-ft side.)

9. ***Familiarize***. Let x represent the height of the file and y represent the width. We make a drawing.

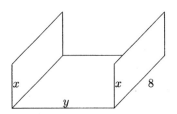

Translate. We have two equations.

$$2x + y = 14$$
$$V = 8xy$$

Carry out. Solve the first equation for y.

$$y = 14 - 2x$$

Substitute for y in the second equation.

$$V = 8x(14 - 2x)$$
$$V = -16x^2 + 112x$$

Completing the square, we get

$$V = -16\left(x - \frac{7}{2}\right)^2 + 196.$$

The maximum function value of 196 occurs when $x = \frac{7}{2}$. When $x = \frac{7}{2}$, $y = 14 - 2 \cdot \frac{7}{2} = 7$.

Check. Check a function value for x less than $\frac{7}{2}$ and for x greater than $\frac{7}{2}$.

$$V(3) = -16 \cdot 3^2 + 112 \cdot 3 = 192$$
$$V(4) = -16 \cdot 4^2 + 112 \cdot 4 = 192$$

Since 196 is greater than these numbers, it looks as though we have a maximum.

State. The file should be $\frac{7}{2}$ in., or 3.5 in., tall.

10. 4 ft by 4 ft

11. ***Familiarize***. We let x and y represent the numbers, and we let P represent their product.

Translate. We have two equations.

$$x + y = 18,$$
$$P = xy$$

Carry out. Solving the first equation for y, we get $y = 18 - x$. Substituting for y in the second equation we get a quadratic function P:

$$P = x(18 - x)$$
$$P = -x^2 + 18x$$

Completing the square, we get

$$P = -(x - 9)^2 + 81.$$

The maximum function value is 81. It occurs when $x = 9$. When $x = 9$, $y = 18 - 9$, or 9.

Check. We can check a function value for x less than 9 and for x greater than 9.

$$P(10) = -10^2 + 18 \cdot 10 = 80$$
$$P(8) = -8^2 + 18 \cdot 8 = 80$$

Since 81 is greater than these numbers, it looks as though we have a maximum.

State. The maximum product of 81 occurs for the numbers 9 and 9.

12. 169; 13 and 13

13. Familiarize. We let x and y represent the two numbers, and we let P represent their product.

Translate. We have two equations.

$$x - y = 8,$$
$$P = xy$$

Carry out. Solve the first equation for x.

$$x = 8 + y$$

Substitute for x in the second equation.

$$P = (8 + y)y$$
$$P = y^2 + 8y$$

Completing the square, we get

$$P = (y + 4)^2 - 16.$$

The minimum function value is -16. It occurs when $y = -4$. When $y = -4$, $x = 8 + (-4)$, or 4.

Check. Check a function value for y less than -4 and for y greater than -4.

$$P(-5) = (-5)^2 + 8(-5) = -15$$
$$P(-3) = (-3)^2 + 8(-3) = -15$$

Since -16 is less than these numbers, it looks as though we have a minimum.

State. The minimum product of -16 occurs for the numbers 4 and -4.

14. $-\dfrac{49}{4}$; $\dfrac{7}{2}$ and $-\dfrac{7}{2}$

15. From the results of Exercises 11 and 12, we might observe that the numbers are -5 and -5 and that the maximum product is 25. We could also solve this problem as follows.

Familiarize. We let x and y represent the two numbers, and we let P represent their product.

Translate. We have two equations.

$$x + y = -10,$$
$$P = xy$$

Carry out. Solve the first equation for y.

$$y = -10 - x$$

Substitute for y in the second equation.

$$P = x(-10 - x)$$
$$P = -x^2 - 10x$$

Completing the square, we get

$$P = -(x + 5)^2 + 25$$

The maximum function value is 25. It occurs when $x = -5$. When $x = -5$, $y = -10 - (-5)$, or -5.

Check. Check a function value for x less than -5 and for x greater than -5.

$$P(-6) = -(-6)^2 - 10(-6) = 24$$
$$P(-4) = -(-4)^2 - 10(-4) = 24$$

Since 25 is greater than these numbers, it looks as though we have a maximum.

State. The maximum product of 25 occurs for the numbers -5 and -5.

16. 36; -6 and -6

17. The data points fall and then rise. The graph appears to represent a quadratic function because the data points approximate a parabola that opens upward.

18. Quadratic; the data approximate a parabola that opens downward.

19. The data points rise, in general. The graph does not appear to represent a quadratic function in which the data points would rise and then fall or vise versa. That is, the data points do not approximate a parabola.

20. Not quadratic; the data do not approximate a parabola.

21. The data appear nearly linear so the graph does not approximate a parabola. A linear function is a better model for this situation than a quadratic function.

22. Quadratic; the data approximate a parabola that opens upward.

23. The data points do not approximate a parabola, so the graph does not appear to represent a quadratic function.

24. Quadratic; the data approximate half a parabola opening upward.

25. The data points resemble the right half of a parabola that opens upward, so a quadratic function $f(x) = ax^2 + bx + c$, $a > 0$, $x \geq 0$, could be used to model the situation.

26. Not quadratic; the data do not approximate a parabola.

27. We look for a function of the form $f(x) = ax^2 + bx + c$. Substituting the data points, we get

$$4 = a(1)^2 + b(1) + c,$$
$$-2 = a(-1)^2 + b(-1) + c,$$
$$13 = a(2)^2 + b(2) + c,$$

or

$$4 = a + b + c,$$
$$-2 = a - b + c,$$
$$13 = 4a + 2b + c.$$

Solving this system, we get

$$a = 2,\ b = 3,\ \text{and}\ c = -1.$$

Therefore the function we are looking for is

$$f(x) = 2x^2 + 3x - 1.$$

8. $f(x) = 3x^2 - x + 2$

9. We look for a function of the form $f(x) = ax^2 + bx + c$. Substituting the data points, we get

$$0 = a(2)^2 + b(2) + c,$$
$$3 = a(4)^2 + b(4) + c,$$
$$-5 = a(12)^2 + b(12) + c,$$

or

$$0 = 4a + 2b + c,$$
$$3 = 16a + 4b + c,$$
$$-5 = 144a + 12b + c.$$

Solving this system, we get

$$a = -\frac{1}{4},\ b = 3,\ c = -5.$$

Therefore the function we are looking for is

$$f(x) = -\frac{1}{4}x^2 + 3x - 5.$$

10. $f(x) = -\frac{1}{3}x^2 + 5x - 12$

11. a) **Familiarize.** We look for a function of the form $A(s) = as^2 + bs + c$, where $A(s)$ represents the number of nighttime accidents (for every 200 million km) and s represents the travel speed (in km/h).

Translate. We substitute the given values of s and $A(s)$.

$$400 = a(60)^2 + b(60) + c,$$
$$250 = a(80)^2 + b(80) + c,$$
$$250 = a(100)^2 + b(100) + c,$$

or

$$400 = 3600a + 60b + c,$$
$$250 = 6400a + 80b + c,$$
$$250 = 10{,}000a + 100b + c.$$

Carry out. Solving the system of equations, we get

$$a = \frac{3}{16},\ b = -\frac{135}{4},\ c = 1750.$$

Check. Recheck the calculations.

State. The function

$$A(s) = \frac{3}{16}s^2 - \frac{135}{4}s + 1750 \text{ fits the data.}$$

b) Find $A(50)$.

$$A(50) = \frac{3}{16}(50)^2 - \frac{135}{4}(50) + 1750 = 531.25$$

About 531 accidents occur at 50 km/h.

32. a) $A(s) = 0.05x^2 - 5.5x + 250$

b) 100 accidents occur

33. **Familiarize.** Think of a coordinate system placed on the drawing in the text with the origin at the point where the arrow is released. Then three points on the arrow's parabolic path are $(0, 0)$, $(63, 27)$, and $(126, 0)$. We look for a function of the form $h(d) = ad^2 + bd + c$, where $h(d)$ represents the arrow's height and d represents the distance the arrow has traveled horizontally.

Translate. We substitute the values given above for d and $h(d)$.

$$0 = a \cdot 0^2 + b \cdot 0 + c,$$
$$27 = a \cdot 63^2 + b \cdot 63 + c,$$
$$0 = a \cdot 126^2 + b \cdot 126 + c$$

or

$$0 = c,$$
$$27 = 3969a + 63b + c,$$
$$0 = 15{,}876a + 126b + c$$

Carry out. Solving the system of equations, we get

$$a \approx -0.0068,\ b \approx 0.8571,\ \text{and}\ c = 0.$$

Check. Recheck the calculations.

State. The function $h(d) = -0.0068d^2 + 0.8571d$ expresses the arrow's height as a function of the distance it has traveled horizontally.

34. a) $P(d) = \frac{1}{64}d^2 + \frac{5}{16}d + \frac{5}{2}$

b) \$9.94

35. a) Enter the data and then use the quadratic regression feature. We have $D(x) = -0.0082833093x^2 + 0.8242996891x + 0.2121786608$.

b) $D(70) \approx 17.325$, so we estimate that the river is about 17.325 ft deep 70 ft from the left bank.

36. a) $W(x) = 0.0111428571x^2 - 0.6637142857x + 21.42857143$, where x is the number of years after 1960.

b) 16.1%

37. a) Enter the data and then use the quadratic regression feature. We have $c(x) = 261.875x^2 - 882.5642857x + 2134.571429$, where x is the number of years after 1992.

b) In 2004, $x = 2004 - 1992 = 12$.

$c(12) \approx 29,254$, so we estimate that about 29,254 cars will be fueled by electricity in 2004.

38. a) $h(x) = 1.577142857x^2 - 11.08571429x + 50.82857143$, where x is the number of years after 1960.

b) \$3439

39. *Writing Exercise*

40. *Writing Exercise*

41.
$$\frac{x}{x^2 + 17x + 72} - \frac{8}{x^2 + 15x + 56}$$
$$= \frac{x}{(x + 8)(x + 9)} - \frac{8}{(x + 8)(x + 7)}$$
$$= \frac{x}{(x + 8)(x + 9)} \cdot \frac{x + 7}{x + 7} - \frac{8}{(x + 8)(x + 7)} \cdot \frac{x + 9}{x + 9}$$
$$= \frac{x(x + 7) - 8(x + 9)}{(x + 8)(x + 9)(x + 7)}$$
$$= \frac{x^2 + 7x - 8x - 72}{(x + 8)(x + 9)(x + 7)}$$
$$= \frac{x^2 - x - 72}{(x + 8)(x + 9)(x + 7)} = \frac{(x - 9)(x + 8)}{(x + 8)(x + 9)(x + 7)}$$
$$= \frac{x - 9}{(x + 9)(x + 7)}$$

42. $\dfrac{(x - 3)(x + 1)}{(x - 7)(x + 3)}$

43. $5x - 9 < 31$
$$5x < 40$$
$$x < 8$$
The solutions set is $\{x | x < 8\}$, or $(-\infty, 8)$.

44. $\{x | x \geq 10\}$, or $[10, \infty)$

45. First find the slope.
$$m = \frac{46,782 - 23,505}{20 - 4} = \frac{23,277}{16} = 1454.8125.$$
Now use the point-slope equation. We will use the point $(4, 23,505)$.
$$y - y_1 = m(x - x_1)$$
$$y - 23,505 = 1454.8125(x - 4)$$
$$y - 23,505 = 1454.8125x - 5819.25$$
$$y = 1454.8125x + 17,685.75$$

46. $r(x) = 1514.092857x + 17,505.2381$, where x is the number of years after 1980.

47. *Writing Exercise*

48. *Writing Exercise*

49. Familiarize. We make a drawing and label it.

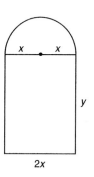

The perimeter of the semicircular portion of the window is $\frac{1}{2} \cdot 2\pi x$, or πx. The perimeter of the rectangular portion is $y + 2x + y$, or $2x + 2y$. The area of the semicircular portion of the window is $\frac{1}{2} \cdot \pi x^2$, or $\frac{\pi}{2}x^2$. The area of the rectangular portion is $2xy$.

Translate. We have two equations, one giving the perimeter of the window and the other giving the area.
$$\pi x + 2x + 2y = 24,$$
$$A = \frac{\pi}{2}x^2 + 2xy$$

Carry out. Solve the first equation for y.
$$\pi x + 2x + 2y = 24$$
$$2y = 24 - \pi x - 2x$$
$$y = 12 - \frac{\pi x}{2} - x$$
Substitute for y in the second equation.
$$A = \frac{\pi}{2}x^2 + 2x\left(12 - \frac{\pi x}{2} - x\right)$$
$$A = \frac{\pi}{2}x^2 + 24x - \pi x^2 - 2x^2$$
$$A = -2x^2 - \frac{\pi}{2}x^2 + 24x$$
$$A = -\left(2x + \frac{\pi}{2}\right)x^2 + 24x$$
Completing the square, we get
$$A = -\left(2 + \frac{\pi}{2}\right)\left(x^2 + \frac{24}{-\left(2 + \frac{\pi}{2}\right)}x\right)$$
$$A = -\left(2 + \frac{\pi}{2}\right)\left(x^2 - \frac{48}{4 + \pi}x\right)$$
$$A = -\left(2 + \frac{\pi}{2}\right)\left(x - \frac{24}{4 + \pi}\right)^2 + \left(\frac{24}{4 + \pi}\right)^2$$
The maximum function value occurs when

$x = \dfrac{24}{4+\pi}$. When $x = \dfrac{24}{4+\pi}$,

$y = 12 - \dfrac{\pi}{2}\left(\dfrac{24}{4+\pi}\right) - \dfrac{24}{4+\pi} =$

$\dfrac{48+12\pi}{4+\pi} - \dfrac{12\pi}{4+\pi} - \dfrac{24}{4+\pi} = \dfrac{24}{4+\pi}.$

Check. Recheck the calculations.

State. The radius of the circular portion of the window and the height of the rectangular portion should each be $\dfrac{24}{4+\pi}$ ft.

50. The length of the piece used to form the circle is $\dfrac{36\pi}{4+\pi}$ in. and the length of the piece used to form the square is $\dfrac{144}{4+\pi}$ in.

51. **Familiarize**. Let x represent the number of trees added to an acre. Then $20 + x$ represents the total number of trees per acre and $40 - x$ represents the corresponding yield per tree. Let T represent the total yield per acre.

Translate. Since total yield is number of trees times yield per tree we have the following function for total yield per acre.

$T(x) = (20+x)(40-x)$

$T(x) = -x^2 + 20x + 800$

Carry out. Completing the square, we get

$T(x) = -(x-10)^2 + 900.$

The maximum function value of 900 occurs when $x = 10$. When $x = 10$, the number of trees per acre is $20 + 10$, or 30.

Check. We check a function value for x less than 10 and for x greater than 10.

$T(9) = (20+9)(40-9) = 899$

$T(11) = (20+11)(40-11) = 899$

Since 900 is greater than these numbers, it looks as though we have a maximum.

State. The grower should plant 30 trees per acre.

52. $15

53. **Familiarize**. We want to find the maximum value of a function of the form $h(t) = at^2 + bt + c$ that fits the following data.

Time (sec)	Height (ft)
0	0
3	0
3 + 2, or 5	−64

Translate. Substitute the given values for t and $h(t)$.

$0 = a(0)^2 + b(0) + c,$

$0 = a(3)^2 + b(3) + c,$

$-64 = a(5)^2 + b(5) + c,$

or

$0 = c,$

$0 = 9a + 3b + c,$

$-64 = 25a + 5b + c.$

Carry out. Solving the system of equations, we get $a = -6.4$, $b = 19.2$, $c = 0$. The function $h(t) = -6.4t^2 + 19.2t$ fits the data.

Completing the square, we get

$h(t) = -6.4(t-1.5)^2 + 14.4.$

The maximum function value of 14.4 occurs at $t = 1.5$.

Check. Recheck the calculations. Also check a function value for t less than 1.5 and for t greater than 1.5.

$h(1) = -6.4(1)^2 + 19.2(1) = 12.8$

$h(2) = -6.4(2)^2 + 19.2(2) = 12.8$

Since 14.4 is greater than these numbers, it looks as though we have a maximum.

State. The maximum height above the cliff is 14.4 ft. The maximum height above sea level is $64 + 14.4$, or 78.4 ft.

54. 158 ft

Exercise Set 8.9

---

1. We see that $p(x) = 0$ when $x = -4$ or $x = \dfrac{3}{2}$, and $p(x) < 0$ between -4 and $\dfrac{3}{2}$. The solution set of the inequality is $\left[-4, \dfrac{3}{2}\right]$.

2. $\left(-4, -\dfrac{2}{3}\right)$

3. $x^4 + 12x > 3x^3 + 4x^2$

$x^4 - 3x^3 - 4x^2 + 12x > 0$

From the graph we see that $p(x) > 0$ on $(-\infty, -2) \cup (0, 2) \cup (3, \infty)$. This is the solution set of the inequality.

4. $(-\infty, -3] \cup \{0\} \cup [2, \infty)$

5. $\dfrac{x-1}{x+2} < 3$

$\dfrac{x-1}{x+2} - 3 < 0$

We see that $r(x) < 0$ on $\left(-\infty, -\dfrac{7}{2}\right) \cup (-2, \infty)$. This is the solution set of the inequality.

6. $(-\infty, -4] \cup (5, \infty)$

7. $(x + 4)(x - 3) < 0$

We solve the related equation.

$$(x + 4)(x - 3) = 0$$
$$x + 4 = 0 \quad or \quad x - 3 = 0$$
$$x = -4 \quad or \qquad x = 3$$

The numbers -4 and 3 divide the number line into 3 intervals.

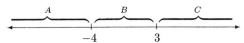

We graph $p(x) = (x + 4)(x - 3)$ in the window $[-10, 10, -15, 5]$ and determine the sign of the function in each interval.

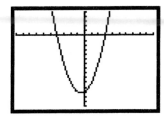

We see that $p(x) < 0$ in interval B, or in $(-4, 3)$. Thus, the solution set of the inequality is $(-4, 3)$, or $\{x | -4 < x < 3\}$.

8. $(-\infty, -2) \cup (5, \infty)$, or $\{x | x < -2 \text{ or } x > 5\}$

9. $(x + 7)(x - 2) \geq 0$

The solutions of $(x + 7)(x - 2) = 0$ are -7 and 2. They divide the number line into three intervals as shown:

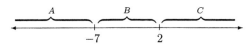

We graph $p(x) = (x + 7)(x - 2)$ in the window $[-10, 10, -25, 5]$, Yscl $= 5$.

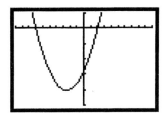

We see that $p(x) \geq 0$ in intervals A and C, or in $(-\infty, -7) \cup (2, \infty)$. We also know that $p(-7) = 0$ and $p(2) = 0$. Thus, the solution set of the inequality is $(-\infty, -7] \cup [2, \infty)$, or $\{x | x \leq -7 \text{ or } x \geq 2\}$.

10. $[-4, 1]$, or $\{x | -4 \leq x \leq 1\}$

11. $x^2 - x - 2 < 0$

$(x + 1)(x - 2) < 0$ Factoring

The solutions of $(x + 1)(x - 2) = 0$ are -1 and 2. They divide the number line into three intervals as shown:

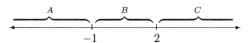

We enter $p(x) = x^2 - x - 2$ on a graphing calculator and try a test number in each interval. We choose -2 from interval A, 0 from B, and 3 from C.

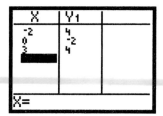

We see that $p(x) < 0$ in interval B, so the solution set is $(-1, 2)$, or $\{x | -1 < x < 2\}$.

12. $(-2, 1)$, or $\{x | -2 < x < 1\}$

13. $\qquad 25 - x^2 \geq 0$

$(5 - x)(5 + x) \geq 0$

The solutions of $(5 - x)(5 + x) = 0$ are 5 and -5. Graph $p(x) = 25 - x^2$ in the window $[-10, 10, -10, 30]$, Yscl $= 5$.

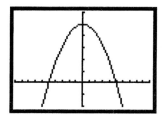

We see that $p(x) > 0$ in $(-5, 5)$; also $p(-5) = 0$ and $p(5) = 0$. The solution set is $[-5, 5]$, or $\{x | -5 \leq x \leq 5\}$.

14. $[-2, 2]$, or $\{x | -2 \leq x \leq 2\}$

15. $x^2 + 4x + 4 < 0$

$(x + 2)^2 < 0$

Observe that $(x + 2)^2 \geq 0$ for all values of z. Thus, the solution set is $\emptyset$.

The graph of $p(x) = x^2 + 4x + 4$ confirms this.

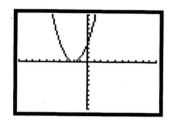

16. ∅

17.
$$x^2 - 4x < 12$$
$$x^2 - 4x - 12 < 0$$
$$(x - 6)(x + 2) < 0$$

The solutions of $(x - 6)(x + 2) = 0$ are 6 and -2. Graph $p(x) = x^2 - 4x - 12$ in the window $[-10, 10, -20, 5]$, Yscl $= 5$.

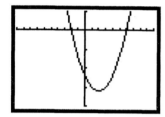

We see that $p(x) < 0$ in the interval $(-2, 6)$ or $\{x | -2 < x < 6\}$. This is the solution set of the inequality.

18. $(-\infty, -4) \cup (-2, \infty)$, or $\{x | x < -4 \text{ or } x > -2\}$

19. $3x(x + 2)(x - 2) < 0$

The solutions of $3x(x + 2)(x - 2) = 0$ are 0, -2, and 2. Graph $p(x) = 3x(x + 2)(x - 2)$ in the window $[-5, 5, -10, 10]$.

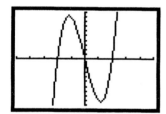

We see that $p(x) < 0$ on $(-\infty, -2) \cup (0, 2)$, or $\{x | x < -2 \text{ or } x > 2\}$. This is the solution set of the inequality.

20. $(-1, 0) \cup (1, \infty)$, or $\{x | -1 < x < 0 \text{ or } x > 1\}$

21. $(x + 3)(x - 2)(x + 1) > 0$

The solutions of $(x + 3)(x - 2)(x + 1) = 0$ are -3, 2, and -1. Graph $p(x) = (x + 3)(x - 2)(x - 1)$ in the window $[-5, 5, -10, 10]$.

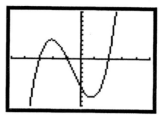

We see that $p(x) > 0$ on $(-3, -1) \cup (2, \infty)$, or $\{x | -3 < x < -1 \text{ or } x > 2\}$. This is the solution set of the inequality.

22. $(-\infty, -2) \cup (1, 4)$, or $\{x | x < -2 \text{ or } 1 < x < 4\}$

23. $(x + 3)(x + 2)(x - 1) < 0$

The solutions of $(x + 3)(x + 2)(x - 1) = 0$ are -3, -2, and 1. Graph $p(x) = (x + 3)(x + 2)(x - 1)$ in the window $[-5, 5, -10, 10]$.

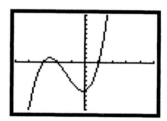

We see that $p(x) < 0$ in $(-\infty, -3) \cup (-2, 1)$, or $\{x | x < -3 \text{ or } -2 < x < 1\}$. This is the solution set of the inequality.

24. $(-\infty, -1) \cup (2, 3)$, or $\{x | x < -1 \text{ or } 2 < x < 3\}$

25. $4.32x^2 - 3.54x - 5.34 \le 0$

Graph $p(x) = 4.32x^2 - 3.54x - 5.34$ in the window $[-5, 5, -10, 10]$.

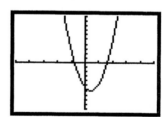

Using the Zero feature we find that $p(x) = 0$ when $x \approx -0.78$ and when $x \approx 1.59$. Also observe that $p(x) < 0$ on the interval $(-0.78, 1.59)$. Thus, the solution set of the inequality is $[-0.78, 1.59]$, or $\{x | -0.78 \le x \le 1.59\}$.

26. $(-\infty, -0.21] \cup [2.47, \infty)$, or $\{x | x \le -0.21 \text{ or } x \ge 2.47\}$

27. $x^3 - 2x^2 - 5x + 6 < 0$

Graph $p(x) = x^3 - 2x^2 - 5x + 6$ in the window $[-5, 5, -10, 10]$.

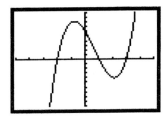

Using the Zero feature we find that $p(x) = 0$ when $x = -2$, when $x = 1$, and when $x = 3$. Then we see that $p(x) < 0$ on $(-\infty, -2) \cup (1, 3)$, or $\{x | x < -2 \ or \ 1 < x < 3\}$. This is the solution set of the inequality.

28. $(-2, 1) \cup (1, \infty)$, or $\{x | -2 < x < 1 \ or \ x > 1\}$

29. $\dfrac{1}{x+3} < 0$

We write the related equation by changing the $<$ symbol to $=$:

$$\frac{1}{x+3} = 0$$

We solve the related equation.

$$(x+3) \cdot \frac{1}{x+3} = (x+3) \cdot 0$$
$$1 = 0$$

The related equation has no solution.

Next we find the values that make the denominator 0 by setting the denominate equal to 0 and solving:

$$x + 3 = 0$$
$$x = -3$$

We use -3 to divide the number line into two intervals as shown:

$$\overset{\displaystyle A \qquad\qquad\qquad B}{\underset{\displaystyle -3}{\longleftrightarrow\!}}$$

We try a test number in each interval. Enter $y = \dfrac{1}{x+3}$ on a graphing calculator and use the Table feature set in ASK mode. We try -4 in interval A and 0 in B.

X	Y1
-8	-1
0	.14286
X=	

We see that $y < 0$ in interval A, so the solution set of the inequality is $(-\infty, -3)$ or $\{x | x < -3\}$.

30. $(-4, \infty)$, or $\{x | x > -4\}$

31. $\dfrac{x+1}{x-5} \geq 0$

Graph $r(x) = \dfrac{x+1}{x-5}$ using DOT mode in the window $[-5, 15, -5, 5]$.

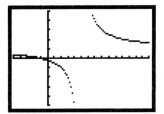

Using the Zero feature we find that $r(x) = 0$ when $x = -1$. Also observe that $r(x) > 0$ on the intervals $(-\infty, -1)$ and $(5, \infty)$. Thus, the solution set of the inequality is $(-\infty, -1] \cup (5, \infty)$, or $\{x | x \leq 1 \ or \ x > 5\}$.

32. $(-5, 2]$, or $\{x | -5 < x \leq 2\}$

33. $\dfrac{3x+2}{2x-4} \leq 0$

Graph $r(x) = \dfrac{3x+2}{2x-4}$ using DOT mode in the standard window.

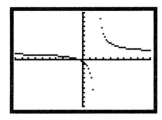

Using the Zero feature we find that $r(x) = 0$ when $x = -0.\overline{6}$, or $-\dfrac{2}{3}$. Also observe that $r(x) < 0$ on $\left(-\dfrac{2}{3}, 2\right)$. Thus, the solution set is $\left[-\dfrac{2}{3}, 2\right)$.

34. $\left(-\infty, -\dfrac{3}{4}\right) \cup \left[\dfrac{5}{2}, \infty\right)$, or $\left\{x \middle| x < -\dfrac{3}{4} \ or \ x \geq \dfrac{5}{2}\right\}$

35. $\dfrac{x+1}{x+6} > 1$

$$\frac{x+1}{x+6} - 1 > 0$$

If $r(x) = \dfrac{x+1}{x+6} - 1$, the solution set of the inequality is all values of x for which $r(x) > 0$.

First we solve $r(x) = 0$.

$$\frac{x+1}{x+6} - 1 = 0$$

$$(x+6)\left(\frac{x+1}{x+6} - 1\right) = (x+6) \cdot 0$$

$$(x+6)\left(\frac{x+1}{x+6}\right) - (x+6) \cdot 1 = 0$$

$$x + 1 - x - 6 = 0$$

$$-5 = 0$$

This equation has no solution.

Find the values that make the denominator 0.

$$x + 6 = 0$$

$$x = -6$$

Use −6 to divide the number line into intervals.

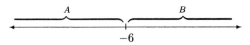

Enter $y = r(x)$ on a graphing calculator and evaluate a test number in each interval. We test −7 and 0.

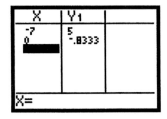

We see that $r(x) > 0$ in interval A. The solution set is $(-\infty, -6)$, or $\{x|x < -6\}$.

36. $(-\infty, 2)$, or $\{x|x < 2\}$

37. $\dfrac{(x-2)(x+1)}{x-5} \le 0$

Solve the related equation.

$$\frac{(x-2)(x+1)}{x-5} = 0$$

$$(x-2)(x+1) = 0$$

$$x = 2 \ or \ x = -1$$

Find the values that make the denominator 0.

$$x - 5 = 0$$

$$x = 5$$

Use the numbers 2, −1, and 5 to divide the number line into intervals as shown:

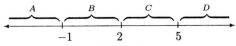

Enter $r(x) = \dfrac{(x-2)(x+1)}{x-5}$ and evaluate a test number in each interval. We test −2, 0, 3, and 6.

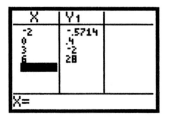

We see that $r(x) < 0$ in intervals A and C. From above we also know that $r(x) = 0$ when $x = 2$ or $x = -1$. Thus, the solution set is $(-\infty, -1] \cup [2, 5)$, or $\{x|x \le -1 \ or \ 2 \le x < 5\}$.

38. $[-4, -3) \cup [1, \infty)$, or $\{x| -4 \le x < -3 \ or \ x \ge 1\}$

39. $\dfrac{x}{x+3} \ge 0$

Graph $r(x) = \dfrac{x}{x+3}$ using DOT mode in the window $[-10, 10, -5, 5]$.

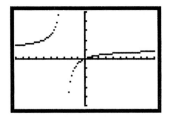

Using the Zero feature we find that $r(x) = 0$ when $x = 0$. Also observe that $r(x) > 0$ in the interval $(-\infty, -3)$ and in $(0, \infty)$. Then the solution set is $(-\infty, -3) \cup [0, \infty)$, or $\{x|x < -3 \ or \ x \ge 0\}$.

40. $(0, 2]$, or $\{x|0 < x \le 2\}$

41. $\dfrac{x-5}{x} < 1$

$$\frac{x-5}{x} - 1 < 0$$

Let $r(x) = \dfrac{x-5}{x} - 1$ and solve $r(x) = 0$.

$$\frac{x-5}{x} - 1 = 0$$

$$x\left(\frac{x-5}{x} - 1\right) = x \cdot 0$$

$$x\left(\frac{x-5}{x}\right) - x \cdot 1 = 0$$

$$x - 5 - x = 0$$

$$-5 = 0$$

This equation has no solution.

Find the values that make the denominator 0.

$$x = 0$$

Use the number 0 to divide the number line into two intervals as shown.

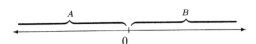

Enter $y = r(x)$ in a graphing calculator and evaluate a test number in each interval. We test -1 and 1.

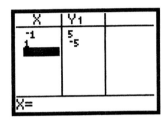

We see that $r(x) < 0$ in interval B. Thus, the solution set is $(0, \infty)$, or $\{x | x > 0\}$.

42. $(1, 2)$, or $\{x | 1 < x < 2\}$

43. $\dfrac{x-1}{(x-3)(x+4)} \leq 0$

Solve the related equation.

$$\frac{x-1}{(x-3)(x+4)} = 0$$

$$x - 1 = 0$$

$$x = 1$$

Find the values that make the denominator 0.

$$(x-3)(x+4) = 0$$

$$x = 3 \ or \ x = -4$$

Use the numbers 1, 3, and -4 to divide the number line into intervals as shown:

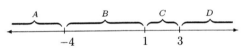

Enter $r(x) = \dfrac{x-1}{(x-3)(x+4)}$ in a graphing calculator and evaluate a test point in each interval. We test -5, 0, 2, and 4.

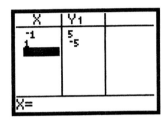

We see that $r(x) < 0$ in intervals A and C. From above we also know that $r(x) = 0$ when $x = 1$. Thus, the solution set is $(-\infty, -4) \cup [1, 3)$, or $\{x | x < -4 \ or \ 1 \leq x < 3\}$.

44. $(-7, -2] \cup (2, \infty)$, or $\{x | -7 < x \leq -2 \ or \ x > 2\}$

45. $4 < \dfrac{1}{x}$

$$4 - \frac{1}{x} < 0$$

Graph $r(x) = 4 - \dfrac{1}{x}$ in the window $[-10, 10, -5, 10]$.

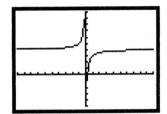

Using the Zero feature we find that $r(x) = 0$ when $x = 0.25$. Observe that $r(x) < 0$ on $(0, 0.25)$, or $\left(0, \dfrac{1}{4}\right)$, or $\left\{x \Big| 0 < x < \dfrac{1}{4}\right\}$. This is the solution set of the inequality.

46. $(-\infty, 0) \cup \left[\dfrac{1}{5}, \infty\right)$, or $\left\{x \Big| x < 0 \ or \ x \geq \dfrac{1}{5}\right\}$

47. *Writing Exercise*

48. *Writing Exercise*

49. $(2a^3 b^2 c^4)^3 = 2^3 (a^3)^3 (b^2)^3 (c^4)^3 = 8a^{3 \cdot 3} b^{2 \cdot 3} c^{4 \cdot 3} = 8a^9 b^6 c^{12}$

50. $25a^8 b^{14}$

51. $2^{-5} = \dfrac{1}{2^5} = \dfrac{1}{32}$

52. $\dfrac{1}{81}$

53. $f(x) = 3x^2$

$f(a+1) = 3(a+1)^2 = 3(a^2 + 2a + 1) = 3a^2 + 6a + 3$

54. $5a + 7$

55. *Writing Exercise*

56. *Writing Exercise*

57. $x^2 + 2x < 5$

$$x^2 + 2x - 5 < 0$$

Using the quadratic formula, we find that the solutions of the related equation are $x = -1 \pm \sqrt{6}$. Graph $p(x) = x^2 + 2x - 5$ in the standard window. Observe that $p(x) < 0$ on $(-1 - \sqrt{6}, -1 + \sqrt{6})$, or $\{x | -1 - \sqrt{6} < x < -1 + \sqrt{6}\}$. This is the solution set of the inequality. This can also be expressed as $(-3.24, 1.24)$, or $\{x | -3.24 < x < 1.24\}$.

58. $(-\infty, \infty)$, or the set of all real numbers

59. $x^4 + 3x^2 \leq 0$

$x^2(x^2 + 3) \leq 0$

$x^2 = 0$ for $x = 0$, $x^2 > 0$ for $x \neq 0$, $x^2 + 3 > 0$ for all x

The solution set is $\{0\}$.

60. $(-\infty, 0.25] \cup [2.5, \infty)$, or $\{x | x \leq 0.25 \text{ or } x \geq 2.5\}$

61. a) $-3x^2 + 630x - 6000 > 0$

$x^2 - 210x + 2000 < 0$ Multiplying by $-\dfrac{1}{3}$

$(x - 200)(x - 10) < 0$

The solutions of $f(x) = (x - 200)(x - 10) = 0$ are 200 and 10. They divide the number line as shown:

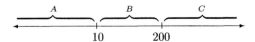

Enter $p(x) = x^2 - 210x + 2000$ in a graphing calculator and evaluate a test point in each interval. We test 9, 11, and 201.

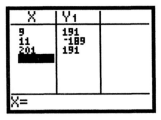

We see that $p(x) < 0$ in interval B, so the company makes a profit for values of x such that $10 < x < 200$, or for values of x in the interval $(10, 200)$, or in the set $\{x | 10 < x < 200\}$.

b) See part (a). Keep in mind that x must be nonnegative since negative numbers have no meaning in this application.

The company loses money for values of x such that $0 \leq x < 10$ or $x > 200$, or for values of x in the interval $[0, 10) \cup (200, \infty)$, or in the set $\{x | 0 \leq x < 10 \text{ or } x > 200\}$.

62. a) $\{t | 0 \text{ sec} < t < 2 \text{ sec}\}$

b) $\{t | t > 10 \text{ sec}\}$

63. We find values of n such that $N \geq 66$ *and* $N \leq 300$.

For $N \geq 66$:

$\dfrac{n(n - 1)}{2} \geq 66$

$n(n - 1) \geq 132$

$n^2 - n - 132 \geq 0$

$(n - 12)(n + 11) \geq 0$

The solutions of $f(n) = (n - 12)(n + 11) = 0$ are 12 and -11. They divide the number line as shown:

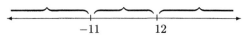

However, only positive values of n have meaning in this exercise so we need only consider the intervals shown below:

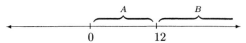

Enter $p(x) = x^2 - x - 132$ in a graphing calculator and evaluate a test point in each interval. We test 1 and 13.

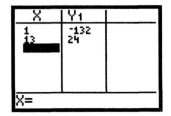

We see that $p(x) > 0$ in interval B. From above we also know that $p(12) = 0$, so the solution set for this inequality is $[12, \infty)$.

For $N \leq 300$:

$\dfrac{n(n - 1)}{2} \leq 300$

$n(n - 1) \leq 600$

$n^2 - n - 600 \leq 0$

$(n - 25)(n + 24) \leq 0$

The solutions of $f(n) = (n - 25)(n + 24) = 0$ are 25 and -24. They divide the number line as shown:

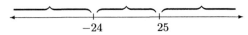

However, only positive values of n have meaning in this exercise so we need only consider the intervals shown below:

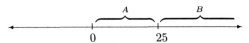

Enter $p(x) = x^2 - x - 600$ in a graphing calculator and evaluate a test point in each interval. We test 1 and 26.

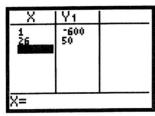

We see that $p(x) < 0$ in interval A. From above we also know that $p(25) = 0$, so the solution set for this inequality is $(0, 25]$. Then $66 \leq N \leq 300$ for $[12, \infty) \cap (0, 25]$, or on $[12, 25]$. We can express the solution set as $\{n | n \text{ is an integer } and \ 12 \leq n \leq 25\}$.

64. $\{n | n \text{ is an integer } and \ 9 \leq n \leq 23\}$

65. From the graph we determine the following:

$f(x)$ has no zeros.

The solutions $f(x) < 0$ are $(-\infty, 0)$, or $\{x | x < 0\}$.

The solutions of $f(x) > 0$ are $(0, \infty)$, or $\{x | x > 0\}$.

66. $f(x) = 0$ for $x = 0 \ or \ x = 1$;

$f(x) < 0$ for $(0, 1)$, or $\{x | 0 < x < 1\}$;

$f(x) > 0$ for $(1, \infty)$, or $\{x | x > 1\}$

67. From the graph we determine the following:

The solutions of $f(x) = 0$ are -2, 1, and 3.

The solution of $f(x) < 0$ is $(-\infty, -2) \cup (1, 3)$, or $\{x | x < -2 \ or \ 1 < x < 3\}$.

The solution of $f(x) > 0$ is $(-2, 1) \cup (3, \infty)$, or $\{x | -2 < x < 1 \ or \ x > 3\}$.

68. $f(x) = 0$ for -2, 1, 2, and 3;

$f(x) < 0$ for $(-2, 1) \cup (2, 3)$, or $\{x | -2 < x < 1 \ or \ 2 < x < 3\}$;

$f(x) > 0$ for $(-\infty, -2) \cup (1, 2) \cup (3, \infty)$, or $\{x | x < -2 \ or \ 1 < x < 2 \ or \ x > 3\}$.

69. a) Enter the data and then use the quadratic regression feature. We have $w(x) = 1388.888889x^2 - 14,900x + 73,800$, where x is the number of years after 1994.

b) Graph $y_1 = w(x)$ and $y_2 = 50,000$. We choose the window $[0, 15, 0, 100,000]$, $\text{Yscl} = 10,000$.

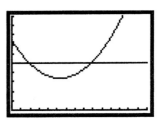

We use the Intersect feature to find that the first coordinates of the points of intersection of the graphs are approximately 2 and 9. Observe that the graph of y_1 lies above the graph of y_2 to the left of $x = 2$ and to the right of $x = 9$. Then the number of welfare cases is greater than 50,000 from 1994 to about 2 years after 1994, or from 1994 to 1996, and also after about 9 years after 1994, or after 2003.

Chapter 9

Exponential and Logarithmic Functions

1. $(f \circ g)(1) = f(g(1)) = f(2 \cdot 1 + 1)$

$\qquad = f(3) = 3^2 + 3$

$\qquad = 9 + 3 = 12$

$(g \circ f)(1) = g(f(1)) = g(1^2 + 3)$

$\qquad = g(4) = 2 \cdot 4 + 1 = 9$

$(f \circ g)(x) = f(g(x)) = f(2x + 1)$

$\qquad = (2x + 1)^2 + 3$

$\qquad = 4x^2 + 4x + 1 + 3$

$\qquad = 4x^2 + 4x + 4$

$(g \circ f)(x) = g(f(x)) = g(x^2 + 3)$

$\qquad = 2(x^2 + 3) + 1$

$\qquad = 2x^2 + 6 + 1$

$\qquad = 2x^2 + 7$

2. $-7; 4; 2x^2 - 9; 4x^2 + 4x - 4$

3. $(f \circ g)(x) = f(g(1)) = f(5 \cdot 1^2 + 2)$

$\qquad = f(7) = 3 \cdot 7 - 1$

$\qquad = 21 - 1 = 20$

$(g \circ f)(1) = g(f(1)) = g(3 \cdot 1 - 1)$

$\qquad = g(2) = 5 \cdot 2^2 + 2$

$\qquad = 5 \cdot 4 + 2 = 20 + 2 = 22$

$(f \circ g)(x) = f(g(x)) = f(5x^2 + 2)$

$\qquad = 3(5x^2 + 2) - 1$

$\qquad = 15x^2 + 6 - 1$

$\qquad = 15x^2 + 5$

$(g \circ f)(x) = g(f(x)) = g(3x - 1)$

$\qquad = 5(3x - 1)^2 + 2$

$\qquad = 5(9x^2 - 6x + 1) + 2$

$\qquad = 45x^2 - 30x + 5 + 2$

$\qquad = 45x^2 - 30x + 7$

4. $31; 27; 48x^2 - 24x + 7; 12x^2 + 15$

5. $(f \circ g)(1) = f(g(1)) = f\left(\dfrac{1}{1^2}\right)$

$\qquad = f(1) = 1 + 7 = 8$

$(g \circ f)(1) = g(f(1)) = g(1 + 7)$

$\qquad = g(8) = \dfrac{1}{8^2} = \dfrac{1}{64}$

$(f \circ g)(x) = f(g(x))$

$\qquad = f\left(\dfrac{1}{x^2}\right) = \dfrac{1}{x^2} + 7$

$(g \circ f)(x) = g(f(x))$

$\qquad = g(x + 7) = \dfrac{1}{(x + 7)^2}$

6. $\dfrac{1}{9}; 3; \dfrac{1}{(x + 2)^2}; \dfrac{1}{x^2} + 2$

7. Since $(y_1 \circ y_2)(-3) = y_1(y_2(-3))$, we first find $y_2(-3)$. Locate -3 in the x-column and then move across to the y_2-column to find that $y_2(-3) = 1$. Now we have $y_1(y_2(-3)) = y_1(1)$. Locate 1 in the x-column and then move across to the y_1-column to find that $y_1(1) = 8$. Thus, $(y_1 \circ y_2)(-3) = 8$.

8. Not defined

9. Since $(y_1 \circ y_2)(-1) = y_1(y_2(-1))$, we first find $y_2(-1)$. Locate -1 in the x-column and then move across to the y_2-column to find that $y_2(-1) = -3$. Now we have $y_1(y_2(-1)) = y_1(-3)$. Locate -3 in the x-column and then move across to the y_1-column to find that $y_1(-3) = -4$. Thus, $(y_1 \circ y_2)(-1) = -4$.

10. 6

11. Since $(y_2 \circ y_1)(1) = y_2(y_1(1))$, we first find $y_1(1)$. Locate 1 in the x-column and then move across to the y_1-column to find that $y_1(1) = 8$. Now we have $y_2(y_1(1)) = y_2(8)$. However, y_2 is not defined for $x = 8$, so $(y_2 \circ y_1)(1)$ is not defined.

12. 8

13. Since $(f \circ g)(2) = f(g(2))$, we first find $g(2)$. Locate 2 in the x-column and then move across to the $g(x)$-column to find that $g(2) = 5$. Now we have $f(g(2)) = f(5)$. Locate 5 in the x-column and then move across to the $f(x)$-column to find that $f(5) = 4$. Thus, $(f \circ g)(2) = 4$.

14. Not defined

15. To find $f(g(3))$ we first find $g(3)$. Locate 3 in the x-column and then move across to the $g(x)$-column to find that $g(3) = 8$. Now we have $f(g(3)) = f(8)$. However, $f(x)$ is not defined for $x = 8$, so $f(g(3))$ is not defined.

16. 5

17. $h(x) = (7 + 5x)^2$

This is $7 + 5x$ raised to the second power, so the two most obvious functions are $f(x) = x^2$ and $g(x) = 7 + 5x$.

18. $f(x) = x^2$, $g(x) = 3x - 1$

19. $h(x) = \sqrt{2x + 7}$

We have $2x + 7$ and take the square root of their expression, so the two most obvious functions are $f(x) = \sqrt{x}$ and $g(x) = 2x + 7$.

20. $f(x) = \sqrt{x}$, $g(x) = 5x + 2$

21. $h(x) = \dfrac{2}{x - 3}$

This is 2 divided by $x - 3$, so two functions that can be used are $f(x) = \dfrac{2}{x}$ and $g(x) = x - 3$.

22. $f(x) = x + 4$, $g(x) = \dfrac{3}{x}$

23. $h(x) = \dfrac{1}{\sqrt{7x + 2}}$

This is the reciprocal of the square root of $7x + 2$. Two functions that can be used are $f(x) = \dfrac{1}{\sqrt{x}}$ and $g(x) = 7x + 2$.

24. $f(x) = \sqrt{x} - 3$, $g(x) = x - 7$

25. $h(x) = \dfrac{1}{\sqrt{3x}} + \sqrt{3x}$

This is the reciprocal of the square root of $3x$ plus the square root of $3x$. Two functions that can be used are $f(x) = \dfrac{1}{x} + x$ and $g(x) = \sqrt{3x}$.

26. $f(x) = \dfrac{1}{x} - x$, $g(x) = \sqrt{2x}$

27. The graph of $f(x) = x - 5$ is shown below.

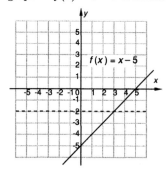

Since there is no horizontal line that crosses the graph more than once, the function is one-to-one.

28. Yes

29. $f(x) = x^2 + 1$

Observe that the graph of this function is a parabola that opens up. Thus, there are many horizontal lines that cross the graph more than once, so the function is not one-to-one. We can also draw the graph as shown below.

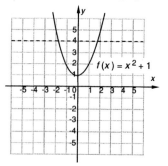

There are many horizontal lines that cross the graph more than once. In particular, the line $y = 4$ crosses the graph more than once. The function is not one-to-one.

30. No

31. The graph of $g(x) = x^3$ is shown below.

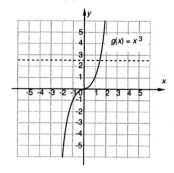

Since no horizontal line crosses the graph more than once, the function is one-to-one.

32. Yes

33. The graph of $g(x) = |x|$ is shown below.

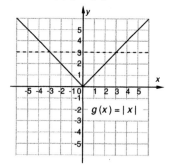

There are many horizontal lines that cross the graph more than once. In particular, the line $y = 3$ crosses the graph more than once. The function is not one-to-one.

34. No

35. a) The function $f(x) = x - 4$ is a linear function that is not constant, so it passes the horizontal-line test. Thus, f is one-to-one.

b) Replace $f(x)$ by y: $y = x - 4$

Interchange x and y: $x = y - 4$

Solve for y: $x + 4 = y$

Replace y by $f^{-1}(x)$: $f^{-1}(x) = x + 4$

36. a) Yes

b) $f^{-1}(x) = x + 2$

37. a) The function $f(x) = 3 + x$ is a linear function that is not constant, so it passes the horizontal-line test. Thus, f is one-to-one.

b) Replace $f(x)$ by y: $y = 3 + x$

Interchange x and y: $x = 3 + y$

Solve for y: $y = x - 3$

Replace y by $f^{-1}(x)$: $f^{-1}(x) = x - 3$

38. a) Yes

b) $f^{-1}(x) = x - 9$

39. a) The function $g(x) = x + 5$ is a linear function that is not constant, so it passes the horizontal-line test. Thus, g is one-to-one.

b) Replace $g(x)$ by y: $y = x + 5$

Interchange x and y: $x = y + 5$

Solve for y: $x - 5 = y$

Replace y by $g^{-1}(x)$: $g^{-1}(x) = x - 5$

40. a) Yes

b) $g^{-1}(x) = x - 8$

41. a) The function $f(x) = 4x$ is a linear function that is not constant, so it passes the horizontal-line test. Thus, f is one-to-one.

b) Replace $f(x)$ by y: $y = 4x$

Interchange x and y: $x = 4y$

Solve for y: $\dfrac{x}{4} = y$

Replace y by $f^{-1}(x)$: $f^{-1}(x) = \dfrac{x}{4}$

42. a) Yes

b) $f^{-1}(x) = \dfrac{x}{7}$

43. a) The function $g(x) = 4x - 1$ is a linear function that is not constant, so it passes the horizontal-line test. Thus, g is one-to-one.

b) Replace $g(x)$ by y: $y = 4x - 1$

Interchange variables: $x = 4y - 1$

Solve for y: $x + 1 = 4y$

$$\dfrac{x+1}{4} = y$$

Replace y by $g^{-1}(x)$: $g^{-1}(x) = \dfrac{x+1}{4}$

44. a) Yes

b) $g^{-1}(x) = \dfrac{x+6}{4}$

45. a) The graph of $h(x) = 5$ is shown below. The horizontal line $y = 5$ crosses the graph more than once, so the function is not one-to-one.

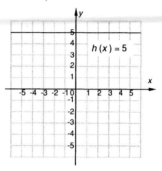

46. a) No

47. a) The graph of $f(x) = \dfrac{1}{x}$ is shown below. It passes the horizontal-line test, so the function is one-to-one.

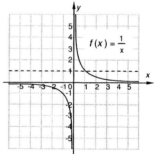

b) Replace $f(x)$ by y: $y = \dfrac{1}{x}$

Interchange x and y: $x = \dfrac{1}{y}$

Solve for y: $xy = 1$

$$y = \dfrac{1}{x}$$

Replace y by $f^{-1}(x)$: $f^{-1}(x) = \dfrac{1}{x}$

48. a) Yes

b) $f^{-1}(x) = \dfrac{3}{x}$

49. a) The function $f(x) = \dfrac{2x+1}{3} = \dfrac{2}{3}x + \dfrac{1}{3}$ is a linear function that is not constant, so it passes the horizontal-line test. Thus, f is one-to-one.

b) Replace $f(x)$ by y: $y = \dfrac{2x+1}{3}$

Interchange x and y: $x = \dfrac{2y+1}{3}$

Solve for y: $3x = 2y + 1$

$3x - 1 = 2y$

$\dfrac{3x-1}{2} = y$

Replace y by $f^{-1}(x)$: $f^{-1}(x) = \dfrac{3x-1}{2}$

50. a) Yes

b) $f^{-1}(x) = \dfrac{5x-2}{3}$

51. a) The graph of $f(x) = x^3 - 5$ is shown below. It passes the horizontal-line test, so the function is one-to-one.

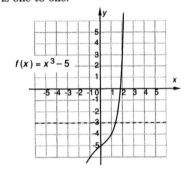

b) Replace $f(x)$ by y: $y = x^3 - 5$

Interchange x and y: $x = y^3 - 5$

Solve for y: $x + 5 = y^3$

$\sqrt[3]{x+5} = y$

Replace y by $f^{-1}(x)$: $f^{-1}(x) = \sqrt[3]{x+5}$

52. a) Yes

b) $f^{-1}(x) = \sqrt[3]{x-2}$

53. a) The graph of $g(x) = (x-2)^3$ is shown below. It passes the horizontal-line test, so the function is one-to-one.

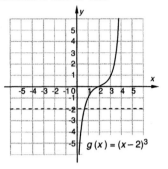

b) Replace $g(x)$ by y: $y = (x-2)^3$

Interchange x and y: $x = (y-2)^3$

Solve for y: $\sqrt[3]{x} = y - 2$

$\sqrt[3]{x} + 2 = y$

Replace y by $g^{-1}(x)$: $g^{-1}(x) = \sqrt[3]{x} + 2$

54. a) Yes

b) $g^{-1}(x) = \sqrt[3]{x} - 7$

55. a) The graph of $f(x) = \sqrt{x}$ is shown below. It passes the horizontal-line test, so the function is one-to-one.

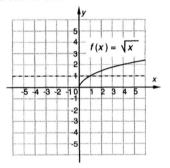

b) Replace $f(x)$ by y: $y = \sqrt{x}$ (Note that $f(x) \geq 0$.)

Interchange x and y: $x = \sqrt{y}$

Solve for y: $x^2 = y$

Replace y by $f^{-1}(x)$: $f^{-1}(x) = x^2,\ x \geq 0$

56. a) Yes

b) $f^{-1}(x) = x^2 + 1,\ x \geq 0$

57. a) The graph of $f(x) = 2x^2 + 1$, $x \geq 0$, is shown below. It passes the horizontal-line test, so the function is one-to-one.

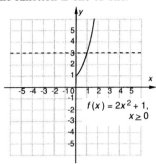

b) Replace $f(x)$ by y: $y = 2x^2 + 1$

Interchange x and y: $x = 2y^2 + 1$

Solve for y: $x - 1 = 2y^2$

$$\frac{x-1}{2} = y^2$$

$$\sqrt{\frac{x-1}{2}} = y$$

(We take the principal square root since $y \geq 0$.)

Replace y by $f^{-1}(x)$: $f^{-1}(x) = \sqrt{\frac{x-1}{2}}$

58. a) Yes

b) $f^{-1}(x) = \sqrt{\frac{x+2}{3}}$

59. First graph $f(x) = \frac{1}{3}x - 2$. Then graph the inverse function by reflecting the graph of $f(x) = \frac{1}{3}x - 2$ across the line $y = x$. The graph of the inverse function can also be found by first finding a formula for the inverse, substituting to find function values, and then plotting points.

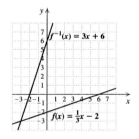

60.

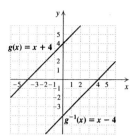

61. Follow the procedure described in Exercise 59 to graph the function and its inverse.

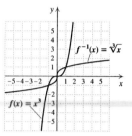

62.

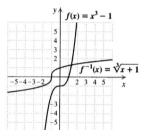

63. Use the procedure described in Exercise 59 to graph the function and its inverse.

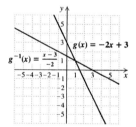

64.

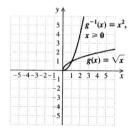

65. Use the procedure described in Exercise 59 to graph the function and its inverse.

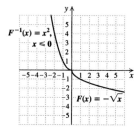

66.

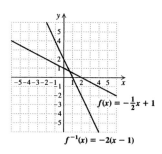

67. Use the procedure described in Exercise 59 to graph the function and its inverse.

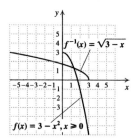

68.

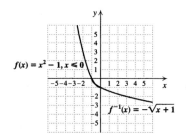

69. We check to see that $f^{-1} \circ f(x) = x$ and $f \circ f^{-1}(x) = x$.

a) $f^{-1} \circ f(x) = f^{-1}(f(x)) = f^{-1}\left(\frac{4}{5}x\right) =$

$\frac{5}{4} \cdot \frac{4}{5}x = x$

b) $f \circ f^{-1}(x) = f(f^{-1}(x)) = f\left(\frac{5}{4}x\right) =$

$\frac{4}{5} \cdot \frac{5}{4}x = x$

70. a) $f^{-1} \circ f(x) = 3\left(\frac{x+7}{3}\right) - 7 = x + 7 - 7 = x$

b) $f \circ f^{-1}(x) = \frac{(3x-7)+7}{3} = \frac{3x}{3} = x$

71. We check to see that $f^{-1} \circ f(x) = x$ and $f \circ f^{-1}(x) = x$.

a) $f^{-1} \circ f(x) = f^{-1}(f(x)) = f^{-1}\left(\frac{1-x}{x}\right) =$

$\frac{1}{\dfrac{1-x}{x}+1} = \frac{1}{\dfrac{1-x}{x}+1} \cdot \frac{x}{x} = \frac{x}{1-x+x} =$

$\frac{x}{1} = x$

b) $f \circ f^{-1}(x) = f(f^{-1}(x)) = f\left(\frac{1}{x+1}\right) =$

$\frac{1 - \dfrac{1}{x+1}}{\dfrac{1}{x+1}} = \frac{1 - \dfrac{1}{x+1}}{\dfrac{1}{x+1}} \cdot \frac{x+1}{x+1} =$

$\frac{x+1-1}{1} = \frac{x}{1} = x$

72. a) $f^{-1} \circ f(x) = \sqrt[3]{x^3 - 5 + 5} = \sqrt[3]{x^3} = x$

b) $f \circ f^{-1}(x) = (\sqrt[3]{x+5})^3 - 5 = x + 5 - 5 = x$

73. Let $y_1 = f(x)$, $y_2 = g(x)$, $y_3 = y_1(y_2)$, and $y_4 = y_2(y_1)$. A table of values shows that $y_3 \neq x$ nor is $y_4 = x$, so $f(x)$ and $g(x)$ are not inverses of each other.

74. Yes

75. Let $y_1 = f(x)$, $y_2 = g(x)$, $y_3 = y_1(y_2)$, and $y_4 = y_2(y_1)$. A table of values shows that $y_3 = x$ and $y_4 = x$ for any value of x, so $f(x)$ and $g(x)$ are inverses of each other.

76. No

77. (1) C; (2) D; (3) B; (4) A

78. (1) D; (2) C; (3) B; (4) A

79. a) $f(8) = 8 + 32 = 40$

Size 40 in France corresponds to size 8 in the U.S.

$f(10) = 10 + 32 = 42$

Size 42 in France corresponds to size 10 in the U.S.

$f(14) = 14 + 32 = 46$

Size 46 in France corresponds to size 14 in the U.S.

$f(18) = 18 + 32 = 50$

Size 50 in France corresponds to size 18 in the U.S.

b) The function $f(x) = x + 32$ is a linear function that is not constant, so it passes the horizontal-line test. Thus, f is one-to-one and, hence, has an inverse that is a function. We now find a formula for the inverse.

Replace $f(x)$ by y: $y = x + 32$

Interchange x and y: $x = y + 32$

Solve for y: $x - 32 = y$

Replace y by $f^{-1}(x)$: $f^{-1}(x) = x - 32$

c) $f^{-1}(40) = 40 - 32 = 8$

Size 8 in the U.S. corresponds to size 40 in France.

$f^{-1}(42) = 42 - 32 = 10$

Size 10 in the U.S. corresponds to size 42 in France.

$f^{-1}(46) = 46 - 32 = 14$

Size 14 in the U.S. corresponds to size 46 in France.

$f^{-1}(50) = 50 - 32 = 18$

Size 18 in the U.S. corresponds to size 50 in France.

80. a) 40; 44; 52; 60

b) $f^{-1}(x) = \dfrac{x - 24}{2}$, or $\dfrac{x}{2} - 12$

c) 8; 10; 14; 18

81. *Writing Exercise*

82. *Writing Exercise*

83. $(a^5 b^4)^2 (a^3 b^5) = (a^5)^2 (b^4)^2 (a^3 b^5)$

$\qquad = a^{5 \cdot 2} b^{4 \cdot 2} a^3 b^5$

$\qquad = a^{10} b^8 a^3 b^5$

$\qquad = a^{10+3} b^{8+5}$

$\qquad = a^{13} b^{13}$

84. $x^{10} y^{12}$

85. $27^{4/3} = (3^3)^{4/3} = 3^{3 \cdot \frac{4}{3}} = 3^4 = 81$

86. 125

87. $\qquad x = \dfrac{2}{3} y - 7$

$\qquad x + 7 = \dfrac{2}{3} y$

$\qquad \dfrac{3}{2}(x + 7) = y$

88. $y = \dfrac{10 - x}{3}$

89. *Writing Exercise*

90. *Writing Exercise*

91. Reflect the graph of f across the line $y = x$.

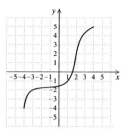

92.

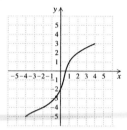

93. From Exercise 80(b), we know that a function that converts dress sizes in Italy to those in the United States is $g(x) = \dfrac{x - 24}{2}$. From Exercise 79(a), we know that a function that converts dress sizes in the United States to those in France is $f(x) = x + 32$. Then a function that converts dress sizes in Italy to those in France is

$h(x) = (f \circ g)(x)$

$h(x) = f\left(\dfrac{x - 24}{2}\right)$

$h(x) = \dfrac{x - 24}{2} + 32$

$h(x) = \dfrac{x}{2} - 12 + 32$

$h(x) = \dfrac{x}{2} + 20.$

94. $h(x) = 2(x - 20)$

95. *Writing Exercise*

96. $((f \circ g) \circ h)(x) = (f \circ g)(h(x))$

$\qquad = f(g(h(x))) = f((g \circ h)(x))$

$\qquad = (f \circ (g \circ h))(x)$

97. Suppose that $h(x) = (f \circ g)(x)$. First note that for $I(x) = x$, $(f \circ I)(x) = f(I(x))$ for any function f.

i) $((g^{-1} \circ f^{-1}) \circ h)(x) = ((g^{-1} \circ f^{-1}) \circ (f \circ g))(x)$

$\qquad = ((g^{-1} \circ (f^{-1} \circ f)) \circ g)(x)$

$\qquad = ((g^{-1} \circ I) \circ g)(x)$

$\qquad = (g^{-1} \circ g)(x) = x$

ii) $(h \circ (g^{-1} \circ f^{-1}))(x) = ((f \circ g) \circ (g^{-1} \circ f^{-1}))(x)$
$$= ((f \circ (g \circ g^{-1})) \circ f^{-1})(x)$$
$$= ((f \circ I) \circ f^{-1})(x)$$
$$= (f \circ f^{-1})(x) = x$$

Therefore, $(g^{-1} \circ f^{-1})(x) = h^{-1}(x)$.

98. (1) C; (2) A; (3) B; (4) D

99. *Writing Exercise.* Observe the following:

$$f(6) = 6 \text{ and } g(6) = 6,$$
$$f(8) = 7 \text{ and } g(7) = 8,$$
$$f(10) = 8 \text{ and } g(8) = 10,$$
$$f(12) = 9 \text{ and } g(9) = 12.$$

It appears that the functions are inverses.

00. $f(x) = \dfrac{1}{2}x + 3,\ g(x) = 2x - 6$

01. $(c \circ f)(n)$ represents the cost of mailing n copies of the book.

02. $(c \circ g)(a)$; it represents the cost of sealant required for a bamboo floor with area a.

03. $R(10) \approx 18$ and $p(18) \approx 22$, so $p(R(10)) \approx 22$ mm of mercury.

04. The pressure in the artery after 10 minutes of bicycling

05. Locate 20 on the vertical axis of the second graph, move across to the curve, and then move down to the horizontal axis to find that $p^{-1}(20) \approx 15$ liters per minute.

06. The rate of blood flow for the heart when the pressure in the artery is 20 mm of mercury

---

Exercise Set 9.2

---

1. The function values increase as x increases, so $a > 1$.

2. $0 < a < 1$

3. The function values decrease as x increases, so $0 < a < 1$.

4. $a > 1$

5. Graph: $y = 2^x$

We compute some function values, thinking of y as $f(x)$, and keep the results in a table.

$$f(0) = 2^0 = 1$$
$$f(1) = 2^1 = 2$$
$$f(2) = 2^2 = 4$$

$$f(-1) = 2^{-1} = \frac{1}{2^1} = \frac{1}{2}$$
$$f(-2) = 2^{-2} = \frac{1}{2^2} = \frac{1}{4}$$

x	y, or $f(x)$
0	1
1	2
2	4
-1	$\dfrac{1}{2}$
-2	$\dfrac{1}{4}$

Next we plot these points and connect them with a smooth curve.

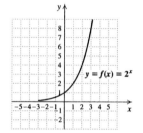

6.

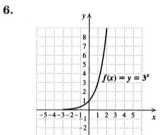

7. Graph: $y = 5^x$

We compute some function values, thinking of y as $f(x)$, and keep the results in a table.

$$f(0) = 5^0 = 1$$
$$f(1) = 5^1 = 5$$
$$f(2) = 5^2 = 25$$

$$f(-1) = 5^{-1} = \frac{1}{5^1} = \frac{1}{5}$$
$$f(-2) = 5^{-2} = \frac{1}{5^2} = \frac{1}{25}$$

x	y, or $f(x)$
0	1
1	5
2	25
-1	$\dfrac{1}{5}$
-2	$\dfrac{1}{25}$

Next we plot these points and connect them with a smooth curve.

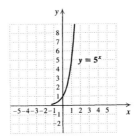

8.

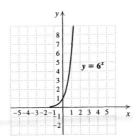

9. Graph: $y = 2^x + 3$

We compute some function values, thinking of y as $f(x)$, and keep the results in a table.

$$f(-4) = 2^{-4} + 3 = \frac{1}{2^4} + 3 = \frac{1}{16} + 3 = 3\frac{1}{16}$$

$$f(-2) = 2^{-2} + 3 = \frac{1}{2^2} + 3 = \frac{1}{4} + 3 = 3\frac{1}{4}$$

$$f(0) = 2^0 + 3 = 1 + 3 = 4$$

$$f(1) = 2^1 + 3 = 2 + 3 = 5$$

$$f(2) = 2^2 + 3 = 4 + 3 = 7$$

x	y, or $f(x)$
-4	$3\frac{1}{16}$
-2	$3\frac{1}{4}$
0	4
1	5
2	7

Next we plot these points and connect them with a smooth curve.

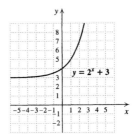

10.

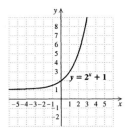

11. Graph: $y = 3^x - 1$

We compute some function values, thinking of y as $f(x)$, and keep the results in a table.

$$f(-3) = 3^{-3} - 1 = \frac{1}{3^3} - 1 = \frac{1}{27} - 1 = -\frac{26}{27}$$

$$f(-1) = 3^{-1} - 1 = \frac{1}{3} - 1 = -\frac{2}{3}$$

$$f(0) = 3^0 - 1 = 1 - 1 = 0$$

$$f(1) = 3^1 - 1 = 3 - 1 = 2$$

$$f(2) = 3^2 - 1 = 9 - 1 = 8$$

x	y, or $f(x)$
-3	$-\frac{26}{27}$
-1	$-\frac{2}{3}$
0	0
1	2
2	8

Next we plot these points and connect them with a smooth curve.

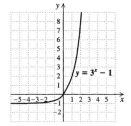

12.

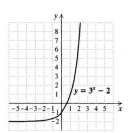

13. Graph: $y = 2^{x-1}$

We construct a table of values, thinking of y as $f(x)$. Then we plot the points and connect them with a smooth curve.

$$f(0) = 2^{0-1} = 2^{-1} = \frac{1}{2}$$

$$f(-1) = 2^{-1-1} = 2^{-2} = \frac{1}{2^2} = \frac{1}{4}$$

$$f(-2) = 2^{-2-1} = 2^{-3} = \frac{1}{2^3} = \frac{1}{8}$$

$$f(1) = 2^{1-1} = 2^0 = 1$$

$$f(2) = 2^{2-1} = 2^1 = 2$$

$$f(3) = 2^{3-1} = 2^2 = 4$$

$$f(4) = 2^{4-1} = 2^3 = 8$$

x	y, or $f(x)$
0	$\frac{1}{2}$
-1	$\frac{1}{4}$
-2	$\frac{1}{8}$
1	1
2	2
3	4
4	8

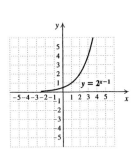

14.

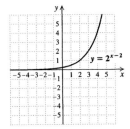

15. Graph: $y = 2^{x+3}$

We construct a table of values, thinking of y as $f(x)$. Then we plot the points and connect them with a smooth curve.

$$f(-4) = 2^{-4+3} = 2^{-1} = \frac{1}{2}$$

$$f(-2) = 2^{-2+3} = 2$$

$$f(-1) = 2^{-1+3} = 2^2 = 4$$

$$f(0) = 2^{0+3} = 2^3 = 8$$

x	y, or $f(x)$
-4	$\frac{1}{2}$
-2	2
-1	4
0	8

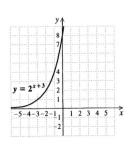

16.

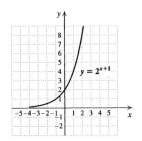

17. Graph: $y = \left(\frac{1}{5}\right)^x$

We construct a table of values, thinking of y as $f(x)$. Then we plot the points and connect them with a smooth curve.

$$f(0) = \left(\frac{1}{5}\right)^0 = 1$$

$$f(1) = \left(\frac{1}{5}\right)^1 = \frac{1}{5}$$

$$f(2) = \left(\frac{1}{5}\right)^2 = \frac{1}{25}$$

$$f(-1) = \left(\frac{1}{5}\right)^{-1} = \frac{1}{\frac{1}{5}} = 5$$

$$f(-2) = \left(\frac{1}{5}\right)^{-2} = \frac{1}{\frac{1}{25}} = 25$$

x	y, or $f(x)$
0	1
1	$\frac{1}{5}$
2	$\frac{1}{25}$
-1	5
-2	25

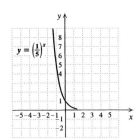

18.

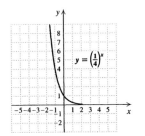

19. Graph: $y = \left(\frac{1}{2}\right)^x$

We construct a table of values, thinking of y as $f(x)$. Then we plot the points and connect them with a smooth curve.

$f(0) = \left(\frac{1}{2}\right)^0 = 1$

$f(1) = \left(\frac{1}{2}\right)^1 = \frac{1}{2}$

$f(2) = \left(\frac{1}{2}\right)^2 = \frac{1}{4}$

$f(3) = \left(\frac{1}{2}\right)^3 = \frac{1}{8}$

$f(-1) = \left(\frac{1}{2}\right)^{-1} = \frac{1}{\left(\frac{1}{2}\right)^1} = \frac{1}{\frac{1}{2}} = 2$

$f(-2) = \left(\frac{1}{2}\right)^{-2} = \frac{1}{\left(\frac{1}{2}\right)^2} = \frac{1}{\frac{1}{4}} = 4$

$f(-3) = \left(\frac{1}{2}\right)^{-3} = \frac{1}{\left(\frac{1}{2}\right)^3} = \frac{1}{\frac{1}{8}} = 8$

x	y, or $f(x)$
0	1
1	$\frac{1}{2}$
2	$\frac{1}{4}$
3	$\frac{1}{8}$
−1	2
−2	4
−3	8

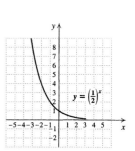

20.

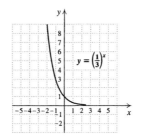

21. Graph: $y = 2^{x-3} - 1$

We construct a table of values, thinking of y as $f(x)$. Then we plot the points and connect them with a smooth curve.

$f(0) = 2^{0-3} - 1 = 2^{-3} - 1 = \frac{1}{8} - 1 = -\frac{7}{8}$

$f(1) = 2^{1-3} - 1 = 2^{-2} - 1 = \frac{1}{4} - 1 = -\frac{3}{4}$

$f(2) = 2^{2-3} - 1 = 2^{-1} - 1 = \frac{1}{2} - 1 = -\frac{1}{2}$

$f(3) = 2^{3-3} - 1 = 2^0 - 1 = 1 - 1 = 0$

$f(4) = 2^{4-3} - 1 = 2^1 - 1 = 2 - 1 = 1$

$f(5) = 2^{5-3} - 1 = 2^2 - 1 = 4 - 1 = 3$

$f(6) = 2^{6-3} - 1 = 2^3 - 1 = 8 - 1 = 7$

x	y, or $f(x)$
0	$-\frac{7}{8}$
1	$-\frac{3}{4}$
2	$-\frac{1}{2}$
3	0
4	1
5	3
6	7

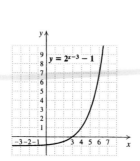

22.

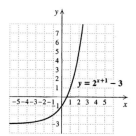

23. Graph: $y = 1.7^x$

We use a graphing calculator.

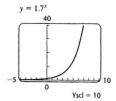

24.

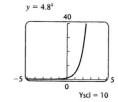

25. Graph: $y = 0.15^x$

We use a graphing calculator.

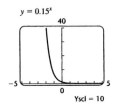

$y = 0.15^x$

26.

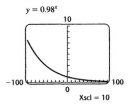

$y = 0.98^x$

27. Graph: $x = 3^y$

We can find ordered pairs by choosing values for y and then computing values for x.

For $y = 0$, $x = 3^0 = 1$.

For $y = 1$, $x = 3^1 = 3$.

For $y = 2$, $x = 3^2 = 9$.

For $y = 3$, $x = 3^3 = 27$.

For $y = -1$, $x = 3^{-1} = \dfrac{1}{3^1} = \dfrac{1}{3}$.

For $y = -2$, $x = 3^{-2} = \dfrac{1}{3^2} = \dfrac{1}{9}$.

For $y = -3$, $x = 3^{-3} = \dfrac{1}{3^3} = \dfrac{1}{27}$.

x	y
1	0
3	1
9	2
27	3
$\dfrac{1}{3}$	-1
$\dfrac{1}{9}$	-2
$\dfrac{1}{27}$	-3

 (1) Choose values for y.

 (2) Compute values for x.

We plot the points and connect them with a smooth curve.

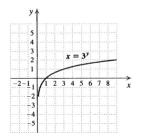

$x = 3^y$

28.

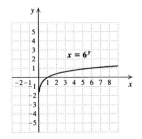

$x = 6^y$

29. Graph: $x = 2^{-y} = \left(\dfrac{1}{2}\right)^y$

We can find ordered pairs by choosing values for y and then computing values for x. Then we plot these points and connect them with a smooth curve.

For $y = 0$, $x = \left(\dfrac{1}{2}\right)^0 = 1$.

For $y = 1$, $x = \left(\dfrac{1}{2}\right)^1 = \dfrac{1}{2}$.

For $y = 2$, $x = \left(\dfrac{1}{2}\right)^2 = \dfrac{1}{4}$.

For $y = 3$, $x = \left(\dfrac{1}{2}\right)^3 = \dfrac{1}{8}$.

For $y = -1$, $x = \left(\dfrac{1}{2}\right)^{-1} = \dfrac{1}{\dfrac{1}{2}} = 2$.

For $y = -2$, $x = \left(\dfrac{1}{2}\right)^{-2} = \dfrac{1}{\dfrac{1}{4}} = 4$.

For $y = -3$, $x = \left(\dfrac{1}{2}\right)^{-3} = \dfrac{1}{\dfrac{1}{8}} = 8$.

x	y
1	0
$\dfrac{1}{2}$	1
$\dfrac{1}{4}$	2
$\dfrac{1}{8}$	3
2	-1
4	-2
8	-3

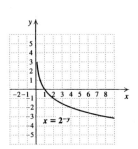

$x = 2^{-y}$

30.

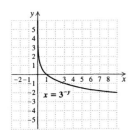

31. Graph: $x = 5^y$

We can find ordered pairs by choosing values for y and then computing values for x. Then we plot these points and connect them with a smooth curve.

For $y = 0$, $x = 5^0 = 1$.

For $y = 1$, $x = 5^1 = 5$.

For $y = 2$, $x = 5^2 = 25$.

For $y = -1$, $x = 5^{-1} = \dfrac{1}{5}$.

For $y = -2$, $x = 5^{-2} = \dfrac{1}{25}$.

x	y
1	0
5	1
25	2
$\dfrac{1}{5}$	-1
$\dfrac{1}{25}$	-2

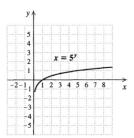

32.

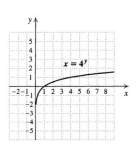

33. Graph: $x = \left(\dfrac{3}{2}\right)^y$

We can find ordered pairs by choosing values for y and then computing values for x. Then we plot these points and connect them with a smooth curve.

For $y = 0$, $x = \left(\dfrac{3}{2}\right)^0 = 1$.

For $y = 1$, $x = \left(\dfrac{3}{2}\right)^1 = \dfrac{3}{2}$.

For $y = 2$, $x = \left(\dfrac{3}{2}\right)^2 = \dfrac{9}{4}$.

For $y = 3$, $x = \left(\dfrac{3}{2}\right)^3 = \dfrac{27}{8}$.

For $y = -1$, $x = \left(\dfrac{3}{2}\right)^{-1} = \dfrac{1}{\frac{3}{2}} = \dfrac{2}{3}$.

For $y = -2$, $x = \left(\dfrac{3}{2}\right)^{-2} = \dfrac{1}{\frac{9}{4}} = \dfrac{4}{9}$.

For $y = -3$, $x = \left(\dfrac{3}{2}\right)^{-3} = \dfrac{1}{\frac{27}{8}} = \dfrac{8}{27}$.

x	y
1	0
$\dfrac{3}{2}$	1
$\dfrac{9}{4}$	2
$\dfrac{27}{8}$	3
$\dfrac{2}{3}$	-1
$\dfrac{4}{9}$	-2
$\dfrac{8}{27}$	-3

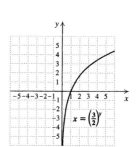

34.

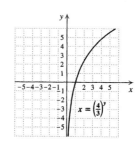

35. Graph $y = 3^x$ (see Exercise 2) and $x = 3^y$ (see Exercise 21) using the same set of axes.

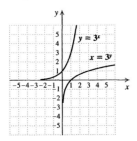

36.

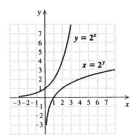

37. Graph $y = \left(\frac{1}{2}\right)^x$ (see Exercise 13) and $x = \left(\frac{1}{2}\right)^y$ (see Exercise 23) using the same set of axes.

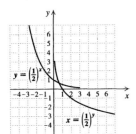

38.

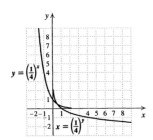

39. $y = \left(\frac{5}{2}\right)^x$ is an exponential function of the form $y = a^x$ with $a > 1$, so y-values will increase as x-values increase. Also, observe that when $x = 0$, $y = 1$. Thus, graph (d) corresponds to this equation.

40. (e)

41. For $x = \left(\frac{2}{5}\right)^y$, when $y = 0$, $x = 1$. The only graph that contains the point $(1,0)$ is (f). This graph corresponds to the given equation.

42. (a)

43. $y = \left(\frac{2}{5}\right)^{x-2}$ is an exponential function of the form $y = a^x$ with $0 < a < 1$, so y-values will decrease as x-values increase. Also, observe that when $x = 2$, $y = 1$. Thus, graph (c) corresponds to the given equation.

44. (b)

45. a) In 2004, $t = 2004 - 1975 = 29$.
$$P(29) = 4(1.0164)^{29} \approx 6.4$$
The world population will be about 6.4 billion in 2004.
In 2008, $t = 2008 - 1975 = 33$.
$$P(33) = 4(1.0164)^{33} \approx 6.8$$
The world population will be about 6.8 billion in 2008.
In 2012, $t = 2012 - 1975 = 37$.
$$P(37) = 4(1.0164)^{37} \approx 7.3$$
The world population will be about 7.3 billion in 2012.

b)

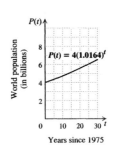

46. a) 4243; 6000; 8485; 12,000; 24,000

b)

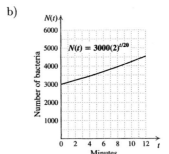

47. a) In 1930, $t = 1930 - 1900 = 30$.

$$P(t) = 150(0.960)^t$$
$$P(30) = 150(0.960)^{30}$$
$$\approx 44.079$$

In 1930, about 44.079 thousand, or 44,079, humpback whales were alive.

In 1960, $t = 1960 - 1900 = 60$.

$$P(t) = 150(0.960)^t$$
$$P(60) = 150(0.960)^{60}$$
$$\approx 12.953$$

In 1960, about 12.953 thousand, or 12,953, humpback whales were alive.

b) Plot the points found in part (a), (30, 44,079) and (60, 12,953) and additional points as needed and graph the function.

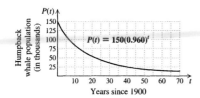

48. a) About 8706; about 13,163

b)

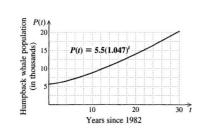

49. a) Substitute for t.

$$N(0) = 250,000\left(\frac{2}{3}\right)^0 = 250,000 \cdot 1 = 250,000;$$

$$N(1) = 250,000\left(\frac{2}{3}\right)^1 = 250,000 \cdot \frac{2}{3} = 166,667;$$

$$N(4) = 250,000\left(\frac{2}{3}\right)^4 = 250,000 \cdot \frac{16}{81} \approx 49,383;$$

$$N(10) = 250,000\left(\frac{2}{3}\right)^{10} = 250,000 \cdot \frac{1024}{59,049} \approx 4335$$

b) We use the function values computed in part (a) to draw the graph of the function. Note that the axes are scaled differently because of the large function values.

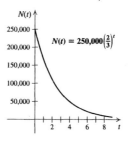

50. a) 5200; 4160; 3328; 1703.94; $558.35

b)

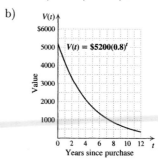

51. a) In 1985, $t = 1985 - 1985 = 0$.

$N(0) = 0.3(1.4477)^0 = 0.3(1) = 0.3$ million, or 300,000

In 1995, $t = 1995 - 1985 = 10$.

$N(10) = 0.3(1.4477)^{10} \approx 12.1$ million

In 2005, $t = 2005 - 1985 = 20$.

$N(20) = 0.3(1.4477)^{20} \approx 490.6$ million

In 2010, $t = 2010 - 1985 = 25$.

$N(25) = 0.3(1.4477)^{25} \approx 3119.5$ million, or 3.1195 billion

b) We use the function values computed in part (a) to draw the graph of the function. Note that the axes are scaled differently because of the large function values.

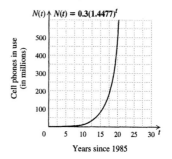

52. a) $454,354,240$ cm^2; $525,233,501,400$ cm^2

b)

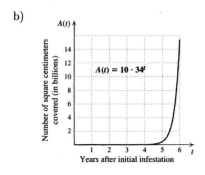

53. *Writing Exercise*

54. *Writing Exercise*

55. $5^{-2} = \dfrac{1}{5^2} = \dfrac{1}{25}$

56. $\dfrac{1}{32}$

57. $1000^{2/3} = (10^3)^{2/3} = 10^{3 \cdot \frac{2}{3}} = 10^2 = 100$

58. $\dfrac{1}{125}$

59. $\dfrac{10a^8 b^7}{2a^2 b^4} = \dfrac{10}{2} a^{8-2} b^{7-4} = 5a^6 b^3$

60. $6x^4 y$

61. *Writing Exercise*

62. *Writing Exercise*

63. Since the bases are the same, the one with the larger exponent is the larger number. Thus $\pi^{2.4}$ is larger.

64. $8^{\sqrt{3}}$

65. Graph: $y = 2^x + 2^{-x}$

Construct a table of values, thinking of y as $f(x)$. Then plot these points and connect them with a curve.

$f(0) = 2^0 + 2^{-0} = 1 + 1 = 2$

$f(1) = 2^1 + 2^{-1} = 2 + \dfrac{1}{2} = 2\dfrac{1}{2}$

$f(2) = 2^2 + 2^{-2} = 4 + \dfrac{1}{4} = 4\dfrac{1}{4}$

$f(3) = 2^3 + 2^{-3} = 8 + \dfrac{1}{8} = 8\dfrac{1}{8}$

$f(-1) = 2^{-1} + 2^{-(-1)} = \dfrac{1}{2} + 2 = 2\dfrac{1}{2}$

$f(-2) = 2^{-2} + 2^{-(-2)} = \dfrac{1}{4} + 4 = 4\dfrac{1}{4}$

$f(-3) = 2^{-3} + 2^{-(-3)} = \dfrac{1}{8} + 8 = 8\dfrac{1}{8}$

x	y, or $f(x)$
0	2
1	$2\dfrac{1}{2}$
2	$4\dfrac{1}{4}$
3	$8\dfrac{1}{8}$
-1	$2\dfrac{1}{2}$
-2	$4\dfrac{1}{4}$
-3	$8\dfrac{1}{8}$

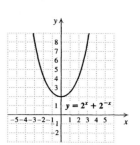

66.

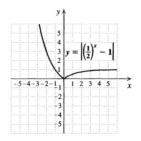

67. Graph: $y = |2^x - 2|$

We construct a table of values, thinking of y as $f(x)$. Then plot these points and connect them with a curve.

$f(0) = |2^0 - 2| = |1 - 2| = |-1| = 1$

$f(1) = |2^1 - 2| = |2 - 2| = |0| = 0$

$f(2) = |2^2 - 2| = |4 - 2| = |2| = 2$

$f(3) = |2^3 - 2| = |8 - 2| = |6| = 6$

$f(-1) = |2^{-1} - 2| = \left|\dfrac{1}{2} - 2\right| = \left|-\dfrac{3}{2}\right| = \dfrac{3}{2}$

$f(-3) = |2^{-3} - 2| = \left|\dfrac{1}{8} - 2\right| = \left|-\dfrac{15}{8}\right| = \dfrac{15}{8}$

$f(-5) = |2^{-5} - 2| = \left|\dfrac{1}{32} - 2\right| = \left|-\dfrac{63}{32}\right| = \dfrac{63}{32}$

x	y, or $f(x)$
0	1
1	0
2	2
3	6
-1	$\dfrac{3}{2}$
-3	$\dfrac{15}{8}$
-5	$\dfrac{63}{32}$

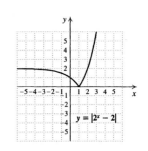

68.

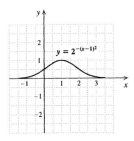

69. Graph: $y = |2x^2 - 1|$

We construct a table of values, thinking of y as $f(x)$. Then we plot these points and connect them with a curve.

$$f(0) = |2^{0^2} - 1| = |1 - 1| = 0$$
$$f(1) = |2^{1^2} - 1| = |2 - 1| = 1$$
$$f(2) = |2^{2^2} - 1| = |16 - 1| = 15$$
$$f(-1) = |2^{(-1)^2} - 1| = |2 - 1| = 1$$
$$f(-2) = |2^{(-2)^2} - 1| = |16 - 1| = 15$$

x	y, or $f(x)$
0	0
1	1
2	15
-1	1
-2	15

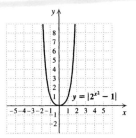

70.

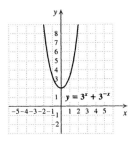

71. $y = 3^{-(x-1)}$ $\qquad$ $x = 3^{-(y-1)}$

x	y
0	3
1	1
2	$\frac{1}{3}$
3	$\frac{1}{9}$
-1	9

x	y
3	0
1	1
$\frac{1}{3}$	2
$\frac{1}{9}$	3
9	-1

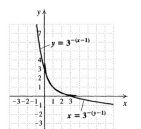

72.

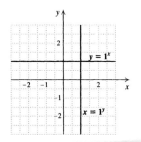

73. Enter the data points $(0, 171)$, $(1, 421)$, and $(2, 1099)$ and then use the exponential regression feature of the graphing calculator to find an exponential function that models the data.

$A(t) = 169.3393318(2.535133248)^t$, where $A(t)$ is total U.S. sales, in millions of dollars, t years after 1997.

In 2005, $t = 2005 - 1997 = 8$.

$A(8) = 169.3393318(2.535133248)^8 \approx \$288,911.0615$ million, or $\$288,911,061,500$

74. 19 words per minute; 66 words per minute; 110 words per minute

75. *Writing Exercise*

Exercise Set 9.3

1. $\log_{10} 100$ is the power to which we raise 10 to get 100. Since $10^2 = 100$, $\log_{10} 100 = 2$.

2. 3

3. $\log_2 8$ is the power to which we raise 2 to get 8. Since $2^3 = 8$, $\log_2 8 = 3$.

4. 4

5. $\log_3 81$ is the power to which we raise 3 to get 81. Since $3^4 = 81$, $\log_3 81 = 4$.

6. 3

7. $\log_4 \frac{1}{16}$ is the power to which we raise 4 to get $\frac{1}{16}$. Since $4^{-2} = \frac{1}{16}$, $\log_4 \frac{1}{16} = -2$.

8. -1

9. Since $7^{-1} = \frac{1}{7}$, $\log_7 \frac{1}{7} = -1$.

10. -2

11. Since $5^4 = 625$, $\log_5 625 = 4$.

12. 3

13. Since $6^1 = 6$, $\log_6 6 = 1$.

14. 0

15. Since $8^0 = 1$, $\log_8 1 = 0$.

16. 1

17. $\log_9 9^7$ is the power to which we raise 9 to get 9^7. Clearly, this power is 7, so $\log_9 9^7 = 7$.

18. 10

19. Since $10^{-1} = \frac{1}{10} = 0.1$, $\log_{10} 0.1 = -1$.

20. -2

21. Since $9^{1/2} = 3$, $\log_9 3 = \frac{1}{2}$.

22. $\frac{1}{2}$

23. Since $9 = 3^2$ and $(3^2)^{3/2} = 3^3 = 27$, $\log_9 27 = \frac{3}{2}$.

24. $\frac{3}{2}$

25. Since $1000 = 10^3$ and $(10^3)^{2/3} = 10^2 = 100$,
$\log_{1000} 100 = \frac{2}{3}$.

26. $\frac{2}{3}$

27. Since $\log_5 7$ is the power to which we raise 5 to get 7, then 5 raised to this power is 7. That is, $5^{\log_5 7} = 7$.

28. 13

29. Graph: $y = \log_{10} x$

The equation $y = \log_{10} x$ is equivalent to $10^y = x$. We can find ordered pairs by choosing values for y and computing the corresponding x-values.

For $y = 0$, $x = 10^0 = 1$.
For $y = 1$, $x = 10^1 = 10$.
For $y = 2$, $x = 10^2 = 100$.

For $y = -1$, $x = 10^{-1} = \frac{1}{10}$.

For $y = -2$, $x = 10^{-2} = \frac{1}{100}$.

x, or 10^y	y
1	0
10	1
100	2
$\frac{1}{10}$	-1
$\frac{1}{100}$	-2

$\uparrow$ (1) Select y.

(2) Compute x.

We plot the set of ordered pairs and connect the points with a smooth curve.

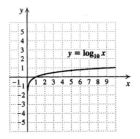

30.

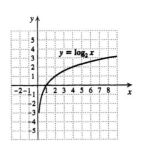

31. Graph: $y = \log_3 x$

The equation $y = \log_3 x$ is equivalent to $3^y = x$. We can find ordered pairs by choosing values for y and computing the corresponding x-values.

For $y = 0$, $x = 3^0 = 1$.
For $y = 1$, $x = 3^1 = 3$.
For $y = 2$, $x = 3^2 = 9$.

For $y = -1$, $x = 3^{-1} = \frac{1}{3}$.

For $y = -2$, $x = 3^{-2} = \frac{1}{9}$.

x, or 3^y	y
1	0
3	1
9	2
$\frac{1}{3}$	-1
$\frac{1}{9}$	-2

We plot the set of ordered pairs and connect the points with a smooth curve.

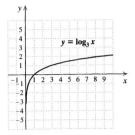

32.

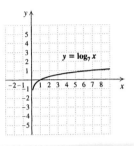

33. Graph: $f(x) = \log_6 x$

Think of $f(x)$ as y. Then $y = \log_6 x$ is equivalent to $6^y = x$. We find ordered pairs by choosing values for y and computing the corresponding x-values. Then we plot the points and connect them with a smooth curve.

For $y = 0$, $x = 6^0 = 1$.

For $y = 1$, $x = 6^1 = 6$.

For $y = 2$, $x = 6^2 = 36$.

For $y = -1$, $x = 6^{-1} = \dfrac{1}{6}$.

For $y = -2$, $x = 6^{-2} = \dfrac{1}{36}$.

x, or 6^y	y
1	0
6	1
36	2
$\dfrac{1}{6}$	-1
$\dfrac{1}{36}$	-2

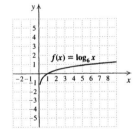

34.

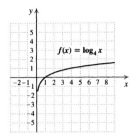

35. Graph: $f(x) = \log_{2.5} x$

Think of $f(x)$ as y. Then $y = \log_{2.5} x$ is equivalent to $2.5^y = x$. We construct a table of values, plot these points and connect them with a smooth curve.

For $y = 0$, $x = 2.5^0 = 1$.

For $y = 1$, $x = 2.5^1 = 2.5$.

For $y = 2$, $x = 2.5^2 = 6.25$.

For $y = 3$, $x = 2.5^3 = 15.625$.

For $y = -1$, $x = 2.5^{-1} = 0.4$.

For $y = -2$, $x = 2.5^{-2} = 0.16$.

x, or 2.5^y	y
1	0
2.5	1
6.25	2
15.625	3
0.4	-1
0.16	-2

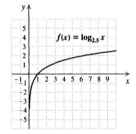

36.

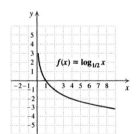

37. Graph $f(x) = 3^x$ (see Exercise Set 9.2, Exercise 6) and $f^{-1}(x) = \log_3 x$ (see Exercise 31 above) on the same set of axes.

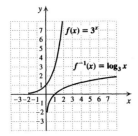

38.

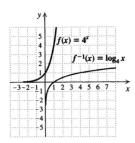

39. $\log 4 \approx 0.6021$

40. 0.6990

41. $\log 13,400 \approx 4.1271$

42. 4.9689

43. $\log 0.527 \approx -0.2782$

44. -0.3072

45. $10^{2.3} \approx 199.5262$

46. 1.4894

47. $10^{-2.9523} \approx 0.0011$

48. $79{,}104.2833$

49. $10^{0.0012} \approx 1.0028$

50. 0.0001

51.

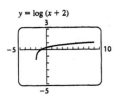

52.

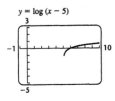

53.

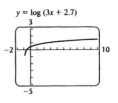

54.

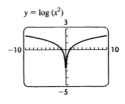

55.

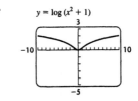

56.

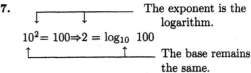

57.

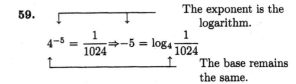

The exponent is the logarithm.

$10^2 = 100 \Rightarrow 2 = \log_{10} 100$

The base remains the same.

58. $4 = \log_{10} 10{,}000$

59.

The exponent is the logarithm.

$4^{-5} = \dfrac{1}{1024} \Rightarrow -5 = \log_4 \dfrac{1}{1024}$

The base remains the same.

60. $-3 = \log_5 \dfrac{1}{125}$

61. $16^{3/4} = 8$ is equivalent to $\dfrac{3}{4} = \log_{16} 8$.

62. $\dfrac{1}{3} = \log_8 2$

63. $10^{0.4771} = 3$ is equivalent to $0.4771 = \log_{10} 3$.

64. $0.3010 = \log_{10} 2$

65. $p^k = 3$ is equivalent to $k = \log_p 3$.

66. $n = \log_m r$

67. $p^m = V$ is equivalent to $m = \log_p V$.

68. $t = \log_Q x$

69. $e^3 = 20.0855$ is equivalent to $3 = \log_e 20.0855$.

70. $2 = \log_e 7.3891$

71. $e^{-4} = 0.0183$ is equivalent to $-4 = \log_e 0.0183$.

72. $-2 = \log_e 0.1353$

73.

$$t = \log_3 8 \Rightarrow 3^t = 8$$

The base remains the same. The logarithm is the exponent.

74. $7^h = 10$

75.

$$\log_5 25 = 2 \Rightarrow 5^2 = 25$$

The logarithm is the exponent. The base remains the same.

76. $6^1 = 6$

77. $\log_{10} 0.1 = -1$ is equivalent to $10^{-1} = 0.1$.

78. $10^{-2} = 0.01$

79. $\log_{10} 7 = 0.845$ is equivalent to $10^{0.845} = 7$.

80. $10^{0.4771} = 3$

81. $\log_c m = 8$ is equivalent to $c^8 = m$.

82. $b^{23} = n$

83. $\log_t Q = r$ is equivalent to $t^r = Q$.

84. $m^a = P$

85. $\log_e 0.25 = -1.3863$ is equivalent to $e^{-1.3863} = 0.25$.

86. $e^{-0.0111} = 0.989$

87. $\log_r T = -x$ is equivalent to $r^{-x} = T$.

88. $c^{-w} = M$

89. $\log_3 x = 2$

$3^2 = x$ Converting to an exponential equation

$9 = x$ Computing 3^2

90. 64

91. $\log_x 64 = 3$

$x^3 = 64$ Converting to an exponential equation

$x = 4$ Taking cube roots

92. 5

93. $\log_5 25 = x$

$5^x = 25$ Converting to an exponential equation

$5^x = 5^2$

$x = 2$ The exponents must be the same.

94. 4

95. $\log_4 16 = x$

$4^x = 16$ Converting to an exponential equation

$4^x = 4^2$

$x = 2$ The exponents must be the same.

96. 3

97. $\log_x 7 = 1$

$x^1 = 7$ Converting to an exponential equation

$x = 7$ Simplifying x^1

98. 8

99. $\log_9 x = 1$

$9^1 = x$ Converting to an exponential equation

$9 = x$ Simplifying 9^1

100. 1

101. $\log_3 x = -2$

$3^{-2} = x$ Converting to an exponential equation

$\dfrac{1}{9} = x$ Simplifying

102. $\dfrac{1}{2}$

103. $\log_{32} x = \dfrac{2}{5}$

$32^{2/5} = x$ Converting to an exponential equation

$(2^5)^{2/5} = x$

$4 = x$

104. 4

105. *Writing Exercise*

106. *Writing Exercise*

107. $\dfrac{x^{12}}{x^4} = x^{12-4} = x^8$

108. a^{12}

109. $(a^4b^6)(a^3b^2) = a^{4+3}b^{6+2} = a^7b^8$

110. x^5y^{12}

111. $\dfrac{\dfrac{3}{x} - \dfrac{2}{xy}}{\dfrac{2}{x^2} + \dfrac{1}{xy}}$

The LCD of all the denominators is x^2y. We multiply numerator and denominator by the LCD.

$$\frac{\dfrac{3}{x} - \dfrac{2}{xy}}{\dfrac{2}{x^2} + \dfrac{1}{xy}} \cdot \frac{x^2y}{x^2y} = \frac{\left(\dfrac{3}{x} - \dfrac{2}{xy}\right)x^2y}{\left(\dfrac{2}{x^2} + \dfrac{1}{xy}\right)x^2y}$$

$$= \frac{\dfrac{3}{x} \cdot x^2y - \dfrac{2}{xy} \cdot x^2y}{\dfrac{2}{x^2} \cdot x^2y + \dfrac{1}{xy} \cdot x^2y}$$

$$= \frac{3xy - 2x}{2y + x}, \text{ or}$$

$$\frac{x(3y - 2)}{2y + x}$$

112. $\dfrac{x + 2}{x + 1}$

113. *Writing Exercise*

114. *Writing Exercise*

115. Graph: $y = \left(\dfrac{3}{2}\right)^x$ Graph: $y = \log_{3/2} x$, or
$$x = \left(\dfrac{3}{2}\right)^y$$

x	y, or $\left(\dfrac{3}{2}\right)^x$
0	1
1	$\dfrac{3}{2}$
2	$\dfrac{9}{4}$
3	$\dfrac{27}{8}$
-1	$\dfrac{2}{3}$
-2	$\dfrac{4}{9}$

x, or $\left(\dfrac{3}{2}\right)^y$	y
1	0
$\dfrac{3}{2}$	1
$\dfrac{9}{4}$	2
$\dfrac{27}{8}$	3
$\dfrac{2}{3}$	-1
$\dfrac{4}{9}$	-2

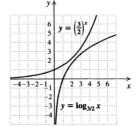

116.

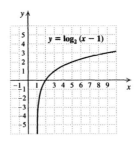

117. Graph: $y = \log_3 |x + 1|$

x	y
0	0
2	1
8	2
-2	0
-4	1
-9	2

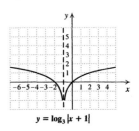

118. $\dfrac{1}{9}, 9$

119. $\log_{125} x = \dfrac{2}{3}$

$\qquad 125^{2/3} = x$

$\qquad (5^3)^{2/3} = x$

$\qquad 5^2 = x$

$\qquad 25 = x$

120. 6

121. $\log_8(2x+1) = -1$

$\qquad 8^{-1} = 2x+1$

$\qquad \dfrac{1}{8} = 2x+1$

$\qquad 1 = 16x+8 \qquad$ Multiplying by 8

$\qquad -7 = 16x$

$\qquad -\dfrac{7}{16} = x$

122. $-25, 4$

123. Let $\log_{1/4} \dfrac{1}{64} = x$. Then

$\qquad \left(\dfrac{1}{4}\right)^x = \dfrac{1}{64}$

$\qquad \left(\dfrac{1}{4}\right)^x = \left(\dfrac{1}{4}\right)^3$

$\qquad x = 3.$

Thus, $\log_{1/4} \dfrac{1}{64} = 3.$

124. -2

125. $\log_{81} 3 \cdot \log_3 81$

$\qquad = \dfrac{1}{4} \cdot 4 \qquad \left(\log_{81} 3 = \dfrac{1}{4}, \ \log_3 81 = 4\right)$

$\qquad = 1$

126. 0

127. $\log_2(\log_2(\log_4 256))$

$\qquad = \log_2(\log_2 4) \qquad (\log_4 256 = 4)$

$\qquad = \log_2 2 \qquad\qquad (\log_2 4 = 2)$

$\qquad = 1$

128. Let $b = 0$, $x = 1$, and $y = 2$. Then $0^1 = 0^2$, but $1 \neq 2$. Let $b = 1$, $x = 1$, and $y = 2$. Then $1^1 = 1^2$, but $1 \neq 2$.

129. *Writing Exercise*

Exercise Set 9.4

1. $\log_3 (81 \cdot 27) = \log_3 81 + \log_3 27$ Using the product rule

2. $\log_2 16 + \log_2 32$

3. $\log_4 (64 \cdot 16) = \log_4 64 + \log_4 16$ Using the product rule

4. $\log_5 25 + \log_5 125$

5. $\log_c rst$

$\quad = \log_c r + \log_c s + \log_c t$ Using the product rule

6. $\log_t 3 + \log_t a + \log_t b$

7. $\log_a 5 + \log_a 14 = \log_a (5 \cdot 14)$ Using the product rule

The result can also be expressed as $\log_a 70$.

8. $\log_b (65 \cdot 2)$, or $\log_b 130$

9. $\log_c t + \log_c y = \log_c (t \cdot y)$ Using the product rule

10. $\log_t (H \cdot M)$

11. $\log_a r^8 = 8 \log_a r$ Using the power rule

12. $5 \log_b t$

13. $\log_c y^6 = 6 \log_c y$ Using the power rule

14. $7 \log_{10} y$

15. $\log_b C^{-3} = -3 \log_b C$ Using the power rule

16. $-5 \log_c M$

17. $\log_2 \dfrac{53}{17} = \log_2 53 - \log_2 17$ Using the quotient rule

18. $\log_3 23 - \log_3 9$

19. $\log_b \dfrac{m}{n} = \log_b m - \log_b n$ Using the quotient rule

20. $\log_a y - \log_a x$

21. $\log_a 15 - \log_a 3$

$\quad = \log_a \dfrac{15}{3},$ Using the quotient rule

$\quad$ or $\log_a 5$

22. $\log_b \dfrac{42}{7}$, or $\log_b 6$

23. $\log_b 36 - \log_b 4$

$= \log_b \dfrac{36}{4},$ Using the quotient rule

or $\log_b 9$

24. $\log_a \dfrac{26}{2},$ or $\log_a 13$

25. $\log_a 7 - \log_z 18 = \log_a \dfrac{7}{18}$ Using the quotient rule

26. $\log_b \dfrac{5}{13}$

27. $\log_a x^5 y^7 z^6$

$= \log_a x^5 + \log_a y^7 + \log_a z^6$ Using the product rule

$= 5 \log_a x + 7 \log_a y + 6 \log_a z$ Using the power rule

28. $\log_a x + 4 \log_a y + 3 \log_a z$

29. $\log_b \dfrac{xy^2}{z^3}$

$= \log_b xy^2 - \log_b z^3$ Using the quotient rule

$= \log_b x + \log_b y^2 - \log_b z^3$ Using the product rule

$= \log_b x + 2 \log_b y - 3 \log_b z$ Using the power rule

30. $2 \log_b x + 5 \log_b y - 4 \log_b w - 7 \log_b z$

31. $\log_a \dfrac{x^4}{y^3 z}$

$= \log_a x^4 - \log_a y^3 z$ Using the quotient rule

$= \log_a x^4 - (\log_a y^3 + \log_a z)$ Using the product rule

$= \log_a x^4 - \log_a y^3 - \log_a z$ Removing parentheses

$= 4 \log_a x - 3 \log_a y - \log_a z$ Using the power rule

32. $4 \log_a x - \log_a y - 2 \log_a z$

33. $\log_b \dfrac{xy^2}{wz^3}$

$= \log_b xy^2 - \log_b wz^3$ Using the quotient rule

$= \log_b x + \log_b y^2 - (\log_b w + \log_b z^3)$ Using the product rule

$= \log_b x + \log_b y^2 - \log_b w - \log_b z^3$ Removing parentheses

$= \log_b x + 2 \log_b y - \log_b w - 3 \log_b z$ Using the power rule

34. $2 \log_b w + \log_b x - 3 \log_b y - \log_b z$

35. $\log_a \sqrt{\dfrac{x^7}{y^5 z^8}}$

$= \log_a \left(\dfrac{x^7}{y^5 z^8} \right)^{1/2}$

$= \dfrac{1}{2} \log_a \dfrac{x^7}{y^5 z^8}$ Using the power rule

$= \dfrac{1}{2}(\log_a x^7 - \log_a y^5 z^8)$ Using the quotient rule

$= \dfrac{1}{2}\left[\log_a x^7 - (\log_a y^5 + \log_a z^8) \right]$ Using the product rule

$= \dfrac{1}{2}(\log_a x^7 - \log_a y^5 - \log_a z^8)$ Removing parentheses

$= \dfrac{1}{2}(7 \log_a x - 5 \log_a y - 8 \log_a z)$ Using the power rule

36. $\dfrac{1}{3}(4 \log_c x - 3 \log_c y - 2 \log_c z)$

37. $\log_a \sqrt[3]{\dfrac{x^6 y^3}{a^2 z^7}}$

$= \log_a \left(\dfrac{x^6 y^3}{a^2 z^7} \right)^{1/3}$

$= \dfrac{1}{3} \log_a \dfrac{x^6 y^3}{a^2 z^7}$ Using the power rule

$= \dfrac{1}{3}(\log_a x^6 y^3 - \log_a a^2 z^7)$ Using the quotient rule

$= \dfrac{1}{3}[\log_a x^6 + \log_a y^3 - (\log_a a^2 + \log_a z^7)]$ Using the product rule

$= \dfrac{1}{3}(\log_a x^6 + \log_a y^3 - \log_a a^2 - \log_a z^7)$ Removing parentheses

$= \dfrac{1}{3}(\log_a x^6 + \log_a y^3 - 2 - \log_a z^7)$ 2 is the number to which we raise a to get a^2.

$= \dfrac{1}{3}(6 \log_a x + 3 \log_a y - 2 - 7 \log_a z)$ Using the power rule

38. $\dfrac{1}{4}(8 \log_a x + 12 \log_a y - 3 - 5 \log_a z)$

39. $7 \log_a x + 3 \log_a z$

$= \log_a x^7 + \log_a z^3$ Using the power rule

$= \log_a x^7 z^3$ Using the product rule

40. $\log_b m^2 n^{1/2},$ or $\log_b m^2 \sqrt{n}$

41. $\log_a x^2 - 2\log_a \sqrt{x}$

$= \log_a x^2 - \log_a (\sqrt{x})^2$ Using the power rule

$= \log_a x^2 - \log_a x$ $(\sqrt{x})^2 = x$

$= \log_a \dfrac{x^2}{x}$ Using the quotient rule

$= \log_a x$ Simplifying

42. $\log_a \dfrac{\sqrt{a}}{x}$

43. $\dfrac{1}{2}\log_a x + 5\log_a y - 2\log_a x$

$= \log_a x^{1/2} + \log_a y^5 - \log_a x^2$ Using the power rule

$= \log_a x^{1/2}y^5 - \log_a x^2$ Using the product rule

$= \log_a \dfrac{x^{1/2}y^5}{x^2}$ Using the quotient rule

The result can also be expressed as $\log_a \dfrac{\sqrt{x}y^5}{x^2}$ or as $\log_a \dfrac{y^5}{x^{3/2}}$.

44. $\log_a \dfrac{2x^4}{y^3}$

45. $\log_a(x^2 - 4) - \log_a(x + 2)$

$= \log_a \dfrac{x^2 - 4}{x + 2}$ Using the quotient rule

$= \log_a \dfrac{(x+2)(x-2)}{x+2}$

$= \log_a \dfrac{(\cancel{x+2})(x - 2)}{\cancel{x+2}}$ Simplifying

$= \log_a(x - 2)$

46. $\log_a \dfrac{2}{x - 5}$

47. $\log_b 15 = \log_b (3 \cdot 5)$

$= \log_b 3 + \log_b 5$ Using the product rule

$= 0.792 + 1.161$

$= 1.953$

48. 0.369

49. $\log_b \dfrac{3}{5} = \log_b 3 - \log_b 5$ Using the quotient rule

$= 0.792 - 1.161$

$= -0.369$

50. -0.792

51. $\log_b \dfrac{1}{5} = \log_b 1 - \log_b 5$ Using the quotient rule

$= 0 - 1.161$ $(\log_b 1 = 0)$

$= -1.161$

52. $\dfrac{1}{2}$

53. $\log_b \sqrt{b^3} = \log_b b^{3/2} = \dfrac{3}{2}$ 3/2 is the number to which we raise b to get $b^{3/2}$.

54. 1.792

55. $\log_b 6$

Since 6 cannot be expressed using the numbers 1, 3, and 5, we cannot find $\log_b 6$ using the given information.

56. 2.745

57. $\log_b 75$

$= \log_b(3 \cdot 5^2)$

$= \log_b 3 + \log_b 5^2$ Using the product rule

$= \log_b 3 + 2\log_b 5$ Using the power rule

$= 0.792 + 2(1.161)$

$= 3.114$

58. Cannot be found

59. $\log_t t^9 = 9$ 9 is the power to which we raise t to get t^9.

60. 4

61. $\log_e e^m = m$ m is the power to which we raise e to get e^m.

62. -2

63. $\log_5 125 = 3$ and $\log_5 625 = 4$, so $\log_5 (125 \cdot 625) = 3 + 4 = 7$.

64. 6

65. $\log_2 128 = 7$ and $\log_2 16 = 4$, so $\log_2 \left(\dfrac{128}{16}\right) = 7 - 4 = 3$.

66. 2

67. *Writing Exercise*

68. *Writing Exercise*

69. Graph $f(x) = \sqrt{x} - 3$.

We construct a table of values, plot points, and connect them with a smooth curve. Note that we must choose nonnegative values of x in order for $\sqrt{x}$ to be a real number.

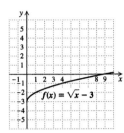

x	$f(x)$
0	-3
1	-2
4	-1
9	0

70.

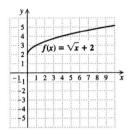

71. Graph $g(x) = \sqrt[3]{x} + 1$.

We construct a table of values, plot points, and connect them with a smooth curve.

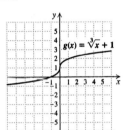

x	$g(x)$
-8	-1
-1	0
0	1
1	2
8	3

72.

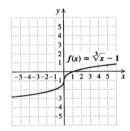

73. $(a^3b^2)^5(a^2b^7) = (a^{3\cdot5}b^{2\cdot5})(a^2b^7) =$
$a^{15}b^{10}a^2b^7 = a^{15+2}b^{10+7} = a^{17}b^{17}$

74. $x^{11}y^6z^8$

75. *Writing Exercise*

76. *Writing Exercise*

77. $\log_a (x^8 - y^8) - \log_a (x^2 + y^2)$

$= \log_a \dfrac{x^8 - y^8}{x^2 + y^2}$

$= \log_a \dfrac{(x^4 + y^4)(x^2 + y^2)(x + y)(x - y)}{x^2 + y^2}$

$= \log_a [(x^4 + y^4)(x^2 - y^2)] \qquad \text{Simplifying}$

$= \log_a (x^6 - x^4y^2 + x^2y^4 - y^6)$

78. $\log_a (x^3 + y^3)$

79. $\log_a \sqrt{1 - s^2}$

$= \log_a (1 - s^2)^{1/2}$

$= \dfrac{1}{2} \log_a (1 - s^2)$

$= \dfrac{1}{2} \log_a [(1 - s)(1 + s)]$

$= \dfrac{1}{2} \log_a (1 - s) + \dfrac{1}{2} \log_a (1 + s)$

80. $\dfrac{1}{2} \log_a (c - d) - \dfrac{1}{2} \log_a (c + d)$

81. $\log_a \dfrac{\sqrt[3]{x^2 z}}{\sqrt[3]{y^2 z^{-2}}}$

$= \log_a \left(\dfrac{x^2 z^3}{y^2} \right)^{1/3}$

$= \dfrac{1}{3} (\log_a x^2 z^3 - \log_a y^2)$

$= \dfrac{1}{3} (2 \log_a x + 3 \log_a z - 2 \log_a y)$

$= \dfrac{1}{3} [2 \cdot 2 + 3 \cdot 4 - 2 \cdot 3]$

$= \dfrac{1}{3} (10)$

$= \dfrac{10}{3}$

82. -2

83. $\log_a x = 2$, so $a^2 = x$.

Let $\log_{1/a} x = n$ and solve for n.

$\log_{1/a} a^2 = n \qquad \text{Substituting } a^2 \text{ for } x$

$\left(\dfrac{1}{a} \right)^n = a^2$

$(a^{-1})^n = a^2$

$a^{-n} = a^2$

$-n = 2$

$n = -2$

Thus, $\log_{1/a} x = -2$ when $\log_a x = 2$.

84. False

85. True; $\log_a(Q + Q^2) = \log_a[Q(1 + Q)] = \log_a Q + \log_a(1 + Q) = \log_a Q + \log_a(Q + 1)$.

86. Graph $y_1 = \log x^2$ and $y_2 = \log x \cdot \log x$ and observe that the graphs do not coincide.

Exercise Set 9.5

1. 1.6094

2. 0.6931

3. 3.9512

4. 3.4012

5. -5.0832

6. -7.2225

7. 96.7583

8. 107.8516

9. 0.7850

10. -0.3939

11. 1.3877

12. 1.8199

13. 15.0293

14. 21.3276

15. 0.0305

16. 0.0714

17. 109.9472

18. 3.4212

19. We will use common logarithms for the conversion. Let $a = 10$, $b = 6$, and $M = 92$ and substitute in the change-of-base formula.

$$\log_b M = \frac{\log_a M}{\log_a b}$$

$$\log_6 92 = \frac{\log_{10} 92}{\log_{10} 6}$$

$$\approx \frac{1.963787827}{0.7781512504}$$

$$\approx 2.5237$$

20. 3.9656

21. We will use common logarithms for the conversion. Let $a = 10$, $b = 2$, and $M = 100$ and substitute in the change-of-base formula.

$$\log_2 100 = \frac{\log_{10} 100}{\log_{10} 2}$$

$$\approx \frac{2}{0.3010}$$

$$\approx 6.6439$$

22. 2.3666

23. We will use natural logarithms for the conversion. Let $a = e$, $b = 7$, and $M = 65$ and substitute in the change-of-base formula.

$$\log_7 65 = \frac{\ln 65}{\ln 7}$$

$$\approx \frac{4.1744}{1.9459}$$

$$\approx 2.1452$$

24. 2.3223

25. We will use natural logarithms for the conversion. Let $a = e$, $b = 0.5$, and $M = 5$ and substitute in the change-of-base formula.

$$\log_{0.5} 5 = \frac{\ln 5}{\ln 0.5}$$

$$\approx \frac{1.6094}{-0.6931}$$

$$\approx -2.3219$$

26. -0.4771

27. We will use common logarithms for the conversion. Let $a = 10$, $b = 2$, and $M = 0.2$ and substitute in the change-of-base formula.

$$\log_2 0.2 = \frac{\log_{10} 0.2}{\log_{10} 2}$$

$$\approx \frac{-0.6990}{0.3010}$$

$$\approx -2.3219$$

28. -3.6439

29. We will use natural logarithms for the conversion. Let $a = e$, $b = \pi$, and $M = 58$ and substitute in the change-of-base formula.

$$\log_\pi 58 = \frac{\ln 58}{\ln \pi}$$

$$\approx \frac{4.0604}{1.1447}$$

$$\approx 3.5471$$

30. 4.6284

31. Graph: $f(x) = e^x$

We find some function values with a calculator. We use these values to plot points and draw the graph.

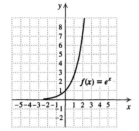

x	e^x
0	1
1	2.7
2	7.4
3	20.1
-1	0.4
-2	0.1

The domain is the set of real numbers and the range is $(0, \infty)$.

32.

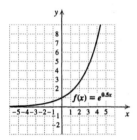

The domain is the set of real numbers and the range is $(0, \infty)$.

33. Graph: $f(x) = e^{-0.4x}$

We find some function values, plot points, and draw the graph.

x	$e^{-0.4x}$
0	1
1	0.67
2	0.45
-1	1.49
-2	2.23
-3	3.32
-4	4.95

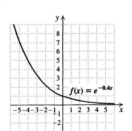

The domain is the set of real numbers and the range is $(0, \infty)$.

34.

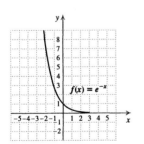

The domain is the set of real numbers and the range is $(0, \infty)$.

35. Graph: $f(x) = e^x + 1$

We find some function values, plot points, and draw the graph.

x	$e^x + 1$
0	2
1	3.72
2	8.39
-1	1.37
-2	1.14

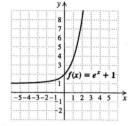

The domain is the set of real numbers and the range is $(1, \infty)$.

36.

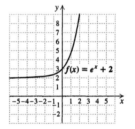

The domain is the set of real numbers and the range is $(2, \infty)$.

37. Graph: $f(x) = e^x - 2$

We find some function values, plot points, and draw the graph.

x	$e^x - 2$
0	-1
1	0.72
2	5.4
-1	-1.6
-2	-1.9

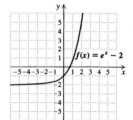

The domain is the set of real numbers and the range is $(-2, \infty)$.

38.

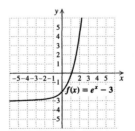

The domain is the set of real numbers and the range is $(-3, \infty)$.

39. Graph: $f(x) = 0.5e^x$

We find some function values, plot points, and draw the graph.

x	$0.5e^x$
0	0.5
1	1.36
2	3.69
-1	0.18
-2	0.07

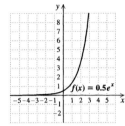

The domain is the set of real numbers and the range is $(0, \infty)$.

40.

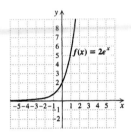

The domain is the set of real numbers and the range is $(0, \infty)$.

41. Graph: $f(x) = 2e^{-0.5x}$

We find some function values, plot points, and draw the graph.

x	$2e^{-0.5x}$
0	2
1	1.21
2	0.74
3	0.45
-1	3.30
-2	5.44
-3	8.96

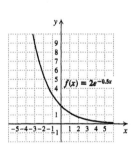

The domain is the set of real numbers and the range is $(0, \infty)$.

42.

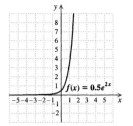

The domain is the set of real numbers and the range is $(0, \infty)$.

43. Graph: $f(x) = e^{x-2}$

We find some function values, plot points, and draw the graph.

x	e^{x-2}
0	0.14
2	1
4	7.39
-1	0.05
-2	0.02

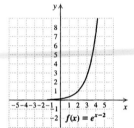

The domain is the set of real numbers and the range is $(0, \infty)$.

44.

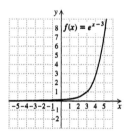

The domain is the set of real numbers and the range is $(0, \infty)$.

45. Graph: $f(x) = e^{x+3}$

We find some function values, plot points, and draw the graph.

x	e^{x+3}
0	20.09
1	54.60
-1	7.39
-3	1
-4	0.37

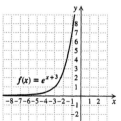

The domain is the set of real numbers and the range is $(0, \infty)$.

46.

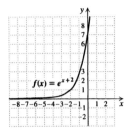

The domain is the set of real numbers and the range is $(0, \infty)$.

47. Graph: $f(x) = 2 \ln x$

x	$2 \ln x$
0.5	-1.4
1	0
2	1.4
3	2.2
4	2.8
5	3.2
6	3.6

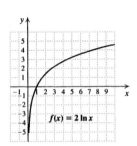

The domain is $(0, \infty)$ and the range is the set of real numbers.

48.

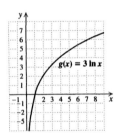

The domain is $(0, \infty)$ and the range is the set of real numbers.

49. Graph: $f(x) = 0.5 \ln x$

x	$0.5 \ln x$
0.5	-0.35
1	0
2	0.35
3	0.55
4	0.69
5	0.80

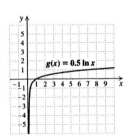

The domain is $(0, \infty)$ and the range is the set of real numbers.

50.

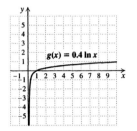

The domain is $(0, \infty)$ and the range is the set of real numbers.

51. Graph: $g(x) = \ln x + 3$

x	$\ln x + 3$
1	3
2	3.69
3	4.10
4	4.39
5	4.61

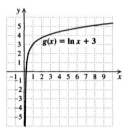

The domain is $(0, \infty)$ and the range is the set of real numbers.

52.

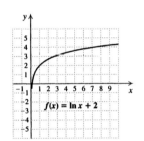

The domain is $(0, \infty)$ and the range is the set of real numbers.

53. Graph: $g(x) = \ln x - 2$

x	$\ln x - 2$
1	-2
2	-1.31
3	-0.90
4	-0.61
5	-0.39

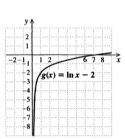

The domain is $(0, \infty)$ and the range is the set of real numbers.

54.

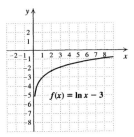

The domain is $(0, \infty)$ and the range is the set of real numbers.

55. Graph: $f(x) = \ln(x + 1)$

We find some function values, plot points, and draw the graph.

x	$\ln(x + 1)$
0	0
1	0.69
2	1.10
4	1.61
6	1.95
-1	Undefined

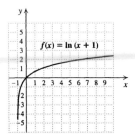

The domain is $(-1, \infty)$ and the range is the set of real numbers.

56.

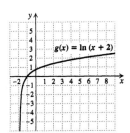

The domain is $(-2, \infty)$ and the range is the set of real numbers.

57. Graph: $g(x) = \ln(x - 3)$

We find some function values, plot points, and draw the graph.

x	$\ln(x - 3)$
3.1	-2.30
4	0
5	0.69
6	1.10
7	1.39

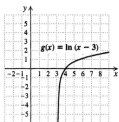

The domain is $(3, \infty)$ and the range is the set of real numbers.

58.

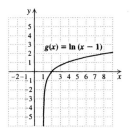

The domain is $(1, \infty)$ and the range is the set of real numbers.

59. We use the change of base formula:
$$f(x) = \frac{\log x}{\log 5} \text{ or } f(x) = \frac{\ln x}{\ln 5}$$

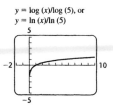

60.

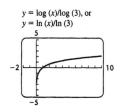

61. We use the change of base formula.
$$f(x) = \frac{\log(x - 5)}{\log 2} \text{ or } f(x) = \frac{\ln(x - 5)}{\ln 2}$$

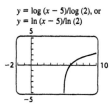

62.

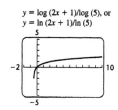

63. We use the change of base formula.

$$f(x) = \frac{\log x}{\log 3} + x \text{ or } f(x) = \frac{\ln x}{\ln 3} + x$$

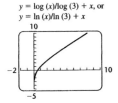
$y = \log (x)/\log (3) + x$, or
$y = \ln (x)/\ln (3) + x$

64.

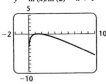
$y = \log (x)/\log (2) - x + 1$, or
$y = \ln (x)/\ln (2) - x + 1$

65. *Writing Exercise*

66. *Writing Exercise*

67.
$$4x^2 - 25 = 0$$
$$(2x + 5)(2x - 5) = 0$$
$$2x + 5 = 0 \quad or \quad 2x - 5 = 0$$
$$2x = -5 \quad or \quad 2x = 5$$
$$x = -\frac{5}{2} \quad or \quad x = \frac{5}{2}$$

The solutions are $-\frac{5}{2}$ and $\frac{5}{2}$.

68. $0, \dfrac{7}{5}$

69.
$$17x - 15 = 0$$
$$17x = 15$$
$$x = \frac{15}{17}$$

The solution is $\dfrac{15}{17}$.

70. $\dfrac{9}{13}$

71. $x^{1/2} - 6x^{1/4} + 8 = 0$

Let $u = x^{1/4}$.

$$u^2 - 6u + 8 = 0 \qquad \text{Substituting}$$
$$(u - 4)(u - 2) = 0$$
$$u = 4 \quad or \quad u = 2$$
$$x^{1/4} = 4 \quad or \quad x^{1/4} = 2$$
$$x = 256 \quad or \quad x = 16 \quad \text{Raising both sides to the fourth power}$$

Both numbers check. The solutions are 256 and 16.

72. $\dfrac{1}{4}, 9$

73. *Writing Exercise*

74. *Writing Exercise*

75. We use the change-of-base formula.

$$\log_6 81 = \frac{\log 81}{\log 6}$$
$$= \frac{\log 3^4}{\log(2 \cdot 3)}$$
$$= \frac{4 \log 3}{\log 2 + \log 3}$$
$$\approx \frac{4(0.477)}{0.301 + 0.477}$$
$$\approx 2.452$$

76. 1.262

77. We use the change-of-base formula.

$$\log_{12} 36 = \frac{\log 36}{\log 12}$$
$$= \frac{\log(2 \cdot 3)^2}{\log(2^2 \cdot 3)}$$
$$= \frac{2 \log(2 \cdot 3)}{\log 2^2 + \log 3}$$
$$= \frac{2(\log 2 + \log 3)}{2 \log 2 + \log 3}$$
$$\approx \frac{2(0.301 + 0.477)}{2(0.301) + 0.477}$$
$$\approx 1.442$$

78. $\ln M = \dfrac{\log M}{\log e}$

79. Use the change-of-base formula with $a = e$ and $b = 10$. We obtain

$$\log M = \frac{\ln M}{\ln 10}.$$

80. $\pm 6.0302 \times 10^{17}$

81.
$$\log(492x) = 5.728$$
$$10^{5.728} = 492x$$
$$\frac{10^{5.728}}{492} = x$$
$$1086.5129 \approx x$$

82. 1.5893

83. $\log 692 + \log x = \log 3450$

$$\log x = \log 3450 - \log 692$$

$$\log x = \log \frac{3450}{692}$$

$$x = \frac{3450}{692}$$

$$x \approx 4.9855$$

84. (a) Domain: $\{x|x > 0\}$, or $(0, \infty)$; range: the set of real numbers;

(b) $[-3, 10, -100, 1000]$, Xscl $= 1$, Yscl $= 100$;

(c)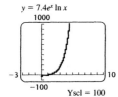

85. (a) Domain: $\{x|x > 0\}$, or $(0, \infty)$; range: $\{y|y < 0.5135\}$, or $(-\infty, 0.5135)$;

(b) $[-1, 5, -10, 5]$;

(c)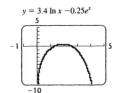

86. (a) Domain: $\{x|x > 2.1\}$, or $(2.1, \infty)$; range: the set of real numbers;

(b) $[-1, 10, -10, 20]$, Xscl $= 1$, Yscl $= 5$;

(c)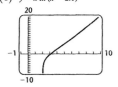

87. (a) Domain $\{x|x > 0\}$, or $(0, \infty)$; range: $\{y|y > -0.2453\}$, or $(-0.2453, \infty)$

(b) $[-1, 5, -1, 10]$;

(c)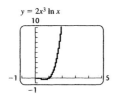

Exercise Set 9.6

1. $2^x = 16$

$$2^x = 2^4$$

$$x = 4 \quad \text{The exponents must be the same.}$$

The solution is 4.

2. 3

3. $3^x = 27$

$$3^x = 3^3$$

$$x = 3 \quad \text{The exponents must be the same.}$$

The solution is 3.

4. 3

5. $2^{x+3} = 32$

$$2^{x+3} = 2^5$$

$$x + 3 = 5$$

$$x = 2$$

The solution is 2.

6. 5

7. $5^{3x} = 625$

$$5^{3x} = 5^4$$

$$3x = 4$$

$$x = \frac{4}{3}$$

The solution is $\frac{4}{3}$.

8. $\frac{3}{2}$

9. $7^{4x} = 1$

Since $a^0 = 1 \ a \neq 0$, then $4x = 0$ and thus $x = 0$. The solution is 0.

10. 0

11. $4^{2x-1} = 64$

$$4^{2x-1} = 4^3$$

$$2x - 1 = 3$$

$$2x = 4$$

$$x = 2$$

The solution is 2.

12. $\frac{5}{2}$

13.
$$3^{x^2} \cdot 3^{3x} = 81$$
$$3^{x^2+3x} = 3^4$$
$$x^2 + 3x = 4$$
$$x^2 + 3x - 4 = 0$$
$$(x+4)(x-1) = 0$$
$$x = -4 \;\; or \;\; x = 1$$

The solutions are -4 and 1.

14. $-3, -1$

15.
$$2^x = 15$$
$$\log 2^x = \log 15$$
$$x \log 2 = \log 15$$
$$x = \frac{\log 15}{\log 2}$$
$$x \approx 3.907$$

The solution is $\log 15 / \log 2$, or approximately 3.907.

16. $\dfrac{\log 19}{\log 2} \approx 4.248$

17.
$$4^{x+1} = 13$$
$$\log 4^{x+1} = \log 13$$
$$(x+1)\log 4 = \log 13$$
$$x + 1 = \frac{\log 13}{\log 4}$$
$$x = \frac{\log 13}{\log 4} - 1$$
$$x \approx 0.850$$

The solution is $\log 13 / \log 4 - 1$ or approximately 0.850.

18. $\dfrac{\log 17}{\log 8} + 1 \approx 2.362$

19.
$$e^t = 100$$

$\ln e^t = \ln 100$	Taking ln on both sides
$t = \ln 100$	Finding the logarithm of the base to a power
$t \approx 4.605$	Using a calculator

20. $\ln 1000 \approx 6.908$

21.
$$e^{-0.07t} + 3 = 3.08$$
$$e^{-0.07t} = 0.08$$

$\ln e^{-0.07t} = \ln 0.08$	Taking ln on both sides
$-0.07t = \ln 0.08$	Finding the logarithm of the base to a power
$t = \dfrac{\ln 0.08}{-0.07}$	
$t \approx 36.082$	

22. $\dfrac{\ln 5}{0.03} \approx 53.648$

23.
$$2^x = 3^{x-1}$$
$$\log 2^x = \log 3^{x-1}$$
$$x \log 2 = (x-1) \log 3$$
$$x \log 2 = x \log 3 - \log 3$$
$$\log 3 = x \log 3 - x \log 2$$
$$\log 3 = x(\log 3 - \log 2)$$
$$\frac{\log 3}{\log 3 - \log 2} = x$$
$$2.710 \approx x$$

24. $\dfrac{\log 3}{\log 5 - \log 3} \approx 2.151$

25.
$$7.2^x - 65 = 0$$
$$7.2^x = 65$$
$$\log 7.2^x = \log 65$$
$$x \log 7.2 = \log 65$$
$$x = \frac{\log 65}{\log 7.2}$$
$$x \approx 2.115$$

26. $\dfrac{\log 87}{\log 4.9} \approx 2.810$

27. $e^{0.5x} - 7 = 2x + 6$

Graph $y_1 = e^{0.5x} - 7$ and $y_2 = 2x + 6$ in a window that shows the points of intersection of the graphs. One good choice is $[-10, 10, -10, 25]$, $Yscl = 5$. Use Intersect to find the first coordinates of the points of intersection. They are the solutions of the given equation. They are about -6.480 and 6.519.

28. -1.873

29.
$$\log_5 x = 3$$

$x = 5^3$	Writing an equivalent exponential equation
$x = 125$	

30. 81

31.
$$\log_4 x = \frac{1}{2}$$

$x = 4^{1/2}$	Writing an equivalent exponential equation
$x = 2$	

32. $\dfrac{1}{8}$

33. $\log x = 3$ The base is 10.

$$x = 10^3$$
$$x = 1000$$

34. 10

35. $2 \log x = -8$

$$\log x = -4 \qquad \text{The base is 10.}$$
$$x = 10^{-4}$$
$$x = \frac{1}{10,000}, \text{ or } 0.0001$$

36. $\dfrac{1}{10,000}$

37. $\ln x = 1$

$$x = e \approx 2.718$$

38. $e^2 \approx 7.389$

39. $5 \ln x = -15$

$$\ln x = -3$$
$$x = e^{-3} \approx 0.050$$

40. $e^{-1} \approx 0.368$

41. $\log_2(8 - 6x) = 5$

$$8 - 6x = 2^5$$
$$8 - 6x = 32$$
$$-6x = 24$$
$$x = -4$$

The answer checks. The solution is -4.

42. 66

43. $\log(x - 9) + \log x = 1 \qquad$ The base is 10.

$$\log_{10}[(x-9)(x)] = 1 \quad \text{Using the product rule}$$
$$x(x - 9) = 10^1$$
$$x^2 - 9x = 10$$
$$x^2 - 9x - 10 = 0$$
$$(x + 1)(x - 10) = 0$$
$$x = -1 \text{ or } x = 10$$

Check: For -1:

$$\frac{\log(x - 9) + \log x = 1}{\log(-1 + 9) + \log(-1) \text{ ? } 1 \qquad \text{FALSE}}$$

For 10:

$$\frac{\log(x - 9) + \log x = 1}{\begin{array}{c} \log(10 - 9) + \log(10) \text{ ? } 1 \\ \log 1 + \log 10 \\ 0 + 1 \\ 1 \end{array} \bigg| 1 \quad \text{TRUE}}$$

The number -1 does not check, because negative numbers do not have logarithms. The solution is 10.

44. 1

45. $\log x - \log(x + 3) = 1 \quad$ The base is 10.

$$\log_{10} \frac{x}{x + 3} = 1 \quad \text{Using the quotient rule}$$
$$\frac{x}{x + 3} = 10^1$$
$$x = 10(x + 3)$$
$$x = 10x + 30$$
$$-9x = 30$$
$$x = -\frac{10}{3}$$

The number $-\dfrac{10}{3}$ does not check. The equation has no solution.

46. $\dfrac{7}{9}$

47. $\log_4(x + 3) - \log_4(x - 5) = 2$

$$\log_4 \frac{x + 3}{x - 5} = 2 \quad \text{Using the quotient rule}$$
$$\frac{x + 3}{x - 5} = 4^2$$
$$\frac{x + 3}{x - 5} = 16$$
$$x + 3 = 16(x - 5)$$
$$x + 3 = 16x - 80$$
$$83 = 15x$$
$$\frac{83}{15} = x$$

The number $\dfrac{83}{15}$ checks. It is the solution.

48. 5

49. $\log_7(x + 2) + \log_7(x + 1) = \log_7 6$

$$\log_7[(x + 2)(x + 1)] = \log_7 6 \quad \text{Using the product rule}$$
$$\log_7(x^2 + 3x + 2) = \log_7 6$$
$$x^2 + 3x + 2 = 6 \qquad \text{Using the property of logarithmic equality}$$
$$x^2 + 3x - 4 = 0$$
$$(x + 4)(x - 1) = 0$$
$$x = -4 \ \text{ or } \ x = 1$$

The number 1 checks, but -4 does not. The solution is 1.

50. 2

51. $\log_3(x+4) + \log_3(x-4) = 2$

$$\log_3[(x+4)(x-4)] = 2$$
$$(x+4)(x-4) = 3^2$$
$$x^2 - 16 = 9$$
$$x^2 = 25$$
$$x = \pm 5$$

The number 5 checks, but -5 does not. The solution is 5.

52. 4

53. $\log_{12}(x+5) - \log_{12}(x-4) = \log_{12} 3$

$$\log_{12} \frac{x+5}{x-4} = \log_{12} 3$$
$$\frac{x+5}{x-4} = 3 \quad \text{Using the property of logarithmic equality}$$
$$x + 5 = 3(x-4)$$
$$x + 5 = 3x - 12$$
$$17 = 2x$$
$$\frac{17}{2} = x$$

The number $\frac{17}{2}$ checks and is the solution.

54. $\frac{17}{4}$

55. $\log_2(x-2) + \log_2 x = 3$

$$\log_2[(x-2)(x)] = 3$$
$$x(x-2) = 2^3$$
$$x^2 - 2x = 8$$
$$x^2 - 2x - 8 = 0$$
$$(x-4)(x+2) = 0$$
$$x = 4 \quad or \quad x = -2$$

The number 4 checks, but -2 does not. The solution is 4.

56. $\frac{2}{5}$

57. $\ln(3x) = 3x - 8$

Graph $y_1 = \ln(3x)$ and $y_2 = 3x - 8$ in a window that shows the points of intersection of the graphs. One good choice is $[-5, 5, -15, 5]$. When we use Intersect in this window we can find only the coordinates of the right-hand point of intersection. They are about $(3.445, 2.336)$, so one solution of the equation is about 3.445. To find the coordinates of the left-hand point of intersection we make the window smaller. One window that is appropriate is $[-1, 1, -15, 5]$. Using Intersect again we find that the other solution of the equation is about 0.0001. (The answer approximated to the nearest thousandth is 0.000, so we express it to the nearest ten-thousandth.)

58. $-0.753, 0.753$

59. Solve $\ln x = \log x$.

Graph $y_1 = \ln x$ and $y_2 = \log x$ in a window that shows the point of intersection of the graphs. One good choice is $[-5, 5, 5, 5]$. Use Intersect to find the first coordinate of the point of intersection. It is the solution of the given equation. It is 1.

60. 1, 100

61. *Writing Exercise*

62. *Writing Exercise*

63. The number of transplants performed has increased at a constant rate, so these data can be modeled better by a linear function than a quadratic function.

64. Quadratic

65. Using the data points $(0, 16)$ and $(10, 24)$, we first find the slope.

$$m - \frac{24 - 16}{10 - 0} = \frac{8}{10} = \frac{4}{5}$$

The y-intercept is $(0, 16)$, so we have $t(x) = \frac{4}{5}x + 16$, where x is the number of years after 1991 and t is in thousands.

66. $p(x) = \frac{1}{6}x^2 + \frac{23}{6}x + 25$

67.
$$\frac{1}{6}x^2 + \frac{23}{6}x + 25 = 5\left(\frac{4}{5}x + 16\right)$$
$$\frac{1}{6}x^2 + \frac{23}{6}x + 25 = 4x + 80$$
$$6\left(\frac{1}{6}x^2 + \frac{23}{6}x + 25\right) = 6(4x + 80)$$
$$x^2 + 23x + 150 = 24x + 480$$
$$x^2 - x - 330 = 0$$

We use the quadratic formula with $a = 1$, $b = -1$, and $c = -330$.

$$x = \frac{-b \pm \sqrt{b^2 - 4ac}}{2a}$$
$$= \frac{-(-1) \pm \sqrt{(-1)^2 - 4 \cdot 1 \cdot (-330)}}{2 \cdot 1}$$
$$= \frac{1 \pm \sqrt{1321}}{2} \approx \frac{1 \pm 36.35}{2}$$
$$x \approx \frac{1 + 36.35}{2} \quad or \quad x \approx \frac{1 - 36.35}{2}$$
$$x \approx 19 \quad\quad or \quad x \approx -18$$

Only 19 has meaning in the original problem. Thus, there will be five times as many patients on the waiting list as there are transplants performed about 19 years after 1991, or in 2010.

68. $t(x) = 0.7571428571x + 15.71428571$, where x is the number of years after 1991 and t is in thousands.

69. Enter the data and then use the quadratic regression feature. We have $p(x) = 0.0892857143x^2 + 4.907142857x + 23.85714286$, where x is the number of years after 1991 and p is in thousands.

70. 2010

71. *Writing Exercise*

72. *Writing Exercise*

73.
$$100^{3x} = 1000^{2x+1}$$
$$(10^2)^{3x} = (10^3)^{2x+1}$$
$$10^{6x} = 10^{6x+1}$$
$$6x = 6x + 1$$
$$0 = 1$$
We get a false equation, so the equation has no solution.

74. $\dfrac{12}{5}$

75.
$$8^x = 16^{3x+9}$$
$$(2^3)^x = (2^4)^{3x+9}$$
$$2^{3x} = 2^{12x+36}$$
$$3x = 12x + 36$$
$$-36 = 9x$$
$$-4 = x$$
The solution is -4.

76. $\sqrt[3]{3}$

77.
$$\log_6 (\log_2 x) = 0$$
$$\log_2 x = 6^0$$
$$\log_2 x = 1$$
$$x = 2^1$$
$$x = 2$$
The solution is 2.

78. -1

79.
$$\log_5 \sqrt{x^2 - 9} = 1$$
$$\sqrt{x^2 - 9} = 5^1$$
$$(\sqrt{x^2 - 9})^2 = 5^2$$
$$x^2 - 9 = 25$$
$$x^2 = 34$$
$$x = \pm\sqrt{34}$$
The solutions are $\pm\sqrt{34}$.

80. $-3, -1$

81.
$$\log (\log x) = 5$$
$$\log x = 10^5$$
$$\log x = 100,000$$
$$x = 10^{100,000}$$
The solution is $10^{100,000}$.

82. $-625, 625$

83.
$$\log x^2 = (\log x)^2$$
$$2 \log x = (\log x)^2$$
$$0 = (\log x)^2 - 2 \log x$$
Let $u = \log x$.
$$0 = u^2 - 2u$$
$$0 = u(u - 2)$$
$$u = 0 \quad or \quad u = 2$$
$$\log x = 0 \quad or \quad \log x = 2 \quad \text{Replacing } u \text{ with } \log x$$
$$x = 10^0 \quad or \quad x = 10^2$$
$$x = 1 \quad or \quad x = 100$$
Both numbers check. The solutions are 1 and 100.

84. $\dfrac{1}{2}$, 5000

85.
$$\log x^{\log x} = 25$$
$$\log x (\log x) = 25 \quad \text{Using the power rule}$$
$$(\log x)^2 = 25$$
$$\log x = \pm 5$$
$$x = 10^5 \quad or \quad x = 10^{-5}$$
$$x = 100,000 \quad or \quad x = \frac{1}{100,000}$$
Both numbers check. The solutions are 100,000 and $\dfrac{1}{100,000}$.

86. $1, \dfrac{\log 5}{\log 3} \approx 1.465$

87.
$$(81^{x-2})(27^{x+1}) = 9^{2x-3}$$
$$[(3^4)^{x-2}][(3^3)^{x+1}] = (3^2)^{2x-3}$$
$$(3^{4x-8})(3^{3x+3}) = 3^{4x-6}$$
$$3^{7x-5} = 3^{4x-6}$$
$$7x - 5 = 4x - 6$$
$$3x = -1$$
$$x = -\frac{1}{3}$$
The solution is $-\dfrac{1}{3}$.

88. $\dfrac{3}{2}$

89. $2^y = 16^{x-3}$ and $3^{y+2} = 27^x$

$2^y = (2^4)^{x-3}$ and $3^{y+2} = (3^3)^x$

$y = 4x - 12$ and $y + 2 = 3x$

$12 = 4x - y$ and $2 = 3x - y$

Solving this system of equations we get $x = 10$ and $y = 28$.
Then $x + y = 10 + 28 = 38$.

90. -3

91. Set $S(x) = D(x)$, and solve for x.

$e^x = 162,755\, e^{-x}$

$e^{2x} = 162,755$ Multiplying by e^x on
                         both sides

$\ln e^{2x} = \ln 162,755$

$2x = \ln 162,755$

$x = \dfrac{\ln 162,755}{2}$

$x \approx 6$

To find the second coordinate of the equilibrium point, find
$S(6)$ or $D(6)$. We will find $S(6)$.

$S(6) = e^6 \approx 403$

The equilibrium point is $(6, \$403)$.

Exercise Set 9.7

1. a) Replace $N(t)$ with 200 and solve for t.

$N(t) = 153(1.37)^t$

$200 = 153(1.37)^t$

$1.3072 \approx (1.37)^t$ Dividing by 153

$\ln 1.3072 \approx \ln(1.37)^t$ Taking the natural
                                  logarithm on both sides

$\ln 1.3072 \approx t \ln 1.37$

$\dfrac{\ln 1.3072}{\ln 1.37} \approx t$

$1 \approx t$

200 million cellular phones would be in use 1 yr after
2002, or in 2003.

b) Replace $N(t)$ with 2(153), or 306, and solve for t.

$306 = 153(1.37)^t$

$2 = (1.37)^t$

$\ln 2 = \ln(1.37)^t$

$\ln 2 = t \ln(1.37)$

$\dfrac{\ln 2}{\ln 1.37} = t$

$2.2 \approx t$

The doubling time is about 2.2 years.

2. a) 13.5 yr

b) 5.7 yr

3. a) Replace $A(t)$ with 40,000 and solve for t.

$A(t) = 29,000(1.08)^t$

$40,000 = 29,000(1.08)^t$

$1.379 \approx (1.08)^t$

$\log 1.379 \approx \log(1.08)^t$

$\log 1.379 \approx t \log 1.08$

$\dfrac{\log 1.379}{\log 1.08} \approx t$

$4.2 \approx t$

The amount due will reach $40,000 after about
4.2 years.

b) Replace $A(t)$ with 2(29,000), or 58,000, and solve
for t.

$58,000 = 29,000(1.08)^t$

$2 = (1.08)^t$

$\log 2 = \log(1.08)^t$

$\log 2 = t \log 1.08$

$\dfrac{\log 2}{\log 1.08} = t$

$9.0 \approx t$

The doubling time is about 9.0 years.

4. a) 3.6 days

b) 0.6 days

5. a) Find $N(41)$.

$N(x) = 600(0.873)^{x-16}$

$N(41) = 600(0.873)^{41-16}$

$= 600(0.873)^{25}$

≈ 20.114

There are about 20.114 thousand, or 20,114,
41-yr-old skateboarders.

b) Substitute 2 for $N(x)$ and solve for x. (Remember
that $N(x)$ is in thousands.)

$2 = 600(0.873)^{x-16}$

$0.0033 \approx (0.873)^{x-16}$

$\log 0.0033 \approx (x - 16) \log 0.873$

$\dfrac{\log 0.0033}{\log 0.873} \approx x - 16$

$\dfrac{\log 0.0033}{\log 0.873} + 16 \approx x$

$58 \approx x$

There are only 2000 skateboarders at age 58.

6. a) 3.5 yr

b) 13.6 yr

7. pH $= -\log[H^+]$

$= -\log[1.3 \times 10^{-5}]$

$\approx -(-4.886057)$ Using a calculator

≈ 4.9

The pH of fresh-brewed coffee is about 4.9.

8. 6.8

9.   pH $= -\log[H^+]$

$7.0 = -\log[H^+]$

$-7.0 = \log[H^+]$

$10^{-7.0} = [H^+]$ Converting to an exponential equation

The hydrogen ion concentration is 10^{-7} moles per liter.

10. 1.58×10^{-8} moles per liter

11. $L = 10 \cdot \log \dfrac{I}{I_0}$

$= 10 \cdot \log \dfrac{3.2 \times 10^{-6}}{10^{-12}}$

$= 10 \cdot \log(3.2 \times 10^6)$

$\approx 10(6.5)$

≈ 65

The intensity of sound in normal conversation is about 65 decibels.

12. 95 dB

13. $L = 10 \cdot \log \dfrac{I}{I_0}$

$105 = 10 \cdot \log \dfrac{I}{10^{-12}}$

$10.5 = \log \dfrac{I}{10^{-12}}$

$10.5 = \log I - \log 10^{-12}$ Using the quotient rule

$10.5 = \log I - (-12)$ $(\log 10^a = a)$

$10.5 = \log I + 12$

$-1.5 = \log I$

$10^{-1.5} = I$ Converting to an exponential equation

$3.2 \times 10^{-2} \approx I$

The intensity of the sound is $10^{-1.5}$ W/m^2, or about 3.2×10^{-2} W/m^2.

14. $10^{-0.9}$ W/m^2

15. a) Substitute 0.06 for k:

$P(t) = P_0\, e^{0.06t}$

b) To find the balance after one year, replace P_0 with 5000 and t with 1. We find $P(1)$:

$P(1) = 5000\, e^{0.06(1)} = 5000\, e^{0.06} \approx$

$5000(1.061836547) \approx \$5309.18$

To find the balance after 2 years, replace P_0 with 5000 and t with 2. We find $P(2)$:

$P(2) = 5000\, e^{0.06(2)} = 5000\, e^{0.12} \approx$

$5000(1.127496852) \approx \$5637.48$

c) To find the doubling time, replace P_0 with 5000 and $P(t)$ with 10,000 and solve for t.

$10{,}000 = 5000\, e^{0.06t}$

$2 = e^{0.06t}$

$\ln 2 = \ln e^{0.06t}$ Taking the natural logarithm on both sides

$\ln 2 = 0.06t$ Finding the logarithm of the base to a power

$\dfrac{\ln 2}{0.06} = t$

$11.6 \approx t$

The investment will double in about 11.6 years.

16. a) $P(t) = P_0 e^{0.05t}$

b) \$1051.27; \$1105.17

c) 13.9 yr

17. a) $P(t) = 288.3 e^{0.013t}$, where $P(t)$ is in millions and t is the number of years after 2002.

b) In 2005, $t = 2005 - 2002 = 3$. Replace t with 3 and compute $P(3)$.

$P(3) = 288.3 e^{0.013(3)}$

$= 288.3 e^{0.039}$

≈ 299.8

The U.S. population in 2005 will be about 299.8 million.

c) Replace $P(t)$ with 325 and solve for t.

$325 = 288.3 e^{0.013t}$

$1.1273 \approx e^{0.031t}$

$\ln 1.1273 \approx \ln e^{0.013t}$

$\ln 1.1273 \approx 0.013t$

$\dfrac{\ln 1.1273}{0.013} \approx t$

$9 \approx t$

The U.S. population will reach 325 million about 9 years after 2002, or in 2011.

18. a) $P(t) = 6.3e^{0.014t}$, where $P(t)$ is in billions and t is the numbers of years after 2002.

b) 6.6 billion

c) 2019

19. a) Replace $N(t)$ with 60,000 and solve for t.

$$60,000 = 3000(2)^{t/20}$$

$$20 = (2)^{t/20}$$

$$\log 20 = \log(2)^{t/20}$$

$$\log 20 = \frac{t}{20} \log 2$$

$$20 \log 20 = t \log 2$$

$$\frac{20 \log 20}{\log 2} = t$$

$$86.4 \approx t$$

There will be 60,000 bacteria after about 86.4 minutes.

b) Replace $N(t)$ with 100,000,000 and solve for t.

$$100,000,000 = 3000(2)^{t/20}$$

$$33,333.333 = (2)^{t/20}$$

$$\log 33,333.333 = \log(2)^{t/20}$$

$$\log 33,333.333 = \frac{t}{20} \log 2$$

$$20 \log 33,333.333 = t \log 2$$

$$\frac{20 \log 33,333.333}{\log 2} = t$$

$$300.5 \approx t$$

About 300.5 minutes would have to pass in order for a possible infection to occur.

c) Replace $P(t)$ with 6000 and solve for t.

$$6000 = 3000(2)^{t/20}$$

$$2 = (2)^{t/20}$$

$$1 = \frac{t}{20} \qquad \text{The exponents must be the same.}$$

$$t = 20$$

The doubling time is 20 minutes.

20. 19.8 years

21. a) Replace a with 1 and compute $N(1)$.

$$N(a) = 2000 + 500 \log a$$

$$N(1) = 2000 + 500 \log 1$$

$$N(1) = 2000 + 500 \cdot 0$$

$$N(1) = 2000$$

2000 units were sold after $1000 was spent.

b) Find $N(8)$.

$$N(8) = 2000 + 500 \log 8$$

$$N(8) \approx 2451.5$$

About 2452 units were sold after $8000 was spent.

c) Using the values we computed in parts (a) and (b) and any others we wish to calculate, we sketch the graph:

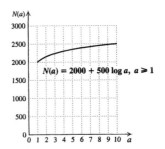

d) Replace $N(a)$ with 5000 and solve for a.

$$5000 = 2000 + 500 \log a$$

$$3000 = 500 \log a$$

$$6 = \log a$$

$$a = 10^6 = 1,000,000$$

$1,000,000 thousand, or $1,000,000,000 would have to be spent.

22. a) 68%

b) 54%; 40%

c)

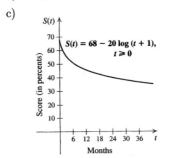

d) 6.9 months

23. a) We use the growth equation $N(t) = N_0 e^{kt}$, where t is the number of years since 1995. In 1995, at $t = 0$, 17 people were infected. We substitute 17 for N_0:

$$N(t) = 17e^{kt}.$$

To find the exponential growth rate k, observe that 1 year later 29 people were infected.

$N(1) = 17e^{k \cdot 1}$ Substituting 1 for t

$29 = 17e^k$ Substituting 29 for $N(1)$

$1.706 \approx e^k$

$\ln 1.706 \approx \ln e^k$

$\ln 1.706 \approx k$

$0.534 \approx k$

The exponential function is $N(t) = 17e^{0.534t}$, where t is the number of years since 1995.

b) In 2001, $t = 2001 - 1995$, or 6. Find $N(6)$.

$N(6) = 17e^{0.534(6)}$

$\qquad = 17e^{3.204}$

$\qquad \approx 418.7$

Approximately 419 people will be infected in 2001.

24. a) $N(t) = 1418\,e^{0.036t}$, where t is the number of years after 1987

b) About 3488 heart transplants

25. We start with the exponential growth equation

$D(t) = D_0\,e^{kt}$, where t is the number of

years after 1995.

Substitute $2\,D_0$ for $D(t)$ and 0.1 for k and solve for t.

$2\,D_0 = D_0\,e^{0.1t}$

$2 = e^{0.1t}$

$\ln 2 = 0.1t$

$\ln 2 = \ln e^{0.1t}$

$\dfrac{\ln 2}{0.1} = t$

$6.9 \approx t$

The demand will be double that of 1998 in $1998 + 7$, or 2005.

26. 2012

27. a) We use the exponential decay equation $W(t) = W_0 e^{-kt}$, where t is the number of years after 1996 and $W(t)$ is in millions of tons. In 1996, at $t = 0$, 17.5 million tons of yard waste were discarded. We substitute 17.5 for W_0.

$W(t) = 17.5e^{-kt}$.

To find the exponential decay rate k, observe that 2 years after 1996, in 1998, 14.5 million tons of yard waste were discarded. We substitute 2 for t and 14.5 for $W(t)$.

$14.5 = 17.5e^{-k \cdot 2}$

$0.8286 \approx e^{-2k}$

$\ln 0.8286 \approx \ln e^{-2k}$

$\ln 0.8286 \approx -2k$

$\dfrac{\ln 0.8286}{-2} \approx k$

$0.094 \approx k$

Then we have $W(t) = 17.5e^{-0.094t}$, where t is the number of years after 1996 and $W(t)$ is in millions of tons.

b) In 2006, $t = 2006 - 1996 = 10$.

$W(10) = 17.5e^{-0.094(10)}$

$\qquad = 17.5e^{-0.94}$

$\qquad \approx 6.8$

In 2006, about 6.8 million tons of yard waste were discarded.

c) 1 ton is equivalent to 0.000001 million tons.

$0.000001 = 17.5e^{-0.094t}$

$5.71 \times 10^{-8} \approx e^{-0.094t}$

$\ln(5.71 \times 10^{-8}) \approx \ln e^{-0.094t}$

$\ln(5.71 \times 10^{-8}) \approx -0.094t$

$\dfrac{\ln(5.71 \times 10^{-8})}{-0.094} \approx t$

$177 \approx t$

Only one ton of yard waste will be discarded about 177 years after 1996, or in 2173.

28. a) $k \approx 0.315$; $M(t) = 5300e^{-0.315t}$, where t is the number of years after 1990

b) 64 cases

c) 2017

29. We will use the function derived in Example 7:

$P(t) = P_0 e^{-0.00012t}$

If the scrolls had lost 22.3% of their carbon-14 from an initial amount P_0, then $77.7\%(P_0)$ is the amount present. To find the age t of the scrolls, we substitute $77.7\%(P_0)$ or $0.777P_0$, for $P(t)$ in the function above and solve for t.

$0.777P_0 = P_0 e^{-0.00012t}$

$0.777 = e^{-0.00012t}$

$\ln 0.777 = \ln e^{-0.00012t}$

$-0.2523 \approx -0.00012t$

$t \approx \dfrac{-0.2523}{-0.00012} \approx 2103$

The scrolls are about 2103 years old.

30. 1654 yr

31. The function $P(t) = P_0 e^{-kt}$, $k > 0$, can be used to model decay. For iodine-131, $k = 9.6\%$, or 0.096. To find the half-life we substitute 0.096 for k and $\frac{1}{2} P_0$ for $P(t)$, and solve for t.

$$\frac{1}{2} P_0 = P_0 e^{-0.096t}, \text{ or } \frac{1}{2} = e^{-0.096t}$$

$$\ln \frac{1}{2} = \ln e^{-0.096t} = -0.096t$$

$$t = \frac{\ln 0.5}{-0.096} \approx \frac{-0.6931}{-0.096} \approx 7.2 \text{ days}$$

32. 11 yr

33. The function $P(t) = P_0 e^{-kt}$, $k > 0$, can be used to model decay. We substitute $\frac{1}{2} P_0$ for $P(t)$ and 1 for t and solve for the decay rate k.

$$\frac{1}{2} P_0 = P_0 e^{-k \cdot 1}$$

$$\frac{1}{2} = e^{-k}$$

$$\ln \frac{1}{2} = \ln e^{-k}$$

$$-0.693 \approx -k$$

$$0.693 \approx k$$

The decay rate is 0.693, or 69.3% per year.

34. 3.15% per year

35. a) We start with the exponential growth equation

$$V(t) = V_0 e^{kt}, \text{ where } t \text{ is the number}$$
of years after 1996.

Substituting 640,500 for V_0, we have

$$V(t) = 640{,}500 e^{kt}.$$

To find the exponential growth rate k, observe that the card sold for $1.1 million, or $1,100,000 in 2000, or 4 years after 1996. We substitute and solve for k.

$$V(5) = 640{,}500 e^{k \cdot 4}$$

$$1{,}100{,}000 = 640{,}500 e^{4k}$$

$$1.7174 \approx e^{4k}$$

$$\ln 1.7174 \approx \ln e^{4k}$$

$$\ln 1.7174 \approx 4k$$

$$\frac{\ln 1.7174}{4} \approx k$$

$$0.135 \approx k$$

Thus, the exponential growth function is $V(t) = 640{,}500 e^{0.135t}$, where t is the number of years after 1996.

b) In 2006, $t = 2006 - 1996 = 10$

$$V(10) = 640{,}500 e^{0.135(10)} \approx 2{,}470{,}681$$

The card's value in 2006 will be about $2.47 million

c) Substitute 2($640,500), or $1,281,000 for $V(t)$ and solve for t.

$$1{,}281{,}000 = 640{,}500 e^{0.135t}$$

$$2 = e^{0.135t}$$

$$\ln 2 = \ln e^{0.135t}$$

$$\ln 2 = 0.135t$$

$$\frac{\ln 2}{0.135} = t$$

$$5.1 \approx t$$

The doubling time is about 5.1 years.

d) Substitute $2,000,000 for $V(t)$ and solve for t.

$$2{,}000{,}000 = 640{,}500 e^{0.135t}$$

$$3.1226 \approx e^{0.135t}$$

$$\ln 3.1226 \approx \ln e^{0.135t}$$

$$\ln 3.1226 \approx 0.135t$$

$$\frac{\ln 3.1226}{0.135} \approx t$$

$$8 \approx t$$

The value of the card will first exceed $2,000,000 about 8 years after 1996, or in 2004.

36. a) $k \approx 0.117$; $V(t) = 58e^{0.117t}$, where t is the number of years after 1987 and $V(t)$ is in millions of dollars

b) About $602.1 million

c) 5.9 yr

d) 24.3 yr

37. Miles per gallon first fell, then rose fairly steeply, and then rose less steeply. This does not fit an exponential model.

38. Yes

39. The ticket price increased from 1980 to 2003 at a rate that makes it appear that an exponential function might fit the data.

40. No

41. a) Enter the data and then use the exponential regression feature. We have

$$p(x) = 6.501242197(1.096109091)^x,$$

where x is the number of years after 1980.

b) $k = \ln b$

$$\approx \ln 1.096109091$$

$$\approx 0.0918, \text{ or } 9.18\%$$

c) In 2006, $t = 2006 - 1980 = 26$. We find $p(26)$ using a table of values or TRACE. If $p(x)$ was copies to the Y = screen as Y_1, we could also enter $Y_1 (26)$.

$$p(26) \approx \$71$$

42. a) $p(x) = 1998.198072(1.291876847)^x$, where x is the number of years after 1995.

   b) 0.256, or 25.6%

   c) \$33,425

43. a) Enter the data points $(0, 50)$, $(10, 100)$, $(20, 150)$, and $(3, 1500)$. (Note that 1.5 billion = 1500 million.) Then use the exponential regression feature. We have

$$P(x) = 37.29665447(1.111922582)^x,$$

   where x is the number of years after 1965 and P is in millions of dollars.

   b) In 2005, $t = 2005 - 1965 = 40$. We find $P(40)$ using a table of values or TRACE. If $P(x)$ was copied to the Y = screen as Y_1, then we could also enter $Y_1(40)$.

$$P(40) \approx 2600$$

   We estimate that the amount spent on new stadium construction in 2005 will be about \$2600 million, or about \$2.6 billion.

44. a) $P(t) = 2.262404 \times 10^{-10} e^{0.0119579321t}$, where P is in billions.

   b) 10.02 billion

45. a) Enter the data and use the exponential regression feature. We get

$$f(x) = 647.6297124(0.5602992676)^x.$$

   b) $f(2.5) \approx 152$, so we estimate that there are 152 decayed, missing, or filled teeth per 100 patients if the fluoride count of the water is 2.5 ppm.

46. a) $P(t) = 0.5434943782 e^{0.369163743t}$, where t is the number of years after 1985 and P is in millions.

   b) 874 million subscribers

47. *Writing Exercise*

48. *Writing Exercise*

49. Graph $y = x^2 - 8x$.

   First we find the vertex.

$$-\frac{b}{2a} = -\frac{-8}{2 \cdot 1} = 4$$

   When $x = 4$, $y = 4^2 - 8 \cdot 4 = 16 - 32 = -16$.

   The vertex is $(4, -16)$ and the axis of symmetry is $x = 4$. We plot a few points on either side of the vertex and graph the parabola.

x	y
4	-16
0	0
2	-12
5	-15
6	-12

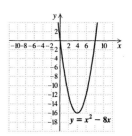

50.

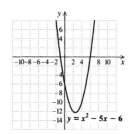

51. Graph $f(x) = 3x^2 - 5x - 1$

   First we find the vertex.

$$-\frac{b}{2a} = -\frac{-5}{2 \cdot 3} = \frac{5}{6}$$

$$f\left(\frac{5}{6}\right) = 3\left(\frac{5}{6}\right)^2 - 5 \cdot \frac{5}{6} - 1 = -\frac{37}{12}$$

   The vertex is $\left(\frac{5}{6}, -\frac{37}{12}\right)$ and the axis of symmetry is $x = \frac{5}{6}$. We plot a few points on either side of the vertex and graph the parabola.

x	$f(x)$
$\frac{5}{6}$	$-\frac{37}{12}$
0	-1
-1	7
2	1
3	11

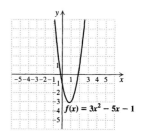

52.

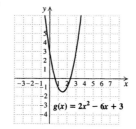

53.
$$x^2 - 8x = 7$$
$$x^2 - 8x + 16 = 7 + 16 \qquad \text{Adding } \left[\frac{1}{2}(-8)\right]^2$$
$$(x - 4)^2 = 23$$
$$x - 4 = \pm\sqrt{23}$$
$$x = 4 \pm \sqrt{23}$$

The solutions are $4 \pm \sqrt{23}$.

54. $-5 \pm \sqrt{31}$

55. *Writing Exercise*

56. *Writing Exercise*

57. We will use the exponential growth equation $V(t) = V_0 e^{kt}$, where t is the number of years after 2001 and $V(t)$ is in millions of dollars. We substitute 21 for $V(t)$, 0.05 for k, and 9 for t and solve for V_0.
$$21 = V_0 e^{0.05(9)}$$
$$21 = V_0 e^{0.45}$$
$$\frac{21}{e^{0.45}} = V_0$$
$$13.4 \approx V_0$$

George Steinbrenner needs to invest \$13.4 million at 5% interest compounded continuously in order to have \$21 million to pay Derek Jeter in 2010.

58. About 80,922 yr or, with rounding of decay rate, about 80,792 yr

59. From Exercises 1 and 17 we know that in 2002 there were 153 million cellular phones in use in the U.S. and the population of the U.S. was 288.3 million. Then the percentage of U.S. residents owning a cellular phone in 2002 was
$$\frac{153}{288.3} \approx 0.5306971904 \approx 53.06971904\%.$$
In 2003, we have $N(1) \approx 209.6$ and $P(1) \approx 292.1$. Then the percentage of U.S. residents owning a cellular phone in 2003 was $\frac{209.6}{292.1} \approx 0.7175967135 \approx 71.75967135\%.$
Now we find a function that models the percentage of U.S. residents owning a cellular phone t years after 2002.
$$P(t) = P_0 e^{kt}$$
$$71.75967135 = 53.06971904 e^{k \cdot 1}$$
$$\frac{71.75967135}{53.06971904} = e^k$$
$$\ln\left(\frac{71.75967135}{53.06971904}\right) = \ln e^k$$
$$\ln\left(\frac{71.75967135}{53.06971904}\right) = k$$
$$0.302 \approx k$$

Then we have $P(t) = 53e^{0.302t}$, rounding P_0, where t is the number of years after 2002 and P is a percent. (Note that this assumes that each resident owns no more than one cellular phone.)

60. *Writing Exercise*

61. a)

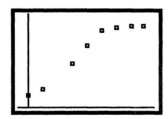

The data appear to be growing exponentially, particularly through 1980.

b) We substitute. First we use the point $(0, 35.0)$.
$$35.0 = ab^0$$
$$35.0 = a$$
Next we use the function $f(x) = 35b^x$ and the point $(50, 90.5)$.
$$90.5 = 35b^{50}$$
$$2.5857 \approx b^{50}$$
$$\log 2.5857 \approx \log b^{50}$$
$$\log 2.5857 \approx 50 \log b$$
$$\frac{\log 2.5857}{50} \approx \log b$$
$$0.0083 \approx \log b$$
$$b \approx 10^{0.0083}$$
$$b \approx 1.0192$$
We have $f(x) = 35.0(1.0192)^x$.

c) $f(90) = 35.0(1.0192)^{90} \approx 194\%$

This estimate does not make sense, since it predicts that more than 100% of U.S. households will have telephones in 2010.

d) $f(x) = \dfrac{102.4604343}{1 + 2.25595242e^{-0.0483573334x}}$

e) $f(90) \approx 99.6\%$

This estimate makes sense.

62. a) $f(x) = \dfrac{19.11252764}{1 + 4.175198683e^{-0.2798470918x}}$

b) \$18.82 billion

Chapter 10

Sequences, Series, and the Binomial Theorem

1. $a_n = 5n - 2$

$a_1 = 5 \cdot 1 - 2 = 3,$

$a_2 = 5 \cdot 2 - 2 = 8,$

$a_3 = 5 \cdot 3 - 2 = 13,$

$a_4 = 5 \cdot 4 - 2 = 18;$

$a_{10} = 5 \cdot 10 - 2 = 48;$

$a_{15} = 5 \cdot 15 - 2 = 73$

2. 5, 7, 9, 11; 23; 33

3. $a_n = \dfrac{n}{n+1}$

$a_1 = \dfrac{1}{1+1} = \dfrac{1}{2},$

$a_2 = \dfrac{2}{2+1} = \dfrac{2}{3},$

$a_3 = \dfrac{3}{3+1} = \dfrac{3}{4},$

$a_4 = \dfrac{4}{4+1} = \dfrac{4}{5};$

$a_{10} = \dfrac{10}{10+1} = \dfrac{10}{11};$

$a_{15} = \dfrac{15}{15+1} = \dfrac{15}{16}$

4. 3, 6, 11, 18; 102; 227

5. $a_n = n^2 - 2n$

$a_1 = 1^2 - 2 \cdot 1 = -1,$

$a_2 = 2^2 - 2 \cdot 2 = 0,$

$a_3 = 3^2 - 2 \cdot 3 = 3,$

$a_4 = 4^2 - 2 \cdot 4 = 8;$

$a_{10} = 10^2 - 2 \cdot 10 = 80;$

$a_{15} = 15^2 - 2 \cdot 15 = 195$

6. $0, \dfrac{3}{5}, \dfrac{4}{5}, \dfrac{15}{17}; \dfrac{99}{101}; \dfrac{112}{113}$

7. $a_n = n + \dfrac{1}{n}$

$a_1 = 1 + \dfrac{1}{1} = 2,$

$a_2 = 2 + \dfrac{1}{2} = 2\dfrac{1}{2},$

$a_3 = 3 + \dfrac{1}{3} = 3\dfrac{1}{3},$

$a_4 = 4 + \dfrac{1}{4} = 4\dfrac{1}{4};$

$a_{10} = 10 + \dfrac{1}{10} = 10\dfrac{1}{10};$

$a_{15} = 15 + \dfrac{1}{15} = 15\dfrac{1}{15}$

8. $1, -\dfrac{1}{2}, \dfrac{1}{4}, -\dfrac{1}{8}; -\dfrac{1}{512}; \dfrac{1}{16,384}$

9. $a_n = (-1)^n n^2$

$a_1 = (-1)^1 1^2 = -1,$

$a_2 = (-1)^2 2^2 = 4,$

$a_3 = (-1)^3 3^2 = -9,$

$a_4 = (-1)^4 4^2 = 16;$

$a_{10} = (-1)^{10} 10^2 = 100;$

$a_{15} = (-1)^{15} 15^2 = -225$

10. $-4, 5, -6, 7; 13; -18$

11. $a_n = (-1)^{n+1}(3n - 5)$

$a_1 = (-1)^{1+1}(3 \cdot 1 - 5) = -2,$

$a_2 = (-1)^{2+1}(3 \cdot 2 - 5) = -1,$

$a_3 = (-1)^{3+1}(3 \cdot 3 - 5) = 4,$

$a_4 = (-1)^{4+1}(3 \cdot 4 - 5) = -7;$

$a_{10} = (-1)^{10+1}(3 \cdot 10 - 5) = -25;$

$a_{15} = (-1)^{15+1}(3 \cdot 15 - 5) = 40$

12. 0, 7, −26, 63; 999; −3374

13. $a_n = 2n - 5$

$a_7 = 2 \cdot 7 - 5 = 14 - 5 = 9$

14. 26

15. $a_n = (3n + 1)(2n - 5)$

$a_9 = (3 \cdot 9 + 1)(2 \cdot 9 - 5) = 28 \cdot 13 = 364$

16. 400

17. $a_n = (-1)^{n-1}(3.4n - 17.3)$

$a_{12} = (-1)^{12-1}[3.4(12) - 17.3] = -23.5$

18. $-37,916,508.16$

19. $a_n = 3n^2(9n - 100)$

$a_{11} = 3 \cdot 11^2(9 \cdot 11 - 100) = 3 \cdot 121(-1) = -363$

20.   9680

21. $a_n = \left(1 + \dfrac{1}{n}\right)^2$

$a_{20} = \left(1 + \dfrac{1}{20}\right)^2 = \left(\dfrac{21}{20}\right)^2 = \dfrac{441}{400}$

22. $\dfrac{2744}{3375}$

23.   1, 3, 5, 7, 9, . . .

These are odd integers, so the general term could be $2n-1$.

24. $2n$

25.   1, −1, 1, −1, . . .

1 and −1 alternate, beginning with 1, so the general term could be $(-1)^{n+1}$.

26. $(-1)^n$

27.   −1, 2, −3, 4, . . .

These are the first four natural numbers, but with alternating signs, beginning with a negative number. The general term could be $(-1)^n \cdot n$.

28. $(-1)^{n+1} \cdot n$

29.   −2, 6, −18, 54, . . .

We can see a pattern if we write the sequence as

$-1 \cdot 2 \cdot 1,\ 1 \cdot 2 \cdot 3,\ -1 \cdot 2 \cdot 9,\ 1 \cdot 2 \cdot 27, \ldots$

The general term could be $(-1)^n 2(3)^{n-1}$.

30. $5n - 7$

31. $\dfrac{1}{2}, \dfrac{2}{3}, \dfrac{3}{4}, \dfrac{4}{5}, \dfrac{5}{6}, \ldots$

These are fractions in which the denominator is 1 greater than the numerator. Also, each numerator is 1 greater than the preceding numerator. The general term could be $\dfrac{n}{n + 1}$.

32. $n(n + 1)$

33.   5, 25, 125, 625, . . .

This is powers of 5, so the general term could be 5^n.

34. 4^n

35.   −1, 4, −9, 16, . . .

This is the squares of the first four natural numbers, but with alternating signs, beginning with a negative number. The general term could be $(-1)^n \cdot n^2$.

36. $(-1)^{n+1} \cdot n^2$

37.   1, −2, 3, −4, 5, −6, . . .

$S_7 = 1 - 2 + 3 - 4 + 5 - 6 + 7 = 4$

38.   −8

39.   2, 4, 6, 8, . . .

$S_5 = 2 + 4 + 6 + 8 + 10 = 30$

40. $\dfrac{5269}{3600}$

41. $\displaystyle\sum_{k=1}^{5} \dfrac{1}{2k} = \dfrac{1}{2 \cdot 1} + \dfrac{1}{2 \cdot 2} + \dfrac{1}{2 \cdot 3} + \dfrac{1}{2 \cdot 4} + \dfrac{1}{2 \cdot 5}$

$= \dfrac{1}{2} + \dfrac{1}{4} + \dfrac{1}{6} + \dfrac{1}{8} + \dfrac{1}{10}$

$= \dfrac{60}{120} + \dfrac{30}{120} + \dfrac{20}{120} + \dfrac{15}{120} + \dfrac{12}{120}$

$= \dfrac{137}{120}$

42. $\dfrac{6508}{3465}$

43. $\displaystyle\sum_{k=0}^{4} 3^k = 3^0 + 3^1 + 3^2 + 3^3 + 3^4$

$= 1 + 3 + 9 + 27 + 81$

$= 121$

44. $\sqrt{9} + \sqrt{11} + \sqrt{13} + \sqrt{15} \approx 13.7952$

45. $\displaystyle\sum_{k=1}^{8} \dfrac{k}{k + 1} = \dfrac{1}{1 + 1} + \dfrac{2}{2 + 1} + \dfrac{3}{3 + 1} + \dfrac{4}{4 + 1} +$

$\dfrac{5}{5 + 1} + \dfrac{6}{6 + 1} + \dfrac{7}{7 + 1} + \dfrac{8}{8 + 1}$

$= \dfrac{1}{2} + \dfrac{2}{3} + \dfrac{3}{4} + \dfrac{4}{5} + \dfrac{5}{6} + \dfrac{6}{7} + \dfrac{7}{8} + \dfrac{8}{9}$

$= \dfrac{15,551}{2520}$

46. $-\dfrac{1}{4} + 0 + \dfrac{1}{6} + \dfrac{2}{7} = \dfrac{17}{84}$

47. $\displaystyle\sum_{k=1}^{8} (-1)^{k+1} 2^k = (-1)^2 2^1 + (-1)^3 2^2 + (-1)^4 2^3 +$

$(-1)^5 2^4 + (-1)^6 2^5 + (-1)^7 2^6 +$

$(-1)^8 2^7 + (-1)^9 2^8$

$= 2 - 4 + 8 - 16 + 32 - 64 +$

$128 - 256$

$= -170$

48. $-4^2 + 4^3 - 4^4 + 4^5 - 4^6 + 4^7 - 4^8 = -52,432$

49. $\displaystyle\sum_{k=0}^{5} (k^2 - 2k + 3)$

$= (0^2 - 2 \cdot 0 + 3) + (1^2 - 2 \cdot 1 + 3) +$

$(2^2 - 2 \cdot 2 + 3) + (3^2 - 2 \cdot 3 + 3) +$

$(4^2 - 2 \cdot 4 + 3) + (5^2 - 2 \cdot 5 + 3)$

$= 3 + 2 + 3 + 6 + 11 + 18$

$= 43$

50. $4 + 2 + 2 + 4 + 8 + 14 = 34$

51. $\displaystyle\sum_{k=3}^{5} \frac{(-1)^k}{k(k+1)} = \frac{(-1)^3}{3(3+1)} + \frac{(-1)^4}{4(4+1)} + \frac{(-1)^5}{5(5+1)}$

$= \dfrac{-1}{3 \cdot 4} + \dfrac{1}{4 \cdot 5} + \dfrac{-1}{5 \cdot 6}$

$= -\dfrac{1}{12} + \dfrac{1}{20} - \dfrac{1}{30}$

$= -\dfrac{4}{60} = -\dfrac{1}{15}$

52. $\dfrac{3}{8} + \dfrac{4}{16} + \dfrac{5}{32} + \dfrac{6}{64} + \dfrac{7}{128} = \dfrac{119}{128}$

53. $\dfrac{2}{3} + \dfrac{3}{4} + \dfrac{4}{5} + \dfrac{5}{6} + \dfrac{6}{7}$

This is a sum of fractions in which the denominator is one greater than the numerator. Also, each numerator is 1 greater than the preceding numerator. Sigma notation is

$$\sum_{k=1}^{5} \frac{k+1}{k+2}.$$

54. $\displaystyle\sum_{k=1}^{5} 3k$

55. $1 + 4 + 9 + 16 + 25 + 36$

This is the sum of the squares of the first six natural numbers. Sigma notation is

$$\sum_{k=1}^{6} k^2.$$

56. $\displaystyle\sum_{k=1}^{5} \frac{1}{k^2}$

57. $4 - 9 + 16 - 25 + \ldots + (-1)^n n^2$

This is a sum of terms of the form $(-1)^k k^2$, beginning with $k = 2$ and continuing through $k = n$. Sigma notation is

$$\sum_{k=2}^{n} (-1)^k k^2.$$

58. $\displaystyle\sum_{k=3}^{n} (-1)^{k+1} k^2$

59. $5 + 10 + 15 + 20 + 25 + \ldots$

This is a sum of multiples of 5, and it is an infinite series. Sigma notation is

$$\sum_{k=1}^{\infty} 5k.$$

60. $\displaystyle\sum_{k=1}^{\infty} 7k$

61. $\dfrac{1}{1 \cdot 2} + \dfrac{1}{2 \cdot 3} + \dfrac{1}{3 \cdot 4} + \dfrac{1}{4 \cdot 5} + \ldots$

This is a sum of fractions in which the numerator is 1 and the denominator is a product of two consecutive integers. The larger integer in each product is the smaller integer in the succeeding product. It is an infinite series. Sigma notation is

$$\sum_{k=1}^{\infty} \frac{1}{k(k+1)}.$$

62. $\displaystyle\sum_{k=1}^{\infty} \frac{1}{k(k+1)^2}$

63. *Writing Exercise*

64. *Writing Exercise*

65. $\dfrac{7}{2}(a_1 + a_7) = \dfrac{7}{2}(8 + 14) = \dfrac{7}{2} \cdot 22 = 77$

66. 23

67. $(x + y)^3$

$= (x + y)(x + y)^2$

$= (x + y)(x^2 + 2xy + y^2)$

$= x(x^2 + 2xy + y^2) + y(x^2 + 2xy + y^2)$

$= x^3 + 2x^2 y + xy^2 + x^2 y + 2xy^2 + y^3$

$= x^3 + 3x^2 y + 3xy^2 + y^3$

68. $a^3 - 3a^2 b + 3ab^2 - b^3$

69. $(2a - b)^3$

$= (2a - b)(2a - b)^2$

$= (2a - b)(4a^2 - 4ab + b^2)$

$= 2a(4a^2 - 4ab + b^2) - b(4a^2 - 4ab + b^2)$

$= 8a^3 - 8a^2 b + 2ab^2 - 4a^2 b + 4ab^2 - b^3$

$= 8a^3 - 12a^2 b + 6ab^2 - b^3$

70. $8x^3 + 12x^2 y + 6xy^2 + y^3$

71. *Writing Exercise*

72. *Writing Exercise*

73. $a_1 = 1$, $a_{n+1} = 5a_n - 2$

$a_1 = 1$

$a_2 = 5 \cdot 1 - 2 = 3$

$a_3 = 5 \cdot 3 - 2 = 13$

$a_4 = 5 \cdot 13 - 2 = 63$

$a_5 = 5 \cdot 63 - 2 = 313$

$a_6 = 5 \cdot 313 - 2 = 1563$

74. 0, 3, 12, 147, 21,612, 467,078,547

75. Find each term by multiplying the preceding term by 2:

1, 2, 4, 8, 16, 32, 64, 128, 256, 512, 1024,

2048, 4096, 8192, 16,384, 32,768, 65,536

76. $5200, $3900, $2925, $2193.75, $1645.31, $1233.98,

$925.49, $694.12, $520.59, $390.44

77. $a_n = (-1)^n$

This sequence is of the form $-1, 1, -1, 1, \ldots$. Each pair of terms adds to 0. S_{100} has 50 such pairs, so $S_{100} = 0$. S_{101} consists of the 50 pairs in S_{100} that add to 0 as well as a_{101}, or -1, so $S_{101} = -1$.

78. $\dfrac{3}{2}, \dfrac{3}{2}, \dfrac{9}{8}, \dfrac{3}{4}, \dfrac{15}{32}; \dfrac{171}{32}$

79. $a_n = i^n$

$a_1 = i^1 = i$

$a_2 = i^2 = -1$

$a_3 = i^3 = i^2 \cdot i = -1 \cdot i = -i$

$a_4 = i^4 = (i^2)^2 = (-1)^2 = 1$

$a_5 = i^5 = (i^2)^2 \cdot i = (-1)^2 \cdot i = 1 \cdot i = i$

$S_5 = i - 1 - i + 1 + i = i$

80. $\{x | x = 4n - 1,$ where n is a natural number$\}$

81. Enter $y_1 = 14x^4 + 6x^3 + 416x^2 - 655x - 1050$. Then scroll through a table of values. We see that $y_1 = 6144$ when $x = 11$, so the 11th term of the sequence is 6144.

82. 1225 handshakes

Exercise Set 10.2

1. 2, 6, 10, 14, . . .

$a_1 = 2$

$d = 4$　　$(6 - 2 = 4, 10 - 6 = 4, 14 - 10 = 4)$

2. $a_1 = 1.06$, $d = 0.06$

3. 6, 2, −2, −6, . . .

$a_1 = 6$

$d = -4$　　$(2 - 6 = -4, -2 - 2 = -4,$
　　　　　　　$-6 - (-2) = -4)$

4. $a_1 = -9$, $d = 3$

5. $\dfrac{3}{2}, \dfrac{9}{4}, 3, \dfrac{15}{4}, \ldots$

$a_1 = \dfrac{3}{2}$

$d = \dfrac{3}{4}$　　$\left(\dfrac{9}{4} - \dfrac{3}{2} = \dfrac{3}{4}, 3 - \dfrac{9}{4} = \dfrac{3}{4}\right)$

6. $a_1 = \dfrac{3}{5}$, $d = -\dfrac{1}{2}$

7. $5.12, $5.24, $5.36, $5.48, . . .

$a_1 = \$5.12$

$d = \$0.12$　$(\$5.24 - \$5.12 = \$0.12, \$5.36 -$
　　　　　　$\$5.24 = \$0.12, \$5.48 - \$5.36 =$
　　　　　　$\$0.12)$

8. $a_1 = \$214$, $d = -\$3$

9. 3, 7, 11, . . .

$a_1 = 3$, $d = 4$, and $n = 12$

$a_n = a_1 + (n - 1)d$

$a_{12} = 3 + (12 - 1)4 = 3 + 11 \cdot 4 = 3 + 44 = 47$

10. 0.57

11. 7, 4, 1, . . .

$a_1 = 7$, $d = -3$, and $n = 17$

$a_n = a_1 + (n - 1)d$

$a_{17} = 7 + (17 - 1)(-3) = 7 + 16(-3) =$
　　　　$7 - 48 = -41$

12. $-\dfrac{17}{3}$

13. $1200, $964.32, $728.64, . . .

$a_1 = \$1200$, $d = \$964.32 - \$1200 = -\$235.68$,

and $n = 13$

$a_n = a_1 + (n - 1)d$

$a_{13} = \$1200 + (13 - 1)(-\$235.68) =$

　　　　$\$1200 + 12(-\$235.68) = \$1200 - \$2828.16 =$

　　　　$-\$1628.16$

14. $7941.62

15. $a_1 = 3$, $d = 4$

$a_n = a_1 + (n-1)d$

Let $a_n = 107$, and solve for n.

$107 = 3 + (n-1)(4)$

$107 = 3 + 4n - 4$

$107 = 4n - 1$

$108 = 4n$

$27 = n$

The 27th term is 107.

16. 33rd

17. $a_1 = 7$, $d = -3$

$a_n = a_1 + (n-1)d$

$-296 = 7 + (n-1)(-3)$

$-296 = 7 - 3n + 3$

$-306 = -3n$

$102 = n$

The 102nd term is -296.

18. 46th

19. $a_n = a_1 + (n-1)d$

$a_{17} = 2 + (17-1)5$ Substituting 17 for n,
                         2 for a_1, and 5 for d

$= 2 + 16 \cdot 5$

$= 2 + 80$

$= 82$

20. -43

21. $a_n = a_1 + (n-1)d$

$33 = a_1 + (8-1)4$ Substituting 33 for a_8,
                       8 for n, and 4 for d

$33 = a_1 + 28$

$5 = a_1$

(Note that this procedure is equivalent to subtracting d from a_8 seven times to get a_1: $33 - 7(4) = 33 - 28 = 5$)

22. -54

23. $a_n = a_1 + (n-1)d$

$-76 = 5 + (n-1)(-3)$ Substituting -76 for
                           a_n, 5 for a_1, and -3
                           for d

$-76 = 5 - 3n + 3$

$-76 = 8 - 3n$

$-84 = -3n$

$28 = n$

24. 39

25. We know that $a_{17} = -40$ and $a_{28} = -73$. We would have to add d eleven times to get from a_{17} to a_{28}. That is,

$-40 + 11d = -73$

$11d = -33$

$d = -3$.

Since $a_{17} = -40$, we subtract d sixteen times to get to a_1.

$a_1 = -40 - 16(-3) = -40 + 48 = 8$

We write the first five terms of the sequence:

$8, 5, 2, -1, -4$

26. $a_1 = \dfrac{1}{3}$; $d = \dfrac{1}{2}$; $\dfrac{1}{3}, \dfrac{5}{6}, \dfrac{4}{3}, \dfrac{11}{6}, \dfrac{7}{3}$

27. $a_{13} = 13$ and $a_{54} = 54$

Observe that for this to be true, $a_1 = 1$ and $d = 1$.

28. $a_1 = 2$, $d = 2$

29. $1 + 5 + 9 + 13 + \ldots$

Note that $a_1 = 1$, $d = 4$, and $n = 20$. Before using the formula for S_n, we find a_{20}:

$a_{20} = 1 + (20-1)4$ Substituting into
                           the formula for a_n

$= 1 + 19 \cdot 4$

$= 77$

Then

$S_{20} = \dfrac{20}{2}(1 + 77)$ Using the formula for S_n

$= 10(78)$

$= 780$.

30. -210

31. The sum is $1 + 2 + 3 + \ldots + 249 + 250$. This is the sum of the arithmetic sequence for which $a_1 = 1$, $a_n = 250$, and $n = 250$. We use the formula for S_n.

$S_n = \dfrac{n}{2}(a_1 + a_n)$

$S_{300} = \dfrac{250}{2}(1 + 250) = 125(251) = 31,375$

32. 80,200

33. The sum is $2 + 4 + 6 + \ldots + 98 + 100$. This is the sum of the arithmetic sequence for which $a_1 = 2$, $a_n = 100$, and $n = 50$. We use the formula for S_n.

$S_n = \dfrac{n}{2}(a_1 + a_n)$

$S_{50} = \dfrac{50}{2}(2 + 100) = 25(102) = 2550$

34. 2500

35. The sum is $6+12+18+ \ldots +96+102$. This is the sum of the arithmetic sequence for which $a_1 = 6$, $a_n = 102$, and $n = 17$. We use the formula for S_n.

$$S_n = \frac{n}{2}(a_1 + a_n)$$

$$S_{17} = \frac{17}{2}(6 + 102) = \frac{17}{2}(108) = 918$$

36. 34,036

37. Before using the formula for S_n, we find a_{20}:

$$a_{20} = 4 + (20 - 1)5 \qquad \text{Substituting into the formula for } a_n$$

$$= 4 + 19 \cdot 5 = 99$$

Then

$$S_{20} = \frac{20}{2}(4 + 99) \qquad \text{Using the formula for } S_n$$

$$= 10(103) = 1030.$$

38. -1200

39. *Familiarize.* We want to find the fifteenth term and the sum of an arithmetic sequence with $a_1 = 14$, $d = 2$, and $n = 15$. We will first use the formula for a_n to find a_{15}. This result is the number of marchers in the last row. Then we will use the formula for S_n to find S_{15}. This is the total number of marchers.

Translate. Substituting into the formula for a_n, we have

$$a_{15} = 14 + (15 - 1)2.$$

Carry out. We first find a_{15}.

$$a_{15} = 14 + 14 \cdot 2 = 42$$

Then use the formula for S_n to find S_{15}.

$$S_{15} = \frac{15}{2}(14 + 42) = \frac{15}{2}(56) = 420$$

Check. We can do the calculations again. We can also do the entire addition.

$$14 + 16 + 18 + \cdots + 42.$$

State. There are 42 marchers in the last row, and there are 420 marchers altogether.

40. 3; 210

41. *Familiarize.* We go from 50 poles in a row, down to six poles in the top row, so there must be 45 rows. We want the sum $50 + 49 + 48 + \ldots + 6$. Thus we want the sum of an arithmetic sequence. We will use the formula $S_n = \frac{n}{2}(a_1 + a_n)$.

Translate. We want to find the sum of the first 45 terms of an arithmetic sequence with $a_1 = 50$ and $a_{45} = 6$.

Carry out. Substituting into the formula for S_n, we have

$$S_{45} = \frac{45}{2}(50 + 6)$$

$$= \frac{45}{2} \cdot 56 = 1260$$

Check. We can do the calculation again, or we can do the entire addition:

$$50 + 49 + 48 + \ldots + 6.$$

State. There will be 1260 poles in the pile.

42. $49.60

43. *Familiarize.* We want to find the sum of an arithmetic sequence with $a_1 = \$600$, $d = \$100$, and $n = 20$. We will use the formula for a_n to find a_{20}, and then we will use the formula for S_n to find S_{20}.

Translate. Substituting into the formula for a_n, we have

$$a_{20} = 600 + (20 - 1)(100).$$

Carry out. We first find a_{20}.

$$a_{20} = 600 + 19 \cdot 100 = 600 + 1900 = 2500$$

Then we use the formula for S_n to find S_{20}.

$$S_{20} = \frac{20}{2}(600 + 2500) = 10(3100) = 31,000$$

Check. We can do the calculation again.

State. They save $31,000 (disregarding interest).

44. $10,230

45. *Familiarize.* We want to find the sum of an arithmetic sequence with $a_1 = 20$, $d = 2$, and $n = 19$. We will use the formula for a_n to find a_{19}, and then we will use the formula for S_n to find S_{19}.

Translate. Substituting into the formula for a_n, we have

$$a_{19} = 20 + (19 - 1)(2).$$

Carry out. We find a_{19}.

$$a_{19} = 20 + 18 \cdot 2 = 56$$

Then we use the formula for S_n to find S_{19}.

$$S_{19} = \frac{19}{2}(20 + 56) = 722$$

Check. We can do the calculation again.

State. There are 722 seats.

46. $462,500

47. *Writing Exercise*

48. *Writing Exercise*

49.
$$\frac{3}{10x} + \frac{2}{15x}, \text{ LCD is } 30x$$

$$= \frac{3}{10x} \cdot \frac{3}{3} + \frac{2}{15x} \cdot \frac{2}{2}$$

$$= \frac{9}{30x} + \frac{4}{30x}$$

$$= \frac{13}{30x}$$

50. $\dfrac{23}{36t}$

51.

The logarithm is the exponent.

$\log_a P = k \qquad a^k = P$

The base does not change.

52. $e^a = t$

53.
$$4^{3x} = 8^{x+2}$$
$$(2^2)^{3x} = (2^3)^{x+2}$$
$$2^{6x} = 2^{3x+6}$$
$$6x = 3x + 6 \qquad \text{Equating exponents}$$
$$3x = 6$$
$$x = 2$$

The solution is 2.

54. 5

55. *Writing Exercise*

56. *Writing Exercise*

57. $a_1 = 1$, $d = 2$, $n = n$
$$a_n = 1 + (n-1)2 = 1 + 2n - 2 = 2n - 1$$
$$S_n = \frac{n}{2}[1 + (2n-1)] = \frac{n}{2} \cdot 2n = n^2$$

Thus, the formula $S_n = n^2$ can be used to find the sum of the first n consecutive odd numbers starting with 1.

58. 3, 5, 7

59.
$$a_1 = \$8760$$
$$a_2 = \$8760 + (-\$798.23) = \$7961.77$$
$$a_3 = \$8760 + 2(-\$798.23) = \$7163.54$$
$$a_4 = \$8760 + 3(-\$798.23) = \$6365.31$$
$$a_5 = \$8760 + 4(-\$798.23) = \$5567.08$$
$$a_6 = \$8760 + 5(-\$798.23) = \$4768.85$$
$$a_7 = \$8760 + 6(-\$798.23) = \$3970.62$$
$$a_8 = \$8760 + 7(-\$798.23) = \$3172.39$$
$$a_9 = \$8760 + 8(-\$798.23) = \$2374.16$$
$$a_{10} = \$8760 + 9(-\$798.23) = \$1575.93$$

60. \$51,679.65

61. See the answer section in the text.

62. a) $a_t = \$5200 - \$512.50t$

    b) \$5200, \$4687.50, \$4175, \$3662.50, \$3150, \$1612.50, \$1100

    c) $a_0 = \$5200$, $a_t = a_{t-1} - \$512.50$

63. Each integer from 501 through 750 is 500 more than the corresponding integer from 1 through 250. There are 250 integers from 501 through 750, so their sum is the sum of the integers from 1 to 250 plus $250 \cdot 500$. From Exercise 31, we know that the sum of the integers from 1 through 250 is 31,375. Thus, we have

$$31,375 + 250 \cdot 500, \text{ or } 156,375.$$

64. Arithmetic; $a_n = 150 - 0.75(n-1)$, where $n = 1$ corresponds to age 20, $n = 2$ corresponds to age 21, and so on

65. We graph the data points, where the first coordinate 1 represents 1998, 2 represents 1999, and so on.

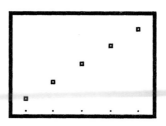

The points appear to lie on a straight line, so this could be the graph of an arithmetic sequence. The general term is

$$a_n = n + 102,$$

where $n = 1$ corresponds to 1998, $n = 2$ corresponds to 1999, and so on.

66. Not arithmetic

67. We graph the data points, where the first coordinate 1 represents 1997, 2 represents 1998, and so on.

The points do not lie on a straight line, so this is not the graph of an arithmetic sequence.

Exercise Set 10.3

1. 7, 14, 28, 56, . . .

$$\frac{14}{7} = 2, \quad \frac{28}{14} = 2, \quad \frac{56}{28} = 2$$

$$r = 2$$

2. 3

3. $5, -5, 5, -5, \ldots$

$\dfrac{-5}{5} = -1, \; \dfrac{5}{-5} = -1, \; \dfrac{-5}{5} = -1$

$r = -1$

4. 0.1

5. $\dfrac{1}{2}, -\dfrac{1}{4}, \dfrac{1}{8}, -\dfrac{1}{16}, \ldots$

$\dfrac{-\frac{1}{4}}{\frac{1}{2}} = -\dfrac{1}{4} \cdot \dfrac{2}{1} = -\dfrac{2}{4} = -\dfrac{1}{2}$

$\dfrac{\frac{1}{8}}{-\frac{1}{4}} = \dfrac{1}{8} \cdot \left(-\dfrac{4}{1}\right) = -\dfrac{4}{8} = -\dfrac{1}{2}$

$\dfrac{-\frac{1}{16}}{\frac{1}{8}} = -\dfrac{1}{16} \cdot \dfrac{8}{1} = -\dfrac{8}{16} = -\dfrac{1}{2}$

$r = -\dfrac{1}{2}$

6. -2

7. $75, 15, 3, \dfrac{3}{5}, \ldots$

$\dfrac{15}{75} = \dfrac{1}{5}, \; \dfrac{3}{15} = \dfrac{1}{5}, \; \dfrac{\frac{3}{5}}{3} = \dfrac{3}{5} \cdot \dfrac{1}{3} = \dfrac{1}{5}$

$r = \dfrac{1}{5}$

8. $-\dfrac{1}{3}$

9. $\dfrac{1}{m}, \dfrac{3}{m^2}, \dfrac{9}{m^3}, \dfrac{27}{m^4}, \ldots$

$\dfrac{\frac{3}{m^2}}{\frac{1}{m}} = \dfrac{3}{m^2} \cdot \dfrac{m}{1} = \dfrac{3}{m}$

$\dfrac{\frac{9}{m^3}}{\frac{3}{m^2}} = \dfrac{9}{m^3} \cdot \dfrac{m^2}{3} = \dfrac{3}{m}$

$\dfrac{\frac{27}{m^4}}{\frac{9}{m^3}} = \dfrac{27}{m^4} \cdot \dfrac{m^3}{9} = \dfrac{3}{m}$

$r = \dfrac{3}{m}$

10. $\dfrac{m}{5}$

11. $3, 6, 12, \ldots$

$a_1 = 3, n = 7, \text{ and } r = \dfrac{6}{3} = 2$

We use the formula $a_n = a_1 r^{n-1}$.

$a_7 = 3 \cdot 2^{7-1} = 3 \cdot 2^6 = 3 \cdot 64 = 192$

12. $131,072$

13. $5, 5\sqrt{2}, 10, \ldots$

$a_1 = 5, n = 9, \text{ and } r = \dfrac{5\sqrt{2}}{5} = \sqrt{2}$

$a_n = a_1 r^{n-1}$

$a_9 = 5(\sqrt{2})^{9-1} = 5(\sqrt{2})^8 = 5 \cdot 16 = 80$

14. $108\sqrt{3}$

15. $-\dfrac{8}{243}, \dfrac{8}{81}, -\dfrac{8}{27}, \ldots$

$a_1 = -\dfrac{8}{243}, \; n = 10, \text{ and } r = \dfrac{\frac{8}{81}}{-\frac{8}{243}} =$

$\dfrac{8}{81}\left(-\dfrac{243}{8}\right) = -3$

$a_n = a_1 r^{n-1}$

$a_{10} = -\dfrac{8}{243}(-3)^{10-1} = -\dfrac{8}{243}(-3)^9 =$

$-\dfrac{8}{243}(-19,683) = 648$

16. $2,734,375$

17. $\$1000, \$1080, \$1166.40, \ldots$

$a_1 = \$1000, \; n = 12, \text{ and } r = \dfrac{\$1080}{\$1000} = 1.08$

$a_n = a_1 r^{n-1}$

$a_{12} = \$1000(1.08)^{12-1} \approx \$1000(2.331638997) \approx$ $\$2331.64$

18. $\$1967.15$

19. $1, 3, 9, \ldots$

$a_1 = 1 \text{ and } r = \dfrac{3}{1}, \text{ or } 3$

$a_n = a_1 r^{n-1}$

$a_n = 1(3)^{n-1} = 3^{n-1}$

20. $a_n = 5^{3-n}$

21. $1, -1, 1, -1, \ldots$

$a_1 = 1 \text{ and } r = \dfrac{-1}{1} = -1$

$a_n = a_1 r^{n-1}$

$a_n = 1(-1)^{n-1} = (-1)^{n-1}$

22. $a_n = 2^n$

23. $\dfrac{1}{x}, \dfrac{1}{x^2}, \dfrac{1}{x^2}, \ldots$

$a_1 = \dfrac{1}{x}$ and $r = \dfrac{\frac{1}{x^2}}{\frac{1}{x}} = \dfrac{1}{x^2} \cdot \dfrac{x}{1} = \dfrac{1}{x}$

$a_n = a_1 r^{n-1}$

$a_n = \dfrac{1}{x}\left(\dfrac{1}{x}\right)^{n-1} = \dfrac{1}{x} \cdot \dfrac{1}{x^{n-1}} = \dfrac{1}{x^{1+n-1}} = \dfrac{1}{x^n}$

24. $a_n = 5\left(\dfrac{m}{2}\right)^{n-1}$

25. $6 + 12 + 24 + \ldots$

$a_1 = 6$, $n = 7$, and $r = \dfrac{12}{6} = 2$

$S_n = \dfrac{a_1(1 - r^n)}{1 - r}$

$S_7 = \dfrac{6(1 - 2^7)}{1 - 2} = \dfrac{6(1 - 128)}{-1} = \dfrac{6(-127)}{-1} = 762$

26. 10.5

27. $\dfrac{1}{18} - \dfrac{1}{6} + \dfrac{1}{2} - \ldots$

$a_1 = \dfrac{1}{18}$, $n = 7$, and $r = \dfrac{-\frac{1}{6}}{\frac{1}{18}} = -\dfrac{1}{6} \cdot \dfrac{18}{1} = -3$

$S_n = \dfrac{a_1(1 - r^n)}{1 - r}$

$S_7 = \dfrac{\frac{1}{18}\left[1 - (-3)^7\right]}{1 - (-3)} = \dfrac{\frac{1}{18}(1 + 2187)}{4} = \dfrac{\frac{1}{18}(2188)}{4} =$

$\dfrac{1}{18}(2188)\left(\dfrac{1}{4}\right) = \dfrac{547}{18}$

28. 7.7777

29. $1 + x + x^2 + x^3 + \ldots$

$a_1 = 1$, $n = 8$, and $r = \dfrac{x}{1}$, or x

$S_n = \dfrac{a_1(1 - r^n)}{1 - r}$

$S_8 = \dfrac{1(1 - x^8)}{1 - x} = \dfrac{(1 + x^4)(1 - x^4)}{1 - x} =$

$\dfrac{(1 + x^4)(1 + x^2)(1 - x^2)}{1 - x} =$

$\dfrac{(1 + x^4)(1 + x^2)(1 + x)(1 - x)}{1 - x} =$

$(1 + x^4)(1 + x^2)(1 + x)$

30. $\dfrac{1 - x^{20}}{1 - x^2}$

31. $\$200, \$200(1.06), \$200(1.06)^2, \ldots$

$a_1 = \$200$, $n = 16$, and $r = \dfrac{\$200(1.06)}{\$200} = 1.06$

$S_n = \dfrac{a_1(1 - r^n)}{1 - r}$

$S_{16} = \dfrac{\$200[1 - (1.06)^{16}]}{1 - 1.06} \approx$

$\dfrac{\$200(1 - 2.540351685)}{-0.06} \approx \$5134.51$

32. $\$60,893.30$

33. $16 + 4 + 1 + \ldots$

$|r| = \left|\dfrac{4}{16}\right| = \left|\dfrac{1}{4}\right| = \dfrac{1}{4}$, and since $|r| < 1$, the series does have a sum.

$S_\infty = \dfrac{a_1}{1 - r} = \dfrac{16}{1 - \frac{1}{4}} = \dfrac{16}{\frac{3}{4}} = 16 \cdot \dfrac{4}{3} = \dfrac{64}{3}$

34. 16

35. $7 + 3 + \dfrac{9}{7} + \ldots$

$|r| = \left|\dfrac{3}{7}\right| = \dfrac{3}{7}$, and since $|r| < 1$, the series does have a sum.

$S_\infty = \dfrac{a_1}{1 - r} = \dfrac{7}{1 - \frac{3}{7}} = \dfrac{7}{\frac{4}{7}} = 7 \cdot \dfrac{7}{4} = \dfrac{49}{4}$

36. 48

37. $3 + 15 + 75 + \ldots$

$|r| = \left|\dfrac{15}{3}\right| = |5| = 5$, and since $|r| \not< 1$ the series does not have a sum.

38. No

39. $4 - 6 + 9 - \dfrac{27}{2} + \ldots$

$|r| = \left|\dfrac{-6}{4}\right| = \left|-\dfrac{3}{2}\right| = \dfrac{3}{2}$, and since $|r| \not< 1$ the series does not have a sum.

40. -4

41. $0.43 + 0.0043 + 0.000043 + \ldots$

$|r| = \left|\dfrac{0.0043}{0.43}\right| = |0.01| = 0.01$, and since $|r| < 1$, the series does have a sum.

$S_\infty = \dfrac{a_1}{1 - r} = \dfrac{0.43}{1 - 0.01} = \dfrac{0.43}{0.99} = \dfrac{43}{99}$

42. $\dfrac{37}{99}$

43. $\$500(1.02)^{-1} + \$500(1.02)^{-2} + \$500(1.02)^{-3} + \ldots$

$|r| = \left|\dfrac{\$500(1.02)^{-2}}{\$500(1.02)^{-1}}\right| = |(1.02)^{-1}| = (1.02)^{-1}$, or

$\dfrac{1}{1.02}$, and since $|r| < 1$, the series does have a sum.

$S_\infty = \dfrac{a_1}{1-r} = \dfrac{\$500(1.02)^{-1}}{1 - \left(\dfrac{1}{1.02}\right)} = \dfrac{\dfrac{\$500}{1.02}}{\dfrac{0.02}{1.02}} =$

$\dfrac{\$500}{1.02} \cdot \dfrac{1.02}{0.02} = \$25,000$

44. $\$12,500$

45. $0.7777\ldots = 0.7 + 0.07 + 0.007 + 0.0007 + \ldots$

This is an infinite geometric series with $a_1 = 0.7$.

$|r| = \left|\dfrac{0.07}{0.7}\right| = |0.1| = 0.1 < 1$, so the series has a sum.

$S_\infty = \dfrac{a_1}{1-r} = \dfrac{0.7}{1 - 0.1} = \dfrac{0.7}{0.9} = \dfrac{7}{9}$

Fractional notation for $0.7777\ldots$ is $\dfrac{7}{9}$.

46. $\dfrac{2}{9}$

47. $8.3838\ldots = 8.3 + 0.083 + 0.00083 + \ldots$

This is an infinite geometric series with $a_1 = 8.3$.

$|r| = \left|\dfrac{0.083}{8.3}\right| = |0.01| = 0.01 < 1$, so the series has a sum.

$S_\infty = \dfrac{a_1}{1-r} = \dfrac{8.3}{1 - 0.01} = \dfrac{8.3}{0.99} = \dfrac{830}{99}$

Fractional notation for $8.3838\ldots$ is $\dfrac{830}{99}$.

48. $\dfrac{740}{99}$

49. $0.15151515\ldots = 0.15 + 0.0015 + 0.000015 + \ldots$

This is an infinite geometric series with $a_1 = 0.15$.

$|r| = \left|\dfrac{0.0015}{0.15}\right| = |0.01| = 0.01 < 1$, so the series has a sum.

$S_\infty = \dfrac{a_1}{1-r} = \dfrac{0.15}{1 - 0.01} = \dfrac{0.15}{0.99} = \dfrac{15}{99} = \dfrac{5}{33}$

Fractional notation for $0.15151515\ldots$ is $\dfrac{5}{33}$.

50. $\dfrac{4}{33}$

51. *Familiarize.* In one year, the population will be $100,000 + 0.03(100,000)$, or $(1.03)100,000$. In two years, the population will be $(1.03)100,000 + 0.03(1.03)100,000$, or $(1.03)^2 100,000$. Thus the populations form a geometric sequence:

$100,000, \quad (1.03)100,000, \quad (1.03)^2 100,000, \ldots$

The population in 15 years will be the 16th term of the sequence.

Translate. We will use the formula $a_n = a_1 r^{n-1}$ with $a_1 = 100,000$, $r = 1.03$, and $n = 16$:

$a_{16} = 100,000(1.03)^{16-1}$

Carry out. We calculate to obtain $a_{16} \approx 155,797$.

Check. We can do the calculation again.

State. In 15 years the population will be about $155,797$.

52. About 24 years

53. *Familiarize.* The rebound distances form a geometric sequence:

$$\dfrac{1}{4} \times 20, \quad \left(\dfrac{1}{4}\right)^2 \times 20, \quad \left(\dfrac{1}{4}\right)^3 \times 20, \ldots,$$

or $\quad 5, \quad \dfrac{1}{4} \times 5, \quad \left(\dfrac{1}{4}\right)^2 \times 5, \ldots$

The height of the 6th rebound is the 6th term of the sequence.

Translate. We will use the formula $a_n = a_1 r^{n-1}$, with $a_1 = 5$, $r = \dfrac{1}{4}$, and $n = 6$:

$$a_6 = 5\left(\dfrac{1}{4}\right)^{6-1}$$

Carry out. We calculate to obtain $a_6 = \dfrac{5}{1024}$.

Check. We can do the calculation again.

State. It rebounds $\dfrac{5}{1024}$ ft the 6th time.

54. $6\dfrac{2}{3}$ ft

55. *Familiarize.* The amounts owed at the beginning of successive years form a geometric sequence:

$\$15,000, \quad (1.085)\$15,000, \quad (1.085)^2\$15,000,$

$(1.085)^3\$15,000, \ldots$

The amount to be repaid at the end of 13 years is the amount owed at the beginning of the 14th year.

Translate. We use the formula $a_n = a_1 r^{n-1}$ with $a_1 = 15,000$, $r = 1.085$, and $n = 14$:

$a_{14} = 15,000(1.085)^{14-1}$

Carry out. We calculate to obtain $a_{14} \approx 43,318.94$.

Check. We can do the calculation again.

State. At the end of 13 years, $\$43,318.94$ will be repaid.

56. 2710 fruit flies

57. We have a geometric sequence

$5000, 5000(0.96), 5000(0.96)^2, \ldots$

where the general term $5000(0.96)^n$ represents the number of fruit flies remaining alive after n minutes. We find the value of n for which the general term is 1800.

$$1800 = 5000(0.96)^n$$
$$0.36 = (0.96)^n$$
$$\log 0.36 = \log(0.96)^n$$
$$\log 0.36 = n \log 0.96$$
$$\frac{\log 0.36}{\log 0.96} = n$$
$$25 \approx n$$

It will take about 25 minutes for only 1800 fruit flies to remain alive.

58. $213,609.57

59. *Familiarize.* The lengths of the falls form a geometric sequence:

$556, \left(\dfrac{3}{4}\right)556, \left(\dfrac{3}{4}\right)^2 556, \left(\dfrac{3}{4}\right)^3 556, \ldots$

The total length of the first 6 falls is the sum of the first six terms of this sequence. The heights of the rebounds also form a geometric sequence:

$\left(\dfrac{3}{4}\right)556, \left(\dfrac{3}{4}\right)^2 556, \left(\dfrac{3}{4}\right)^3 556, \ldots,$ or

$417, \left(\dfrac{3}{4}\right)417, \left(\dfrac{3}{4}\right)^2 417, \ldots$

When the ball hits the ground for the 6th time, it will have rebounded 5 times. Thus the total length of the rebounds is the sum of the first five terms of this sequence.

Translate. We use the formula $S_n = \dfrac{a_1(1 - r^n)}{1 - r}$ twice, once with $a_1 = 556$, $r = \dfrac{3}{4}$, and $n = 6$ and a second time with $a_1 = 417$, $r = \dfrac{3}{4}$, and $n = 5$.

$D = $ Length of falls $+$ length of rebounds

$$= \frac{556\left[1 - \left(\frac{3}{4}\right)^6\right]}{1 - \frac{3}{4}} + \frac{417\left[1 - \left(\frac{3}{4}\right)^5\right]}{1 - \frac{3}{4}}.$$

Carry out. We use a calculator to obtain $D \approx 3100.35$.

Check. We can do the calculations again.

State. The ball will have traveled about 3100.35 ft.

60. 3892 ft

61. *Familiarize.* The heights of the stack form a geometric sequence:

$0.02, 0.02(2), 0.02(2^2), \ldots$

The height of the stack after it is doubled 10 times is given by the 11th term of this sequence.

Translate. We have a geometric sequence with $a_1 = 0.02$, $r = 2$, and $n = 11$. We use the formula

$$a_n = a_1 r^{n-1}.$$

Carry out. We substitute and calculate.

$$a_{11} = 0.02(2^{11-1})$$
$$a_{11} = 0.02(1024) = 20.48$$

Check. We can do the calculation again.

State. The final stack will be 20.48 in. high.

62. $2,684,354.55

63. The points lie on a straight line, so this is the graph of an arithmetic sequence.

64. Geometric

65. The points lie on the graph of an exponential function, so this is the graph of a geometric series.

66. Arithmetic

67. The points lie on the graph of an exponential function, so this is the graph of a geometric series.

68. Arithmetic

69. *Writing Exercise*

70. *Writing Exercise*

71.
$$(x + y)(x^2 + 2xy + y^2)$$
$$= x(x^2 + 2xy + y^2) + y(x^2 + 2xy + y^2)$$
$$= x^3 + 2x^2y + xy^2 + x^2y + 2xy^2 + y^3$$
$$= x^3 + 3x^2y + 3xy^2 + y^3$$

72. $a^3 - 3a^2b + 3ab^2 - b^3$

73.
$$5x - 2y = -3, \quad (1)$$
$$2x + 5y = -24 \quad (2)$$

Multiply Eq. (1) by 5 and Eq. (2) by 2 and add.

$$\begin{array}{r} 25x - 10y = -15 \\ 4x + 10y = -48 \\ \hline 29x \quad\quad\; = -63 \end{array}$$

$$x = -\frac{63}{29}$$

Substitute $-\dfrac{63}{29}$ for x in the second equation and solve for y.

$$2\left(-\frac{63}{29}\right) + 5y = -24$$

$$-\frac{126}{29} + 5y = -24$$

$$5y = -\frac{570}{29}$$

$$y = -\frac{114}{29}$$

The solution is $\left(-\frac{63}{29}, -\frac{114}{29}\right)$.

74. $(-1, 2, 3)$

75. *Writing Exercise*

76. *Writing Exercise*

77. $x^2 - x^3 + x^4 + x^5 + \ldots$

This is a geometric series with $a_1 = x^2$ and $r = -x$.

$$S_n = \frac{a_1(1-r^n)}{1-r} = \frac{x^2[1-(-x)^n]}{1-(-x)} = \frac{x^2[1-(-x)^n]}{1+x}$$

78. $\dfrac{1-x^n}{1-x}$

79. The length of a side of the first square is 16 cm. The length of a side of the next square is the length of the hypotenuse of a right triangle with legs 8 cm and 8 cm, or $8\sqrt{2}$ cm. The length of a side of the next square is the length of the hypotenuse of a right triangle with legs $4\sqrt{2}$ cm and $4\sqrt{2}$ cm, or 8 cm. The areas of the squares form a sequence:

$$(16)^2, \ (8\sqrt{2})^2, \ (8)^2, \ldots, \ \text{or}$$
$$256, \ 128, \ 64, \ldots.$$

This is a geometric sequence with $a_1 = 256$ and $r = \frac{1}{2}$.

We find the sum of the infinite geometric series $256 + 128 + 64 + \ldots$.

$$S_\infty = \frac{256}{1-\frac{1}{2}} = \frac{256}{\frac{1}{2}} = 512 \text{ cm}^2$$

80. Let $a_1 = 0.9$ and $r = 0.1$. Then $S_\infty = \dfrac{0.9}{1-0.1} = \dfrac{0.9}{0.9} = 1$.

Exercise Set 10.4

1. $8! = 8 \cdot 7 \cdot 6 \cdot 5 \cdot 4 \cdot 3 \cdot 2 \cdot 1 = 40,320$

2. $362,880$

3. $10! = 10 \cdot 9 \cdot 8 \cdot 7 \cdot 6 \cdot 5 \cdot 4 \cdot 3 \cdot 2 \cdot 1 = 3,628,800$

4. $39,916,800$

5. $\dfrac{7!}{4!} = \dfrac{7 \cdot 6 \cdot 5 \cdot 4!}{4!} = 7 \cdot 6 \cdot 5 = 210$

6. 56

7. $\dfrac{10!}{7!} = \dfrac{10 \cdot 9 \cdot 8 \cdot 7!}{7!} = 10 \cdot 9 \cdot 8 = 720$

8. 3024

9. $\dbinom{8}{2} = \dfrac{8!}{(8-2)!2!} = \dfrac{8!}{6!2!} = \dfrac{8 \cdot 7 \cdot 6!}{6! \cdot 2 \cdot 1} = \dfrac{8 \cdot 7}{2} = 4 \cdot 7 = 28$

10. 35

11. $\dbinom{10}{6} = \dfrac{10!}{(10-6)!6!} = \dfrac{10!}{4!6!} = \dfrac{10 \cdot 9 \cdot 8 \cdot 7 \cdot 6!}{4 \cdot 3 \cdot 2 \cdot 6!} = \dfrac{10 \cdot 9 \cdot 8 \cdot 7}{4 \cdot 3 \cdot 2} = 10 \cdot 3 \cdot 7 = 210$

12. 126

13. $\dbinom{20}{18} = \dfrac{20!}{(20-18)!18!} = \dfrac{20!}{2!18!} = \dfrac{20 \cdot 19 \cdot 18!}{2 \cdot 1 \cdot 18!} = \dfrac{20 \cdot 19}{2} = 10 \cdot 19 = 190$

14. 4060

15. $\dbinom{35}{2} = \dfrac{35!}{(35-2)!2!} = \dfrac{35!}{33!2!} = \dfrac{35 \cdot 34 \cdot 33!}{33! \cdot 2 \cdot 1} = \dfrac{35 \cdot 34}{2} = 35 \cdot 17 = 595$

16. 780

17. Expand $(m+n)^5$.

Form 1: The expansion of $(m+n)^5$ has $5+1$, or 6 terms. The sum of the exponents in each term is 5. The exponents of m start with 5 and decrease to 0. The last term has no factor of m. The first term has no factor of n. The exponents of n start in the second term with 1 and increase to 5. We get the coefficients from the 6th row of Pascal's triangle.

```
            1
         1     1
      1     2     1
   1     3     3     1
1     4     6     4     1
1  5    10    10    5    1
```

$$(m+n)^5 = 1 \cdot m^5 + 5 \cdot m^4 n^1 + 10 \cdot m^3 \cdot n^2 +$$
$$10 \cdot m^2 \cdot n^3 + 5 \cdot m \cdot n^4 + 1 \cdot n^5$$
$$= m^5 + 5m^4 n + 10m^3 n^2 + 10m^2 n^3 +$$
$$5mn^4 + n^5$$

Form 2: We have $a = m$, $b = n$, and $n = 5$.

$$(m+n)^5 = \binom{5}{0}m^5 + \binom{5}{1}m^4n + \binom{5}{2}m^3n^2 +$$
$$\binom{5}{3}m^2n^3 + \binom{5}{4}mn^4 + \binom{5}{5}n^5$$
$$= \frac{5!}{5!0!}m^5 + \frac{5!}{4!1!}m^4n + \frac{5!}{3!2!}m^3n^2 +$$
$$\frac{5!}{2!3!}m^2n^3 + \frac{5!}{1!4!}mn^4 + \frac{5!}{0!5!}m^5$$
$$= m^5 + 5m^4n + 10m^3n^2 + 10m^2n^3 +$$
$$5mn^4 + n^5$$

18. $a^4 - 4a^3b + 6a^2b^2 - 4ab^3 + b^4$

19. Expand $(x-y)^6$.

Form 1: The expansion of $(x-y)^6$ has $6+1$, or 7 terms. The sum of the exponents in each term is 6. The exponents of x start with 6 and decrease to 0. The last term has no factor of x. The first term has no factor of $-y$. The exponents of $-y$ start in the second term with 1 and increase to 6. We get the coefficients from the 7th row of Pascal's triangle.

$$\begin{array}{ccccccccccccc} & & & & & & 1 & & & & & & \\ & & & & & 1 & & 1 & & & & & \\ & & & & 1 & & 2 & & 1 & & & & \\ & & & 1 & & 3 & & 3 & & 1 & & & \\ & & 1 & & 4 & & 6 & & 4 & & 1 & & \\ & 1 & & 5 & & 10 & & 10 & & 5 & & 1 & \\ 1 & & 6 & & 15 & & 20 & & 15 & & 6 & & 1 \end{array}$$

$$(x-y)^6 = 1 \cdot x^6 + 6 \cdot x^5 \cdot (-y) + 15 \cdot x^4 \cdot (-y)^2 +$$
$$20 \cdot x^3 \cdot (-y)^3 + 15 \cdot x^2 \cdot (-y)^4 +$$
$$6 \cdot x \cdot (-y)^5 + 1 \cdot (-y)^6$$
$$= x^6 - 6x^5y + 15x^4y^2 - 20x^3y^3 +$$
$$15x^2y^4 - 6xy^5 + y^6$$

Form 2: We have $a = x$, $b = -y$, and $n = 6$.

$$(x-y)^6 = \binom{6}{0}x^6 + \binom{6}{1}x^5(-y) + \binom{6}{2}x^4(-y)^2 +$$
$$\binom{6}{3}x^3(-y)^3 + \binom{6}{4}x^2(-y)^4 +$$
$$\binom{6}{5}x(-y)^5 + \binom{6}{6}(-y)^6$$
$$= \frac{6!}{6!0!}x^6 + \frac{6!}{5!1!}x^5(-y) + \frac{6!}{4!2!}x^4y^2 +$$
$$\frac{6!}{3!3!}x^3(-y^3) + \frac{6!}{2!4!}x^2y^4 + \frac{6!}{1!5!}x(-y^5) +$$
$$\frac{6!}{0!6!}y^6$$
$$= x^6 - 6x^5y + 15x^4y^2 - 20x^3y^3 +$$
$$15x^2y^4 - 6xy^5 + y^6$$

20. $p^7 + 7p^6q + 21p^5q^2 + 35p^4q^3 + 35p^3q^4 + 21p^2q^5 + 7pq^6 + q^7$

21. Expand $(x^2 - 3y)^5$.

We have $a = x^2$, $b = -3y$, and $n = 5$.

Form 1: We get the coefficients from the 6th row of Pascal's triangle. From Exercise 17 we know that the coefficients are

$$\begin{array}{cccccc} 1 & 5 & 10 & 10 & 5 & 1. \end{array}$$
$$(x^2 - 3y)^5 = 1 \cdot (x^2)^5 + 5 \cdot (x^2)^4 \cdot (-3y) +$$
$$10 \cdot (x^2)^3 \cdot (-3y)^2 + 10 \cdot (x^2)^2 \cdot (-3y)^3 +$$
$$5 \cdot (x^2) \cdot (-3y)^4 + 1 \cdot (-3y)^5$$
$$= x^{10} - 15x^8y + 90x^6y^2 - 270x^4y^3 +$$
$$405x^2y^4 - 243y^5$$

Form 2:

$$(x^2 + 3y)^5 = \binom{5}{0}(x^2)^5 + \binom{5}{1}(x^2)^4(-3y) +$$
$$\binom{5}{2}(x^2)^3(-3y)^2 + \binom{5}{3}(x^2)^2(-3y)^3 +$$
$$\binom{5}{4}x^2(-3y)^4 + \binom{5}{5}(-3y)^5$$
$$= \frac{5!}{5!0!}x^{10} + \frac{5!}{4!1!}x^8(-3y) + \frac{5!}{3!2!}x^6(9y^2) +$$
$$\frac{5!}{2!3!}x^4(-27y^3) + \frac{5!}{1!4!}x^2(81y^4) +$$
$$\frac{5!}{0!5!}(-243y^5)$$
$$= x^{10} - 15x^8y + 90x^6y^2 - 270x^4y^3 +$$
$$405x^2y^4 - 243y^5$$

22. $2187c^7 - 5103c^6d + 5103c^5d^2 - 2835c^4d^3 + 945c^3d^4 - 189c^2d^5 + 21cd^6 - d^7$

23. Expand $(3c - d)^6$.

We have $a = 3c$, $b = -d$, and $n = 6$.

Form 1: We get the coefficients from the 7th row of Pascal's triangle. From Exercise 19 we know that the coefficients are

$$\begin{array}{ccccccc} 1 & 6 & 15 & 20 & 15 & 6 & 1. \end{array}$$
$$(3c - d)^6 = 1 \cdot (3c)^6 + 6 \cdot (3c)^5 \cdot (-d) +$$
$$15 \cdot (3c)^4 \cdot (-d)^2 + 20 \cdot (3c)^3 \cdot (-d)^3 +$$
$$15 \cdot (3c)^2 \cdot (-d)^4 + 6 \cdot (3c) \cdot (-d)^5 +$$
$$1 \cdot (-d)^6$$
$$= 3^6c^6 - 6 \cdot 3^5c^5d + 15 \cdot 3^4c^4d^2 -$$
$$20 \cdot 3^3c^3d^3 + 15 \cdot 3^2c^2d^4 - 6 \cdot 3cd^5 + d^6$$
$$= 729c^6 - 6 \cdot 243c^5d + 15 \cdot 81c^4d^2 -$$
$$20 \cdot 27c^3d^3 + 15 \cdot 9c^2d^4 - 6 \cdot 3cd^5 + d^6$$
$$= 729c^6 - 1458c^5d + 1215c^4d^2 - 540c^3d^3 +$$
$$135c^2d^4 - 18cd^5 + d^6$$

Form 2:

$$(3c - d)^6 = \binom{6}{0}(3c)^6 + \binom{6}{1}(3c)^5(-d) +$$

$$\binom{6}{2}(3c)^4(-d)^2 + \binom{6}{3}(3c)^3(-d)^3 +$$

$$\binom{6}{4}(3c)^2(-d)^4 + \binom{6}{5}(3c)(-d)^5 +$$

$$\binom{6}{6}(-d)^6$$

$$= \frac{6!}{6!0!}(729c^6) + \frac{6!}{5!1!}(243c^5)(-d) +$$

$$\frac{6!}{4!2!}(81c^4)(d^2) + \frac{6!}{3!3!}(27c^3)(-d^3) +$$

$$\frac{6!}{2!4!}(9c^2)(d^4) + \frac{6!}{1!5!}(3c)(-d^5) +$$

$$\frac{6!}{0!6!}d^6$$

$$= 729c^6 - 1458c^5d + 1215c^4d^2 - 540c^3d^3 +$$

$$135c^2d^4 - 18cd^5 + d^6$$

24. $t^{-12} + 12t^{-10} + 60t^{-8} + 160t^{-6} + 240t^{-4} + 192t^{-2} + 64$

25. Expand $(x - y)^3$.

We have $a = x$, $b = -y$, and $n = 3$.

Form 1: We get the coefficients from the 4th row of Pascal's triangle.

$$1$$
$$1 \quad 1$$
$$1 \quad 2 \quad 1$$
$$1 \quad 3 \quad 3 \quad 1$$

$$(x - y)^3$$
$$= 1 \cdot x^3 + 3x^2(-y) + 3x(-y)^2 + 1 \cdot (-y)^3$$
$$= x^3 - 3x^2y + 3xy^2 - y^3$$

Form 2:

$$(x - y)^3$$

$$= \binom{3}{0}x^3 + \binom{3}{1}x^2(-y) + \binom{3}{2}x(-y)^2 +$$

$$\binom{3}{3}(-y)^3$$

$$= \frac{3!}{3!0!}x^3 + \frac{3!}{2!1!}x^2(-y) + \frac{3!}{1!2!}xy^2 +$$

$$\frac{3!}{0!3!}(-y^3)$$

$$= x^3 - 3x^2y + 3xy^2 - y^3$$

26. $x^5 - 5x^4y + 10x^3y^2 - 10x^2y^3 + 5xy^4 - y^5$

27. Expand $\left(x + \dfrac{2}{y}\right)^9$.

We have $a = x$, $b = \dfrac{2}{y}$, and $n = 9$.

Form 1: We get the coefficients from the 10th row of Pascal's triangle.

$$1$$
$$1 \quad 1$$
$$1 \quad 2 \quad 1$$
$$1 \quad 3 \quad 3 \quad 1$$
$$1 \quad 4 \quad 6 \quad 4 \quad 1$$
$$1 \quad 5 \quad 10 \quad 10 \quad 5 \quad 1$$
$$1 \quad 6 \quad 15 \quad 20 \quad 15 \quad 6 \quad 1$$
$$1 \quad 7 \quad 21 \quad 35 \quad 35 \quad 21 \quad 7 \quad 1$$
$$1 \quad 8 \quad 28 \quad 56 \quad 70 \quad 56 \quad 28 \quad 8 \quad 1$$
$$1 \quad 9 \quad 36 \quad 84 \quad 126 \quad 126 \quad 84 \quad 36 \quad 9 \quad 1$$

$$\left(x + \frac{2}{y}\right)^9 = 1 \cdot x^9 + 9x^8\left(\frac{2}{y}\right) + 36x^7\left(\frac{2}{y}\right)^2 +$$

$$84x^6\left(\frac{2}{y}\right)^3 + 126x^5\left(\frac{2}{y}\right)^4 +$$

$$126x^4\left(\frac{2}{y}\right)^5 + 84x^3\left(\frac{2}{y}\right)^6 +$$

$$36x^2\left(\frac{2}{y}\right)^7 + 9x\left(\frac{2}{y}\right)^8 + 1 \cdot \left(\frac{2}{y}\right)^9$$

$$= x^9 + \frac{18x^8}{y} + \frac{144x^7}{y^2} + \frac{672x^6}{y^3} +$$

$$\frac{2016x^5}{y^4} + \frac{4032x^4}{y^5} + \frac{5376x^3}{y^6} +$$

$$\frac{4608x^2}{y^7} + \frac{2304x}{y^8} + \frac{512}{y^9}$$

Form 2:

$$\left(x - \frac{2}{y}\right)^9$$

$$= \binom{9}{0}x^9 + \binom{9}{1}x^8\left(\frac{2}{y}\right) + \binom{9}{2}x^7\left(\frac{2}{y}\right)^2 +$$

$$\binom{9}{3}x^6\left(\frac{2}{y}\right)^3 + \binom{9}{4}x^5\left(\frac{2}{y}\right)^4 +$$

$$\binom{9}{5}x^4\left(\frac{2}{y}\right)^5 + \binom{9}{6}x^3\left(\frac{2}{y}\right)^6 +$$

$$\binom{9}{7}x^2\left(\frac{2}{y}\right)^7 + \binom{9}{8}x\left(\frac{2}{y}\right)^8 +$$

$$\binom{9}{9}\left(\frac{2}{y}\right)^9$$

$$= \frac{9!}{9!0!}x^9 + \frac{9!}{8!1!}x^8\left(\frac{2}{y}\right) + \frac{9!}{7!2!}x^7\left(\frac{4}{y^2}\right) +$$

$$\frac{9!}{6!3!}x^6\left(\frac{8}{y^3}\right) + \frac{9!}{5!4!}x^5\left(\frac{16}{y^4}\right) +$$

$$\frac{9!}{4!5!}x^4\left(\frac{32}{y^5}\right) + \frac{9!}{3!6!}x^3\left(\frac{64}{y^6}\right) +$$

$$\frac{9!}{2!7!}x^2\left(\frac{128}{y^7}\right) + \frac{9!}{1!8!}x\left(\frac{256}{y^8}\right) +$$

$$\frac{9!}{0!9!}\left(\frac{512}{y^9}\right)$$

$$= x^9 + 9x^8\left(\frac{2}{y}\right) + 36x^7\left(\frac{4}{y^2}\right) + 84x^6\left(\frac{8}{y^3}\right) +$$

$$126x^5\left(\frac{16}{y^4}\right) + 126x^4\left(\frac{32}{y^5}\right) + 84x^3\left(\frac{64}{y^6}\right) +$$

$$36x^2\left(\frac{128}{y^7}\right) + 9x\left(\frac{256}{y^8}\right) + \frac{512}{y^9}$$

$$= x^9 + \frac{18x^8}{y} + \frac{144x^7}{y^2} + \frac{672x^6}{y^3} +$$

$$\frac{2016x^5}{y^4} + \frac{4032x^4}{y^5} + \frac{5376x^3}{y^6} +$$

$$\frac{4608x^2}{y^7} + \frac{2304x}{y^8} + \frac{512}{y^9}$$

28. $19,683s^9 + \dfrac{59,049s^8}{t} + \dfrac{78,732s^7}{t^2} + \dfrac{61,236s^6}{t^3} +$

$$\frac{30,618s^5}{t^4} + \frac{10,206s^4}{t^5} + \frac{2268s^3}{t^6} + \frac{324s^2}{t^7} + \frac{27s}{t^8} + \frac{1}{t^9}$$

29. Expand $(a^2 - b^3)^5$.

We have $a = a^2$, $b = -b^3$, and $n = 5$.

Form 1: We get the coefficient from the 6th row of Pascal's triangle. From Exercise 17 we know that the coefficients are

1   5   10   10   5   1.

$$(a^2 - b^3)^5$$
$$= 1 \cdot (a^2)^5 + 5(a^2)^4(-b^3) + 10(a^2)^3(-b^3)^2 +$$
$$10(a^2)^2(-b^3)^3 + 5(a^2)(-b^3)^4 + 1 \cdot (-b^3)^5$$
$$= a^{10} - 5a^8b^3 + 10a^6b^6 - 10a^4b^9 +$$
$$5a^2b^{12} - b^{15}$$

Form 2:
$$(a^2 - b^3)^5$$
$$= \binom{5}{0}(a^2)^5 + \binom{5}{1}(a^2)^4(-b^3) +$$
$$\binom{5}{2}(a^2)^3(-b^3)^2 + \binom{5}{3}(a^2)^2(-b^3)^3 +$$
$$\binom{5}{4}(a^2)(-b^3)^4 + \binom{5}{5}(-b^3)^5$$
$$= \frac{5!}{5!0!}a^{10} + \frac{5!}{4!1!}a^8(-b^3) + \frac{5!}{3!2!}a^6(b^6) +$$
$$\frac{5!}{2!3!}a^4(-b^9) + \frac{5!}{1!4!}a^2(b^{12}) + \frac{5!}{0!5!}(-b^{15})$$
$$= a^{10} - 5a^8b^3 + 10a^6b^6 - 10a^4b^9 +$$
$$5a^2b^{12} - b^{15}$$

30. $x^{15} - 10x^{12}y + 40x^9y^2 - 80x^6y^3 + 80x^3y^4 - 32y^5$

31. Expand $(\sqrt{3} - t)^4$.

We have $a = \sqrt{3}$, $b = -t$, and $n = 4$.

Form 1: We get the coefficients from the 5th row of Pascal's triangle.

```
            1
          1   1
        1   2   1
      1   3   3   1
    1   4   6   4   1
```

$$(\sqrt{3} - t)^4 = 1 \cdot (\sqrt{3})^4 + 4(\sqrt{3})^3(-t) +$$
$$6(\sqrt{3})^2(-t)^2 + 4(\sqrt{3})(-t)^3 + 1 \cdot (-t)^4$$
$$= 9 - 12\sqrt{3}t + 18t^2 - 4\sqrt{3}t^3 + t^4$$

Form 2:
$$(\sqrt{3} - t)^4 = \binom{4}{0}(\sqrt{3})^4 + \binom{4}{1}(\sqrt{3})^3(-t) +$$
$$\binom{4}{2}(\sqrt{3})^2(-t)^2 + \binom{4}{3}(\sqrt{3})(-t)^3 +$$
$$\binom{4}{4}(-t)^4$$
$$= \frac{4!}{4!0!}(9) + \frac{4!}{3!1!}(3\sqrt{3})(-t) +$$
$$\frac{4!}{2!2!}(3)(t^2) + \frac{4!}{1!3!}(\sqrt{3})(-t^3) +$$
$$\frac{4!}{0!4!}(t^4)$$
$$= 9 - 12\sqrt{3}t + 18t^2 - 4\sqrt{3}t^3 + t^4$$

32. $125 + 150\sqrt{5}\,t + 375t^2 + 100\sqrt{5}\,t^3 + 75t^4 + 6\sqrt{5}\,t^5 + t^6$

33. Expand $(x^{-2} + x^2)^4$.

We have $a = x^{-2}$, $b = x^2$, and $n = 4$.

Form 1: We get the coefficients from the fifth row of Pascal's triangle. From Exercise 31 we know that the coefficients are

$$1 \quad 4 \quad 6 \quad 4 \quad 1.$$
$$(x^{-2} + x^2)^4$$
$$= 1 \cdot (x^{-2})^4 + 4(x^{-2})^3(x^2) + 6(x^{-2})^2(x^2)^2 +$$
$$\quad 4(x^{-2})(x^2)^3 + 1 \cdot (x^2)^4$$
$$= x^{-8} + 4x^{-4} + 6 + 4x^4 + x^8$$

Form 2:
$$(x^{-2} + x^2)^4$$
$$= \binom{4}{0}(x^{-2})^4 + \binom{4}{1}(x^{-2})^3(x^2) +$$
$$\quad \binom{4}{2}(x^{-2})^2(x^2)^2 + \binom{4}{3}(x^{-2})(x^2)^3 +$$
$$\quad \binom{4}{4}(x^2)^4$$
$$= \frac{4!}{4!0!}(x^{-8}) + \frac{4!}{3!1!}(x^{-6})(x^2) + \frac{4!}{2!2!}(x^{-4})(x^4) +$$
$$\quad \frac{4!}{1!3!}(x^{-2})(x^6) + \frac{4!}{0!4!}(x^8)$$
$$= x^{-8} + 4x^{-4} + 6 + 4x^4 + x^8$$

34. $x^{-3} - 6x^{-2} + 15x^{-1} - 20 + 15x - 6x^2 + x^3$

35. Find the 3rd term of $(a + b)^6$.

First, we note that $3 = 2 + 1$, $a = a$, $b = b$, and $n = 6$. Then the 3rd term of the expansion of $(a + b)^6$ is

$$\binom{6}{2}a^{6-2}b^2, \text{ or } \frac{6!}{4!2!}a^4b^2, \text{ or } 15a^4b^2.$$

36. $21x^2y^5$

37. Find the 12th term of $(a - 3)^{14}$.

First, we note that $12 = 11 + 1$, $a = a$, $b = -3$, and $n = 14$. Then the 12th term of the expansion of $(a - 3)^{14}$ is

$$\binom{14}{11}a^{14-11} \cdot (-3)^{11} = \frac{14!}{3!11!}a^3(-177,147)$$
$$= 364a^3(-177,147)$$
$$= -64,481,508a^3$$

38. $67,584x^2$

39. Find the 5th term of $(2x^3 - \sqrt{y})^8$.

First, we note that $5 = 4 + 1$, $a = 2x^3$, $b = -\sqrt{y}$, and $n = 8$. Then the 5th term of the expansion of $(2x^3 - \sqrt{y})^8$ is

$$\binom{8}{4}(2x^3)^{8-4}(-\sqrt{y})^4$$
$$= \frac{8!}{4!4!}(2x^3)^4(-\sqrt{y})^4$$
$$= 70(16x^{12})(y^2)$$
$$= 1120x^{12}y^2$$

40. $\dfrac{35c^3}{b^8}$

41. The expansion of $(2u - 3v^2)^{10}$ has 11 terms so the 6th term is the middle term. Note that $6 = 5 + 1$, $a = 2u$, $b = -3v^2$, and $n = 10$. Then the 6th term of the expansion of $(2u - 3v^2)^{10}$ is

$$\binom{10}{5}(2u)^{10-5}(-3v^2)^5$$
$$= \frac{10!}{5!5!}(2u)^5(-3v^2)^5$$
$$= 252(32u^5)(-243v^{10})$$
$$= -1,959,552u^5v^{10}$$

42. $30x\sqrt{x}$, $30x\sqrt{3}$

43. The 9th term of $(x - y)^8$ is the last term, y^8.

44. $-b^9$

45. *Writing Exercise*

46. *Writing Exercise*

47. $\log_2 x + \log_2(x - 2) = 3$
$$\log_2 x(x - 2) = 3$$
$$x(x - 2) = 2^3$$
$$x^2 - 2x = 8$$
$$x^2 - 2x - 8 = 0$$
$$(x - 4)(x + 2) = 0$$
$$x = 4 \text{ or } x = -2$$

Only 4 checks. It is the solution.

48. $\dfrac{5}{2}$

49. $e^t = 280$
$$\ln e^t = \ln 280$$
$$t = \ln 280$$
$$t \approx 5.6348$$

50. ± 5

51. *Writing Exercise*

52. *Writing Exercise*

53. Consider a set of 5 elements, $\{A, B, C, D, E\}$. List all the subsets of size 3:

$\{A, B, C\}, \{A, B, D\}, \{A, B, E\}, \{A, C, D\},$
$\{A, C, E\}, \{A, D, E\}, \{B, C, D\}, \{B, C, E\},$
$\{B, D, E\}, \{C, D, E\}.$

There are exactly 10 subsets of size 3 and $\binom{5}{3} = 10$, so

there are exactly $\binom{5}{3}$ ways of forming a subset of size 3 from a set of 5 elements.

54. $\binom{5}{n}(0.325)^{5-2}(0.675)^2 \approx 0.156$

55. Find the sixth term of $(0.15 + 0.85)^8$:

$\binom{8}{5}(0.15)^{8-5}(0.85)^5 = \frac{8!}{3!5!}(0.15)^3(0.85)^5 \approx 0.084$

56. $\binom{5}{2}(0.325)^3(0.675)^2 + \binom{5}{3}(0.325)^2(0.675)^3 +$

$\binom{5}{4}(0.325)(0.675)^4 + \binom{5}{5}(0.675)^5 \approx 0.959$

57. Find and add the 7th through the 9th terms of $(0.15 + 0.85)^9$:

$\binom{8}{6}(0.15)^2(0.85)^6 + \binom{8}{7}(0.15)(0.85)^7 +$

$\binom{8}{8}(0.85)^8 \approx 0.89$

58. $\binom{n}{n-r} = \frac{n!}{[n-(n-r)!](n-r)!} = \frac{n!}{r!(n-r)!} = $
$\binom{n}{r}$

59. The $(r+1)$st term of $\left(\frac{3x^2}{2} - \frac{1}{3x}\right)^{12}$ is

$\binom{12}{r}\left(\frac{3x^2}{2}\right)^{12-r}\left(-\frac{1}{3x}\right)^r$. In the term which does not contain x, the exponent of x in the numerator is equal to the exponent of x in the denominator.

$2(12 - r) = r$
$24 - 2r = r$
$24 = 3r$
$8 = r$

Find the $(8+1)$st, or 9th term:

$\binom{12}{8}\left(\frac{3x^2}{2}\right)^4\left(-\frac{1}{3x}\right)^8 = \frac{12!}{4!8!}\left(\frac{3^4x^8}{2^4}\right)\left(\frac{1}{3^8x^8}\right) = \frac{55}{144}$

60. $-4320x^6y^{9/2}$

61. $\dfrac{\binom{5}{3}(p^2)^2\left(-\frac{1}{2}p\sqrt[3]{q}\right)^3}{\binom{5}{2}(p^2)^3\left(-\frac{1}{2}p\sqrt[3]{q}\right)^2} = \dfrac{-\frac{1}{8}p^7q}{\frac{1}{4}p^8\sqrt[3]{q^2}} = $

$\dfrac{-\frac{1}{8}p^7q}{\frac{1}{4}p^8q^{2/3}} = -\frac{1}{8}\cdot\frac{4}{1}\cdot p^{7-8}\cdot q^{1-2/3} = $

$-\frac{1}{2}p^{-1}q^{1/3} = -\dfrac{\sqrt[3]{q}}{2p}$

62. $-\dfrac{35}{x^{1/6}}$

63. The degree of $(x^2 + 3)^4$ is the degree of $(x^2)^4 = x^8$, or 8.

64. $x^7 + 7x^6y + 21x^5y^2 + 35x^4y^3 + 35x^3y^4 + 21x^2y^5 + 7xy^6 + y^7$